DIGITAL
SIGNAL PROCESSING
IN VLSI

ANALOG DEVICES TECHNICAL REFERENCE BOOKS

Published by Prentice Hall
        Analog-Digital Conversion Handbook
        Digital Signal-Processing in VLSI

Published by Analog Devices
        Nonlinear Circuits Handbook
        Transducer Interfacing Handbook
        Synchro & Resolver Conversion

DIGITAL
SIGNAL PROCESSING
IN VLSI

by
Richard J. Higgins
Georgia Institute of Technology

PRENTICE HALL, Englewood Cliffs, NJ 07632

Published by Prentice-Hall, Inc.
A Division of Simon & Schuster
Englewood Cliffs, New Jersey 07632

Printed in the United States of America

10 9 8 7 6 5 4 3 2

ISBN 0-13-212887-X

Prentice-Hall International (UK) Limited, *London*
Prentice-Hall of Australia Pty. Limited, *Sydney*
Prentice-Hall Canada Inc., *Toronto*
Prentice-Hall Hispanoamericana, S.A., *Mexico*
Prentice-Hall of India Private Limited, *New Delhi*
Prentice-Hall of Japan, Inc., *Tokyo*
Simon & Schuster Asia Pte. Ltd., *Singapore*
Editora Prentice-Hall do Brasil, Ltda., *Rio de Janeiro*

Table of Contents

v

Preface

Do it digitally. These days, this precept goes far beyond personal computers. Inside practically any measuring instrument, communications or audio product, traditional electronic circuits are being emulated in software and firmware. The software drives a new generation of specialized microprocessors whose architecture is optimized to carry out digital signal-processing (DSP) algorithms, typified by the multiply-accumulate computation sequence of digital filters and spectral analysis.

With the gap between the analog and digital worlds bridged by economical analog-to-digital and digital-to-analog converters, the translation of signal processing functions to digital terms is further driven by advantages of stability, programmability, enhanced performance, as well as some digital signal processing algorithms and tricks that have no analog equivalent.

DSP computing processes and processors are having an impact similar to the effect the microprocessor had on computing. Talking calculators, smart measuring instruments, sophisticated music synthesizers, translating machines and seeing aids for the blind are coming into vogue, and even DSP-based smart toys (pioneered by TI's "Speak and Spell") are commonplace.

This book is for users and potential users of DSP who need to massage signals or create systems and are more interested in its applications than in theory (as represented by the classic DSP treatises).

A successful user of DSP techniques, I was driven to DSP by necessity as well as inclination. My lab measures the properties of electronic materials and microelectronic devices. Pulling information out of data by spectral analysis or real-time signal processing gave us a competitive edge in our research. In the mid-1970's I was introduced to digital filter concepts by a former student, Hal Alles, then at Bell Laboratories. Hal's group had constructed a high-performance digital filter and demonstrated it with real-time examples, including speech and music synthesis. While my lab had long been computer-massaging sampled data, I was impressed to see how well you could implement circuit functions by difference equations and solve them *in real time* with compact digital hardware.

This book is for a technical audience that needs to understand DSP algorithms and the special-purpose DSP hardware ICs and software tools developed to carry them out efficiently. Electrical engineers and scientists who are familiar with signals and their manipulation through analog means or through traditional digital computer analysis need to learn about the new tools available through IC chips optimized for DSP algorithms. Computer engineers and computer scientists with good microprocessor background need to learn about real-world signals and signal processing basics in order to design with DSP chips without violating fundamental DSP laws. Readers who have been deterred from DSP by treatises that begin mathematically with the z-transform will find here an introduction to the basics by an intuitive approach, with illustrations based on familiar examples.

In combining theory and practice, the book is a user guide that seeks to follow the tradition established by the *Analog-Digital Conversion Handbook* and other useful handbooks published by Analog Devices. The book is organized in two halves: fundamental DSP basics (Chapters 1-4) and their application (Chapters 5-8). Although the fundamentals half has been subjected to the scrutiny of several DSP experts, our approach is heuristic, with key ideas and rules summarized for rapid use, often without full proofs. The applications half is more motivational than exhaustive, with chip design, software issues, and applications illustrated by selected examples to point the reader to the potential for his or her own application.

Here is a list of the chapters and a brief description of their contents, to establish a perspective on the book's flow:

Chapter 1, Real-World Signal Processing: An introductory survey of digital signal processing, to fill in missing background and provide motivational examples. Builds bridges for the practicing EE via DSP implementations of analog circuit functions, and for the computer expert via software implementation of simple digital filter functions.

Chapter 2, Sampled Signals and Systems: Convolution theorem in Fourier transforms; information theorem and spectral limits in sampled data; the discrete Fourier transform; poles, zeros, and system stability; the z-transform. Provides the equivalent of the first course in DSP for readers without that background, and a ready-reference summary for others.

Chapter 3, The Discrete Fourier Transform and the Fast Fourier-Transform Algorithm: Relation to the continuous transform and to Fourier series; DFT limits (samples, window size); "fast Fourier transform" (FFT) algorithms and speed enhancement; pipelining and parallelism enhancements; avoiding DFT pitfalls: aliasing, leakage, picket fence effect; critical comparison of window weighting functions; spectral analysis examples including decimation (zoom), correlation, convolution; Fourier filtering.

Chapter 4, Digital Filters: Comparison of digital filter types; finite impulse-response (FIR) filter background and design methods; finite wordlength errors; infinite impulse-response (IIR) filter background and design methods; mapping analog to digital designs; CAD design methods for both FIR and IIR filter types.

Chapter 5, The Bridge to VLSI: Tradeoffs between speed, accuracy, and costs; parallelism and pipelining; quantization error, with special emphasis on FFTs and digital filter output error; analog input and output requirements (dynamic range, throughput).

Chapter 6, Real Hardware: Arithmetic-logic units; barrel shifters; address generators; instruction sequencers); system alternatives (single- and multiple-chip processors; bit- and Word-Slice configurations); comparison of alternatives in terms of performance, as well as cost of hardware and software development. Illustration with present day systems (Texas Instruments TMS-320 series; Analog Devices 2100 series); comparison of fixed- and floating-point systems; microprogrammable systems.

Chapter 7, Software Development for the DSP System: Development tools for translating and debugging DSP source code; target system specification; the assembler; the linking loader; instruction-level simulation; circuit emulation. DSP microprocessor applications, illustrated via an FIR filter module and a speech recognition system. Brief comparative introduction of software development (the meta-assembler) for microprogrammed DSP systems.

Chapter 8, DSP Applications: Major elements of a DSP system, using digital transmit-receive (or modulate-detect) prototype; digital detection alternatives (signal averaging, correlation, coherent detection, spectral estimation); digital heterodyning, decimation, and interpolation. Real time detection (examples of matched filters, coherent detection or lock-in, spectrum analyzer, spectral analysis and estimation). Modeling in real time; examples from telecommunications and speech; comparison of digital encoding and data-compression alternatives. Real-time signal generation and music synthesis: fast function generation; musical instrument analysis and synthesis methods; fm synthesizer example. Image processing: machine vision, image enhancement; medical image processing (examples from CAT, PET, and MRI image reconstruction)

ACKNOWLEDGMENTS

This book began during a two year stay in Boston (1983-85), where I divided my time between semiconductor research at the M.I.T. Francis Bitter National Magnet Laboratory and consulting at Analog Devices. I would like to thank the staff of the National Magnet Laboratory for their hospitality during this stimulating period.

The staff of the DSP Division of Analog Devices patiently bore with my many questions. In particular, I would single out Matt Johnson and Bob Fine, who gave generously of their advice, clear explanations, and feedback based on their contact as application engineers with real-world DSP users.

While any book is ultimately the author's responsibility, this book benefited greatly from a succession of technical commentaries at all stages of its development, particularly from Ted Dintersmith and Paul Toldalagi, both critical readers with an eye for clarity. Special credit is due to Analog Devices' super-editor Dan Sheingold, who worked over the author's some-times turgid prose with an eye for crystal clarity based on his many years' experience in editing Analog Devices' superb technical publications. His contribution was enormous.

Richard J. Higgins

Atlanta, Georgia

DIGITAL
SIGNAL PROCESSING
IN VLSI

Chapter One

Real-World
Signal Processing

1.1 INTRODUCTION

1.1.1 CHAPTER OVERVIEW

Digital signal-processing (DSP) technology is moving inexorably into measurement, control, graphics, data analysis, and data communications. It is extending the computational capabilities of microcomputers to equal or exceed those of much larger machines, and helping larger machines to perform hitherto unheard of prodigies of number crunching. Finally, DSP circuitry both supplements and supplants analog circuits in signal-conditioning roles.

This chapter, an introductory survey of digital signal processing, is intended to get all our readers on somewhat the same footing. We anticipate that our readers will have diverse backgrounds. Some will have strong analog background but little experience with sampled data; we hope to put at rest the concern that more is lost than gained in sampling or amplitude quantization. Some will have strong digital hardware background but little analog experience; we will introduce you to the principles that underlie filter system behavior. Some with a computer-science background may have little experience with signal-processing hardware or with signals; we will show you what computers, augmented by signal processing technology, can accomplish in the real world. Many other readers already know some DSP from having used the techniques with minicomputers in data acquisition, but may want to extend their knowledge and capabilities to real-time applications of DSP with specialized microprocessors and components.

No previous DSP experience is assumed, except for some general EE circuit background and a firm resolve not to panic at the prospect of brief exposure to mathematical tools necessary for description—even if not for analysis. This introduction, light on the formal math, will emphasize the results you will be able to obtain.

There are five main topics to be covered in this chapter:

1. What is "real-world DSP", and what are typical applications?
2. What is to be gained in going digital? What are the tradeoffs?
3. DSP programmability = flexibility, compared to analog solutions.
4. How to estimate the tradeoffs between hardware and software solutions.
5. VLSI gives new freedom: custom-tailored architecture.

These themes are introduced through spectral analysis, which places time and frequency on equal footing as equivalent representations of information; and through commonly used filters, which enhance signal-to-noise. The benefits are clear: DSP promises flexibility, programmability, multiplexing, performance which is predictable, accurate, repeatable, and adaptable in real time. New DSP issues are understandable: sampling and aliasing; quantization; throughput limitations; program-development effort replacing circuit development. Two familiar examples are brought into digital form: the moving-average filter leads to the finite impulse response (FIR) concept; the first-order low-pass introduces digital feedback or infinite impulse response (IIR). Digital audio applications will be surveyed here and recur throughout the book, since they illustrate DSP computations in an undeniably real-time environment.

This chapter, a survey of possibilities, introduces ideas to be met later in more depth. Formal mathematics is minimized, though we employ some key tools: transfer functions and filter coefficients. Where possible, the analog circuit counterparts will be shown, and typical DSP hardware and software will be introduced.

The chapter begins with a review of signals and analog signal processing at (1.2). Spectral analysis is introduced (1.3) by familiar examples, e.g., vibration and resonance. (1.4) What are the limitations of analog signal processing? (1.5) How well can we approximate signals when they are sampled? We survey sampling ideas such as bit rate and quantization. (1.6) The first digital filter is one you already know—the moving average. We develop the moving average in time and also in the frequency domain, as a function of number of "taps" or averaged points. (1.7) The familiar analog first-order low-pass filter will introduce recursive systems (i.e., feedback). This will be compared with a nonrecursive approximation, to show how the number of taps needed to simulate the long-time response can be reduced. While mathematically more complicated, recursive systems are fast and good for adaptive simulation, for example modeling of the vocal tract. (1.8) An in-depth example, digital audio, will leave the reader (we hope) ready to dive into the more challenging DSP background that follows.

1.1.2 WHAT IS REAL-WORLD DSP?

The term usually implies:

1. Contact with analog (continuous) signals as inputs or outputs. Real-world DSP in its simplest form is analog signal processing done digitally.

2. Real-time signal processing as opposed to off-line. The computation keeps pace with the input and output signals. The off-line processing of seismic data from magnetic tape, for example, is not real-time; but filtering an audio signal while it is happening *is* real-time.

An example of a real-world DSP system is shown in Figure 1.1a. Signals are translated from the analog domain to the digital domain via an analog-to-digital

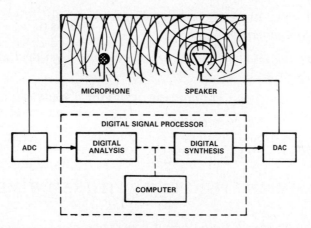

a. A real-world digital signal processing system collects data from the analog domain via an a/d converter and provides it to the analog domain via a d/a converter. This example includes a loudspeaker and microphone as transducers for studying room acoustics in the time and frequency domains.

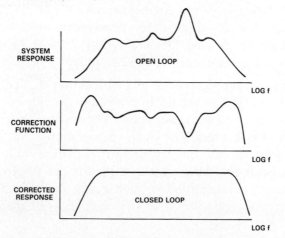

b. Closed-loop active enhancement of imperfect components (speaker, walls of room) in a digital feedback loop can improve the response curve.

Figure 1.1. Example of digital signal processing.

converter (A/D, ADC), and returned to the analog domain via a digital-to-analog converter (D/A, DAC). Computations embracing both analysis and synthesis are done in binary form in digital hardware and software.

The same system hardware can perform quite diverse applications. For example, with a computer-generated signal producing stimuli via the DAC-plus-speaker and a calibrated microphone-plus-ADC reading the response, the speaker characteristics can be measured; with changes to the program, the acoustics of the room can be characterized. A further program change—closing the feedback loop in software—can optimize the filtering by equalization, as shown in Figure 1.1b, to compensate dynamically for limitations in the speaker or in the room acoustics.

Perhaps 75% of signal processing tasks fall into three categories: *convolution*, *correlation*, and *transformation*.

- *Convolution* filters; it enhances signal-to-noise by selecting frequency bands to pass or suppress.

- *Correlation* compares, suppressing random events and amplifying repeated ones.

- *Transformation* finds frequency content, the characteristic pitch and harmonic-content "fingerprint".

1.1.3 MOS-VLSI SIGNAL PROCESSORS WITH REAL-TIME CAPABILITY

Inexpensive programmable digital signal-processing chips now have the speed and accuracy adequate to process signals in real time. They range from dedicated devices, such as multipliers and multiplier/accumulators, to families of components that are used to form high-performance DSP systems (for example, AMD's Bit-Slice™ and Analog Devices Word-Slice®, to single-chip processors. Examples of single-chip processors available from major manufacturers include: Analog Devices ADSP-2100, Nippon Electric Co. NEC 7720, and Texas Instruments TMS 320 family.

Some of these are in NMOS technology, with power dissipation measurable in watts. However, CMOS has recently evolved to provide competitive speeds with one-tenth as much power dissipation. These chips are quite complex, a characteristic of VLSI (very-large-scale integration). VLSI gives device designers the freedom to custom-design elaborate chip architectures, nevertheless—in practice—the internal complexity that it leads to (apparent in the photomicrograph of an example of such a chip shown in Figure 1.2a) is of no more concern to typical users than the complexity of their own brains or cell structure.

The development of such compact signal processors has significance comparable to the introduction of microcomputers: computing was no longer solely the domain of computer specialists. However, before we can make best

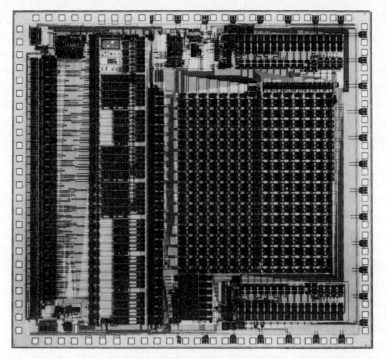

Figure 1.2. VLSI digital signal processing chip. This example is a floating-point CMOS multiplier, capable of executing a full 32 by 32-bit multiply in only 100 ns. The chip can operate programmably in various formats, from fixed point to single- or double-precision IEEE-standard floating point. More-recent members of the family have speeds up to 40 MFLOPS (millions of floating-point operations per second). (Analog Devices)

use of them, we must first understand the fundamentals of digital signal processing—the purpose of this book. Inexpensive, easily available DSP processors can contribute to numerous applications in a variety of fields, for example:

- Instrumentation and Measurement: Spectrum analyzers, Correlators, Signal averagers, Coherent detectors, Filters

- Communications: Voice telecommunications, Data communications, Modems, Encryption

- Digital audio: Speech and music generation, Signal generation, Speech recognition

- Graphics: Computer-graphic art, films, and CAD; Image enhancement and reconstruction, Medical imaging

- Navigation: Radar, Sonar

- Control: Robotics, Machine vision, Guidance, Decision-making

- Seismic investigations

- Computation: Microcomputer and mainframe accelerators; Workstations for science and engineering; Array processors

While these applications have been developing for some years with minicomputers or fixed-purpose hardware, cost and throughput limitations have limited the number of users. Consider digital speech recognition. This application combines digital filters and spectrum analyzers with phoneme recognition algorithms. Putting all this on an integrated circuit chip can bring speech recognition out of the expensive mainframe and into the world—for example, a door-lock which recognizes when you speak that it is you who wants to come in.

1.2 REVIEW OF SIGNALS AND SIGNAL PROCESSING

This section uses the familiar example of analog filters, which selectively enhance signal frequencies and attenuate noise, to introduce some of the nomenclature of signal processing.

1.2.1 ENHANCEMENT OF SIGNAL TO NOISE

Why use filters? Filters are typically used to pick out signals of interest from noise, by making use of their differing frequency characteristics. In Figure 1.3a, if the data consists of a rapidly varying waveform, biased by a slowly varying background and swamped by a low-frequency oscillation, the signal can be passed through a high-pass filter, with the result shown in the bottom trace of Figure 1.3b. On the other hand, if the data is the slow variation being swamped by high frequency noise, pass the signal through a low-pass filter to get the upper trace.

1.2.2 SYSTEM MODELS AND THE TRANSFER FUNCTION

A model of the "system"—in Figure 1.1, it is the room, and perhaps also the speaker and mike—is the starting point for characterizing its behavior. A system's response to a varying input signal, $s(t)$, with a frequency spectrum $S(f)$, can be described, essentially interchangeably, by the response $r(t)$ in the time domain (as a time history) or $R(f)$ in the frequency domain (as a frequency spectrum), as illustrated by Figure 1.4. Key system characteristics operate on the signal to produce the response to the stimulus. In the frequency domain, the operation can be expressed as a simple product; the ratio of response to stimulus is called the *transfer function*, $H(f)$.

$$H(f) = \frac{R(f)}{S(f)} \tag{1.1}$$

We will use lower-case letters for time-domain properties and upper-case letters for the frequency domain; the two domains are intimately connected via the Fourier transform (Chapter 2). The box is a model of the system which must be a convincing, adequate, and self-consistent description of system behavior. We observe what comes out (response) in response to what goes in (stimulus), without looking inside.

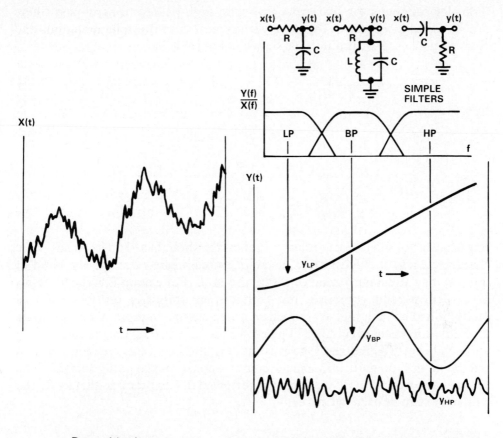

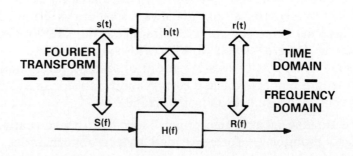

a. Data with a broad range
of spectral content.

b. Low-pass, band-pass, and high-pass
filters can separate out information contained in various frequency bands.

Figure 1.3. Filtering example.

Figure 1.4. A system's transfer function, $h(t)$ and $H(f)$, characterizes its response $r(t)$
in the time domain and $R(f)$ in the frequency domain, to respective stimuli, $s(t)$ and
$S(f)$. The two domains are related by the Fourier transform.

Transfer functions for first-order low-pass, high-pass, and band-pass filters are sketched in Figure 1.3b, along with typical circuit implementations and their responses to the input signal shown in Figure 1.3a.

$$H(f)_{\text{LP}} = \frac{1}{1 + j\omega RC} \qquad \omega = 2\pi f, \; j = \sqrt{-1} \qquad (1.2a)$$

$$H(f)_{\text{HP}} = \frac{j\omega RC}{1 + j\omega RC} \qquad\qquad\qquad (1.2b)$$

$$H(f)_{\text{BP}} = \frac{1}{1 + j\dfrac{R}{\omega_0 L}\left[\dfrac{\omega}{\omega_0} - \dfrac{\omega_0}{\omega}\right]} \qquad \omega_0 = \frac{1}{\sqrt{LC}} \qquad (1.2c)$$

It is usually not sufficient to select a filter off the shelf, based on its steady-state characteristics, and plug it in. Instabilities manifested outside the band of interest may affect performance within the band. For example, if the filter has a correct in-band response, but contains an amplifier on the brink of oscillation at a much higher frequency, a step change in input or a noise spike can cause unexpected ringing (Figure 1.5); this may produce significant errors at much lower frequencies. One must understand how filter characteristics in the frequency domain influence system response in the time domain. The analog background to filter stability is reviewed in Chapter 2 so that its digital equivalent can be used in Chapter 4.

1.3 SPECTRA: TIME AND FREQUENCY AS COMPLEMENTARY DOMAINS

Why compute spectra? Electrical systems and the signals they deal with have measurable responses as time histories in the time domain, but they can be transformed for study in the frequency domain to yield further information. In a police radar, for example, the echo time yields the distance to the target, and the frequency shift (Doppler shift) yields the target speed. Because the data handled by a computer may not always start out as a time series, we can replace the terms "time domain" and "frequency domain" by "data domain" and "spectral domain" when working with systems where time is not the variable. For example, the static pattern of loading forces and deflections along the length of a bridge has a corresponding spectrum of "frequencies" inversely related to distances along the bridge.

These spectra may act as "fingerprints," specific to a given entity, which can be used to monitor performance and detect abnormalities. Significant applications are in evaluation of safety in physical structures, observing significant shifts in spectra of structures in the human body for investigating the presence of diseases, and, in research, establishing functional relationships in the presence of seemingly patternless data—to name just a

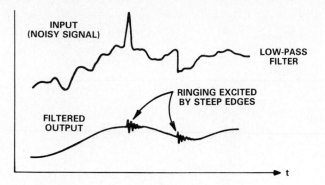

a. A poorly designed filter, even a low-pass filter, may ring when hit by a sudden change of input.

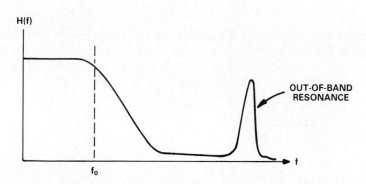

b. The ringing is related to an unanticipated out-of-band peak in the transfer function, $H(f)$.

Figure 1.5. Example of unanticipated filter problem.

few. We shall see that improved algorithms, such as the fast Fourier transform, allow spectral analysis for large amounts of data to be performed quite rapidly (in "real time").

1.3.1 TIME AND FREQUENCY: TWO WAYS TO STUDY A SYSTEM.

Consider the example of acoustic response. The speaker and microphone of Figure 1.1a can probe room acoustics in either the time or the frequency domain. In the time domain method, a sudden pulse (a click or pop) is caused to be emitted from the speaker. The response (Figure 1.6a), measured by the microphone, is a series of echoes from the walls, decaying with time. This is the impulse response. The amplitude of the pulse should be kept small enough not to overload the system, i.e., to keep it *linear* (a linear system is one in which, twice the stimulus gives twice the response, and the sums of the responses to two stimuli are equal to the responses to the sums of the stimuli). While the impulse response is in principle adequate to predict the time response to any waveform, system characteristics may show up more directly when viewed as a function of frequency.

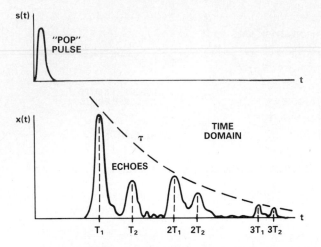

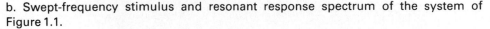

a. A sudden stimulus from the speaker, and the resulting decaying echo response measured by the microphone.

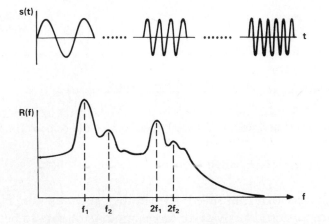

b. Swept-frequency stimulus and resonant response spectrum of the system of Figure 1.1.

Figure 1.6. Stimulus and response in the time and frequency domains.

Alternatively, one can increase the frequency of an oscillator in small increments over a wide range (i.e., *sweep* it) and measure the resonances in the steady-state response as a function of frequency, as in Figure 1.6b. The same speaker/microphone combination is used, but the synthesis/analysis processes are changed. The response in the frequency domain is called the frequency spectrum of the system. Analog instrumentation techniques are different for the time and frequency domains. But the same hardware of Figure 1.1a, assuming it has adequate performance characteristics, can do either time-domain or frequency-domain probing (with a change in the software) via the digital approach.

System characteristics in the two domains are related. Each echo period, T_0, is related to a resonant frequency, $f_n = 1/T_n$. You have probably observed the phenomenon when singing in a resonant hard-surfaced room—such as a shower. If, for example, the room is about 4 feet square and 8 feet high, there are two modes of echo, with delay times of roughly 4 ft/1100 ft/s = 3.6 ms, and 8/1100 = 7.2 ms, corresponding to a fundamental frequency of about 138 Hz, and its second harmonic at 275 Hz—the vicinity of C below middle C and middle C. Considering this basic relationship between acoustic delay and resonant frequency, it should not be surprising that one's voice is sonorous in the shower, especially when the richness of harmonics in the human voice is taken into account. The time and frequency points of view are equivalent and complementary, and—as we will show—connected by the Fourier transform (Chapter 2).

1.3.2 SYSTEM PROTOTYPES: THE SECOND ORDER SYSTEM AND THE DELAY LINE

Two simple systems will occur often in different incarnations.

(a) The second-order system trades energy back and forth at a characteristic rate between "potential" energy (a spring constant) and "kinetic" energy (an inertia).

(b) The delay line stores energy in free wave propagation, bouncing it back and forth between reflecting "walls." There are no forces except at the walls, and the modes of motion are set by geometry—sizes of the system.

Some familiar examples are:

Second Order System	*Delay Line*
LC resonator	Transmission line
Spring and mass	Echo chamber

The guitar string, the quartz crystal, the laser cavity are similar examples of these prototypes; the dual views of time and frequency response for such systems are compared in Figure 1.7.

The impulse response of the second-order system is a damped sine wave (Figure 1.7), of period T_0 and decay time τ_{decay}. T_0 is set by the rate at which energy is traded back and forth between potential and kinetic. There is always a potential energy reservoir (the capacitance C, the spring constant k, the elasticity, ϵ, of the string. . .) and an inertia (the inductance L, the mass, m, of the object or density, ρ, of the string. . .). The characteristic angular frequency, ω_0, is:

$$2\pi f_0 = \sqrt{\frac{1}{LC}} = \sqrt{\frac{k}{m}} = \sqrt{\frac{\epsilon}{\rho}} \qquad (1.3)$$

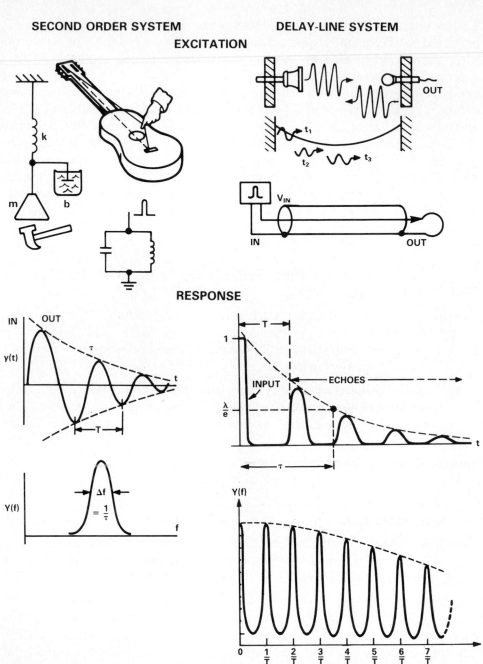

Figure 1.7. Second-order systems, typical time response, and the corresponding frequency response: (left) 2nd order resonator; (right) Echo or reverberative system.

The frequency response of the 2nd-order system displays a peak having half-power width Δf at the natural frequency f_0. The width of the peak specifies the "Q" of the system; $Q = f_0/\Delta f$.

The impulse response and frequency response are two ways of looking at the same system. The output parameters are related.

$$f_0 = \frac{1}{T_0} \tag{1.4a}$$

$$Q = \frac{4\pi}{\tau_{\text{decay}}} \tag{1.4b}$$

The impulse response of an ideal delay-line system is a series of echoes whose spacing is a measure of geometry (velocity of propagation and distance traveled), and whose decay rate is a measure of material properties, such as losses in the walls. This system has two characteristics to be encountered often: *delay* and *feedback*. The frequency response of a delay-line system has a resonance whenever a standing wave fits between the boundaries.

Exercise 1.1. A guitar string is a 2nd-order system, said to be linear. Yet, any guitar player knows how to excite a harmonic response. Isn't that a nonlinearity?

Answer: The normal "pluck" contains the fundamental and its harmonics, since it forces the string into a triangular shape before release. To excite a harmonic response, the player rests a finger where the fundamental oscillation has a peak, so only the harmonics can be excited. This is a form of filtering in the data domain: the finger forces a characteristic mode to zero amplitude.

Exercise 1.2. In a reverberative system, the amplitude of a tone decays slowly after the tone input has stopped. A reverberation can be implemented by electronic circuits, either analog or digital.

Draw a block diagram for a delay-line reverberative system. Available boxes:

Delay line, length T
Summing junction
Attenuator, passes a fraction β.

By imagining an input impulse, show how delay time plus attenuator fraction combine to give the reverb time of the closed-loop system.

Answer: Compare your result to the reverb modeled in the next Chapter.

1.3.3 THE SPECTRUM AND SPECTRAL FINGERPRINTS

Every signal has a spectrum. For each independent variable (such as time) in the data domain, a corresponding variable with inverse dimensions (such as frequency $=$ (time)$^{-1}$) becomes the independent variable in the spectral domain. The *spectrum* (plural *spectra*) is the function of the inverse variable that corresponds to a particular dependent variable in the original data. The spectrum, extracted from the data by the *Fourier transform*, is computed as the sum of products of data and sine waves at all frequencies and has a peak whenever the two contain a common frequency. The spectrum has two components, *magnitude* and *phase*; often, when spectral curves are being discussed (as in these pages), you will find only the magnitude plotted, but—as you will find in the next few chapters—the phase is essential to determining the system effects of a spectrum.

While both domains contain (or serve as *maps* to represent) the same information, one or the other may prove more useful in signal processing or in compact representation of information. For example, a single peak in the frequency domain represents a sine wave over infinite time (see Figure 1.8a). A more-complex sinusoidal signal may have numerous peaks, whose relative amplitudes and positions identify the signal waveform uniquely; for example, a sine wave linearly modulated by another sine wave has three peaks (funda-

SIGNAL–SPECTRUM PAIRS

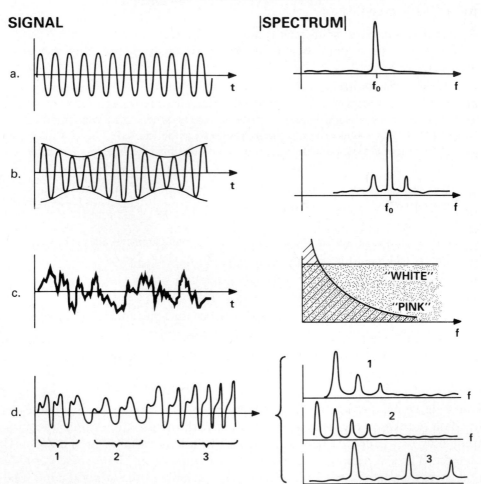

Figure 1.8. Signal-magnitude spectrum pairs: sine wave, beat pattern, white and pink noise, and speech.

mental and sidebands), as in (b). A non-repetitive noise signal has a broadband spectrum (c), "white" if its spectral density is equal everywhere and "pink" if it varies inversely with frequency. Random noise may appear to have spectral peaks if examined over a short time interval, but the spectrum's shape averages out (appears stationary) as time goes on. The spectral structure in speech and music (d) is highly nonstationary over long time intervals, but over short periods is stationary with well-defined spectral "fingerprints" which give speech phonemes and musical instruments their distinctive sounds.

Examples are many: Spectral fingerprints of sounds uttered by creatures, also called *voice prints* or *sonographs*, have long been known to birdlovers. The spectral fingerprint can diagnose an impending failure of a mechanical part, since the vibration spectrum changes as fatigue weakens the part or a crack develops. If the normal vibrational spectra of structures or rotating machines—or geological faults—were periodically monitored in appropriate ways, disasters could be averted.

A *spectrum analyzer* is an instrument that automatically determines the spectrum of a set of data. A portable spectrum analyzer is a straightforward application for DSP. As sizes and prices shrink—as they seem to in modern electronics—it could, in concept, become a normal part of bridges, cars, airplanes, etc. Hospitals use similar pattern comparison to monitor cardiac anomalies in intensive care units; the patient's recent heartbeat history, stored in memory, is compared by the microprocessor with the current heartbeat; anomalies signal the possible onset of a problem.

Spectral analysis in *real time* (i.e., while the process under study is going on, rather than *off-line*) has become practical for increasingly faster processes because of both hardware improvements and powerful algorithms. For example, you will learn in a later chapter that, while the conventional Fourier transform requires N^2 multiplications—each requiring a certain amount of time, however brief—the *fast Fourier transform* (FFT) algorithm reduces this to $(N/2)(\log_2(N))$, a factor-of-200 reduction if $N = 1,000$.

Since a DSP process can be laid out on a single chip, researchers are experimenting with special-purpose processors implanted under the skin to replace a damaged ear canal by direct stimulus to the brain, using the finite set of frequency signals which normally come from various portions of the ear canal—a biological Fourier analyzer.

The transformation from time to frequency is reversible. If the spectrum as a function of frequency is "run backwards" through the spectrum analyzer (i.e., by computing the *inverse* transform), the original signal is, in principle, reconstructed as a function of time. Since the spectrum is defined as extending from minus infinity to plus infinity, it is necessary to limit the portion of the spectrum to be transformed back. However, if we limit the spectrum, by multiplying it by a carefully chosen *bandpass window* and then performing what is called the inverse transform, only the portion of the signal lying in that frequency band is reconstructed, just like the filtered waveforms of Figure 1.3. Spectral analysis and filtering are thus closely connected, as will be seen in depth in Chapter 3.

1.4 LIMITATIONS OF ANALOG SYSTEMS

Why process signals digitally? The main answers are *predictability* and *flexibility*. Although many of the real-world quantities that we want to process have measurement accuracies and resolutions considerably worse than those of the best analog signal-processing components, the analog variables may have wide dynamic range. Although we may not require end results to great accuracy, the end result of a chain of analog computations is subject to their accumulated errors, including unpredictable *drift* and *noise*. Thus, limited dynamic range and finite signal-to-noise of analog components hamper analog processing. In addition, analog systems are programmed by wiring configurations and gain setting (and perhaps calibration of a given setup). If changes must be made (effectively) in the interconnections and gains—and especially if recalibration would be required in an analog system each time changes are made—the flexibility of digital software makes it much easier to perform the whole process digitally.

1.4.1 DYNAMIC RANGE EXAMPLE: ANALOG VS. DIGITAL AUDIO

The dynamic range of analog recordings is limited by the 1-part-in-1,000 range from smallest to largest excursions in the record groove, or in the range from noise level to maximum magnetization of magnetic tape, as shown in Figure 1.9. In the digital version, only 1's and 0's are encoded, with dynamic range extended to 1 part in 65,536 if the signal is recorded with 16 bits per measurement. Dynamic range is often measured in *decibels* (dB), proportional to the logarithm of the dynamic range.

Dynamic Range (dB) $= 20 \log_{10}$ (largest /smallest discernible signal) (1.5)*

(signals expressed in voltage or current). The realistic analog and digital limits in this example are thus:

$$\begin{array}{ll} \text{(analog)} & \text{1 part in } 1{,}000 = 60\,\text{dB} \\ \text{(digital)} & \text{1 part in } 65{,}536 = 96\,\text{dB} \end{array}$$

The relationship between number of bits, M, recorded digitally and the dynamic range DR in dB is

$$\text{DR(dB)} = 20 \log(2^M) = 20\,M \log(2) = 6.02\,M \qquad (1.6)$$

or about 6 dB per bit.

---

*The decibel is defined as $10\log_{10}$ (power ratio). Since power goes as the square of voltage or current (for constant impedance), the log of voltage or current ratios can be expressed in dB, with a factor of 20 instead of 10.

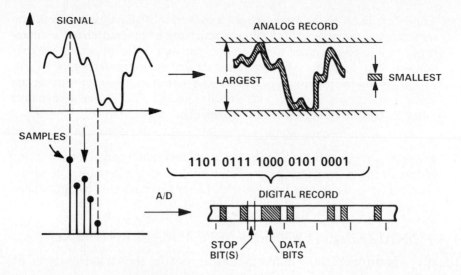

Figure 1.9. An analog record of a signal is limited by the dynamic range of the recording medium. A digitized record is limited only by the number of bits which can be encoded and stored in the time and space available.

The dynamic range of digital processing and storage is limited only by the speed of the processor, the capacity of the storage medium, and roundoff errors in arithmetical processing. While several hours of information can fit on an analog tape, only a short record could fit if stored directly in digital form. High-fidelity audio requires digitization at about 50 kHz, or 20 microseconds per sample. If the information is stored as 16-bit words, then for each million words of storage, the record could run for a time, t, equal to

$$t = [10^6 \text{ 16-bit words}] \times [20 \,\mu\text{s/word}]$$
$$= 20 \text{ seconds}$$

It's not surprising, then, that compact disks have memory capacities approaching a gigabyte. Digital compression techniques can reduce storage requirements by several orders of magnitude, making the storage demands of analog and compressed digital nearly comparable.

Example: Dynamic-range scaling is one way to handle wide-range signals with limited range channels. For example, the human ear can sense a range of more than 100 dB, from "silence" to painfully loud sounds. Yet, the individual sensors in the ear have only about a 60 dB range from smallest to largest movement. How do we do it? Recent biomedical research suggests that more than one sensor must be involved. How many would it take if the extra information were encoded in binary format?

$$100 \text{ dB} - 60 \text{ dB} = 40 \text{ dB}$$

which is close to the dynamic range of 7 binary sensors ($2^7 = 32 \cong 40$ dB) acting in concert for large-amplitude signals

Those familiar with the multiplying d/a converter (M-DAC) will note the resemblance to the wide scale of voltages which can be covered when the binary input of a fixed-reference converter sets the M-DAC's full-scale reference voltage, V_{ref}, for any value from full scale to 0 with a resolution of 1 LSB: $2^{-N} V_{ref}$; and the multiplying DAC's binary input sets the gain value at any value from 0 to nearly unity, with a resolution of 2^{-M}. The overall dynamic range (barring noise and other analog errors) is:

$$\text{Dynamic Range} = 2^N \times 2^M = 2^{N+M} \qquad (1.7)$$

where M is the number of bits on the variable reference and N is the number of bits in the gain or scale factor. This kind of dynamic-range scaling can keep arrays at the largest possible scale factor for maximum computational accuracy.

1.4.2 ANALOG FILTER TRADEOFFS: IDEALS AND REALITY

Ideally, a bandpass filter would have the properties shown in Figure 1.10:
  passband width as narrow or as wide as desired;
  complete flatness within the passband;
  rolloff as steep as desired outside the passband;
  infinite attenuation or zero leakage in the stopband;
  zero phase shift within the pass-band.

Where the passband must be wide, it makes sense to treat a bandpass filter essentially as a composite of separately designed low-pass and high-pass filters, and to optimize the rolloff characteristics of each. However, when the passband is relatively narrow, compared to the center frequency, a resonant filter may be a better prototype, as exemplified by the R-L-C circuit in Figure 1.3, with overall characteristics as shown in Figure 1.10.

The 2nd-order section (the LC filter and its equivalent) illustrates typical realities, and demonstrates the imperfectness of analog filters. The transfer function sketched in Figure 1.10 has the form:

$$|H(f)| = \frac{1}{\sqrt{1 + \zeta^2 \left(\frac{f}{f_0} - \frac{f_0}{f}\right)^2}} \qquad (1.8a)$$

$$\tan \phi = \zeta \left(\frac{f}{f_0} - \frac{f_0}{f}\right) \qquad (1.8b)$$

The phase shift covers a range of 180° and passes through zero at resonance. Lack of flatness in the pass-band gives an unfaithful output, especially with respect to phase. If the signal is a sine wave whose frequency f is just slightly off the center frequency of the bandpass, a serious phase shift will be introduced—lagging by up to 90° for f above f_0 and leading by up to 90° for f below f_0. A signal comprising a group of frequencies within the passband

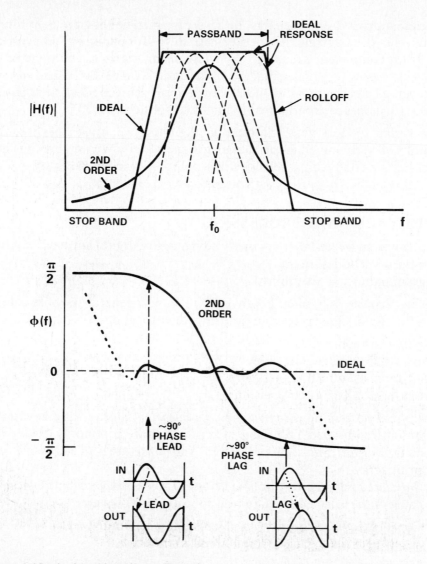

Figure 1.10. An ideal bandpass filter is characterized by its pass-band, stopband, and rolloff rate. A real filter may represent a less-than-satisfactory compromise, as in this 2nd-order example. Classic improved filters stagger-tune (dashed lines) a series of 2nd order filters to approximate the ideal.

will emerge from the 2nd-order bandpass distorted in waveform, due to the frequency-dependent phase shift.

In addition, the rolloff is slow, asymptotic to f and $1/f$ below and above f_0, which gives poor rejection of frequencies outside the passband. Narrowing the bandwidth by increasing ζ makes tuning more difficult and aggravates the phase problem. A practical approach is to use a higher-order transfer function—or a number of second-order sections with related center frequencies.

Practical bandpass design staggers the center frequencies of a group of filters to flatten the pass-band and lessen the phase shift. But parameter adjustment depends on component accuracy, and noise accumulates as the number of sections increases. We will see that the digital equivalent of this biquad section serves well as a component of filters which approach the ideal sketched in Figure 1.10; it does so programmably and with high accuracy.

But in order to deal with information digitally, we must abandon the continuity of behavior inherent in analog signals and devices, and consider approaches to establishing relationships point-by-discrete point. We now outline the elements of dealing with discretely sampled data.

1.5 DIGITAL SIGNAL PROCESSING

Earlier, it was suggested that a primary advantage of digital approaches could be expressed by the notion that:
Programmability = Flexibility

Exploring that thesis in more depth, we will show a general-purpose configuration for digital signal processing and introduce these key DSP issues:

Sampling and aliasing
Quantization and dynamic range
Windowing and information theory
Throughput and data compression

Digital signal processing is *computationally intensive*; it requires many multiplications and additions. To a range of approaches are commonly used to implement DSP algorithms. They represent various degrees of hardware-software optimization.

1. *Software with general-purpose microprocessor* develops a bottleneck because of the time required for multiplication in software.

2. Coprocessors, multipliers, and multiplier/accumulators as *on-the-bus microperipherals* offload computation-intensive tasks.

3. *Hardware-intensive solutions*—e.g., processor systems with parallel operation; microcoded Bit-, Byte-, and Word-Slice^{TM} systems— minimize the number of cycles to execute a DSP program.

4. The *programmable single-chip digital signal processor* combines software flexibility with DSP hardware power.

While (1) and (2) are moderate extensions of conventional microprocessor systems, (3) tends to link the most hardware to achieve the fastest performance at highest cost, while (4) unbottlenecks conventional microprocessor architecture for high DSP throughput at moderate cost. Specific performance comparisons are drawn in Section 1.7.4.

When digital processing replaces analog circuits, the time spent on software development expands. Necessary DSP development tools include assemblers to translate program statements into binary code, simulators to test design performance in software, and emulators to cycle controllably through system operation. Although the front-end investment may be substantial, the payoff is enormous.

1.5.1 FLEXIBILITY: KEY ADVANTAGE TO DSP

Figure 1.11 compares block diagrams for solving problems employing analog and digital signal-processing approaches. A solution to a signal processing problem using analog circuit hardware looks different for each problem. By contrast, the same system components appear in nearly every DSP block diagram; what changes is the program.

The digital configuration shown in Figure 1.11 is fairly general. The analog to digital (a/d) and digital to analog (d/a) converter connect the sampled-data processor to real-world analog continuous-valued signals. The digital components in the center resemble a typical microcomputer system, but with some enhancements. In addition to the procesing hardware, there are two memory banks. One, for temporary data storage, must be alterable (RAM); the other is for fixed coefficients—equivalent to the circuit component parameters in the analog circuit—and for program steps, which configure the circuit function. This second memory may be alterable (RAM), permanent (ROM), or semi-permanent (EPROM).

A separate multiplier or multiplier/accumulator (and high-speed ALU) speeds DSP programs, especially if its inputs are connected on the same bus as the memories to avoid the bottleneck of passing through the ALU. The hardware ALU and multiplier(-accumulator) enhances system throughput in two ways. In addition to the direct speed increase provided by fast multiplication in hardware, both inputs of the MAC and/or ALU may be loaded in one step; the next inputs can be loaded while the previous product is being calculated (this is known as *pipelining*). Finally, since DSP algorithms can involve jumping through address space to comb out selected data samples, an address generator (also programmable), prepares the next address to be selected while computation is going on. This eases the load on the ALU, since it is not needed for address calculations.

A hardware multiplier computes the product of two n-bit digital inputs, x and y, in a short time— for example, from less than 25 to 150 nanoseconds. The multiplication is performed asynchronously. To relieve the overhead on the microprocessor, multiplier hardware (such as the 16-bit × 16-bit ADSP-1016A; see Figure 1.12a) includes registers that store not only the x and y inputs and z outputs, but also control and format information. The product may have twice as many bits as either input, and the user has control over whether the less-significant half is truncated, rounded, or returned for the increased accuracy of double precision.

Since many filter and spectral-analysis operations involve a multiplication fol-

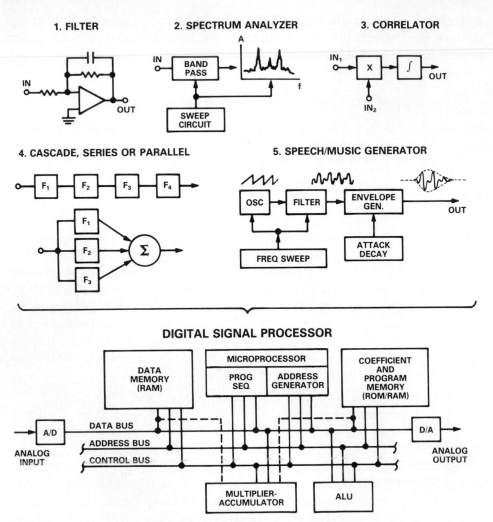

Figure 1.11. General-purpose DSP architecture compared to a variety of analog circuits it can replace.

lowed by addition to an existing sum (e.g., Equation 1.11 in Section 1.6.1), an on-chip accumulator is a natural extension to a multiplier chip; a typical example is the 16×16 ADSP-1010A multiplier-accumulator (MAC), shown in Figure 1.12b. Combining these functions can avoid an extra program step.

Signal processing problems have been solved successfully for years with software programs, using increasingly faster machines, without hardwired multiplication circuits. The dedicated digital signal-processor, with its real-world connections, is not just a new breed of computer; it is also a subsystem for enhancing computers, with fast hardware and architectural features that speed real-time throughput. The impact of the new hardware is especially great since one can now accomplish with a single chip what earlier required the resources of an expensive general-purpose computer.

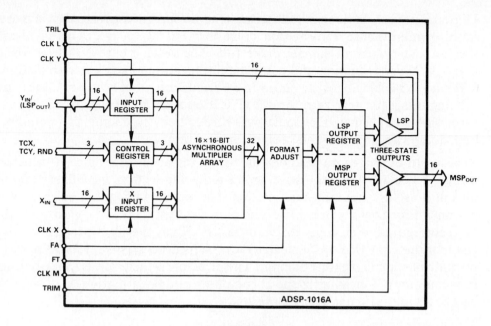

a. Multiplier.

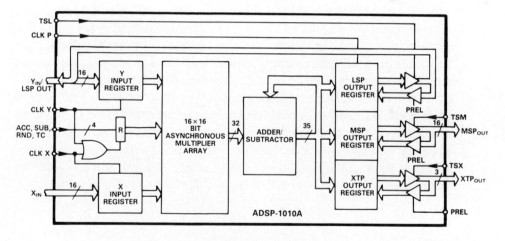

b. Multiplier-accumulator.

Figure 1.12. Examples of high-speed CMOS multiplier architectures.

1.5.2 DSP ISSUES AND TERMINOLOGY

A digital circuit computation may require dozens of clock cycles for a filter or thousands for a spectrum analysis. The subsequently limited *throughput* (data rate) used to be a serious limitation of real-time DSP bandwidth. Now that co-processors, independent multipliers, and improvements in organization (architecture) replace slow software program loops with operations requiring but a few cycles, DSP throughputs have increased substantially, reaching real-time for an increasing number of analog problems.

Digital data transmission is less sensitive to noise. The process allows more faithful long-distance transmission, since noise has less chance of being mistaken for data when all signals make full-scale voltage excursions between 0 and 1. Noise-reduction and confirmation techniques such as redundancy and parity checking boost accuracy in environments where the probability of missing bits is not near-zero. And DSP is itself a key to recovering digital information from signals that must pass through environments that combine high noise with large amounts of signal attenuation.

Real-world DSP is more than just an analog replacement; it makes possible things you couldn't do otherwise. In the following section, we will see examples of a useful filter transfer function that has no analog equivalent. Programmability gives the flexibility inherent in software; the same hardware can be reconfigured to do quite different tasks. Often, the DSP processor is so fast that the same DSP system can handle many processes "simultaneously", by multiplexing in the time domain. The principle is the same as time-sharing in a computer: a megahertz-speed computer can pay attention to dozens of people typing at 10 characters per second.

The key design issues to pay attention to when designing digital systems to handle analog data are:

1. The data must be sampled at discrete intervals, t_s (Figure 1.13a), which must be carefully chosen to insure accurate representation of an analog signal without *aliasing*. A familiar—and extreme—example of what can happen with inadequate sampling rates is analogous to the stroboscopic effect: moviegoers, (who are watching sampled data consisting of stationary frames shown at a high-enough speed to produce the illusion of continuous motion) observe that wagon wheels in older Western movies appear to stop or even go backwards when the rate of rotation of the spokes just matches the film frame speed—the same phenomenon that allows a flashing strobe light to freeze the motion of a rotating machine.

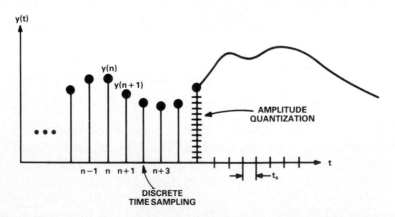

a. Data is sampled in the process of going from the analog to the digital domain.

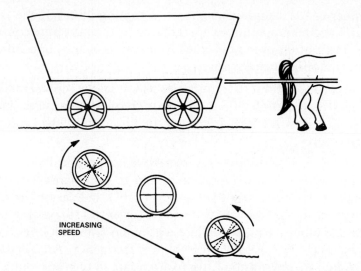

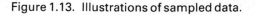

b. The familiar "wagon wheels going backwards" effect of inadequate sampling, called "aliasing" in DSP.

Figure 1.13. Illustrations of sampled data.

2. Analog data is continuous but digital data is quantized with 2^n levels— corresponding to n bits—for each observation. Since an operation such as filtering or spectrum analysis involves a series of computations, care must be taken to minimize the effect of accumulated roundoff error on the accuracy.

For example, a bank which truncates (and keeps!) fractional pennies on your savings interest, rather than rounding (up or down) honestly, will accumulate significant extra revenue when there are many such accounts. For example,

INTEREST COMPUTATION

	Accumulated	Truncated	Rounded	
Month 1	$1.474	$1.47	$1.47	
Month 2	$1.826	$1.82	$1.83	
Subtotal	$3.300	$3.29	$3.30	
Bank keeps:		$0.01	$0.00	from each customer
		$10,000.00	$0.00	from a million customers

Sampling theory is described below and developed in Chapter 2. Error analysis is touched on throughout later chapters.

1.5.3 SAMPLED-DATA CONSIDERATIONS

When a waveform like that shown in Figure 1.13a is measured, enough data samples must be taken to characterize the system adequately. Although the phenomenon may be continuous, the analysis of data in a digital computer is always discrete; the continuous function $y(t)$ becomes the set $y(n\,t_s)$, where

n is the number of the sample and t_s is the sampling interval (the sampling rate, $f_s = 1/t_s$). Continous operations are replaced by their discrete equivalents; for example, integration over t becomes summation over n, and differential equations become difference equations.

Sampling puts an upper limit to frequencies which can be represented. There must be at least two samples of the highest frequency present (including noise), or else *aliasing* will take place. Undesired high-frequency components must be filtered out.

The wagon wheels or strobe light example above is a practical demonstration of *aliasing*. The wheels appear to stop and then go backwards when the rate at which spokes go by (or multiples of it) approaches and passes the sampling frequency. As long as the frequency of spokes (or any data) passing by is less than half the sampling frequency, there will be no ambiguity or alias. This maximum frequency, $f_s/2$, is called the Nyquist frequency. Analog data must be passed through an antialiasing filter to attenuate to zero any signals above $f_s/2$, even if the data itself has no higher frequencies, because even unwanted high-frequency noise can cause aliasing errors in the received signal, appearing at frequencies in the signal band.

The wagon-wheel phenomenon has come back in modern dress with computer-graphic film sequences composed of line segments, which must be blurred to avoid aliased motion. The blurring (electrically equivalent to low-pass filtering) is equivalent to defocusing a lens for moving objects so that spokes or line segments can no longer be resolved when the sampling rate is less than the Nyquist frequency.

Data is usually recorded over a finite observation interval, called the *window*. Because the frequency of a wave can only be measured perfectly over an infinite time interval, the window limits the spectral resolution in the frequency domain. From the point of view of the digital process, only that portion of the event lying within the observation window exists. If the window is not large enough to take in an adequate picture, the results can be misleading, as with the real-life "window" situation pictured in Figure 1.14. Window issues occur both in spectral analysis and in filtering, since the filter shape in the frequency domain affects the fidelity of the output signal. The ringing in the example of Figure 1.5 is an example of an improperly chosen filter window.

1.5.4 THROUGHPUT EXPANSION, DATA COMPRESSION, AND PIPELINING

Information flow often has to be compressed. We do this without knowing it. The amount of information in one visual image seen by the human eyes would, if digitized directly, fill all the brain cells in a matter of seconds, as shown by the following estimate (pixels are picture elements, the unit of picture resolution):

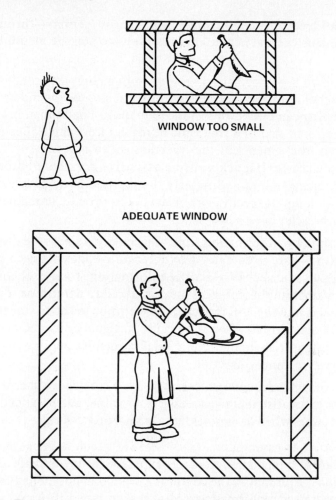

WINDOW TOO SMALL

ADEQUATE WINDOW

Figure 1.14. The "window" in a digitally sampled system must be chosen carefully to give an accurate picture of what is happening.

Info per glance $= (N\text{ Pixels}) \times (\text{Gray scale})^\star \times (N\text{ Colors})$ bits

N Pixels $= [(\text{visual field})/(\text{visual resolution})]^2$ (for solid angle)
$\quad\quad\quad = [(100°)/(0.1°)]^2 = 10^6$

Gray scale (8 bits) $= 2^8 = 256$

M colors $= (2500\text{ Å vision range})/(50\text{ Å resolution}) = 50$

Info per glance $= 10^6 \times 256 \times 50 \cong 10^{10}$ bits

At this rate, the data handling capacity of the 10^{12} neurons in the human brain would be exceeded in no time (i.e., 100 samples). Compression of information

---

*Gray scale is the number of shades of gray the human eye can resolve between white and black. 8 bits is the typical resolution used in digital displays; it is more than adequate.

can occur at all levels in the pipeline, from sensor (retina)—through inter-mediate processors (neural nets to optic nerve)—to storage medium (optical cortex).

For example, if all the information needed from a complex scene is whether a moving object of appropriate size is within reach of an appendage, special-ized preprocessing can cut down storage demands. The human brain is con-siderably larger than that of a frog, but humans rarely catch flies with their tongues (except by accident). If the question is, "was it moving?" the pea-brained frog can answer it quickly, using extensive distributed processing in the pipeline from eye to brain. Similarly, hard-wired human reflex arcs per-form rapid evaluation of specific sensor data, and provide immediate stimuli to appropriate muscles in response.

In the case of the frog, information from adjacent "pixels"—the discrete sen-sors in the retina—can be compared to determine edges between regions of light and shadow, since edges take little information storage. Second, infor-mation on adjacent time samples can be compared to determine if one of the shadow regions is moving. All this takes place prior to the brain, so the frog can relax in the sun until an interrupt occurs to signal dinner. Of course, the process is not quite that simple, but similar principles apply with regard to determining size, distance, direction.

Similarly, to obtain adequate processing speed, for specialized problems, when and how to use distributed processing, pipeline, and parallel processing are important issues when designing digital signal processors.

Data compression is essential in large bandwidth communications. This is because the rate of information transfer, expressed as the bit-rate (in bits per second, or *baud*), is limited by the communications channel. For example, the direct (uncompressed) information-transfer requirements of common com-munications are:

System	Features	Bit-Rate
Computer phone line		300 to 2400
Telephone, analog	Voice-grade	3kHz
Telephone, digital	8kHz, 8 bits	64kHz
Digital audio	16 bits $\times$ 40kHz	640kHz
TV transmission	400 points $\times$ 500 lines $\times$ 30 frames/second	6MHz

While the bandwidth per communication channel can often be set wide enough to handle this (as is now done for phone lines and TV), channel capaci-ty can be saved by compression of information, using algorithms to be developed in later chapters. Typical strategies are to send only differences between sampled values, or to model the information and send only the (changing) parameters of the model.

1.6 NONRECURSIVE FILTERS: THE MOVING AVERAGE AS AN EXAMPLE

Filtering is one of the major applications of DSP. The reader probably knows the most common digital filter already, but by a different name. The moving average (or tapped delay line), familiar through elementary lab experiments, stock-market, or diet plotting, is a prototype of the nonrecursive or *finite impulse-response* (FIR) filter. We begin in the time domain, using this filter to introduce the convolution sum. Then this filter prescription in the time domain is related to the analog transfer function in the frequency domain. We will evaluate the moving-average filter performance as a function of number of points averaged, called taps. We will see a 90-tap FIR filter, unlikely to be implemented by means other than digital.

1.6.1 FINITE IMPULSE-RESPONSE FILTERS

The moving average is familiar from periodic records of one's diet progress, the value of an investment, or averaging noisy data in a freshman physics experiment. The average of nearby points (Figure 1.15) is the simplest way to eliminate noise and look for real trends.

Taking human weight as our example—it fluctuates continuously, but we may sample it at regular intervals by standing on the scale at, say, 7:00 AM every day. To compute our average weight over the last 7 days, on the 100th day of our diet, seeking to average out daily variations during the past week, we will sum our weight on day 100, multiplied by 1/7, with our weight on days 99, 98, 97, 96, 95, and 94, each multiplied by 1/7. We ignore our weight on all previous days and future days (i.e., in effect, we multiply them by 0 instead of by 1/7).

To formalize the process, write what we have done as:

$$y(100) = \tfrac{1}{7}x(100) + \tfrac{1}{7}x(99) + \ldots + \tfrac{1}{7}x(94) \tag{1.9}$$

Figure 1.15a shows a set of 14 such data points, and the averages for the most-recent 7 days. If we want to use a somewhat fancier average, for example, one weighted to be more sensitive to the most-recent data, we may use coefficients, h, that differ somewhat from 1/7— and from one another (but add up to unity).

The corresponding formula for such a filter would be:

$$y(n) = h(0)\,x(n) + h(1)\,x(n-1) + \ldots + h(6)\,x(n-6) \tag{1.10}$$

where each data point, x, is multiplied by a weighting factor, h, numbered in order of time from today (0) to the earliest day (6). We can generalize it to

$$y(n) = \sum_{0}^{N-1} h(m)\,x(n-m) \tag{1.11}$$

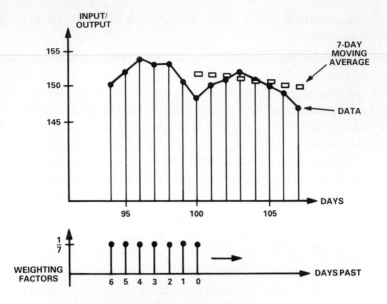

a. Input and response; weighting factors.

| | | 7-DAY WEIGHTS | | | | | | | |
| DAY | WEIGHT | DAY | DAY | DAY | DAY | DAY | DAY | DAY | DAY |
	(LBS)	100	101	102	103	104	105	106	107
94	150	150	–	–	–	–	–	–	–
95	152	152	152	–	–	–	–	–	–
96	154	154	154	154	–	–	–	–	–
97	153	153	153	153	153	–	–	–	–
98	153	153	153	153	153	153	–	–	–
99	151	151	151	151	151	151	151	–	–
100	148	148	148	148	148	148	148	148	–
101	150	–	150	150	150	150	150	150	150
102	151	–	–	151	151	151	151	151	151
103	152	–	–	–	152	152	152	152	152
104	151	–	–	–	–	151	151	151	151
105	150	–	–	–	–	–	150	150	150
106	149	–	–	–	–	–	–	149	149
107	147	–	–	–	–	–	–	–	147
Average = Sum/7		151.6	151.6	151.4	151.1	150.9	150.4	150.1	150.0

b. How the results are computed.

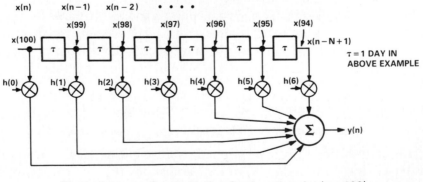

c. Block diagram of computation for the nth point ($n = 100$).

Figure 1.15. Digital filtering example—the moving-average filter combines recent samples.

where N is the number of data points and m is the number assigned to a data point, starting with 0, for the point associated with n. The actual calculation process (Figure 1.15b) shows how the filter moves past the data, taking each set of 7 points in turn.*

The example above is a practical definition of a *finite impulse-response* (FIR) filter. It is *finite*, because the number of steps before the output becomes constant after the input becomes constant is limited to N, and there is no recursion (i.e., feedback)—the output depends only on the input and the coefficients; there are no output terms used in calculating the sum. If the input ceases to change, the output stops changing N steps later. (We will discuss *impulse-response* later on.)

If we are to calculate this sum digitally, we can use the digital circuit representation (Figure 1.15c) involving basic DSP building blocks, the unit delay (τ), representing each day in the example, multiplication (by the weighting factors), and summation. To make a unit-delay, which represents the time between samples in real time, store the data in memory and access each point, in its turn, one clock-cycle later.

In embodying filters of real-time signals, the number of *taps* in a filter (Figure 1.15c) is equal to N, the number of nearby samples which are combined to form an output at any point in time. Recent previous inputs, $x(n-m)$, are multiplied by filter coefficients, $h(m)$, and summed; this is called a *convolution sum*, for reasons that will become clear in later chapters. For the simplest moving average, $h(m) = 1/N$, since the sum is over finite N, the output falls to zero after N samples of zero input, hence the term *finite* impulse response.

What about the frequency response? Although we calculate this filter's response in the time domain (data domain), what does the transfer function look like? How would we relate, for example, the action of a moving average to an analog low-pass filter? How is the response of the moving-average filter affected by the number of taps? The explanation must wait until more background is developed in Chapter 2, but it will be shown there that the transfer function, $H(f)$, of the N-point moving average has the form:

$$H(f) = \frac{\sin(\pi N f t_S)}{N \sin(\pi f t_S)} \tag{1.12}$$

While more taps make more zeros, the amplitudes of the sideband peaks in the stopband are unchanged, so high-frequency components can leak through. As seen in the plot of Figure 1.16a, increasing the number of points averaged narrows the passband but does nothing to attenuate the stopband.

---

*When the coefficients are all equal ($= 1/7$), there is no problem with dividing the sums (instead of the individual terms) by 7, for the sake of simplicity—as was done here. This is only one of the algorithms that can simplify calculation in some cases, especially when DSP hardware is not used. Another one that can be used, where coefficients are equal, is to recognize that, once the first sum is performed, subsequent sums can be computed by simply subtracting the oldest point and adding the newest one.

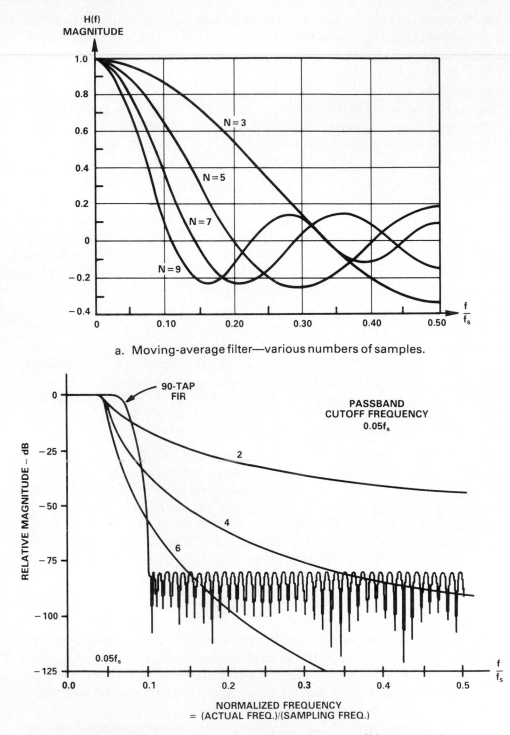

a. Moving-average filter—various numbers of samples.

b. 90-tap FIR filter response compared with sharp-cutoff Chebyshev filters.

Figure 1.16. Frequency response of digital low-pass filters.

Used as a low-pass filter (not its most suitable application), the moving-average filter with equal weighting has narrow bandwidth and poor performance.

1.6.2 DIGITAL FILTERS

A filter's desired sine-wave response can be specified in terms of its transfer function, $H(f)$. To translate this to a set of FIR coefficients, (as in Figure 1.15c), algorithms (to be discussed in later chapters) are available. The set of coefficients, $h(m)$, is identical (but reversed—or *folded*—in time) to the sampled time response of the filter to a single unit impulse, as it propagates down the delay line. Since that response disappears after the $(N - 1)$th tap, the filter has a *finite impulse response* (FIR).

The shape of a digital filter transfer function can thus be custom-tailored to fit individual requirements, and—being programmable—can be changed at will with no hardware changes. In Figure 1.16b, 2nd, 4th, and 6th-order ideal Chebyshev low-pass filter transfer functions (a commonly used analog filter configuration), optimized for 1-dB (12%) in-band error, are compared with a 90-tap (i.e., 90 coefficients and 90 sequential memory locations) digital filter, optimized for *0.024%* error.

There is no practical analog equivalent; this is higher order than is realistic with analog hardware. Since the response is flatter within the passband, the signal is reproduced more faithfully, and phase distortion in the pass-band is negligible. Maximum error in the stop band is at -80 dB, or about ½-LSB at 12-bit resolution. Note that, by a simple choice of appropriate coefficients, lower-order analog filters could be simulated quite accurately—and be free from errors due to component tolerances or variation.

The improvement is illustrated by considering the reduction in harmonic distortion of an imperfect sine wave near the cutoff frequency. The rolloff of an Nth-order filter well beyond cutoff goes as $(f_c/f)^N$, so distortion in the rth harmonic is thus reduced by about $(1/r)^N$. More precisely, reading off the graphs of Fig. 1.16,

Filter	Reduction in harmonic distortion 2nd harmonic	5th harmonic
2nd order	-16 dB	-36 dB
4th order	-36 dB	-68 dB
6th order	-60 dB	-112 dB
90-tap FIR	-80 dB	-80 dB

The 90-tap digital filter, configured for the response shown, reduces the output by 80 dB, or four orders of magnitude, at any frequency above about 2 times the cutoff frequency. This filter is effectively a "brick wall" above its passband.

Exercise 1.3. Do-it-yourself digital filter example. The table of data below represents the behavior of a certain stock at daily intervals over a time period of 10 weeks. Perform digital filter computations to

(a) Smooth the data. Try N-point averages, with N in the range from 8 to 24.

(b) Figure the rate of change. Do a derivative—but do it on the smoothed data of (a) to avoid noise.

(c) Is there a cyclic trend visible which would allow you to decide whether to buy more or sell it off in the near future? Try the following filter:

$$y(n) = \frac{3}{4}y(n-1) + [x(n) - x(n-m)]$$

where $2m$ is (approximately) the number of samples within a suspected trend cycle.

26	27	29	28.625	27.5
27	28	29.25	28	28.375
27.625	28.375	29.25	27.625	28.125
27.75	28.5	29.875	27.25	27
27.5	29	28.625	27	27
28.125	22.625	24.25	23.375	21
28	22.5	24.5	23	21.75
27.125	23.5	24.5	20	20.5
24.75	21.875	23.125	20.5	19.75
23.25	23.75	23.25	21.25	22

Exercise 1.4. The moving-average filter viewed in the frequency domain has the transfer function of Eq. 1.12, Fig. 1.16a. How are the passband width and stopband attenuation related to the number of taps?

(a) Passband width. Roughly: where does the first zero occur? Answer: $f_s/(N-1)$. More conventionally, calculate the -3dB point vs. N.

(b) Stopband attenuation. Calculate the worst-case point, the first sideband peak in Eq. 1.9.

1.7 RECURSIVE FILTERS FROM THEIR ANALOG COUNTERPARTS

This section will show how digital filters can be designed to have fewer coefficients and thus greater speed, in configurations that relate closely to their analog counterparts, by using feedback. A time estimate for the low-pass done in software makes it clear why hardware implementations of multipliers are necessary. The recursive difference equation can be related to the analog system's differential equation and transfer function. The impulse response of the familiar RC filter helps explain the term "convolution sum" as a memory of previous "kicks". Although a recursive system has no exact non-recursive counterpart, an approximate FIR time response can be matched to the response of a recursive filter up to a defined cutoff in the impulse response, and shows how many taps would be needed to simulate infinite-impulse-response filter behavior.

1.7.1 ANALOG FEEDBACK FILTERS AND THEIR RECURSIVE DIGITAL COUNTERPARTS

The circuit of a first-order analog low pass filter is shown in Figure 1.17a alongside its digital counterpart. The analog-computer block diagram, which models the filter and provides the solution of equation 1.2a, demonstrates the recursive nature of the continuous solution. It is similar in form to the block diagram for the digital solution, with continuous time integration (or differentiation, if loop causality is reversed) replaced by time-delay and iterative summation. Both involve the sum of the input and output signals. The feedback paths from the output are called recursive (from *recursion* relations in mathematics), because an output signal plays a role in its own future, as shown by the the difference equation, 1.14. Recursive filters require fewer coefficients and unit time delays—and the associated multiplications—than do nonrecursive filters; this gives them a higher throughput for a given performance specification.

The 1st-order low-pass stopband rejection can be improved upon by adding energy storages; the next step is the 2nd order order low-pass filter; one configuration is depicted in Figure 1.17b. The transfer function, also shown, carries key system properties in its poles (the low-pass transfer function has only poles, no zeros). The 2nd-order low pass is an example of the more general biquadratic ("BIQUAD") transfer function:

$$\frac{Y(s)}{X(s)} = \frac{a + bs + cs^2}{1 + ds + es^2} \tag{1.13}$$

The response of the filter is determined by which coefficients are present, and by their magnitudes. If b and c are zero, the filter is a low-pass; if a and b are zero, it is a high-pass, etc. Similarly (Figure 1.17c), the digital equivalent also has its poles and zeros, controlled by the user via the coefficient values fed to the multipliers to set the characteristic frequency, shape (HP,LP,BP), and sharpness of the response.

Higher-order filters (both analog and digital) can be synthesized from a cascade of similar or dissimilar 2nd order sections to provide predictable 4th-, 6th-, . . . order filtering. The complexity of higher-order analog circuits makes digital replacement attractive, since higher-order digital filters are drift-free and require little or no additional hardware investment, just additional program steps. However, as with analog feedback filters, stability is an important consideration in the design, and accumulated roundoff errors in the digital counterpart can add noise if not carefully anticipated in simulation.

1.7.2 INTRODUCING RECURSIVE FILTERS

The following relationship between input and output is one way to generate a 1st-order low-pass filter:

$$y(n) = \alpha\, x(n) + (1-\alpha)y(n-1) \qquad (1.14)$$

In words, the output of the filter is equal to some fraction of the present input value plus the complement-of-the-fraction times the previous output value. Figure 1.17d tabulates the calculation of an example in which x is a step function from 0 to a constant 1.0, with α equal to 0.3935. The response, computed by a simple BASIC program, is asymptotic to 1, and is virtually identical to the characteristic exponential step response of a continuous RC circuit having a time constant equal to 2 units, also shown. (The solutions would be more nearly identical if α weren't rounded off, i.e., $\alpha = 0.3934693 = e^{-1/2}$.) The constant, α, plays the role of a time constant. Small α means longer time constants—the output is mostly the accumulated memory of past outputs. For large α, it is the most recent inputs that matter. The constant, α, will be related below to the decay time, RC, of the analog version.

It is revealing to follow the response of the 1st-order low-pass to a unit impulse at $n = 1$.

n	$x(n)$	$\alpha x(n)$	$(1-\alpha)\ y(n-1)$	$y(n)$
0	0	0	0	0
1	1	α	0	α
2	0	0	$(1-\alpha)\ \alpha$	$(1-\alpha)\ \alpha$
3	0	0	$(1-\alpha)^2\alpha$	$(1-\alpha)^2\alpha$
.	.	.	.	.
.	.	.	.	.
m	0	0		$(1-\alpha)^{m-1}\alpha$

The output at time n is a constant fraction, $(1-\alpha)$, of the previous output at $(n-1)$, modelling the characteristic decreasing exponential response of the RC circuit (in the example, the complement of the responses in Figure 1.17d). We observe that, following the impulse, the response continues indefinitely and is asymptotic to zero at infinite time, hence the filter has *infinite impulse response* (IIR).

The following general recipe includes both the recursive examples of this section and nonrecursive filters of the previous section.

The difference equation is a key concept in digital signal processing. An important consideration when implementing of either FIR or IIR filters digitally

RECURSIVE OR IIR FILTER

$$y(n) = \sum_{0}^{M-1} a(k)\,x(n-k) + \sum_{0}^{N-1} b(k)\,y(n-k) \qquad (1.15)$$

NONRECURSIVE OR FIR FILTER

$$y(n) = \sum_{0}^{M-1} a(k)\,x(n-k) \qquad (1.16)$$

with difference equations is throughput, discussed in the following section. The four approaches to digital signal-processing (introduced in Section 1.5) are discussed in somewhat greater detail below to compare the bandwidths of signals handled in real time by several common signal-processing algorithms.

1.7.3 THE DIGITAL FILTER IN BLOCK-DIAGRAM FORM

The block diagram for signal flow of Figure 1.17a is constructed by inspection of the difference equation, Equation 1.14. The symbols specify universal building blocks in DSP: summation $(+)$, multiplication $(\times)$, and the unit delay τ_s. The unit-delay operation is a register which passes its contents on after one clock cycle. The generalized delay, τ, is equal to $m\tau_s$ clock cycles.

Recursion is feedback; the analog circuit of Figure 1.17a computes the 'next' output from the present input plus a delayed sample of the output. Recursive filters share the strengths and weakness of any feedback system: feedback enhances signal-processing power but can result in instability. The timing example in Section 1.7.6 will illustrate IIR's attraction: rapid processing, compared to an FIR counterpart. The guidelines to be developed in Chapter 4 will lead the reader past the traps to the rewards of recursive filters—where they are appropriate.

1.7.4 COMPARING SOFTWARE AND HARDWARE-MULTIPLIER HANDLING OF THE 1ST-ORDER LOW-PASS

For a fairer speed comparison between hardware and software approaches, we use the assembly language structure of Table 1.1 and the memory map of Figure 1.18 to carry out the filter of Equation 1.14 in software, with greater speed than in the case of Figure 1.17's BASIC model.

We assume a typical microprocessor architecture, where both input/output data (I/O) and memory data must pass through an accumulator, A. MM specifies the starting address of a buffer in memory for temporary storage of most-recent computations and coefficients. An extra y-location has been allowed so that a high-pass filter or a derivative filter can also be embodied

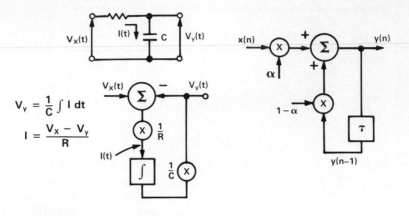

$$V_y = \frac{1}{C} \int I \, dt$$

$$I = \frac{V_x - V_y}{R}$$

a. First-order low pass analog circuit, the equivalent analog-computer diagram, and the recursive digital counterpart.

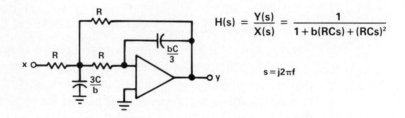

$$H(s) = \frac{Y(s)}{X(s)} = \frac{1}{1 + b(RCs) + (RCs)^2}$$

$$s = j2\pi f$$

b. Second-order low-pass circuit and its transfer function.

$$H(s) = \frac{-Z_1^{-1} Z_4^{-1}}{Z_5^{-1}\left(\sum\limits_{n=1}^{4} Z_n^{-1}\right) + Z_3^{-1} Z_4^{-1}}$$

$$H(z) = C_1 \frac{1 + C_2 z^{-1} + C_3 z^{-2}}{1 + C_4 z^{-1} + C_5 z^{-2}}$$

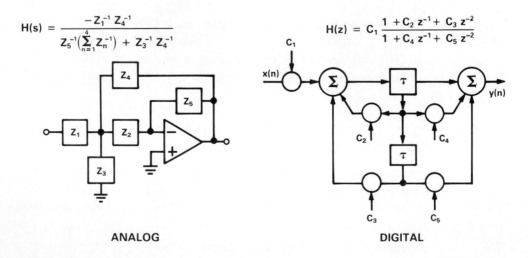

ANALOG   **DIGITAL**

c. Biquad filters.

```
REM  Comparing 1st-order IIR and  unit-lag
DIM x(25): DIM y(25): DIM z(25): DIM w(25)
x(0) = 0: y(0) = 0: z(0) = 0
M = 16: REM number of data points
GOSUB IIR
GOSUB Exponential
GOSUB Printer
END
    IIR:
        A = .3935: REM gain factor
            FOR n = 1 TO M
                x(n) = 1: REM unit step--0 initially, 1 for all n > 0
                y(n) = A * x(n) + (1 - A) * y(n - 1): REM  discrete steps
            NEXT n
        RETURN
    Exponential:
            FOR t = 1 TO M
                z(t) = 1 - EXP(-t/2): REM continuous function
            NEXT t
        RETURN
    Printer:
        PRINT "Comparing IIR and Unit-Lag":PRINT
        PRINT "n or t", "  x(n)";"      y(n)", " z(t)", "  y - z"
        PRINT ,:"          IIR"," Exp":PRINT
        PRINT 0, x(0), y(0), z(0)
            FOR i = 1 TO M
                w(i)  = y(i) - z(i)
                y(i) = INT(1000 * y(i) + .5)/1000: REM round to 3 digits
                z(i) = INT(1000 * z(i) + .5)/1000
                w(i) = INT(100000! * w(i) + .5)/100000!
                PRINT i, x(i);"      "; y(i), z(i), w(i)
            NEXT i
        RETURN
```

COMPARING IIR AND UNIT-LAG

n or t	x(n)	y(n) IIR	z(t) Exp	y - z
0	0	0	0	
1	1	.394	.393	.00003
2	1	.632	.632	.00004
3	1	.777	.777	.00003
4	1	.865	.865	.00003
5	1	.918	.918	.00002
6	1	.95	.95	.00002
7	1	.97	.97	.00001
8	1	.982	.982	.00001
9	1	.989	.989	.00001
10	1	.993	.993	0
11	1	.996	.996	0
12	1	.998	.998	0
13	1	.998	.998	0
14	1	.999	.999	0
15	1	.999	.999	0
16	1	1	1	0

d. Example of recursive low-pass filter response to step function, $\alpha = 0.3935$, $y(n) = 0.3935\, x(n) + 0.6065\, y(n-1)$. Computer program to compute step response and run demonstrates simplicity of recursive low-pass filter; comparison with exponential verifies fidelity to dynamic form.

Figure 1.17. Recursive filtering.

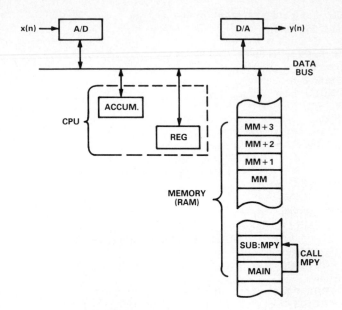

Figure 1.18. Computer register and memory-map structure for the 1st-order recursive filter.

(See Exercises). The structure of a program allows us to estimate the speeds possible in various implementations.

This first-order loop takes 11 instruction steps, each about $1\,\mu s$ long (assuming typically 4 cycles of a 5 MHz clock per instruction) if all instructions were equal, giving a loop time of $11\,\mu s$—or a throughput of about 90 kHz, sufficient for full audio bandwidth. However, all instructions are not equal: the software multiply takes longer. The comparison below shows how coprocessors and fast multipliers reduce multiply time to approach and even equal the time of other operations.

Table 1.1. Program structure for 1st-order low-pass filter

LOOP: Point to buffers
           Input x from A/D to A-register
           Put x into memory
           Multiply α by x. Result returned in memory
           Bring back last $y(n-1)$
           Complement α to get $(1-\alpha)$
           Multiply $(1-\alpha)$ by $y(n-1)$
           Add αx to $(1-\alpha)y(n-1)$
           Store $y(n)$ for next loop
           Output y to D/A
           Go to LOOP

(a) *Multiply in software*.

The multiply operation in software, a sequence of shifts and adds, is carried out in a subroutine call, whose structure is shown in Table 1.2.

The loop must be executed as many times as there are bits in the multiplier, hence the slowness. An estimate is:

> Multiply (MPY) subroutine:
> 6 initialization instructions $\times$ 1 μs/instruction = 6 μs
> (6 mpy loop instr.) $\times$ (1 μs/instr.) $\times$ (16 loops) = 96 μs
> Multiply-time estimate = 102 μs per multiplication

Here is why digital signal processing was not a popular real-time application of microcomputers prior to the advent of digital hardware multipliers:

Execution-time estimate, 1st-order low-pass

Ordinary instructions	9×1 μs/instruction
Multiply subroutine	2×102 μs/multiply
Total	215 μs, or about 0.2 ms
Max sampling frequency	5 kHz

The bandwidth of incoming signals is thus limited to only about 2.5 kHz, even for this, the simplest, most-compact IIR filter. For the many multiplications of a multi-tap FIR filter, bandwidth is even more severely restricted.

Table 1.2. Multiply subroutine structure

SUBROUTINE MPY
    Multiplicand in memory location MM
    Multiplier in location MM + 1
    Put 16-bit result in MM + 2 (least significant half) and
       MM + 3 (most significant half)

MPY: Point H,L to MM
           Get multiplicand
           Get multiplier
           Initialize product to 0
           Counter = 16

LOOP: Partial product (PP) $\times$ 2
           Shift a multiplier bit to carry
           Is this multiplier bit = 1?
           Yes; add multiplicand to PP
           Decrement counter
           Loop for next partial product

           Final product to memory

RETURN

(b) *Coprocessor offloading.*

The coprocessor (the 8087 in an 8086 system, for example) can perform fixed- or floating-point multiplications in about the same time that the 8086 takes for fixed-point, a little more than 20 μs. The real advantage of the coprocessor is in floating point, because it performs a floating-point multiply in about one-tenth the time required by the 8086. A typical execution-time estimate for a first-order low-pass filter would be:

Ordinary instructions	$9 \times 2\,\mu s$/instr
Multiply subroutine	$2 \times 20\,\mu s$/mpy
Total	$58\,\mu s$

The sampling frequency is thus about 16 kHz, a factor of three faster than for software-only.

(c) *Fast multiplier.*

Suppose the multiply subroutine is replaced by a hardware multiplier. Multiply times in typical hardware are from 20 to 150 nanoseconds per multiplication, hundreds of times faster than in software.

If the software overhead is not longer than the equivalent of a CALL statement, the multiply time in the program loop is the same as any other instruction, and the time to complete the loop is

Execution-time estimate, 1st-order low-pass
Loop time	$11 \times 2\,\mu s = 22\,\mu s$

This filter has a sampling throughput rate of the order of 40 kHz (signal bandwidth comfortably covers the audio range).

The comparison becomes even more telling for filters more complex than the 1st-order IIR low-pass, such as biquads, FIRs, or for operations such as spectral analysis—which require many multiplies, even when the fast Fourier transform is used [see Table 1.3].

Table 1.3. Throughput limits of common DSP programs with various multiply implementations.

Method:	Software	Coproc.	Fast MPY
Operation:			
Biquad—Number of multiplies = 5			
Biquad Throughput limit (kHz):	1	8	50
Operation:			
1,024 point FFT—Number of multiplies = 5,120			
FFT Throughput limit (Hz):	1.0	10	1,000

These examples make it apparent that only when multiply time matches other computations does DSP throughput become adequate for real-time full-audio filtering and for spectral analysis at the millisecond speeds adequate for speech processing.

1.7.5 HOW DO DIFFERENCE EQUATIONS RELATE TO THE ANALOG SYSTEM?

While we have established that the performance of the digital 1st-order low-pass emulates that of its analog counterpart (e.g., the data in Fig. 1.17), the large body of analog filter lore makes it desirable to see how the difference equation (Equation 1.14) relates to analog fundamentals. Consider the passive low-pass of Figure 1.17a. Kirchoff's laws for current and voltage between input $x(t)$ and output $y(t)$ yield

$$I(t) = \frac{V_x(t) - V_y(t)}{R} = C \frac{dV_y(t)}{dt} \tag{1.17}$$

A unit impulse input at x gives the familiar exponential decay output.

$$V_y = V_{yo} \, e^{-\frac{t}{RC}} \tag{1.18}$$

Suppose continuous time is replaced with n samples at intervals, t_s, i.e., $t = nt_s$. The differential equation becomes a difference equation

$$\frac{x(n) - y(n)}{R} = C \frac{y(n) - y(n-1)}{t_s} \tag{1.19}$$

Rearrange this and solve for $y(n)$.

$$y(n) = \frac{x(n)}{1 + \frac{RC}{t_s}} + y(n-1) \frac{\frac{RC}{t_s}}{1 + \frac{RC}{t_s}} \tag{1.20}$$

This is precisely the form of Equation 1.14, the original difference equation introduced without proof, if we identify:

$$\alpha = \frac{1}{1 + \frac{RC}{t_s}}, \text{ and } \frac{RC}{t_s} = \frac{1 - \alpha}{\alpha} \tag{1.21}$$

The quantity, RC/t_s, plays the role of a dimensionless decay time-constant for the system. The coefficients, α and $(1 - \alpha)$, which weigh the relative memory of recent events, $y(n-1)$, and new information, $x(n)$, can thus be related to familiar RC filter form (see Exercises).

Equations 1.17 to 1.20 show one way to obtain the difference equation for a filter which has an analog circuit counterpart, i.e., to replace the continuous time interval in the differential equation with discrete samples. While this is useful to illustrate the difference-differential connection, it is by no means the best or easiest method to design a filter. Chapter 4 develops better ways, and Chapter 2 shows how to relate the analog transfer function in the frequency domain to the filter coefficients $h(k)$ in the time domain.

We now review the notion of impulse response: kick the system and see how it responds.

1.7.6 IMPULSE RESPONSE AS A PROBE OF SYSTEM DYNAMICS

Since the filters we have looked at are linear, responses may be superposed. We can think of a sampled continuous waveform as a series of impulses of varied size. The response to any one of these individual impulses over time will be—the impulse response; for example, the IIR filter of Figure 1.17 was shown above to respond with a sampled decreasing exponential asymptotic to zero, while an FIR filter will respond with a finite number of values, depending on the coefficients. The very next impulse will cause another such exponential to start, and the system response at any instant will be the sum of the *first* response to the *latest* impulse, the *second* response from the *previous one*, the *third* response from the *one before that*, etc. This is shown explicitly in the computation table provided in Figure 1.15b.

In other words, if the passive *RC* low-pass filter of Figure 1.17a is hit by a series of kicks (or impulses) of varying strength, the output contains a response from the latest kick, plus the accumulated memory of previous kicks, as shown in Figure 1.19. For a continuous time function, and a filter such as a first-order low-pass, with an exponential response history that goes indefinitely far back in time, the net memory of kicks is the integral of the product

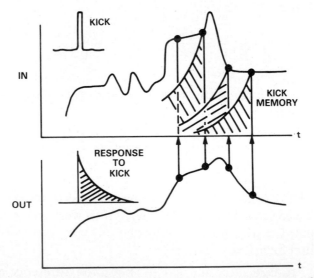

Figure 1.19. The response of a system to a general input is the sum of its memory of previous kicks (considering each sampled input point as an impulse, or "kick"). While a kick *response* decays *forward* in time, the *memory* of a kick decays *backward* in time.

of the impulse time response of the filter and the arbitrary input time function, $x(t)$,

$$y(t) = \int_{-\infty}^{t} \exp\left[-\frac{t-\tau}{RC}\right] x(\tau)\, d\tau \qquad (1.22)$$

Shifting to discrete time samples, this becomes

$$y(nt_s) = \sum_{m=-\infty}^{n} \exp\left[-\frac{nt_s - mt_s}{RC}\right] x(mt_s) \qquad (1.23)$$

$$= \sum_{k=-\infty}^{0} \exp\left[-k\alpha\right] x(n-k)$$

Here, $\alpha^{-1} = RC/t_s$ is a "scaled" time constant, and the second form of the sum comes from a change of variables, $m = n - k$.

The word *convolution* (or "folding back"), is used to describe this "memory of previous kicks" process. A tool which greatly simplifies filter design—and DSP in general—it is developed in Chapter 2. For the 1st-order low-pass IIR, the lower limit (previous history) must go back infinitely. Equation 1.23 is the signature of a recursive or infinite impulse response (IIR) system.

While it is mathematically correct to say that the RC filter remembers a given kick indefinitely, there is a point beyond which the exponential decay approaches the noise level. If we terminate the memory of previous kicks in Equation 1.23 at some point N, the convolution sum looks like the prescription for an FIR filter.

$$y(n) = \sum_{k=0}^{N-1} e^{-\alpha k} x(n-k) \qquad (1.24)$$

N will be identified with the number of taps in the nonrecursive (approximate) filter, as in the moving average example. How many taps does it take for a reasonable approximation nonrecursively? For the example in Figure 1.17, the answer can be seen by inspection, an exponential can be fit with 14 taps for 0.1% error (each 2 samples are equal to one time constant). In general, the decay of an individual kick at a rate, $e^{-\alpha k}$, down to a normalized noise threshold, y_{min}, takes a time (expressed as a multiple of the time constant)

$$N\alpha = -\ln(y_{min}) \qquad (1.25)$$

For $y_{min} = 10^{-3}$ (60 dB), $N\alpha = 7$. For $y_{min} = 10^{-6}$ (120 dB), $N\alpha = 14$. To retain fine structure in time, filter time constants are long compared to the sampling interval, so the number of taps must be increased in proportion (in the example, they are only doubled). The number of taps, N, in the FIR approximation to the simple low-pass is thus

y_{min}	$N/(RC/t_s)$	Minimum N for $RC/t_s =$	
		2	4
10^{-4} (80 dB)	9.2	19	37
10^{-6} (120 dB)	13.8	28	56

Since each tap requires a multiply, this FIR approximation to the first order low-pass runs at least 10 times slower than the direct IIR version.

Other features which govern the choice between recursive and nonrecursive filters will be discussed in Chapter 4.

Exercise 1.5. Rewrite the low-pass program in the text, Table 1.1, to do a derivative operation instead.

Exercise 1.6. Rewrite the low-pass recursive filter program in the text, Table 1.1, to do a high-pass operation instead. If your intuition fails for the moment, we will do it again methodically in Chapter 4.

Exercise 1.7. Multiply time is the bottleneck in software digital signal processing. You can demonstrate this on a personal computer by the following example:

```
(BASIC)
        100   INPUT A,B
        200   C = 0
        300   PRINT "START TIMING"
        400   FOR I = 1 TO 10000
        500   C = A*B
        600   NEXT I
        700   PRINT "DONE"
```

The time between "START" and "DONE" depends upon how the multiply is implemented on a given system, as well as all the other operations in the loop. To estimate the multiply time, change step 500 to 'C = A' and re-run the program. The time difference, divided by the number of repetitions in the loop, is the extra multiply time, typically about a millisecond on personal computers running BASIC software—and would not decrease significantly in shifting from line-by-line interpretive execution to compiled execution.

Exercise 1.8. The difference equation, Eq. 1.14, and the differential equation, Eq. 1.17, for the 1st-order RC low-pass of Fig. 1.17a, become completely equivalent when RC/t_s is large. In this limit, with sampling time short compared to RC, show that the terms in Eq. 1.17 correspond to the superposition of:

(a) The decay of an initial voltage put on the capacitor, $e^{-\frac{t_s}{RC}} y(n-1)$, after a unit of time, t_s, and

(b) The amount of voltage, $\left(1 - e^{-\frac{t_s}{RC}}\right) x(n)$, which builds up on the capacitor, given a step input, assuming no previous charge.

Method: Just expand the exponential above and relate to the terms in Eq. 1.17.

Exercise 1.9. Compare these two familiar situations: (a) A principal amount, P, invested at interest rate, R, per compounding interval, grows as $P(1+R)^N$. (b) Exponential growth of a population P at rate-of-change, $dP/dt = rP$ gives $P = P_0 \exp(rt)$. Show how the two are equivalent as the interval of compounding interest becomes continuous. How are R and r related?

Method: Write the difference equation for compound interest and use the analogy with the RC filter, comparing Eq. 1.14 and 1.17.

1.8 DIGITAL AUDIO EXAMPLE

As an example of the invasion and transformation of a traditional analog area of endeavor by DSP, we end this chapter by looking at the plethora of existing and potential applications for DSP in audio. The key concepts of DSP introduced in this chapter: flexibility, accuracy, predictability, high performance. . ., and the key processes: filtering and spectrum analysis, bring new standards of capability and convenience. Key audio components and functions made easier digitally include:

Digital oscillators, synthesizers
Editor/mixer
Filter/equalization, reverb
Enhancement, restoration
Digital transmission of voice and data
Recording, compression, . . .

1.8.1 DIGITAL AUDIO

Audio signal processing has seen the gradual invasion of digital techniques in recent years. Since digital audio well illustrates the advantages and the techniques common to DSP, we will use it as a recurrent theme in this book, beginning here with an overview.

Signal-to-noise: The enhancement of signal-to-noise and dynamic range digitally was discussed in Section 1.4 as our first analog-digital comparison. Digital audio disks bring concert hall clarity with their 16-bit (96 dB) range. Floating-point concepts extend this further, assigning a few bits to an exponent and the rest to a mantissa. Scaling tests during computation can keep the mantissa always as large as possible to retain maximum precision while not allowing overflow in the accumulated computations. Companding techniques in A/D and D/A conversion and computation further increase dynamic range without excessive word length, and nonlinear (Dolby-like) algorithms reduce noise.

Digital oscillator(s) and Digital synthesizer: How does one design a digital musical instrument? In principle, a room full of oscillators could be tuned in frequency and amplitude to simulate particular instrument spectra, but the adjustments would have to be changed for each note, and also during a note, to get realistic and satisfying sound. Computer music historically replaced the oscillators with digital computations, operating in an off-line mode that was cumbersome for composers and maddening to musicians. The components of a single musical sound or "voice," and their functions, are:

Oscillators or waveform generators set the pitch for each note
Filters alter the harmonics in subtractive synthesis
Multiple oscillators sum harmonics in additive synthesis
Envelope generators assign amplitudes dynamically .
Input devices, such as the keyboard, the string, the wind,
    or even the conductor's baton transduce musicians' wishes in real time.

All are straightforward real-time DSP applications, with sufficient throughput to permit time-multiplexing of dozens of oscillators (enough for a respectable synthesizer) with a single DSP processor. Modern research at computer music labs at Stanford and MIT, for example, not to mention development at numerous companies, is being translated into VLSI chips, with the goal a real-time digital "orchestra."

Digital mixer in real time: A key issue in modern recording is the combination of a dozen or more separate tracks, formerly recorded separately on analog magnetic tape. The analog mixing process is tedious and expensive, and it introduces noise. Let all the "tracks" be stored digitally in memory. Gain control in mixing is just a real-time multiply, and if done in floating point gives noise-free results, much faster and more flexibly (see pitch shifting and tempo changing below).

Digital equalization: An equalizer is an active filter bank with amplitudes adjusted to shape the transfer function over a series of frequency bands. A digital filter is a natural for equalization. The coefficients for the entire filter bank are stored in memory and then read out dynamically to reconfigure the processor as the signal is "switched" through one filter after the other—all with the same signal processor. The concept is illustrated in Figure 1.20. A portion of the signal is stored in a "circular" buffer, continuously being updated, with older data thrown away. The coefficients are loaded, and the filter operates on the buffer and accumulates the result in a second buffer. The next set of filter coefficients is loaded and the process is repeated. As long as the filter calculation is fast compared to the signal speed, many filter bands can be computed in real time.

Digital Reverb (synthetic reverberation): The delay line, implemented digitally, gives concert hall or cathedral ambience to what might have been recorded in a dry studio. Since the reverb parameters are easily changed during the session, an editing pass goes rapidly, and there is little inertia about trying an improvement when the results of each try are stored in memory for easy recall and change. This process is a boon to the composer and arranger, when compared to signal storage on inflexible magnetic tape.

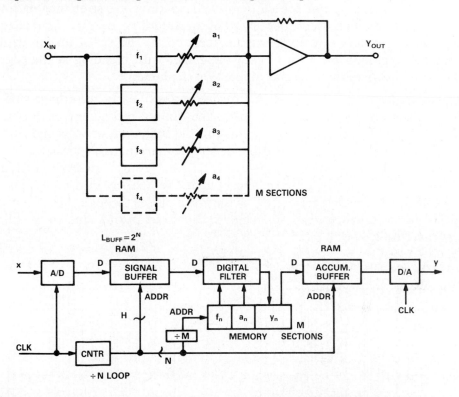

Figure 1.20. Concept of time-multiplexing to replace many identical filters by a single dynamically programmed digital filter. Blocks shown as hardware (e.g., counters) may be replaced by software loops in a program.

Pitch shifting without tempo change: Since a given tone or voice is stored as a waveform in memory, the pitch can be shifted at will by changing the rate of clocking through the waveform buffer addresses. With the envelope-shaping parameters stored independently, a pitch change can be made without changing the tempo at which notes are played. The same concept also applied to stored human sounds (speech, singing), though the variation of pitch without tempo is more delicate and depends upon the storage algorithm—an easy matter when adaptive parametric storage (voice-box modeling) is used, but more difficult when signals stored as uncompressed waveforms must be retrieved.

Tempo change without pitch shifting: When learning a language, it is convenient to hear the words slowed down, but without an overall time-scale change, which would shift the pitch towards the bass; similarly, to do a rapid review of material, a fast playback mode is desirable, but without a treble shift to "Donald Duck" quality. Tempo changes are helpful to analog and acoustic musicians attempting to learn to play a piece along with digital colleagues. This is easy for synthetic sounds: change the clock rate at which notes are read back from memory. For speech, tempo change involves stretching or compressing the phoneme envelope (shape of the mouth vs. time) without changing the vibrational rate of the digital vocal cords. To speed up the tempo requires chopping off a bit of each phoneme, with blending at the edges. To slow it down requires adding a piece to each phoneme.

Digital-audio image restoration: A treasure-house of early recordings suffers from inadequate signal-to-noise due to the limited technology of the time. While digital filtering can remove hiss on the high end, or wow and flutter on the low end, spectral information is lost. An alternative approach, due to Stockham (1975), is shown in Figure 1.21. The "tinny" quality of early recordings is due mostly to sharp "megaphone" resonances of the microphone (visible at left), which are excited as the voice passes through those frequencies. The resonances can be removed by a technique known as blind deconvolution.

Digital Processing in Audio Signals

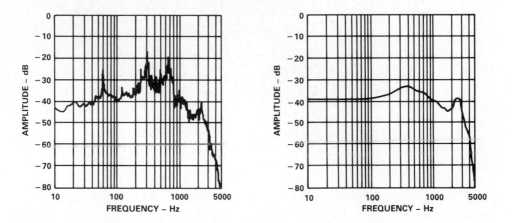

Figure 1.21. Caruso and Bjoerling return in hi fi, thanks to Stockham. Response spectra of cleaned-up noisy low-fi recordings, using adaptive modeling of vocal-tract of the singer and of the distortion in the original recording environment. (Source: Stockham, © 1975 IEEE—see Bibliography)

Certain assumptions about the signal source are necessary: the human voice has spectral capabilities relatively the same today as when the recording was made, and the singer's individual spectrum changes only gradually—if at all—during a recording session. Assumptions about the filtering added in recording are also necessary: the music lasts longer than the filter's impulse response, and the filter did not change during the recording session. Together with a model of the voice extracted from modern recordings, an algorithm can be developed which removes the sharp resonances excited in the microphone, etc., while leaving the essential qualities of the singer with, quoting Stockham, "clarity of expression, texture of the voice, and artistic interest dramatically changed."

Digital telephony and recording: As mentioned in Section 1.5 above, in connection with data compression, dynamic range and signal-to-noise are gained digitally at the cost of high bit-rates: 64 kilobits per second (kb/s) for voice and several hundred kb/s for music, if stored directly Speech coded phonetically by source modeling ("vocoding") can be transmitted at bit rates a factor of 32 slower, down to 2 kHz, but will always sound somewhat machine-like. Improved coding in both the time domain and in the frequency domain is being actively pursued. In the time domain, nonlinear quantization steps (companding), adaptive step size, and locally smaller step size (delta modulation) are key concepts, which, when combined, lead to good sound quality at bit rates down to about 12 kb/s. Frequency-domain methods divide the spectrum into a limited number of bands, and thereby need to transmit less information; or adaptive-filter methods follow dynamically the spectral qualities of speech: formants, pitch, and noise shaping.

Speech Synthesis: Speech synthesis issues are closely related to the storage and transmission issues above, except that the common goal of putting the synthesizer on a chip further restricts the storage capability—and hence the synthesis algorithms. Numerous manufacturers have tried their hand at putting speech algorithms onto silicon; an early consumer product was the Texas Instruments "Speak and Spell", a *tour de force* of linear predictive coding. The TI concept achieves more-human sound than vocoders, yet uses a bit-rate of only 1,200 bits/second. The information per 25-ms interval of speech is stored in 50 bits or less, depending upon whether the sound is voiced or unvoiced. Pitch requires 5-bits, energy uses 4, and a set of ten 4- or 5-bit coefficients establish the parameters of compact two-multiplier lattice filters, which model the vocal tract as a function of time.

Modems: Digital data on audio channels: While not an audio instrument *per se*, a modem is used to ship (and receive) digital information over audio telephone channels. Modem design is a natural DSP application, since coding concepts become more subtle as baud rates go up, and costs must be kept down. The (digital bit)-to-(audio tone) end involves a high-precision digital sine-wave generator of rapidly changeable frequency, spectral shaping by a high-performance FIR filter, and d/a conversion. The receiving end adds a digital demodulator and phase-locked loop to turn frequency and phase information back into binary bits. Throughputs are sufficiently slow for a single DSP processor to time-multiplex all of these operations.

Digital technology is becoming dominant in the recording, playback, and transmission of sound. First came digitally mastered analog records: only the master was stored digitally on tape. Next came digital processing operations in the recording studio, now a relatively standard technique, and advanced by such operations as Lucasfilm (in "Starwars"). Soon after came digital disks in playback, but still through an analog hi-fi.

We can look for examples of future extensions: the all-digital hi fi; controlling phase information to make music move inside your head; adaptations of the audio to fit room acoustics; adaptations of concert hall "brightness" or reverb to compensate for absorption as the number of people in the audience changes. The key is digital flexibility.

Chapter Two

Sampled Signals and Systems

2.1 INTRODUCTION

To lay the foundations for signal processing, it is necessary to review a series of general-purpose tools. The mathematics they require may get a little heavy here, but the tools are powerful once mastered. As with cookbooks, the basic tools and techniques allow a long list of recipes to be carried out simply with impressive results.

The chapter begins (2.2) in the analog or continuous domain, reviewing how a signal in the time (or other data-space) domain is related to its frequency spectrum by the Fourier transform. (2.3) The convolution theorem reduces complicated systems and procedures to a set of simpler ones. (2.4) The information theorem limits how well we can know a system in its two conjugate representations when information in either domain is limited. (2.5) When data is sampled at discrete intervals, the available frequency spectrum is limited. The need to take at least two samples of the highest frequency component in the data fixes the minimum sampling rate. Artifacts must be avoided by anti-alias filtering.

Spectral analysis of sampled data reveals special properties summarized (2.6) in the discrete Fourier transform and the circular convolution. (2.7) To introduce the poles and zeros which govern recursive filter behavior, we review analog stability and the impulse response as basic indicators of system dynamics. The mathematics of discrete-time signal processing brings its own set of tools; they simplify analysis and go beyond what one could do with conventional analog signal processing alone. The z-transform, in particular, plays the role of the Fourier/Laplace transform for *sampled* data. (2.8) The chapter ends with two simple approaches to digital filters, the first from the delay line and the second from numerical analysis. These are less-powerful approaches than those to be considered in Chapter 4, but they illustrate the tools in action in a familiar setting. In this chapter, the digital issues of accuracy and quantization error will generally be ignored, except for a brief example of a dead-zone effect.

This chapter also establishes a background in discretely sampled data that will help to understand the reasons for the components intervening between analog signal and digital process in a data-acquisition system, as exemplified in Figure 2.1. The input signal is filtered by a low-pass "anti-aliasing" filter, sampled at regular, precisely timed intervals, and converted to digital. Digital control may be used to optimize input signal gains, providing a sort of floating-point, so that conversions retain maximum precision as signal levels vary. At the output end, a d/a converter produces a set of analog samples; a "deglitcher" (which may be part of the d/a converter) is generally a sample-hold that provides essentially noise-free transitions between analog levels at the sampled values; it is followed by a low-pass smoothing filter, which filters out frequencies higher than one-half the sampling frequency.

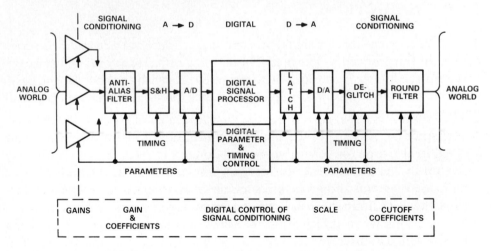

Figure 2.1. Key elements of a sampled-data system.

2.2 REVIEW OF FOURIER SERIES AND FOURIER TRANSFORMS

2.2.1 FOURIER-SERIES ANALYSIS

The signal-spectrum pairs of Chapter 1 (Figure 1.8) have precise interrelationships: the Fourier *Series* is for periodic signals having a period T; the Fourier *transform* is for any finite-energy signal over the infinite interval.

A periodic signal, $y(t)$, with period, T, may be represented by a sum of sines and cosines of frequencies which are multiples of the *fundamental* frequency, $f = 1/T$.

$$y(t) = \frac{A_0}{2} + \sum_{n=1}^{\infty} \left[A_n \cos\left(n\frac{2\pi}{T} t \right) + B_n \sin\left(n\frac{2\pi}{T} t \right) \right] \qquad (2.1)$$

The term, $2\pi/T$, is the fundamental *angular frequency*, ω, in radians per second. The coefficients of each term, or *Fourier coefficients*, A_n and B_n, are obtained by integrating over one period of the fundamental, projecting the components of $y(t)$.

$$A_n = \frac{2}{T} \int_{-T/2}^{+T/2} f(t) \cos\left(n\frac{2\pi}{T} t \right) dt$$

$$(2.2)$$

$$B_n = \frac{2}{T} \int_{-T/2}^{+T/2} f(t) \sin\left(n\frac{2\pi}{T} t \right) dt$$

The terms for each frequency, n/T, may be combined to form a single cosine term, with magnitude, $D_n = \sqrt{(A_n^2 + B_n^2)}$, and phase angle, $\phi_n = \tan^{-1} B_n/A_n)$.

Fourier analysis works because the set of sine/cosine functions is both complete and orthogonal. Completeness guarantees that any periodic $y(t)$ can be represented. Orthogonality guarantees that the Fourier coefficients are independent of one another.⋆

Calculation of Fourier coefficients is greatly simplified for certain types of waveforms if the origin is chosen so as to take advantage of *even-function* (mirror) symmetry (i.e, for $f(t) = f(-t)$, the B_n terms vanish) or *odd-function* symmetry (for $f(t) = -f(-t)$, the A_n terms vanish). Further simplification results if $f(t) = f(t+T/2)$ (*odd coefficients* are zero) or if (excluding the dc bias) $f(t) = -f(t+T/2)$ (*even coefficients*—except for the dc bias, $A_0/2$—are zero).

Example: Square Wave
The Fourier series for a symmetrical biased square wave (Figure 2.2) is a useful prototype.

$$f(t) = +2 \qquad (-T/4 < t < T/4)$$
$$0 \qquad (-T/2 < t < T/4 \text{ and } T/4 < t < T/2) \qquad (2.3)$$

Drawing it as a function with even symmetry, we can see by inspection that the average over the period is 1 unit $= A_0/2$, hence $A_0 = 2$. We need only compute the A_n terms; the B_n terms are all zero. We also note (subtracting

---

⋆The purposes of this book can be served by stating such truths as reminders without rigorous proof. The reader in need of more background will find ample corroboration in the references listed in the Appendix.

the dc bias, $A_0/2$), that $f(t) = -f(t+T/2)$, hence the even coefficients will be zero. (The square wave can also be drawn as an odd function, having only sine terms, and of course with arbitrary phase, giving a mixture of terms.) Eq. 2.2 is used to calculate the terms of the cosine series:

$$A_n = \frac{2}{T} \int_{-T/2}^{+T/2} 2 \cos\left(n\frac{2\pi}{T}t\right) dt = 2 \frac{\sin n\frac{\pi}{2}}{n\frac{\pi}{2}} \tag{2.4}$$

For this case, $A_0 = 2$. For $n = 1, 3, 5$, etc., the expression for the nth odd component, normalized to A_0, is

$$A_n = \frac{2}{n\pi} (-1)^{\frac{n-1}{2}}$$

giving the series,

$$A_n = \frac{2}{\pi}, \ -\frac{2}{3\pi}, \frac{2}{5\pi}, \ -\frac{2}{7\pi}, \text{etc.}$$

$$= 0.637, -0.212, 0.127, -0.091, \text{etc.}$$

The Fourier-series approximation to the square wave shown in Figure 2.2a has ripples and does not make sharp step transitions, as a consequence of the truncation of the series beyond $n = 7$ in this example. With the representation limited to a finite frequency interval or bandwidth, the higher-frequency properties of the ideal square wave are lost. The consequences of bandwidth limiting in the frequency domain on signals in the time domain will be explored in Section 2.4.

Figure 2.2b is a normalized plot of the coefficient amplitudes vs. their number, n; note that each number, n, is proportional to the total angular rotation associated with its frequency component in the period T, hence its angular frequency. Therefore, x, the argument of the $\sin x/x$ envelope, can be thought of as a frequency—a variable having *discrete* values.

Example: Fourier Series for a Pulse
The Fourier-series representation of a pulse series, of amplitude A, period T, and width τ (Figure 2.3a), is closely related to that of the square wave. Noting that the B_n coefficients are zero, and integrating for the A_n coefficients, using Eq. 2.2:

$$A_n = \frac{2}{T} \int_{-\tau/2}^{+\tau/2} A \cos\left(n\frac{2\pi}{T}t\right) dt = 2A \frac{\sin\left(n\frac{\pi\tau}{T}\right)}{n\frac{\pi\tau}{T}} \tag{2.5}$$

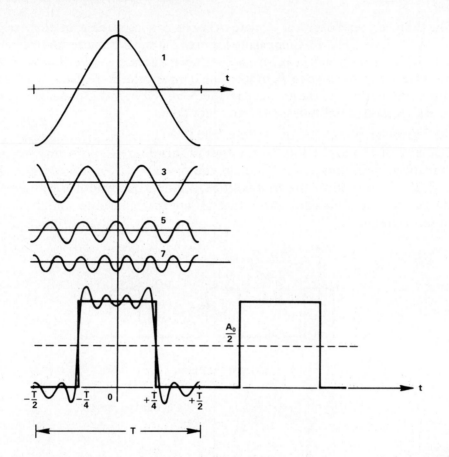

a. The original signal and its representation by a finite sum of terms with amplitudes A_n.

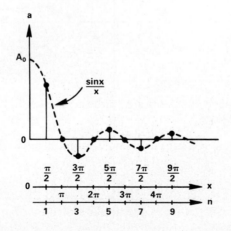

b. Amplitudes of the coefficients (Fourier amplitude spectrum).

Figure 2.2. Fourier analysis of a square wave.

The Fourier series for the square wave is a special case of this, where $\tau/T = \frac{1}{2}$ (Exercise 2.2). Comparing Figures 2.3 and 2.2, more spectral terms fall within each $\sin(x)/x$ peak for a pulse than for a square wave. The narrower the duration, τ, relative to T, in the time domain, the more dense will be the number of terms, and the greater the number of Fourier components that will be of significant amplitude in the frequency domain.

For a given amplitude, A, as τ shrinks and the number of terms increases, the smaller will be the absolute magnitudes in this discrete spectrum; but their *normalized* magnitudes will maintain the discrete $\sin x/x$ relationship. The limit of a narrow pulse, the *delta function*, requires bandwidth ideally extending to infinite frequency—and that A approach infinite amplitude—as τ approaches zero.

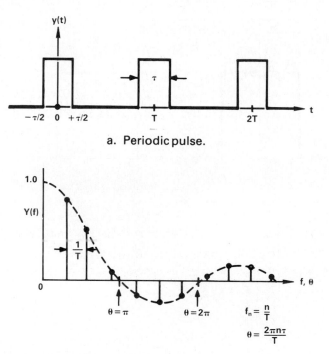

a. Periodic pulse.

b. Its normalized Fourier coefficients are points on a $\sin x/x$ function; the narrower the pulse, the closer the spacing.

Figure 2.3. Fourier analysis of a pulse wave.

2.2.2 THE FOURIER TRANSFORM AND ITS PROPERTIES
The Fourier transform provides the means of converting between the two equivalent representations of information in the data (time) or spectral (frequency) domain. As with other bilateral transformations, such as rectangular to polar coordinates, the transformation works in both directions, and one or the other representation is chosen for convenience in analysis. The Fourier transform is a generalization of the Fourier series to an infinite interval.

$$\text{FT}[y(t)] = Y(f) = \int_{-\infty}^{+\infty} y(t)\, e^{-j2\pi ft}\, dt \qquad (2.6)$$

The transform, $Y(f)$, contains the spectral information necessary to reconstruct the signal by the inverse transform.

$$y(t) = \int_{-\infty}^{+\infty} Y(f)\, e^{j2\pi ft}\, df \qquad (2.7)$$

The time function, $y(t)$, must meet certain conditions or else the integral over infinite limits may diverge. However, for most practical applications, it will not be necessary to deal with these issues. As noted earlier, other sources are available for readers who would dig deeper.

Examples of Fourier Transforms
Pulse: Fourier transforms, even of standard functions, can be difficult to derive. Fortunately, they can usually be looked up in tables. However, it is useful to work out an illustrative example, the *single* pulse of unit amplitude centered at $t = 0$ and of width τ.

$$y(t) = A = 1 \qquad (-\tau/2 < t < +\tau/2)$$

$$0 \qquad (\text{all other values of } t) \qquad (2.8)$$

As a result of choosing the pulse as an even function, only the real part (cosine term, which integrates to sine) is non-vanishing in the result of the integral over the complex exponential (Eq. 2.6).

$$Y(f) = \int_{-\tau/2}^{+\tau/2} e^{-j2\pi ft}\, dt = \int_{-\tau/2}^{+\tau/2} [\cos 2\pi ft - j\sin 2\pi ft]\, dt$$

$$= \frac{1}{2\pi f}\left[\sin\left(2\pi f \frac{\tau}{2}\right) - \sin\left(2\pi f \frac{-\tau}{2}\right)\right]$$

$$= \frac{\sin \pi f \tau}{\pi f} \qquad (\text{for all } f \text{ from } -\infty \text{ to } +\infty) \qquad (2.9a)$$

If the amplitude and width of the pulse are interdependent so that that the pulse has constant area, $a = A\tau$, then

$$Y(f) = a\frac{\sin(\pi f \tau)}{\pi f \tau} \qquad (2.9b)$$

Suppose the pulse occurs, not at $t = 0$ but at a later time, t_0. The delay merely introduces a phase shift in the result. This may be shown explicitly for this example (Exercise 2.3), or more generally by the *shifting theorem*: if the Fourier transform of a function, $y(t)$, is $Y(f)$,

$$\text{FT}\,[y(t-t_0)] = Y(f)\,e^{-j2\pi f t_0}$$

$$= Y(f)\,[\cos{(2\pi f t_0)} - j\sin(2\pi f t_0)] \qquad (2.10)$$

Sine Wave: The Fourier integral of a sine or cosine of frequency f_0 (i.e., $y(t) = \cos(2\pi f_0 t)$ or $y(t) = \sin(2\pi f_0 t)$) is zero for all frequencies except $\pm f_0$. This is expressed by the displacement of a pair of delta functions, or unit samples, by $\pm f_0$, along the frequency axis to the specific values of frequency in the real or imaginary domain (Figure 2.4):

$$\text{FT}\,[\cos(2\pi f_0 t)] = \tfrac{1}{2}\,[\delta(f-f_0) + \delta(f+f_0)] \qquad (2.11)$$

$$\text{FT}\,[\sin(2\pi f_0 t)] = j\tfrac{1}{2}\,[\delta(f-f_0) - \delta(f+f_0)] \qquad (2.12)$$

Exercise 2.9 may help explain the pattern of signs and real/imaginary parts in Equations 2.11 and 2.12 and Figure 2.4, by the use of (a) the inverse Fourier transform and (b) the shifting theorem.

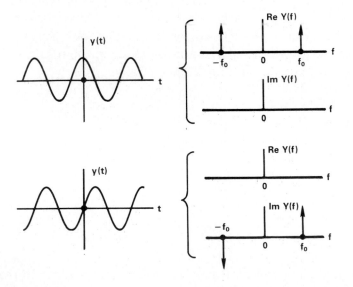

Figure 2.4. Fourier transforms of the cosine and sine waves.

Theorems About Fourier Transforms

The shifting theorem is but one of a number of theorems that make Fourier transforms more useful by easing the mathematical manipulations that are required in employing them to solve problems. A sampling of theorems can be found in Table 2.1.

Table 2.1 Fourier Transform Properties. Here, $x(t)$, $y(t)$ are time functions, and $X(\omega)$, $Y(\omega)$ are the corresponding frequency spectra ($\omega = 2\pi f$). Essentially similar theorems apply to Laplace transforms (Section 2.7).

Theorem	Signal	Spectrum
Dependent Variables	$x(t), y(t)$	$X(\omega), Y(\omega)$
Addition (linearity)	$ax(t) + by(t)$	$aX(\omega) + bY(\omega)$
Time Shifting	$x(t - t_0)$	$e^{-j\omega t_0} X(\omega)$
Freq. Shifting	$e^{j\omega t_0} x(t)$	$X(\omega - \omega_0)$
Time Reversal	$x(-t)$	$X(-\omega)$
Time/Frequency Scaling	$x(t/a)$	$\lvert a \rvert X(a\omega)$
Convolution	$x(t) \star y(t)$ $x(t) y(t)$	$X(\omega) Y(\omega)$ $X(\omega) \star Y(\omega)$
Derivative	$\dfrac{dx(t)}{dt}$ $-j2\pi t\, x(t)$	$j\omega X(\omega)$ $\dfrac{dX(\omega)}{d\omega}$
Integral	$\displaystyle\int_{-\infty}^{t} x(t)\, dt$	$\dfrac{X(\omega)}{j\omega} + X(0)\delta(\omega)$
Complex Conjugate (Real $x(t)$) Equivalent Form:	$x(t) = x^*(t)$	$X(\omega) = X^*(-\omega)$ $\lvert X(\omega) \rvert = \lvert X(-\omega) \rvert$ $\phi(\omega) = \phi(-\omega)$
Even (Real) $x(t)$	$x(t) = x(-t)$	$\operatorname{Im} X(\omega) = 0$
Odd (Real) $x(t)$	$x(t) = -x(-t)$	$\operatorname{Re} X(\omega) = 0$
Parseval's Relation	$\displaystyle\int_{-\infty}^{+\infty} \lvert x(t) \rvert^2 \, dt = \dfrac{1}{2\pi} \int_{-\infty}^{+\infty} \lvert X(\omega) \rvert^2 \, d\omega$	

The theorems follow for the most part from symmetries in the cosine/sine components of the Fourier integral; they demonstrate how symmetry in the data or the spectrum can cut processing time, in the same way as for Fourier series. For example, if $x(t)$ is real, then $X(\omega)$ has even magnitude and odd phase; there is no need to compute the spectrum for negative frequencies. If

$x(t)$ is real and even, then $X(\omega)$ is purely real; if $x(t)$ is real and odd, then $X(\omega)$ is purely imaginary; there is no need to compute both components.

The sampling in Table 2.1 is illustrative. When actually using them, it is helpful for the reader to dig deeper, using some of the indicated references to understand the conditions under which the theorems are useful, their limitations, and their applications. Some of the theorems must be modified for periodic signals whose energy is not finite over infinite time. For example, Parseval's theorem must be modified to make the range of the integral agree with the period of the signal.

Magnitude and Phase

The Fourier transform contains important phase information. This may be represented either by the real and imaginary parts, as in the example of Figure 2.4, or by the magnitude and phase.

$$|Y(f)|^2 = (\text{Re } Y(f))^2 + (\text{Im } Y(f))^2$$

$$\phi(f) \quad = \tan^{-1}\left[\frac{\text{Im } Y(f)}{\text{Re } Y(f)}\right] \tag{2.13}$$

Phase information, often ignored, is sometimes unimportant, as in noise measurements which are plotted as the *noise spectrum* or power spectral density, the first term above. More often, phase information can be quite important. For example, complicated signals are superpositions of simpler ones, and the phase is important. Filters must preserve phase information to avoid distorting waveshapes of signals comprising many components, such as the square wave of Figure 1.1, as we will see in Chapter 4. In machine vision, phases are crucial in defining the *edges* in a picture (like the edge of a square wave) since the edges are where all the components must add up in-phase. An image that has been Fourier-transformed and then inverse-transformed with the phase ignored is unrecognizable. Interestingly, an image similarly processed with inaccurate magnitude information but with the phase preserved is quite recognizable, especially at the edges of objects (see Oppenheim & Willsky, 1983, p. 230).

Phase also has to do with causality. Time reversal $t \rightarrow -t$ for a real-valued signal corresponds in Table 2.1 to a mere flipping of the sign of the phase; since the magnitude is even, it is unchanged. Yet if we play a tape of music or speech backwards, the difference is usually quite apparent. Why? The laws of dissipationless mechanics go equally well backwards and can ignore causality (an orrery or a film of billiard balls running backwards looks fine) but for more complex systems governed by statistical mechanics (where time has a directionality, as in a film of cracking an egg, the discharge of a capacitor through a resistance, or the process of aging), time reversal makes a crucial difference.

2.3 THE CONVOLUTION THEOREM

2.3.1 FOURIER TRANSFORM OF A PRODUCT

Suppose that a frequency function, $Y(f)$, is the product of two functions, $F(f)$ and $G(f)$. Then the inverse Fourier transform, $y(t)$, is related to the FT's of the two functions by a special integral called the *convolution*. If

$$Y(f) = F(f)G(f) \tag{2.14}$$

Then

$$y(t) = \text{FT}[Y(f)] = \int_{-\infty}^{+\infty} f(\tau)g(t-\tau)\,d\tau \tag{2.15}$$

Here,

$$f(t) = \text{FT}[F(f)]$$

$$g(t) = \text{FT}[G(f)] \tag{2.16}$$

The proof follows straightforwardly by change of variables in the Fourier integral of Eq. 2.6 (Exercise 2.4). It is convenient to use a special symbol, $\star$, as shorthand for the convolution. For example,

$$f \star g = \int_{-\infty}^{+\infty} f(\tau)g(t-\tau)\,d\tau \tag{2.17a}$$

2.3.2 PROPERTIES OF CONVOLUTIONS

The convolution has certain general properties.

Convolutions are commutative, i.e., the role of the shifted function can be interchanged without affecting the result:

$$f \star g = g \star f \tag{2.18}$$

Convolutions are also associative and distributive.

$$x \star [y \star z] = [x \star y] \star z \qquad \text{Eq. (2.19)}$$

$$x \star [y + z] = x \star y + x \star z \qquad \text{Eq. (2.20)}$$

The convolution theorem works in either direction, as listed in Table 2.1. The expression for the transform in the frequency domain is:

$$F \star G = \int_{-\infty}^{+\infty} F(\omega_0)G(\omega_0 - \omega)\,d\omega_0 \tag{2.17b}$$

> *The convolution theorem:* The Fourier transform of a *product* of two
> functions is the *convolution* of their individual Fourier transforms.
> Conversely, the Fourier transform of a *convolution* of two functions
> is the *product* of their individual Fourier transforms.

The convolution theorem is central to the design of digital filters, since the
multiplication of frequency transfer functions in filtering to compute frequency response is equivalent to convolution in the time domain to obtain
time response. Convolution also provides great simplification by permitting
the evaluation of complicated spectral analysis problems in terms of simpler
ones, as shown in the following section.

The uses of convolution go far beyond the processing of electrical signals. Although digital electrical signals in computers are the usual vehicle for signal
processing because of the medium's flexibility, physical convolution by analog techniques, as in optical or electrical analog computation, can often be performed at high speed and with low cost. For this reason, it will be helpful to
the reader to seek a physical understanding of convolution that goes beyond
its use as a mathematical tool. A couple of examples may be helpful.

Example: Convolutions with delta functions
The convolution of a function $f(t)$ with a delta function, occurring at t_0, moves
the function over to t_0, the location of the delta function.

$$\delta(t - t_0) \star f(t) = \int_{-\infty}^{+\infty} \delta(\tau - t_0) f(t - \tau) \, d\tau$$

$$= f(t - t_0) \tag{2.21}$$

In order to convey this meaning in physical terms, the function shown in
Figure 2.5a is a two-dimensional image of a three-dimensional object, and
the independent variable is distance, x, rather than time.

Carrying this idea a step further, the convolution of an "object", $f(x,y)$, with
a periodic array of delta functions provides a repeated image of the object at
every point of the array (Figure 2.5b). Thus, a periodic array can be simplified
for mathematical handling by considering the convolution of the lattice-point
set with the basic repeating unit: atoms, molecules, antennas, receivers,
absorbers, etc.

This is the procedure for calculating convolutions, point by point:

(a) For a given point, x, pick up one of the functions, reverse the direction
of x', and slide it over by an amount x.

(b) Multiply by the value of the second function, and integrate over x'.

(c) Repeat for each point x of the convolution function.

The folding of time has already been seen in the "memory of previous kicks"
example of Chapter 1.

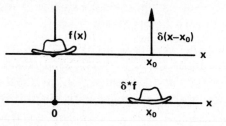

a. Convolution of a function with a delta function displaced by x_0.

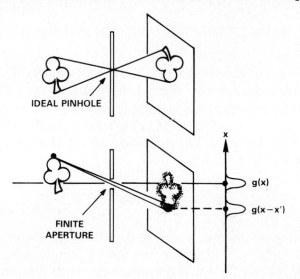

b. Convolution of an object with a periodic array of delta functions.

Figure 2.5. Effects of convolution with delta functions.

What is a convolution? Pinhole camera example
A pinhole camera (Figure 2.6) with a zero-diameter aperture (ignoring for a moment the wave nature of light), like a delta function, ideally forms a perfectly translated (inverted) image, call it $f(x)$, in which each point (or picture-element—pixel) on the object along the x-axis is perfectly reproduced on the screen. If the pinhole has a nonzero diameter, the image is blurred; the image of each point is now a spot, containing a distribution of intensities, call it $g(x)$, rather than a simple unit-point; and some portions of it are superimposed on

Figure 2.6. The blurred image of a pinhole camera with finite diameter hole may be thought of as a convolution.

nearby points. The image of each point is the sum of its own spread-out image and those of portions of nearby spots that overlap it.

Thus, the blurred image is the sum of many such blurred spots (Figure 2.6) shifted by a range of distances, $(x - x')$,—call the function $g(x - x')$—and multiplied by the intensity $f(x')$ at each point x' in the ideal image. The blurred image is the convolution of the true image with blurred spot image of a single point of light.

Note that, again, the independent "time" variable in this case is x. The Fourier transform will still map into the frequency domain, but the dimension of "frequency" is $1/x$, and $\omega = 2\pi/x$. In the frequency domain, the "impulse response" of the pinhole could be expressed by the transform of the spot function, $G(\omega)$, and the transform of the resulting image would be the product of $G(\omega)$ and the transform of the original image, $F(\omega)$.

The German word for convolution, *Faltung* (or *folding*), aptly expresses the operation. Detail in one function is blurred when it is convolved with a second function that is a single broad peak, as in measurements limited by a certain instrumental resolution, and in signal processing over a finite interval. The inverse process of image enhancement follows naturally by a *deconvolution* (See Chapter 8).

The pinhole camera with finite aperture is like a filter, since filtering is also expressed by a convolution. The image of a point light source is the impulse response. The camera builds an image of an arbitrary object (set of points of light) the same way that a filter processes an arbitrary signal input. Image blurring becomes apparent as soon as the size of the hole approaches the scale of spatial structure in the object: the hole is a low-pass filter.

2.3.3 THE CONVOLUTION THEOREM IN FOURIER TRANSFORMS

Diffraction Effects

The pinhole camera of geometrical optics becomes a quite different analog for signal processing if the aperture shrinks to the size of the wavelength of the light. It follows from wave optics that the distant image of a beam of light passing through an aperture or filter is the Fourier transform of the transmission in the plane of the aperture. The same principle applies to the distribution of electric fields radiated by an antenna.

$$A(k) = \int_{-\infty}^{+\infty} a(r)\, e^{-j2\pi kr} dr \qquad (2.22)$$

The transmission of the aperture is $a(r)$: either 1 or 0 for a hole, and an arbitrary function for an optical filter or antenna pattern. Here, k is the wave-vector $2\pi/\lambda$, $kr = (2\pi r/\lambda)\sin\theta$, and θ is the angle from the normal to the plane of the aperture at which the distant image is observed. A portion of an observation sphere distant from the aperture is effectively flat ("far field conditions''), so the variable, $|k| = \sin(\theta/\lambda)$, is effectively the distance from the center of the image plane.

Viewed in this way, the convolution examples which follow will come as no surprise to those familiar with either optics or antennas. A hole larger than the wavelength returns an exact image, as $f(r)$ approaches a delta function. As the hole shrinks to the wavelength of the radiation, the point image spreads out, as with the radiation pattern of an antenna. The correspondence with signals and spectra is as follows:

Signals and Spectra	Diffraction in Optics or Antennas
time t	distance r along aperture
frequency f	wave vector $|k| = \sin\theta/\lambda$
signal $y(t)$	aperture transmission or field distribution $a(r)$
spectrum $Y(f)$	diffraction or radiation pattern $A(k)$

Image enhancement, image reconstruction, or deconvolution of instrumental broadening may be viewed as finding the right extra "filter" to pass the signal through on the way to the output. The examples which follow illustrate the basics.

Example: Spectrum of a Tone Burst
What is the Fourier spectrum of a tone burst of frequency, f_0, with a Gaussian envelope of width 2τ? Write the wave packet as the product of a continuous wave and a Gaussian peak, as shown in Figure 2.7a.

$$y(t) = \left[e^{-\frac{t^2}{2\tau^2}} \right] e^{j2\pi f_0 t_0} \tag{2.23}$$

It will be convenient to use the exponential polar-form representation rather than writing a sine wave (see Eq. 2.10). Since this is a product of two time functions, the Fourier transform is the convolution of individual transforms. The FT of a Gaussian is another Gaussian.

$$\text{FT}\left[e^{-\frac{t^2}{2\tau^2}} \right] = 2\pi\tau^2 e^{-f^2(2\pi\tau^2/2)} \tag{2.24}$$

The FT of the complex exponential in Eq. 2.23 is a delta function located at f_0. Since convolution with a delta function shifts the function over to the delta function position, the Fourier transform of a wave packet is a Gaussian peak centered at the frequency f_0 of the wave, as shown in Figure 2.7a.

Example: Two-slit diffraction
Light or any wave passing through two slits forms an interference pattern well-known to physicists, shown in Figure 2.7b. The pattern can be readily

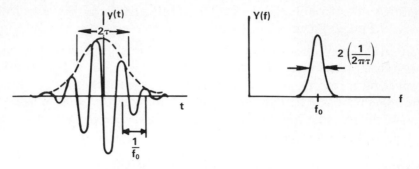

a. Fourier transform of a wave packet or tone burst: the time function and its Fourier transform.

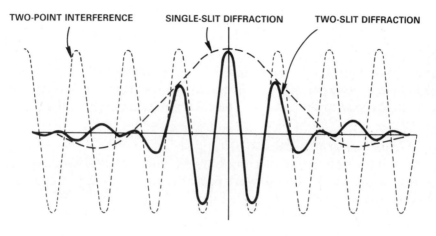

b. Two-slit diffraction.

Figure 2.7. Convolution examples.

understood using the convolution theorem. The two-slits are like the convolution of a single slit with two point-delta-functions. The image is therefore the product of two functions: the sine-wave interference pattern of two points (Exercise 2.5), and the $\sin x/x$ Fourier transform of the slit, which has the form of the pulse's FT, Eq. 2.9.

Example: Diffraction from a Periodic Array

Using the convolution theorem, the pattern of radiation scattered from a periodic array can be separated into the portion due to the periodicity and that due to each scatterer. This can be quite useful in solving for an unknown pattern of scatterers, with signals ranging from sonar to x-ray crystallography. The pattern of scattered radiation $F(k)$, where $k = 2\pi/\lambda$, is the summation of scattering of the wave $e^{-j\mathbf{k}\cdot\mathbf{r}}$ from each point r with amplitude $q(r)$:

$$F(\mathbf{k}) = \int_{-\infty}^{+\infty} q(\mathbf{r})\, e^{-j\mathbf{k}\cdot\mathbf{r}} d\mathbf{r} \qquad (2.25)$$

Here, $\mathbf{k} \cdot \mathbf{r}$ is the dot-product of the two vectors, $\mathbf{k}$ and $\mathbf{r}$. Since $F(\mathbf{k})$ has the form of a Fourier transform, the *inverse* transform will enable the pattern of scatterers in an unknown structure to be determined.

If the objects are arranged in a periodic array, then $q(\mathbf{r})$ has a form analogous to Figure 2.5b:

$$q(\mathbf{r}) = q_a(\mathbf{r}) \star \sum_{\text{LATTICE}} (\mathbf{r} - \mathbf{R_n}) \qquad (2.26)$$

where q_a represents the scattering by a single object. Since this has the form of a convolution, the diffracted image, Eq. 2.25, is separable into a product.

$$F(\mathbf{k}) = Q_a(\mathbf{k}) \sum_{\substack{\text{RECIPROCAL} \\ \text{LATTICE}}} \delta(\mathbf{k} - \mathbf{K_n}) \qquad (2.27)$$

The second term, the FT of the lattice sum of delta functions in Eq. 2.26, determines the *positions* in k-space (or angles of scattering) of the scattered peaks, and measures the lattice structure. The first term, the FT of $q_a(\mathbf{r})$, determines the *amplitudes* of the peaks; it is related to the individual objects doing the scattering. The convolution theorem is thus the key to unraveling complex structures by scattering radiation from them.

2.3.4 RELATED SIGNAL-SPECTRUM THEOREMS: PARSEVAL, CORRELATION, AUTOCORRELATION

Several important signal-processing theorems are closely related to convolution or are readily understood through the convolution theorem.

Parseval's theorem: The total energy of the signal is closely connected to the total energy in the transform. Parseval's theorem states that the the average squared magnitude of the signal (or the product of complex conjugates of the signal) equals the average squared magnitude (or product of complex conjugates) of the transform:

$$\int_{-\infty}^{+\infty} |y(t)|^2 dt = \int_{-\infty}^{+\infty} |Y(f)|^2 df \qquad (2.28)$$

The proof follows from the autocorrelation theorem below. Parseval's theorem has important consequences in the computation of Fourier transforms. Consider the two contrasting examples of Figure 2.8a. In both cases we let the signals be both band-limited in frequency and windowed in time, to avoid the issue of convergence in the above integral; this corresponds to practical signal-processing situations. The upper signal is random noise. The Fourier transform is flat up to the bandwidth, then rolls off. The shaded areas of the magnitude-squared are equal. If the signal has a single strong frequency component (second waveform), the requirement of equal areas in Parseval's theorem forces all the energy into the peaks. A Fourier transform computation must take this correspondence into account in the scaling in order to avoid overflow.

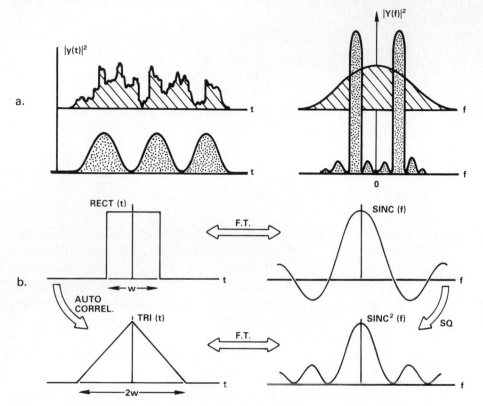

Figure 2.8. Parseval's theorem and autocorrelation. a. Illustration of Parseval's theorem, showing equal areas. b. Illustration of autocorrelation.

Among its applications, *Parseval's theorem* is useful in estimating the propagation of noise through a nonrecursive filter.

Correlation: Two signals may be correlated with one another to any degree from fully identical to completely uncorrelated. That peak on the seismograph may be related to the flash of light detected half-way around the world by a satellite, or it might just have been due to vibrations from the truck that was passing by. Correlation is the search for statistically significant connections between events; it is used extensively for geophysical, sonar, radar, and biomedical signals. Correlation between two signals is measured numerically by the *cross-correlation function, C_{xy}*:

$$C_{xy}(t) = \frac{1}{\sigma_x \sigma_y} \int_{-\infty}^{+\infty} x(\tau) y(t+\tau) \, d\tau \qquad (2.29)$$

The form looks like the convolution, except for the sign of t: there is no time-reversal in the correlation function. Here, σ, which stands for the rms values of the signals, provides normalization. Both x and y are assumed real.* Two signals with related events in time will show a peak in $C_{xy}(t)$. If a stimulus leads to a response, for example, the peak will appear out of the noise when after many repetitions a sufficient number of stimuli accumulate. The location in time is a measure of the delay in the system.

If the two signals contain a common frequency component, a peak in the correlation function's Fourier transform demonstrates the common component in spite of other frequency components in either signal. This follows because the correlation function resembles the transformed convolution, i.e., the product—but with the complex conjugate:

$$FT\, C_{xy}(t) = X(f)\, Y^*(f)$$

The reason is that the correlation may be rewritten as a convolution with a change of sign:

$$C_{xy}(t) = x(t) \star y(-t)$$

by inspection of the definition. The time reversal brings the complex conjugate of the Fourier transform, hence the stated result. If both $X(f)$ and $Y(f)$ contain spectral peaks, only those in common will contribute to their product.

Some correlation-function examples:
   (a) x and y identical functions:
        C_{xy} is a single peak of magnitude unity at $t = 0$.
   (b) x and y have the same form except for a scale factor, c, and a time shift, T:
        C_{xy} is a peak of magnitude unity located at $t = T$.
   (c) x and y have no relationship:
        $C_{xy} = 0$.

Random noise has a zero correlation function with a signal if the summation is carried on long enough. A sample of finite duration may give non-zero C_{xy}, but it tends towards zero as the correlation record length increases. The correlation function can thus bring out the connection otherwise obscured by noise in one or both signals: a peak or other structure emerges above the noise.

The correlator is a key signal-processing element, comparable in importance to the filter and the spectrum analyzer. The *coherent detector*, or *lock-in amplifier*, is a correlator with one input a measure of the stimulus frequency and the other a measure of the response. The *phase-locked loop* is a correlator with the second input generated internally.

---

*For complex signals, C_{xy} is defined with one of the signals written as the complex conjugate; we will ignore this complication (see also Section 8.3).

The Autocorrelation Theorem: The autocorrelation, C_{xx}, of a signal, $y(t)$, (the correlation with itself) equals the Fourier transform of the power spectral density, the term used in statistics for the squared magnitude of the Fourier transform, $Y(f)$ (see Figure 2.8b).

$$C_{xx} = \int_{-\infty}^{+\infty} y(\tau)y(t+\tau)\,d\tau = \int_{-\infty}^{+\infty} |Y(f)|^2 e^{j2\pi ft}\,df \qquad (2.30)$$

If the signal is complex, the first term on the left above is the complex conjugate. The proof follows by a straightforward application of the convolution theorem, together with the symmetry properties of Fourier transform pairs. Phase information is entirely missing from the autocorrelation; the FT is unchanged if the phases are allowed to vary. The autocorrelation function is a tool for the analysis of random noise signals.

2.4 LIMITS TO INFORMATION

2.4.1 TONE-BURSTS AND FREQUENCY UNCERTAINTY

A signal observed over a limited time interval or window has limited spectral definition. This follows because:

> The Fourier spectrum of a wave observed over a finite interval, or window, is the convolution of the true spectrum of the wave with the FT of the window itself.

Consider the sine wave observed over a square window (Figure 2.9), the "tone-burst". The Fourier transform, also shown, has the familiar $\sin(x)/x$ form, from the FT of the square pulse, Eq. 2.9b.* The shorter the window, the wider the peak, hence the less information about the characteristic frequency or pitch. A flute-tone and a drum-beat are both vibrating tones, but the drum has little pitch because the energy input of the beat all occurs in a few cycles of vibration of the drumhead (See Exercises 2.6 and 2.7). This is shown in the Figure; the width of the central lobe is

$$\Delta f = \frac{1}{T_w} \qquad (2.31)$$

Thus, tone bursts which cover only a few cycles of oscillation have little pitch information left in them, since $\Delta f/f_0 \simeq 1$.

2.4.2 THE UNCERTAINTY PRINCIPLE OF INFORMATION

The above tone-burst example illustrates a fundamental property of information. The uncertainty principle of physics has a similar origin in the wave-nature of matter. If I try to determine the *location* of an object to increasingly

---

*It looks different because it is a magnitude plot; all lobes are positive in magnitude.

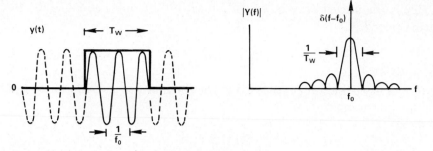

a. A sine wave observed over a finite square window; the Fourier transform has spectral information convolved with the F.T. of the window.

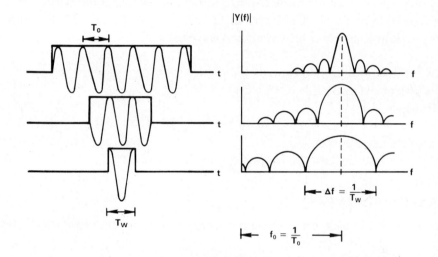

b. Less time in the burst means less spectral definition in the Fourier transform.

Figure 2.9. Transform of sine wave of truncated duration.

finer precision, information about its *velocity* will be lost. Take a box with a hole of size δ_x. A screen at the rear of the box lights up where an object hits it. Firing objects straight at the box does not build up counts in the geometric shadow of the hole, as with the pinhole camera of *geometrical* optics, but builds up a peak which is spread out, like the diffraction of light through the same hole. This thought-experiment is confirmed in reality, and is a consequence of the Heisenberg *uncertainty principle* of particle "waves."★

The uncertainty principle states that

$$\Delta x \Delta k > 1 \qquad (2.32)$$

★Particles act like waves, with momentum, p, and wavelength, λ, related by $p = h/\lambda = h/2\pi k$, where k is the wave vector $2\pi/\lambda$, and h is Planck's constant, which sets the scale for observing wave-particle duality.

Here, Δx is the uncertainty in x-position set by the hole, and Δk, the uncertainty in wave vector, $2\pi/\lambda$, is

$$\Delta k = \Delta(2\pi/\lambda) \tag{2.33}$$

The uncertainty principle, expressed in the form of Eq. 2.32 or by its equivalent in the time domain,

$$\Delta t \Delta f > 1 \tag{2.34}$$

is among the most fundamental theorems of information theory. Observation of any phenomenon over a finite window limits the spectral information obtainable to $\Delta k = 1/\Delta x$, for $x - k$ pairs, or to $\Delta t = 1/\Delta f$, for $t - f$ pairs. This is implicit in our previous examples of the Gaussian wave packet (Figure 2.7a):

Wave packet, width in time $= \sqrt{2}\,\tau$
Width of spectral peak $= 1/(\sqrt{2}\,\tau)$
Product $= 1$

Try it for the tone burst of Figure 2.9, using the width of the first $\sin(x)/x$ peak as a measure of uncertainty in frequency (See Exercise 2.10).

2.5 SAMPLED DATA

"When a tree falls in the forest and no-one is there to hear it, does it make a sound?" (after Kant)

What a signal is doing between samples or outside the window of observation is unknown, but the process of discrete computation forces behavior which had best be understood before the results of a computation are believed. This section explains the consequences of sampling on the spectrum, including aliases and how to avoid them.

2.5.1 SAMPLING, ALIASING, AND ANTI-ALIASING FILTERS

The act of sampling data at discrete intervals limits the frequency range of the spectral information that can be determined. The highest meaningful frequency is half the sampling frequency. The data must be pre-filtered to eliminate components above this value. Even high-frequency noise which may not have been apparent can image down or "alias" if not eliminated by filtering (the antialias filter of Figure 2.1) prior to digitizing the signal. This section will explain the phenomenon and how to control its effects. Aliasing is a general effect, common to the wagon wheels example of Chapter 1, to the vibrations of discrete beads on a string or atoms on a lattice, to the modes of a lumped LC model of a transmission line. For sampling without aliasing, the guideline is as follows:

Sampling Rules

(1) The signal must be sampled at at a frequency, f_s, which is at least twice the rate of the highest frequency component of interest.

(2) Signals containing noise or frequency components greater than $f_s/2$ must be passed through an antialiasing filter prior to digitizing. The filter cutoff frequency and rolloff characteristics are set so as to bring any noise components extending beyond $f_s/2$ below the system resolution, while distorting the signal as little as possible within the passband.

Although time is the usual independent variable in electrical measurement and control systems, and discrete time is the equivalent variable employed in DSP treatment of signals, it may be illustrative to start this discussion of sampled data with a case in which *distance* is the independent variable.

Frequency Spectrum of a Discrete System

A mechanical rod or an electrical transmission line is a system with a *continuous* variation of the mechanical or electrical parameters, such as position or voltage, as a function of distance along its length. For study by numerical methods, it may be decomposed into a discrete (lumped) system, such as the chain of springs and masses or inductors and capacitors shown in Figure 2.10. The positions of the masses (x) or voltages across the capacitors (V) are only defined at discrete positions, n, spaced a units apart (viz., x_n or V_n). While one can make observations anywhere in a continuous rod or transmission line, the only motions one can discuss in the lumped case are the displacements, x_n, at specific intervals, n, or specific voltages, V_n.

In a continuous system that is not at rest, a wave of displacement or voltage is present. At a given instant, its continuously varying value—as a function of x—can be observed at any point; for a lumped system, however, the continuous wave is replaced by a set of discrete points. Whereas in the continuous system, the ability to look anywhere allows any wavelength to be observed, the smallest wavelength that can be sensed in the discrete system is two adjacent points vibrating out of phase with one another. If the separation is a, this minimum wavelength, λ_{min}, is thus $2a$. This minimum wavelength corresponds to a maximum wave vector, k_{max}.

$$|k_{max}| = \frac{2\pi}{\lambda_{min}} = \frac{\pi}{a} \tag{2.35}$$

Note that k_{max} has twice the period of the "sampling" frequency, $1/a$, which corresponds to the wave vector, $2\pi/a$. More than one continuous wave can be drawn through the same discrete set of displacements. For example, in Figure 2.10, call the wave of longest wavelength that can be drawn through the set of points taken from the continuous wave $k = k_0$. An equally good fit to the discrete motions can be drawn with another wave of much shorter wavelength, and wave-vector $k = (2\pi/a) - k_0$. Spectral information in a discrete lattice thus repeats at intervals of the sampling frequency, $2k_{max}$. The sine wave of Figure 2.10 corresponds to delta function modes (Figure 2.4) at $\pm k_0$. Equivalent modes will be found at:

$$2k_{max} \pm k_0, \ 4k_{max} \pm k_o, \ 6k_{max} \pm k_0, \dots \tag{2.36}$$

as seen in Figure 2.10d. For k_0 less than k_{max}, the modes k_0 and $2k_{max} - k_0$ will be disjoint, occurring at clearly different frequencies; however, if k_0 is greater than k_{max}, they will overlap (as will $k_{max} + k_0$ and $2k_{max} - k_0$, etc.) and it will be impossible to distinguish between them.*

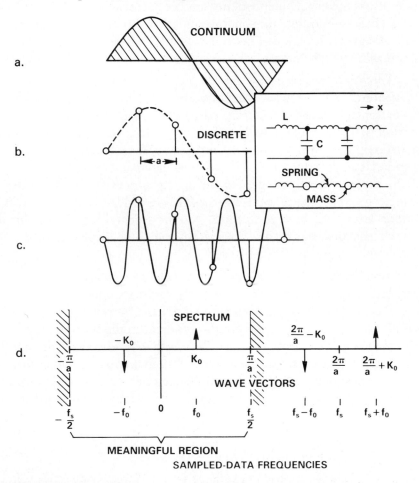

Figure 2.10. A wave passing through a discrete point set such as a line of point masses or lumped LC circuits has more than one possible interpretation: a. The continuous wave. b. Its direct discrete representation. c. An alternative wave which fits the same discrete displacements. d. The ambiguity is represented by showing equivalent points in wave-vector separated by the (repeat-distance)$^{-1}$, or frequencies separated by the sampling frequency.

*This fundamental theorem in sampled data, due to Brillouin, finds its way into lattice filters in transmission-line modeling and is a cornerstone of solid state physics.

Analogously, in the case where the independent variable is time, the same redundancy occurs for signals sampled at discrete intervals of time ($1/f_s$). The spectrum of the signal will be replicated at intervals of f_s. *Any spectral feature at* $\pm f_0$ will find itself repeated at

$$f_s \pm f_0, \ 2f_s \pm f_0, \ 3f_s \pm f_0, \dots \tag{2.37}$$

also shown in Figure 2.10d.

The relationship between sampling and aliasing can now be clearly seen. For a fixed-frequency sine wave sampled at successively decreasing rates, as shown in Figure 2.11, the aliased modes (dashed peaks) move downwards until they pass into the fundamental interval $[-f_0 < f < f_0]$ at $f_0 = f_s/2$. At slower sampling rates, the Fourier spectrum in the region between $-f_0/2$ and $+f_0/2$ will have extra peaks called *aliases*, corresponding to the case in which more than one wave can be drawn through the sampled points. Since the first

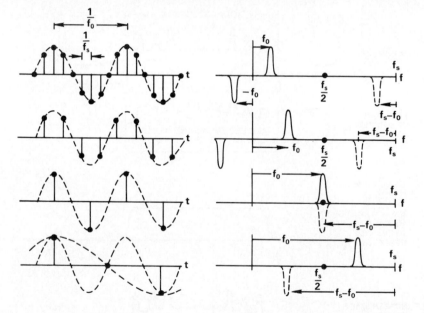

Figure 2.11. Sampling and aliasing. Here the signal frequency is held constant as the sampling rate is decreased; note the shifts in the corresponding peaks in the spectrum (right side), especially the aliases (dashed curves).

alias occurs at the difference between f_s and f_0, aliases can occur at quite low frequencies. If the sampling frequency coincides exactly with a frequency in the signal, the resulting alias will be at zero frequency, i.e., *dc*.

Exercise: The spokes in the filmed wagon wheels of Figure 1.13 stop and then go backwards. Why? A film is a sample of the continuous event. The sampling frequency, f_s, is the frame speed. In Figure 2.11, the peaks are drawn with opposite signs, as for a sine wave. As the aliased peak moves down into the fundamental interval and the "true" peaks move out of it, there is a sign reversal in the peaks which lie within $-f_s$, $+f_s$. The aliased wave has opposite sign, which corresponds to a reversal in the direction of motion.

How to avoid aliasing? Put in a filter which kills frequencies higher than $f_s/2$, as in Figure 2.12. Here, the strong noise component is aliased into the signal

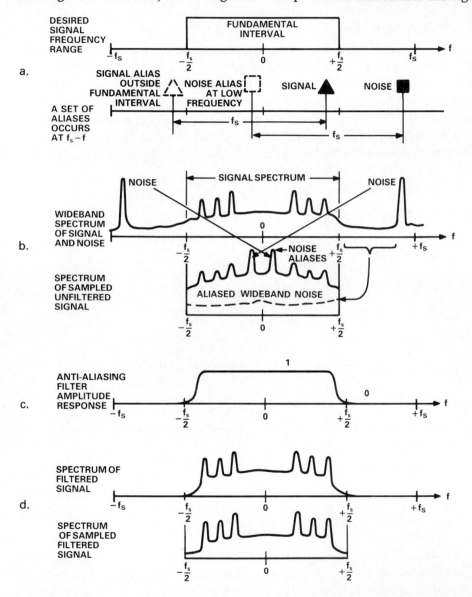

Figure 2.12. Aliases and the effects of an anti-aliasing filter. a. Signal and noise, and their aliases in the neighborhood of the fundamental frequency interval. b. The spectrum of a noisy signal without prior filtering—before and after sampling. c. Amplitude response of anti-aliasing filter. d. Spectrum of filtered signal before and after sampling.

domain when sampled directly. But if the signal is first passed through a low-pass anti-alias filter, frequency components beyond the passband of the low-pass filter are attenuated—and so are their aliases. A component at a frequency in the stop band of the filter is completely attenuated so that there is no alias of that component in the sampled data.

A modern equivalent of the wagon wheels occurs in the computer graphics animation of space ships flying into canyons or past battle cruisers. When the background features have any periodicity, any portion of the background moving faster than half the frame rate will alias. This is eliminated in animated graphics with motional blurring, another word for antialiasing filters, equivalent to letting the shutter stay open part of the time between frames.

> *Exercise*: Stopped motion as in a strobe light can sometimes be seen under ordinary fluorescent illumination but never under light from a heated filament (ordinary light bulb). Why? The fluorescent bulb can respond at the rate of the ac power line driving it, but the heated filament has a low pass filter; its specific heat limits its response time, as you can see by estimating the decay time for light when you turn off the bulb.

2.6 THE FOURIER TRANSFORM OF SAMPLED DATA

2.6.1 SPECIAL FUNCTIONS IN SAMPLED DATA

Discrete data occurs for integer values of an independent variable, n, and nowhere in-between; in discrete-time systems, $t = nt_s$, where $t_s (= 1/f_s)$ is the sampling interval. It is convenient to define some special functions in sampled data (Figure 2.13):

Unit impulse at $t = 0$, $I(nt_s) = 1$ at $n = 0$;   0 everywhere else

Row of unit impulses: 1 for all n

Unit step, $u(n_0)$, at $t = n_0 t_s$: $u(n_0 t_s) = \sum\limits_{k=-\infty}^{n_0} I(kt_s)$

Unit rectangle, $R_M(n)$: $u(nt_s - n_1 t_s) - u(nt_s - (n_2 + 1) t_s), n_2 - n_1 = M$

The unit impulse plays the role of the delta function (its unit amplitude is akin to the delta function's unit area). The properties of the other special functions follow from the unit impulse. The row of unit impulses plays the role of a sampling function: multiply a continuous signal by a row of unit impulses to create a sampled signal. At a given point, $t_n = nt_s$:

$$f(nt_s) = f(t_n) I(nt_s) \qquad (2.38)$$

The unit step helps make the transition to the Laplace transform in sampled data, since it forces the samples to zero before a given point. The unit rectangle, $R_M(n)$, samples over a finite window, Mt_s, the width of the rectangle.

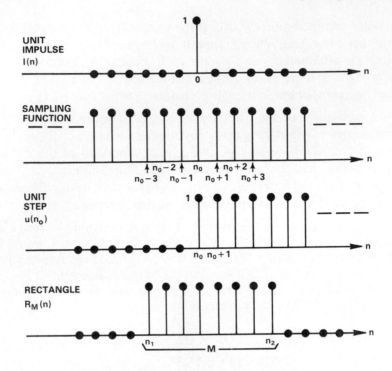

Figure 2.13. Some special functions in sampled signals. From the top: unit impulse, sampling function, unit step, and sampling within a rectangular window.

It is useful to consider the Fourier transforms of some of these functions, as shown in Figure 2.14a.

(a) The Fourier transform of the sampling function is also a row of unit impulses, separated by frequency interval $1/t_s$. Those familiar with delta functions will recognize the reason. From the Fourier integral, Eq. 2.6, the unit impulse series contributes equal amplitudes at each frequency which is a multiple of f_s, and cancels everywhere else.

(b) The Fourier transform of the unit rectangle is, from Eq.2.6, a finite geometric series.

$$\text{FT}[R_M(n)] = \int_{-\infty}^{+\infty} \sum_{n=0}^{M-1} I(nt_s)\, e^{-j2\pi ft}\, dt$$

$$= \sum_{n=0}^{M-1} z_n = \frac{1 - z^M}{1 - z} \tag{2.39}$$

where

$$z = e^{-j2\pi ft_s}$$

It is customary to redefine the rectangular window about its mid-point, so the sum ranges from $-M/2$ to $(+M/2 - 1)$. After a little algebra,

SIGNAL   **SPECTRUM**

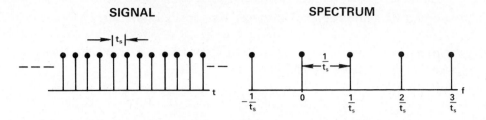

a. The sampling function and its Fourier transform, which is also a row of unit impulses.

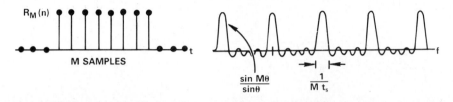

b. The sampling function over a finite window of width M has a series of sin(x)/x-broadened peaks with overlapping tails.

Figure 2.14. Two special signal-spectrum pairs.

$$\text{FT}\,[R(n)] = W(\omega) = e^{j\omega t_s/2}\ \frac{\sin N\omega t_s/2}{\sin \omega t_s/2} \tag{2.40}$$

This has the form of a row of $\sin(x)/x$ functions, as shown in Figure 2.14b. The width of each central peak is $1/(Mt_s)$, and the separation between peaks is $1/(t_s)$.

> *Exercise:* Use the convolution theorem to explain the shape and spacing of $W(\omega)$ above.

As M gets larger, each peak gets narrower, approaching the row of unit impulses of Figure 2.14a in the limit as $M \to \infty$.

2.6.2 SPECTRAL PROPERTIES OF SAMPLED DATA

Given that sampled data is known only at discrete intervals, the computation of the Fourier transform becomes a sum.

$$Y(f) = \sum_{n=-\infty}^{+\infty} y(n)\,e^{-j2\pi nf} \tag{2.41}$$

The time dimension is often suppressed, or normalized to t_s, and written simply as n. Thus written, the frequency, f, is also dimensionless. This somewhat simplifies mathematical expressions. To restore the time axis of measurement for real-time signal processing, replace n by nt_s, and f will come out in units of $1/t_s$.

The sampled-data Fourier transform has some special symmetry properties:

(a) Fourier Transforms of Sampled Data are Periodic in the Frequency Domain
Suppose the (dimensionless) frequency axis is shifted by 1:

$$f \to f + 1$$

The true frequency is shifted by $1/t_s$. Then

$$
\begin{aligned}
Y(f+1) &= \sum_{n=-\infty}^{+\infty} y(n)\, e^{-j2\pi n(f+1)} \\
&= \sum_{n=-\infty}^{+\infty} y(n)\, e^{-j2\pi n}\, e^{-j2\pi nf} \qquad (2.42) \\
&= Y(f)
\end{aligned}
$$

since $e^{j2\pi n} = 1$ for n any integer. The Fourier transform of sampled data is periodic in the interval of the sampling frequency. *All* information is contained in the interval $(-\frac{1}{2}, +\frac{1}{2})$, called the *basic interval*.

(b) The Fourier Inverse for Sampled Data is Computed as a Fourier Series
Since the Fourier transform, $Y(f)$, is periodic, it may be represented as a Fourier series expansion in frequency space, as in Eq. 2.1, but with time and frequency reversed. But because the Fourier series is unique, the coefficients, $y(n)$, must be the same as the original set of sampled data:

$$y(n) = (1/F) \int_{-F/2}^{+F/2} Y(f)\, e^{j2\pi nf}\, df \qquad (2.43)$$

where we write the "interval" F explicitly as a reminder.

(c) Fundamental Intervals in the Spectrum of Sampled Data
The spectrum of sampled data in a finite window thus has three special time intervals, summarized in Figure 2.15a.

> the period of a given spectral component, $1/f_0$;
> the period of the window, $1/f_w$
> the period of sampling, $1/f_s$.

In the frequency domain, there are three corresponding frequencies (b):

> f_o locates the peak in the spectrum
> f_w determines its width
> f_s is the interval of a periodic repetition of the spectrum,

It is often convenient to view the frequency domain of sampled data as a circle which revolves one turn each time f changes by f_s, as in Figure 2.15c. Aliasing is a natural consequence of this picture, since each spectral feature—such as the pair drawn at $\pm f_o$ here—repeats at intervals of f_s.

(d) The Convolution Theorems Work for Sampled Data

The convolution of two functions sampled in the time domain becomes

$$z(n) = \sum_{m=-\infty}^{+\infty} x(m)\,y(n-m) \qquad (2.44)$$

All the general convolution theorems of Fourier transform theory go through unchanged, with the integrals replaced by discrete sums.

However, the convolutions in frequency space will undergo the same circular periodicity shown in Figure 2.15c, called the *circular convolution*. We can imagine the circular convolution as wrapping the two functions around a cylinder (one time-reversed), multiplying and integrating the results, and then rotating one of the functions before computing the next convoluted point.

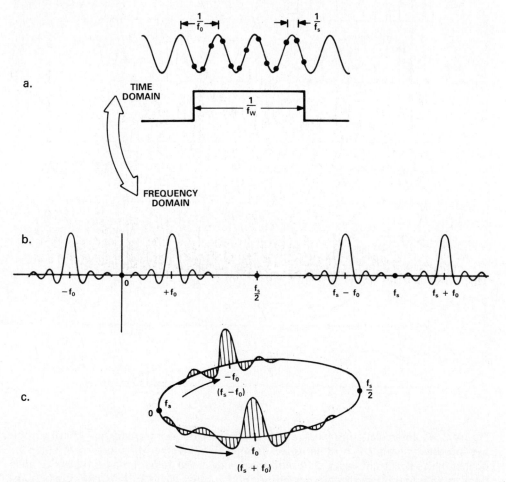

Figure 2.15. Interrelationship between (a) basic time intervals and (b) corresponding frequencies in sampled data over a finite window. The repeating nature of the frequency information is sometimes shown by plotting it over a circular cylinder (c).

2.6.3 SUMMARY OF SPECTRAL FEATURES OF SAMPLED DATA

The interconnection of Fourier transforms and Fourier series, up to now considered as different objects, can be seen graphically in a sequence which gives a preview of the Discrete Fourier Transform of Chapter 3. Consider Figure 2.16.

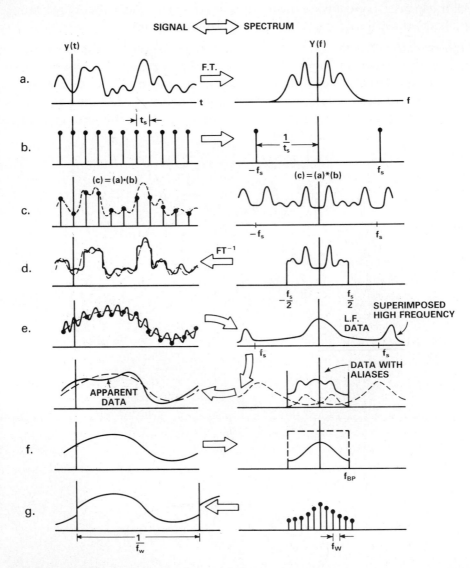

Figure 2.16. Interrelationships of (a) the continuous Fourier transform with (b) sampling impulses give the F.T. of sampled data (c). The inverse (d) is in error if spillover occurs in the spectrum past $f_s/2$. A high frequency in the data (e) may translate or alias back to an apparent low-frequency component, unless (f) a bandpass filter is first applied to the data. Discrete sampling of frequencies (g) forces a periodicity to the data outside the window of observation.

(a) A given continuous function, $y(t)$, has its corresponding continuous Fourier transform.

(b) Suppose $y(t)$ is to be sampled at intervals t_s. The sampling function, a row of unit impulses, has as its Fourier transform also a row of delta functions, at intervals $f_s = 1/t_s$.

(c) The Fourier transform of the sampled data is the convolution of the original FT in (a) with the FT of the sampling function of (b). If spectral content in the spectrum spills over above $f_s/2$, the overlapping contributions will add up.

(d) The inverse transform of $Y(f)$ is the Fourier series representation, $y(nt_s)$, which will have error contributions because of the overlap in (c)'s frequency spectrum.

(e) Worse yet, if $y(t)$ had a noise component at a frequency above $f_s/2$, the back transform will have an error signal or *alias* within the fundamental frequency interval. The alias is very real and troublesome to track down.

(f) Instead, pass the original signal through a lowpass filter prior to doing the transform to limit the spectral content, so no alias terms occur in the transform.

(g) If the frequency spectrum is sampled (as it would be in computations), and an insufficient number of samples are used, discontinuities will be introduced at the boundaries of periods corresponding to the frequency resolution, f_w.

2.6.4 EXAMPLE: SINUSOIDAL SIGNAL GENERATOR

The output of the d/a converter in Figure 2.1 is usually a stair-step approximation to a smooth continuous function of time. The horizontal spacing is the output sampling time, which establishes the nominal step amplitudes.* While the deglitching filter—which is essentially a sample-hold—can remove transient distortion within the DAC's update time, leaving an essentially "clean" (albeit rounded) stairstep, the distortion of a desired smooth function remains unless further filtering is used.

It is illustrative to consider the nature of this distortion for the case of a digital signal generator: The output is the d/a conversion of the clocked output of a sine lookup table from memory (Figure 2.17b)—a sampled, quantized version of the desired clean sinusoidal output.

Among the advantages that digital oscillators have over L-C and R-C-based oscillators are frequency stability (because of the use of a crystal clock) and the ability to be phase-locked to one another (since they can share the same

---

*Because the output is quantized, the actual values of the step amplitudes are also affected by the digital system resolution (2^{-n}, where n is the number of bits) and/or the resolution of the DAC. For the present discussion, the effects of quantized amplitude will be ignored.

clock. A stair-step sine-wave approximation is shown in Figure 2.17a. The approximation is free from *low*-harmonic distortion. For an N sample-per-cycle approximation, harmonics do not show up until $(N-1)$ and $(N+1)$ times the fundamental; their respective amplitudes are $1/(N-1)$ and $1/(N+1)$ of the fundamental amplitude. The next harmonics do not appear until about twice this value of frequency, and their amplitudes are further reduced by about a factor of two. For a 32-sample approximation, the spectrum is:

Frequency (harmonic of f_0)	Amplitude
1	1.000
2	0.000
3	0.000
4	0.000
.	.
.	.
31	0.032
32	0.000
33	0.030
34	0.000
.	.
.	.
63	0.016
64	0.000
65	0.015
66	0.000
67	0.000

Harmonic distortion so widely separated from the fundamental is readily eliminated by simple low-pass filter algorithms.

The explanation is another good application of the convolution theorem. The waveform is equivalent to a point-sampled sine wave "stretched" into stair steps. More precisely, *convolve* the sampled wave with a pulse of width equal to the sampling interval, τ, as in Figure 2.17b. The spectrum in the frequency domain is thus the spectrum of the sampled sine wave *multiplied* by the spectrum of the pulse, as in Figure 2.17c. The spectrum of the sampled sine wave is a set of delta function pairs at $\pm f_0$, repeated at intervals $N f_0$, where N is the number of steps per cycle of the sine wave.

The spectrum of the pulse is $\sin(x)/x$, with $x = 2\pi\, f/(N f_0)$. The zeros of this function occur precisely at intervals of the sampling frequency. The resulting product, shown in Figure 2.17d, has the harmonic amplitudes tabulated above; the calculation is left as Exercise 2.12. Because sampling only repeats the original sine-wave at intervals of f_s, there is no harmonic distortion at frequencies below about Nf_0 for an N-point stair-step. The amplitudes of the harmonics are small because the product terms occur very near the zeros of the $\sin(x)/x$ function.

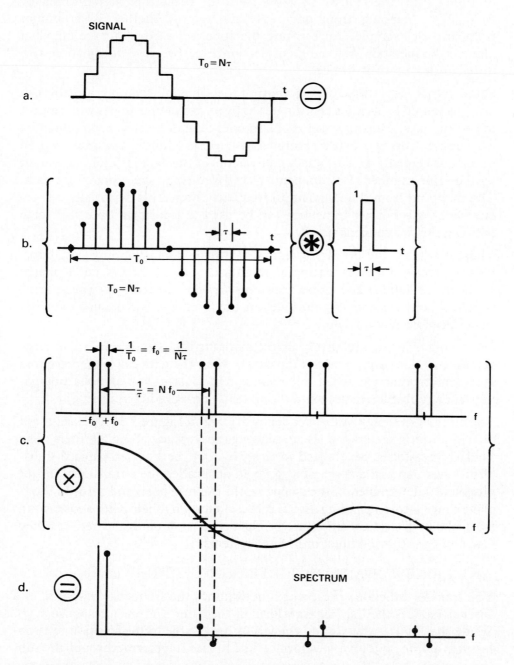

Figure 2.17. Digital oscillator. a. Waveform. b. Interpretation as a convolution. c. Frequency spectrum of the terms in (b). d. Resulting spectrum of the digital oscillator.

2.7 LAPLACE AND z TRANSFORMS

INTRODUCTION: MAPPINGS

Mapping finds use in DSP in going from the domain of digital to analog frequency. Mapping from one representation to another is a familiar problem, for example, representing the spherical earth's surface on a flat plane. Map methods (Mercator's projection, etc.) preserve information with differing degrees of distortion.

Some mappings are two ways of viewing the same information: a vector can be represented by its *x-y-z* coordinates or by its radius and angles with respect to certain axes. This mapping is one-to-one. Signals are also represented by mappings; an array of data measured as voltages as a function of time is stored in memory equally well as binary equivalents of the data for each sample, or by the same number of components of the frequency spectrum of the data. The mapping from time domain to frequency domain is essentially one-to-one, since the Fourier transform can be used to transform back and forth between the two domains at will.

Those mappings preserve information. However, some mappings lose information irretrevably; for example, statistical analysis brings out a mean, standard deviation, and other measures (mode, median, skewness, etc.) which, though representing the data, are not by themselves enough to reconstruct large amounts of data.

On the other hand, a relatively simple mapping can permit retrieval of large amounts of data (i.e., *compress* the data): when the data can be represented as an analytic function; when only *changes* in the data are stored (delta modulation); or key characteristics *model* the system (linear predictive coding).

The circular mapping of the frequency axis, as in Figure 2.15c, is useful not only in plotting spectra but also in plotting the response of digital filters. The Fourier transform, generalized to *complex* ω in the Fourier/Laplace transform,* has a special mapping for sampled data called the z-transform. Filter properties, determined in the analog world by their poles and zeros in the s-plane, have a corresponding digital representation by their poles and zeros in the z-plane. We review first (briefly) the concepts of poles and zeros as they determine the transfer function of filter systems.

2.7.1 REVIEW OF ANALOG FILTER CONCEPTS

The transfer function, $H(s)$, which determines the frequency response of a filter system, is the Laplace transform of the filter's impulse response $h_i(t)$. While the Fourier transform is generally applicable to the solution of problems involving repetitive waveforms, the Laplace transform, through the sub-

---

* We retain the Laplace transform designation for ω complex, even though some authors call it the generalized Fourier transform (FT), because of the large body of filter lore with this terminology. Essentially all of the FT theorems of Table 2.1 apply equally well to the LT. Even though we define the LT over the half-plane, $t = (0, \infty)$, this is only for convenience in connecting with pulse/step signal theory and to avoid unnecessary digression into convergence issues.

stitution of s ($= \sigma + j\omega$) for $j\omega$ in the Fourier transform equation, can handle unique occurrences. For our purposes, the Laplace transform reduces to a one-sided Fourier transform when $s = j\omega$. The relationships are summarized in the box (Equations 2.45).

Laplace Transform Equations:

Direct:

$$F(s) = \int_0^{+\infty} e^{-st} f(t)\, dt$$

Inverse:

$$f(t) = \frac{1}{2\pi j} \int_{c-j\infty}^{c+j\infty} e^{ts} F(s)\, ds$$

where the choice of c involves the domain of convergence (see standard references on Laplace and Fourier transforms).

Transfer Function $H(s) = $ Laplace transform of Impulse response $h(t)$

For any input, $X(s)$, the output $Y(s)$ is

$$Y(s) = H(s) X(s)$$

and the corresponding time response is the convolution of the input $x(t)$ with $h(t)$

$$y(t) = h(t) \star x(t)$$

$$= \int_0^t h(\tau) x(t - \tau)\, d\tau$$

where

$$h(t) = \text{LT}^{-1}[H(s)] = y_i(t),$$

$$H(s) = \text{LT}[h(t)],$$

and

$$X(s) = \text{LT}[x(t)] \tag{2.45}$$

The Laplace transform of the output is related to the input through multiplication with the transfer function, $H(s)$; their time-domain values are related through convolution of the input with the inverse transform of the transfer function, $h(t)$. If the input, $x(t)$, is a delta function, $\delta(t)$, a pulse of zero width but finite area, or *unit impulse* in the continuous time domain), its Laplace transform is unity ($= 1$). When it multiplies the transfer function, the transform of the output is identical to the transfer function; hence, the time response that results from convolution of the delta function with a system having the transfer function, $H(s)$, is $h(t)$, called the *impulse response*.

$$Y_i(s) = \text{LT}[y(t)] = H(s)\, \text{LT}[\delta(t)] \tag{2.46}$$

Since

$$\text{LT}[\delta(t)] = 1 \tag{2.47}$$

then

$$Y_i(s) = LT[y_i(t)] = H(s) = LT[h(t)] \tag{2.48}$$

The inverse transform is

$$y_i(t) = h(t) \tag{2.49}$$

Once the impulse response is known, the response to any other input can be determined by convolution in the time domain—the "memory of previous kicks" idea of Chapter 1.

The transfer function, $H(s)$, contains in simple algebraic form the same information about the system as its differential equation. Consider an analog system describable by a linear constant-coefficient differential equation:

$$\sum_{k=0}^{N} d_k \frac{d^k y(t)}{dt^k} = \sum_{r=0}^{M} c_r \frac{d^r x(t)}{dt^r} \tag{2.50}$$

Take the Laplace transform of both sides. The L.T. of the kth derivative of $f(t)$ brings out s_k (see Table 2.1); the same theorems apply for the L.T. as for the FT:

$$\sum_{k=0}^{N} d_k s^k \, Y(s) = \sum_{r=0}^{M} c_r s^r \, X(s) \tag{2.51}$$

$H(s)$, the transfer function, equal to the ratio, $Y(s)/X(s)$, is thus determined by a ratio of polynomials in s:

$$H(s) = \frac{\displaystyle\sum_{r=0}^{M} c_r s^r}{\displaystyle\sum_{k=0}^{N} d_k s^k} \tag{2.52}$$

Terms in the denominator correspond to feedback of previous *outputs*. Terms in the numerator correspond to feedforward of recent *inputs*. The same parameters, c_k, d_k, which appear in the system differential equation also appear in its transfer function.

Since a polynomial can be factored, $H(s)$ can be expressed as

$$H(s) = A \frac{\displaystyle\prod_{r=0}^{M}\left(1 - \frac{s}{s_{zr}}\right)}{\displaystyle\prod_{k=0}^{N}\left(1 - \frac{s}{s_{pk}}\right)} \tag{2.53}$$

plus an extra polynomial if $M > N$. Here, s_{zr} and s_{pr} locate the zeros and poles of the transfer function and determine the stability of the system. Although it is beyond our scope to build stability background, it may help understanding to list a few facts and two examples.

2.7.2 STABILITY FACTS FOR ANALOG SYSTEMS

(1) The poles of $H(s)$ are the location in the complex plane of the roots of the denominator polynomial; they show up in conjugate pairs having coordinates $(\sigma, +j\omega)$ and $(\sigma, -j\omega)$. They locate in the s-plane plane the "natural frequencies" of the system.

(2) A transfer function which can be written as a ratio of polynomials in s is stable only if all its poles (zeros of the denominator) lie in the left half of the s-plane. Each factored term in the denominator gives a term, $(1 - s/s_p)^{-1}$, which corresponds to an exponential time dependence (see Table 2.2), with $s_p = -b + j\omega$:

$$e^{-(b+jw)t}$$

which does not blow up as long as the real part $(-b)$ is negative.

Nonrecursive filter design manipulates the zeros; there are no feedback terms to give poles. Recursive filters are dominated by the poles of their transfer function. Here are two brief examples:

First-order low-pass filter
The transfer function has the form:

$$H(s) = \frac{1}{1 - \frac{s}{s_1}} = \frac{1}{1 + RCs} \tag{2.54}$$

There is one pole, s_1, on the negative real axis ($s_1 = -1/RC$). The distance of the pole from the origin is inversely proportional to the time constant, RC. The impulse response is, from Table 2.2:

$$y_i(t) = e^{-\frac{t}{RC}} \tag{2.55}$$

Second-order system
Consider a pair of poles, located at $s_p = -b \pm j\omega_0$, as shown in Figure 2.18b. The transfer function is:

$$H(s) = \frac{1}{\left(1 - \frac{s}{s_p}\right)^2} \tag{2.56}$$

which, using Table 2.2, corresponds to a time dependence

$$y(t) = A e^{-bt} \sin(\omega_0 t) \tag{2.57}$$

This is the familiar damped ringing of a resonant system when hit by an impulse.

Table 2.2. Laplace and z-transforms of time functions

Function of time $f(t)$	Laplace Transform $F(s)$	z-Transform $F'(z)$
$\delta(t)$	1	1
$\delta(t - nT)$	e^{-nTs}	z^{-n}
Unit Step (0 to 1), $u(t)$	$\dfrac{1}{s}$	$\dfrac{z}{z-1}$
t	$\dfrac{1}{s^2}$	$\dfrac{Tz}{(z-1)^2}$
t^2	$\dfrac{2!}{s^2}$	$\dfrac{T^2 z(z+1)}{(z-1)^3}$
e^{-at}	$\dfrac{1}{s+a}$	$\dfrac{z}{z-e^{-aT}}$
$\dfrac{1}{(b-a)}\left[e^{-at} - e^{-bt}\right]$	$\dfrac{1}{(s+a)(s+b)}$	$\dfrac{1}{(b-a)}\left[\dfrac{z}{z-e^{-aT}} - \dfrac{z}{z-e^{-bT}}\right]$
$\dfrac{1}{a}\left[(t) - e^{-at}\right]$	$\dfrac{1}{s(s+a)}$	$\dfrac{1}{a}\dfrac{(1-e^{-aT})z}{(z-1)(z-e^{-aT})}$
te^{-at}	$\dfrac{1}{(s+a)^2}$	$\dfrac{Tze^{-aT}}{(z-e^{-aT})^2}$
$\sin at$	$\dfrac{a}{s^2+a^2}$	$\dfrac{z\sin aT}{z^2 - 2z\cos aT + 1}$
$\cos at$	$\dfrac{s}{s^2+a^2}$	$\dfrac{z(z-\cos aT)}{z^2 - 2z\cos aT + 1}$
$\dfrac{1}{b}e^{-at}\sin bt$	$\dfrac{1}{(s+a)^2+b^2}$	$\dfrac{1}{b}\left(\dfrac{ze^{-aT}\sin bT}{z^2 - 2z e^{-at}\cos bt + e^{-2aT}}\right)$
$e^{-at}\cos bt$	$\dfrac{s+a}{(s+a)^2+b^2}$	$\dfrac{z^2 - z e^{-aT}\cos bT}{z^2 - 2z e^{-aT}\cos bT + e^{-2aT}}$

$$Y(s) = \sum_{n=-\infty}^{+\infty} y(n)e^{-sn} \qquad\qquad Y'(z) = \sum_{n=-\infty}^{+\infty} y(n)z^{-n}$$

Example:

If $y(n) = e^{-anT}$, $n \geq 0$, and $y(n) = 0$ for $n < 0$,

then
$$Y(z) = \sum_{n=0}^{+\infty} e^{-anT} z^{-n} = \sum_{n=0}^{+\infty} (e^{-aT} z^{-1})^n$$

$$= \frac{1}{1 - e^{-aT} z^{-1}} = \frac{z}{z - e^{-aT}}$$

2.7.3 POLES, SPECTRAL RESPONSE, AND FILTER BEHAVIOR

The stability of an analog circuit, or its digital filter implementation, is determined by the poles in $H(s)$. Suppose that the poles of Figure 2.18c are moved towards the imaginary axis. What happens to the spectral response of the filter? Consider a 2nd-order bandpass

$$H_{BP}(s) = \frac{Gs}{\left(\dfrac{s}{s_0}\right) + 2\epsilon \dfrac{s}{s_0} + 1} \tag{2.58}$$

Here, ϵ is a dimensionless measure of the horizontal distance, $|b|$, of the pole from the imaginary axis. To calculate the spectral response, substitute $s = j\omega$ and calculate the magnitude of the complex transfer function, as in Figure 2.18c. The sharpness of the peak is a measure of how closely the poles approach the imaginary axis. Since the decay time of Eq. 2.57 is connected with this same quantity, sharper peaks in $H(s)$ are associated with lower damping. The frequency spectrum of a filter may be viewed as a slice of $|H(s)|$ along the axis $s = j\omega$ in the complex s-plane, as in Figure 2.18d. The spectral peaks in Figure 2.18c are cuts which come closer and closer to center of the mountains in the 3-D pole plot.

Undamped behavior can be avoided by proper choice of pole locations. For example, a second-order analog low-pass filter (Figure 2.19a) has more rapid rolloff than the first-order RC (12 dB/octave rather than 6) and will do a better job of smoothing, but it must be properly damped. As damping is reduced, the corner sharpens, then develops a resonant peak in the vicinity of the cutoff frequency. If such a peak is present, a sudden input change can cause the output to ring, as in Figure 2.19c. The means of controlling the ringing can be seen in the transfer function

$$H_{LP}(s) = -\left[\frac{1}{1 + bsRC + (sRC)^2}\right] \tag{2.59}$$

The parameter, b, determined by the capacitance ratio, $b = 3\sqrt{C_2/C_1}$, allows the filter response to be adjusted to eliminate the ringing. Once the frequency, ω_0, and damping, b, have been chosen, and R is fixed, the capacitance values can be chosen using $C_2 = [(\omega_0/R)\sqrt{b/3}]$, and $C_1 = [(\omega_0/R)\sqrt{3/b}]$.

Exercise: How does the damping factor, $b = 3\sqrt{C_2/C_1}$ control ringing?

Solution:
Find the poles and put them where needed. Solve for the roots of the denominator polynomial in Eq. 2.59:

$$s_0 = \left(-\frac{b \pm \sqrt{b^2 - 4}}{2RC}\right)$$

For $b < 2$, s_0 is a complex number, so the poles form a conjugate pair (Figure 2.19d), which indicates overshoot or ringing (c). If $b = 2$, the response is

critically damped. Pick *b* a little smaller to get a sharper corner at the cutoff
frequency, hence a faster initial time response, which approaches the final
value with only a modest amount of ringing, as with the curve marked
$b/2 = 0.707$. This exercise illustrates how poles are used to control filter
behavior.

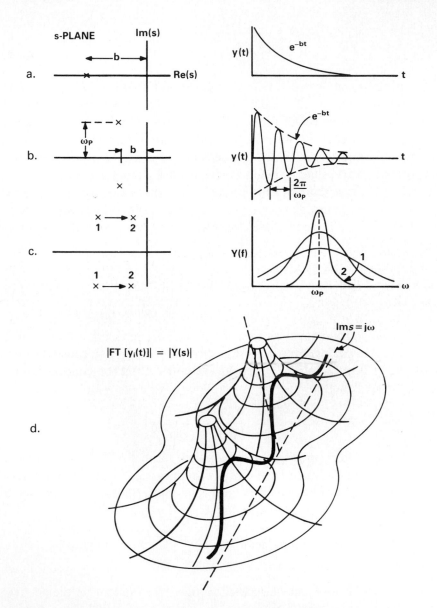

Figure 2.18. Poles and corresponding impulse response for simple analog systems.
a. First order = *RC* low pass. b. Second order = resonant system with damping. c.
Sharpness of spectral response is a function of the closeness of poles to the vertical
axis, Im(*s*) = 0. d. The Fourier spectrum viewed as a slice of H(*s*) along the imaginary
axis, s = *j*ω.

2.7.4 THE *z*-TRANSFORM

The *z*-transform is the equivalent of the Fourier/Laplace transform for sampled data. The sampled function of time, $y(nt_s)$, is

$$y(n_s) = \sum_{k=-\infty}^{\infty} y(t_k)\,\delta(t - nt_s)$$

(2.60)

The Fourier integral becomes a sum, called the *z-transform*:

The *z*-transform: if $y(n) = 0$ for $n < 0$:

$$Y(z) = \sum_{n=0}^{\infty} y(nt_s)\,z^{-n}$$

where

$$z = e^{j\omega t_s}$$

(2.61)

The variable, z, is a mapping of the variable, $j\omega$, and the *z*-transform exists on the entire complex z-plane. The properties of filters and other systems in discrete time are characterized by their poles in the z-plane, just as poles in the complex s-plane characterize continuous-time analog systems. Questions of convergence exist, but they are largely beyond the scope of this chapter (for detailed discussions of this matter, see Oppenheim, A. V., and Schafer, R. W., *Digital Signal Processing*, Englewood Cliffs, NJ: Prentice-Hall (1975)).

The discrete transfer function $H(z)$ follows by taking the *z*-transform of both sides of the difference equations (such as Equations 1.15 and 1.16) that specify the system, and solving for $Y(z)/X(z)$. For a system characterized by a general Nth order difference equation,

$$\sum_{k=0}^{N} a_k\,y(n - k) = \sum_{r=0}^{M} b_r\,x(n - r)$$

The *z*-transform of its transfer function is:

$$H(z) = \frac{Y(z)}{X(z)} = \frac{\displaystyle\sum_{r=0}^{M} b_r z^{-r}}{\displaystyle\sum_{k=0}^{N} a_k z^{-k}}$$

Although similar in form, the coefficients in the sampled $H(z)$ do not equal those in the analog version. To design a digital filter to correspond to an analog filter with a given L.T. transfer function, the mapping $z = e^{st}$ is used. Since

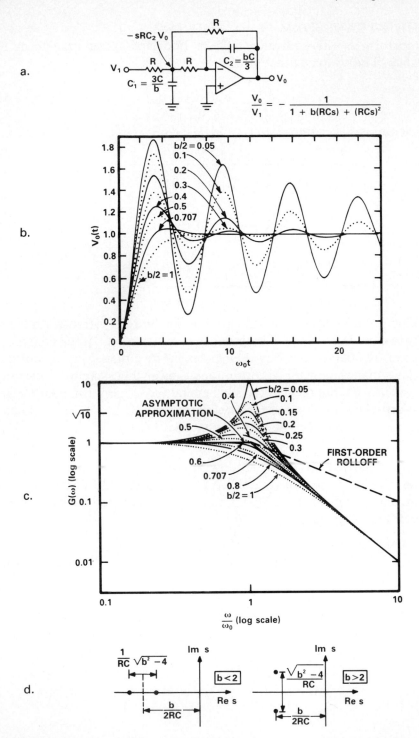

Figure 2.19. Second order low-pass filter. a. Sallen-Key circuit embodiment; b. Transfer function: amplitude response plot.c. Time response to step input. d. Location of poles as a function of the parameter $b = 3\sqrt{C_2/C_1}$.

each z^{-1} element corresponds to a unit time delay, the form of $H(z)$ leads naturally to a block diagram for the digital filter, employing adders, multipliers, and *unit-delay* stages. Procedures for determining the coefficients will be developed in section 4.3.5 (*Impulse Invariance*).

While superficially resembling a Laplace transform, the z-transform has periodic consequences of the discrete sampling. The connection between the Laplace transform and z-transform—and s-plane to z-plane mapping (Figure 2.20) may be summarized as follows:

The Laplace transform, along the frequency axis, $s = j\omega$, is the z-transform evaluated on the unit circle, $|z| = 1$.

(a) The left half of s-plane maps within the unit circle of the z-plane.
(b) A distance, $1/t_s$, along the real frequency axis wraps once around the unit circle in z.

This second property is a consequence of sampling. Filter calculations in the z-plane keep the periodicity visible along the ω axis of Figure 2.20, since the unit circle is traversed once each Nyquist interval.

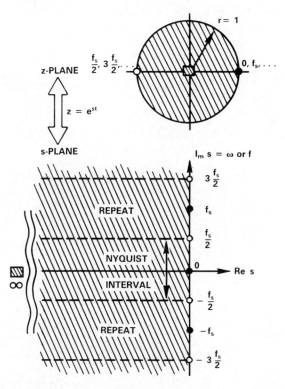

Figure 2.20. Frequency-domain to z-plane mapping. The unit circle in the z plane corresponds to the entire left-half of the s plane, as indicated by the corresponding shaded regions and specifically labeled points in the two domains.

There is an important conceptual consequence for filter modeling: z^{-1} is an operator which plays the role of the *unit delay*. This is a direct consequence of the shifting theorem (see Table 2.3):

> Shifting theorem: $x(n + n_0)$ transforms into $z^{n_0} X(z)$ (2.63)
>
> Therefore, if $n_0 = -1$, and $x(n)$ transforms into $X(z)$,
>
> $x(n - 1)$ transforms into $z^{-1} X(z)$ (2.64)
>
>   and z^{-1} functions as a UNIT DELAY

Thus, multiplication by z^{-1} is equivalent to a unit delay of the function $x(t)$.

Properties of the z-transform: The z-transform has a set of powerful properties (many of which resemble theorems for the Fourier transform). They are summarized in Table 2.3.

Table 2.3 Summary of important z-transform theorems in normalized form, (with $t = nt_s$ written simply as n)

Theorem	Function of n	Function of z
z-transforms are linear	$Ax(n) + By(n)$	$AX(z) + BY(z)$
Shifting theorem	$x(n + n_0)$	$z^{n_0} X(z)$
Multiplication by an exponential	$a^n x(n)$	$X(z/a)$
Multiplication by time	$nx(n)$	$-z \dfrac{dX(z)}{dz}$
Convolution	$\displaystyle\sum_{-\infty}^{+\infty} x(m)\, y(n-m)$	$X(z) \star Y(z)$
	$x(n) \star y(n)$	$\dfrac{1}{j2\pi} \oint X(v)\, Y\left(\dfrac{z}{v}\right) v^{-1} dv$
Initial value theorem	If $x(n) = 0$ for $n < 0,$	$x(0) = \lim\limits_{z \to \infty} X(z)$
Final value theorem	$\lim\limits_{x \to \infty} x(n)$	$\lim\limits_{z \to 1} [X(z)(1 - z^{-1})]$

As a practical matter, as is the case for Fourier and Laplace transforms, the theorems, together with tables of common transforms, usually make it unnecessary to evaluate z-transforms. A selection is given in Table 2.2 (following Eq. 2.57), along with the corresponding Laplace transform.

Even though the z-transform and the Fourier transform seem similar, the form of the functions in the tables is quite different. As a result, the form of transfer functions will be unfamiliar, at least initially.

> *Exercise*: Work out the transfer function $H(z)$ for the 1st-order low-pass. From this, construct the block diagram for the digital filter. Compare your answer with the version developed heuristically in Chapter 1 (Figure 1.17a).

> *Method*: Write the difference equation of the first order low pass. Identify corresponding coefficients in $H(z)$. Construct the filter using adder, multiplier, and unit-delay elements corresponding to those operations present in your $H(z)$.

2.8 SIMPLE DIGITAL FILTERS

These examples illustrate how to use the z-transform. We consider two filter classes: nonrecursive (FIR) and recursive (IIR). For each class we give an example based on the delay line, and an example or two coming from numerical analysis. While not very spectacular as digital filters, these examples show how one can evaluate the performance of discretely sampled digital filters, using the z-transform to go back and forth from discrete time to (digital) frequency domain.

2.8.1 FINITE IMPULSE-RESPONSE (FIR) FILTERS
The delay line
The output of a delay line is the input delayed by a time τ. A long coaxial cable is an analog delay line for high-frequency signals. Delay lines for audio signals require delays approaching a second and, for high fidelity, must store thousands of samples with high precision. Some analog audio "bucket-brigade" delay lines were constructed with charge-coupled devices (CCD) (Figure 2.21). The signal is sampled at the clock rate and is transferred in the form of charge (proportional to the analog signal) along a sequence of MOS capacitors.

As long as the sampling frequency is more than twice the signal bandwidth, little effect of the discrete sampling is heard; however, any errors in the transmission of charge from one capacitor to the next (systematic gain changes or losses of charge) are compounded and can lead to poor performance. It is here that the *digital* bucket brigade has its forte; digital signals can be passed along delay lines indefinitely without error.

Consider the delay line filter of Figure 2.22a. The delayed output is added to the undelayed input, an example of feedforward. Thus, its response to an

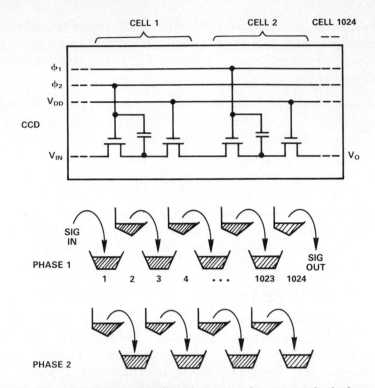

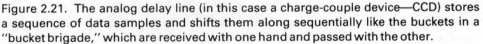

Figure 2.21. The analog delay line (in this case a charge-couple device—CCD) stores a sequence of data samples and shifts them along sequentially like the buckets in a "bucket brigade," which are received with one hand and passed with the other.

impulse input at $n = 0$ will be an impulse at $n = 0$, followed by a second impulse at $n = \tau$. That is, for an impulse input, $x = \delta(t)$, the output is:

$$y_i(t) = \delta(t) + \delta(t - \tau) \qquad (2.65)$$

We retain the familiar terminology $\delta(t)$ for the unit impulse, rather than $I(t)$ as in section 2.6, since we will work with sampled data only. What is the transfer function? *The transfer function, H(s), is the L.T. of the impulse response.* Since the L.T. of $\delta(0) = 1$, and a time shift multiplies the L.T. by an exponential,

$$H(s) = 1 + e^{-s\tau} \qquad (2.66)$$

The frequency response is obtained by the substitution $s = j\omega$. After a little complex algebra, the magnitude and phase shift of $H(s)$ are:

$$|H(f)| = 2\cos\left(\frac{\omega\tau}{2}\right) \qquad (2.67)$$

$$\phi = \frac{\pi f \tau}{2}$$

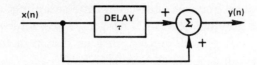

a. Block diagram of filter showing feedforward of undelayed signal.

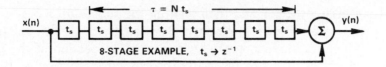

b. Internal structure with N sampled unit-delays.

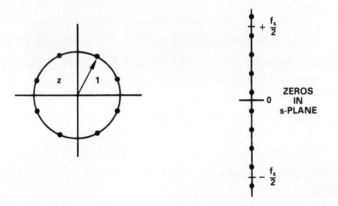

c. Zeros of the filter in (left) the z-plane and (right) the s-plane.

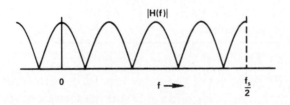

d. Magnitude plot of frequency amplitude response.

Figure 2.22. A delay-line non-recursive filter.

The phase shift of a distortionless delay line is linear in frequency. Most FIR filters try to achieve linear phase behavior.

$|H(f)|$ is zero when the direct and delayed signals summed at the output are 90° out of phase, i.e., at integer multiples of the frequency, $1/(2\tau)$:

$$f = \frac{m}{2\tau} \qquad (2.68)$$

Now consider the consequences of discrete time-sampling, an example of the z-plane ideas of the previous section. The delay, τ, is achieved by a sequence of N unit delays (clock periods) of length t_s as shown in Figure 2.22b. The total delay, for N sections, is

$$\tau = N t_s \qquad\qquad (2.69)$$

Rather than taking the LT and then making the substition $z = e^{st_s}$, we can use the z-transform (Eq. 2.61) directly. By analogy for the LT relationship between $H(s)$ and $y_i(t)$, $H(z)$ can be obtained by inspection of the impulse response, which in turn can be written down by inspection of the filter block diagram. This example will illustrate a general rule to be developed in Chapter 4.

Prescription for obtaining $H(z)$ from FIR-filter impulse response:

Apply a unit impulse input to the filter. Add up the terms contributing to the output with their weightings and delays.

$$\begin{aligned} y_i(t) &= a\delta(t) + b\delta(t-t_s) + c\delta(t-2t_s) + \ldots \\ &= h(t) \end{aligned} \qquad (2.70)$$

Applying the z-transform equation, the transfer function in the z-plane is

$$H(z) = a + bz^{-1} + cz^{-2}\ldots \qquad (2.71)$$

since $\delta(t - Mt_s)$ transforms into z^{-M}, where M is the number of unit delays up to a given point.

Since the impulse response of the delay-line filter with feedforward is given in Eq. 2.62, substitute $\tau = Nt_s$ and apply this prescription, giving:

$$H(z) = 1 + z^{-N} \qquad\qquad (2.72)$$

To see the properties in the z-plane, solve for the zeros of $H(z)$; there are no poles for a nonrecursive filter.

$$0 = 1 + z^{-N}$$

or

$$z^{-N} = -1 = e^{(2m+1)\pi}$$

The values of z for which this is true lie on the unit circle at angles,

$$\phi = \frac{2m+1}{N} \pi \tag{2.73}$$

In the 8-delay example given, the zeros lie at $\pm \pi/8$, $\pm 3\pi/8$, ... $15\pi 8$; there are 8 zeros around the circle with a spacing of $2\pi/N$ between adjacent zeros. The maximum number of zeros is equal to N. Thus, this filter has a series of zeros spaced equally around the unit circle in the z-plane (Figure 2.22c).

What is the frequency response? Transforming back to the s-plane as in Figure 2.22d, N zeros occur within the basic frequency interval, $-(f_s/2) < f < (f_s/2)$, one zero for each stage of delay. Since the number of taps is equal to one plus the number of delays, take away one tap to avoid counting it as a delay. The rule is:

There are at most $M - 1$ zeros available for the M-tap nonrecursive filter in the z-plane, or $(M - 1)/2$ over the frequency interval, $0, f_s/2$.

This can be seen from the polynomial form of the transfer function, Eq. 2.71, which has order $N = M - 1$; when factored, at most $M - 1$ terms of the form $(1 - z/z_o)$ can appear. The name of the game in FIR filter design is to put the zeros where they are needed.

The magnitude of the transfer function, $H(f)$, is plotted in Figure 2.22d. There are a series of zeros at frequencies $(2m + 1)f_s/(2N)$. This is a "comb" filter, which has holes in the spectrum at a sequence of points. Among audio engineers, it is known as a flanger or phaser; it creates holes in the audio spectrum. This example illustrates how the behavior can be predicted by examining the location of zeros (later, also poles) in $H(z)$. If N is odd, the final zero occurs at $\pm f_s/2$; on the circle mapping, since $+f_s/2$ and $-f_s/2$ coincide, the zeros will coincide and be counted as a single zero.

Example: Moving Average Revisited
Consider the running average of Chapter 1, Figure 1.15, for $N = 3$. The output-input relationship is, by inspection:

$$y(n) = \frac{1}{3}[x(n) + x(n-1) + x(n-2)] \tag{2.74}$$

The impulse response is, following the prescription of previous example (see Box above),

$$h(t) = \frac{1}{3}[\delta(t) + \delta(t-t_s) + \delta(t-2t_s)] \tag{2.75}$$

and the z-transform is

$$H(z) = \frac{3}{3}[1 + z^{-1} + z^{-2}] \tag{2.76}$$

$H(z)$ can be written by inspection of the block diagram of unit delays. Conversely, given the desired form of $H(z)$, the circuit can be drawn by inspection. The mapping back to the analog domain depends upon the prescription connecting the domains (Chapter 4). With the mapping, $z = e^{st}$, the transfer function $H(f)$ follows from $H(z)$ above (with some trigonometry):

$$|H(f)|^2 = \frac{1}{9}[3 + 4\cos(\omega t_s) + 2\cos(2\omega t_s)] \qquad (2.77)$$

and is plotted in Figure 2.23. Comparing the two forms illustrates that $H(z)$ and $H(f)$ need not have the same form. However, the plot makes sense: $H(f)$ repeats at intervals of the sampling frequency, and looks like a low-pass filter for frequencies less than $1/(2t_s)$. Trends slower than $f_s/4$ get through with little attenuation, while noise which lasts less than 3 samples is attenuated by the 3-point moving average.

2.8.2 NUMERICAL ANALYSIS FILTERS

Suppose that it is important to detect *when* a new trend (edge, shoulder, change in slope) starts in a signal which is obscured by noise, as in Figure 2.24a. The moving average filter is a natural one to try. Here are some questions that come up in relation to design in the time domain:

(a) What is the optimum number of points—5, 20, or 100?

(b) Is there something to be gained by making the endpoints of the filter-in-time fall to zero trapezoidally, for example with extra endpoints of half-value as in Figure 2.24b?

(c) If the data has structure which might be lost in averaging many points, what is to be gained by a least-squares fit of each point and its neighbors to a polynomial?

(d) The moving average can be combined with a differencing filter, as in Figure 2.24c; this should make it easier to spot a trend, since it measures the rate of change of the moving average—and inflections (second differences) can be used to anticipate trend changes.

A quick look in the frequency domain will convince us why the "cut and try in the signal domain" approach is of limited use, only adequate for simple functions of very limited bandwidth and dynamic range. We include it to make contact with classic numerical analysis. For more details, see Hanning (1983).

The moving average filter in time is the same as the sampled unit rectangle function defined in Section 2.6.1, times a weighting factor $1/N$. The transfer function (Eq. 2.39) was obtained there by a finite geometric series; the variable was labeled z (by no accident). The frequency response in the range 0 to $f_s/2$ has already been discussed in Chapter 1 (Figure 1.16a). That figure shows that increasing the number of points narrows the peak but does nothing to reduce the sidelobes. The filter is a low-pass of very narrow bandwidth.

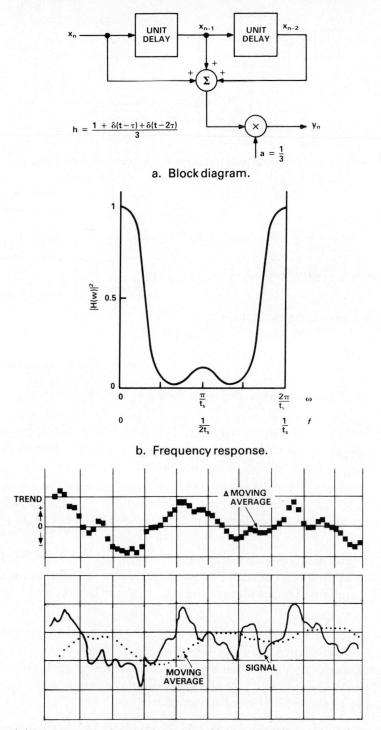

a. Block diagram.

$$h = \frac{1 + \delta(t - \tau) + \delta(t - 2\tau)}{3}$$

b. Frequency response.

c. Filtered data example of multiple-section filter, with differencing above to observe trend of the average (see Figure 2.24c).

Figure 2.23. Moving-average filter.

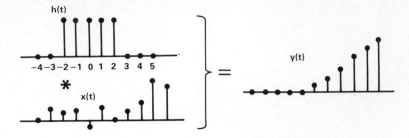

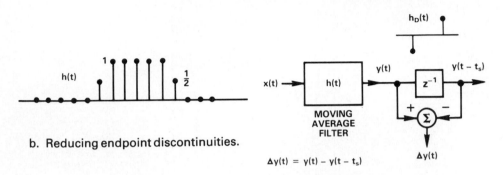

a. The filter, $h(t)$, convolved with data, $x(t)$, gives a smoothed output, $y(t)$.

b. Reducing endpoint discontinuities.

MOVING
AVERAGE
FILTER

$\Delta y(t) = y(t) - y(t - t_s)$

c. Trend-spotting differencing filter.

Figure 2.24. Nonrecursive moving-average filters.

How many zeros fall between 0 and $f_s/2$? A straightforward application of Eq. 2.73 shows that at most $M/2$ zeros fall within the Nyquist interval $[0, f_s/2]$: 2 for 4- or 5-tap filters, 4 for 8- or 9-tap filters, etc.

> *Exercise*: Calculate the bandwidth of the N-point moving average as a function of N. Procedure: let the bandwidth be defined (a) as the frequency at which the response is down by 3 dBV ($= -20 \log_{10} |H(f)|$, or $H(f) = 0.707$), and (b) as the location of the first zero.

Some improvement in low-pass filtering bandwidth using analytic solutions can be made using a least-square fit to a polynomial (See Hamming, 1983). The transfer function has a wider central lobe than the linear fit but no better sidelobes.

Consider forcing the averaging function to zero in two steps rather than one, as in Figure 2.24b. Since the filter function can be expressed as the sum of a 5-point rectangle plus extra half-size impulses added at each end, the FT must be the sum of the $\sin x/x$ with the FT of an impulse pair at $\pm N/2$, which is just $\cos \omega N$. The sum (after some trigonometry), is

$$H(w) = \frac{\sin (N-1)\frac{\omega}{2}}{(N-1) \sin\left(\frac{\omega}{2}\right)} \cos\left(\frac{\omega}{2}\right) \tag{2.78}$$

Exercise: show how Eq. 2.78 follows by the *z*-transform method followed by the substitution $z = e^{sT_s}$.

The extra cosine attenuates the sidelobes somewhat, as seen in Figure 2.25, and brings them to zero with zero derivatives at $f_s/2$. Something has been gained. Yet, by comparison with the filters of Chapter 4, this is crude. Viewed on a dB (i.e., logarithmic) scale, none of these filters would look acceptable.

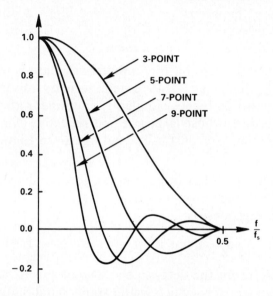

Figure 2.25. Modified straight-line windows (as in Figure 2.24b) improve moving-average filter performance. Compare with the frequency response of the moving average, Figure 1.16a.

There are, however, two useful clues gained from this look at numerical analysis: (1) Avoiding end-point discontinuities improves filter performance; (2) A quest for intuition or performance improvement from design in the data domain is likely to be unproductive. Go into the frequency domain and tailor the filter there.

Exercise: Differencing filter. Consider the first-order difference filter of Figure 2.24c. Does it make a good differentiator? Show its transfer function.

Method: Recall how we added endpoint smoothing delta functions to the moving window above. Write the *z*-transform of the filter

$$h(n) = \delta(n) - \delta(n-1) \tag{2.79}$$

It is convenient first to shift the phase of this by half a sampling interval so $h(n)$ is antisymmetric. This is equivalent to estimating the derivative of the data at the midpoint between samples. Then use trigonometry to evaluate $|H(f)|$. Answer: $H(f) = \sin(\pi f/f_s)$.

2.8.3 INFINITE IMPULSE-RESPONSE (IIR) FILTERS

Infinite impulse-response and its attributes will be introduced by the delay-line recursive filter, extensively used in digital audio, followed by a brief excursion into numerical analysis: trapezoidal integration viewed as recursive filtering, and a 1st order system which illustrates dead-band behavior with finite-wordlength arithmetic.

Recursive Filters from Delay Lines

As in section 2.8.1, the discussion of recursive filters begins with the delay line; instead of nonrecursive filters with delay-line *feedforward*, we will here consider filters with feed*back* (hence *recursive*).

A familiar example of stable infinite impulse-response is reverberation in a good concert hall, which may last for several seconds. Modern digital reverb units recreate acoustic ambiance simulating anything in the range from that of a baroque cathedral, whose large size and reflective stone walls give 3 to 5-second reverberation, to acoustically dead spaces (small rooms with absorptive walls).

The recursive delay line filter in its simplest form is shown in Figure 2.26. The delay line feeds back τ seconds later a fraction, g, of its delayed input, so the unit-impulse response is:

$$h(t) = \delta(t-\tau) + g\delta(t-2\tau) + g^2\delta(t-3\tau) + g^3\delta(t-4\tau) + \dots \qquad (2.80)$$

Define a z-transform for the delay line as a whole, using capital $Z = e^{s\tau}$ to distinguish it from $z = e^{st_s}$—which would apply to individual delay elements if τ were a sampled delay line. This Z-transform is

$$H(Z) = Z^{+1} + gZ^{-2} + g^2Z^{-3} + g^3Z^{-4} + \dots \qquad (2.81)$$

which is the power series expansion of

$$H(Z) = \frac{Z^{-1}}{1 - gZ^{-1}} \qquad (2.82)$$

Comparing the form of this with the original block diagram of Figure 2.26a, we note that the term in the numerator corresponds to the feedforward and the denominator to the feedback term.

> *Exercise*: From the block diagram of Figure 2.26a, write the recursive relationship between output and input data samples, and verify the form of $H(Z)$ given in Eq. 2.82.

> Method: take the Z-transform of the recursive form for $y(n)$ and then solve for $Y(Z)/X(Z)$.

To get the frequency response, make the substitution $Z = e^{s\tau}$:

$$|H(\omega)| = \frac{1}{\sqrt{1 + g^2 - 2g\cos\omega\tau}} \qquad (2.83)$$

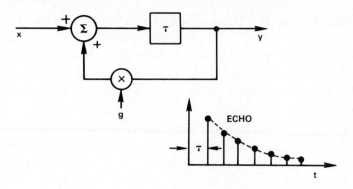

a. Echo generator, with simple weighted feedback.

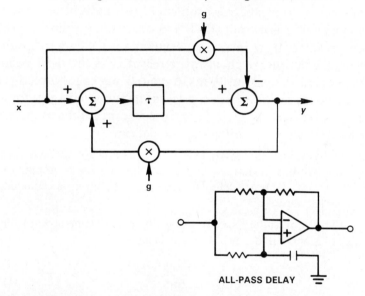

b. Reverb(eration) filter, which emulates an analog all-pass delay circuit (shown).

Figure 2.26. Examples of recursive delay-line filters.

This is a bandpass comb filter, since $H(\omega)$ peaks at frequencies which are multiples of $1/\tau$. For g close to 1, the peaks are much sharper than those of the nonrecursive comb filter of Figure 2.22d. This is the digital version of the response of a resonant cavity, with a resonance at any frequency where the delayed echos add in phase with the input.

One can create a more natural-sounding reverberance by subtracting from the output a portion of the input, as shown in Figure 2.26b. The coefficients have been chosen to give a frequency response of unit magnitude, i.e.,

$$H(Z) = Z^{-1}\frac{1 - gZ^{+1}}{1 - gZ^{-1}} \tag{2.84}$$

with $|H(Z)| = 1$ on the unit circle in the Z-plane. (For further information see Blesser and Katz, 1978.) This *all-pass* delay creates a *frequency-independent* reverberance. The analog counterpart of this reverb circuit is the *"all-pass delay"* also shown in Figure 2.26b, whose transfer function is also of unit magnitude but with frequency-dependent phase shift. Note the similarity in the structure of the block diagram. There is both feedback (the two resistors) and feedforward (the RC passes a signal to the plus input of the op amp), just as in the digital version.

These reverb circuits illustrate the dramatic effect of recursion; once an input has been applied to excite the system, the reverb time can be several orders of magnitude longer than the delay time, τ, giving effectively much longer memory than the number of taps. On the other hand, for a very long decay, g must be close to 1; the difference of small numbers increases sensitivity to coefficient errors (a typical error problem of recursive filters). In digital systems, the choice of τ is a tradeoff. If τ is made long, the increased number of samples per interval, τ, increases, requiring more data memory. In addition, there is a practical psychoacoustic reason: τ is the basic echo time and must be held to less than about 10 ms to avoid being heard directly as an echo. On the other hand, making τ short also has its disadvantages: for a given reverb time, g is pushed closer to 1, with worsening stability problems. The tradeoff is illustrated by the following calculation.

> *Exercise*: what value must g have for $\tau = 1$ ms and a reverberation time t_r of 1 second? Here, t_r is defined as the time for an output voltage to decay to 1/1,000 (or 60 dB) of its original value after the input is removed. How many stages are required?
>
> Answer: The amplitude response decays by g each time around the loop. The value of g for 60 dB decay in 1 second is set by
>
> $$\frac{\tau}{t_r} = \frac{20 \log_{10} g}{60}$$
>
> $$\log g = 3 \times 10^{-3}$$
>
> $$g = 0.993$$

With g this close to 1, small accumulated errors (the digital equivalent of component value drift) can shift the pole closer to instability. Practical reverberation units reduce g slightly but superpose several independent all-pass networks with different delay times.

The number of stages or taps necessary to avoid losing fidelity is related to the sampling inside the delay line. The sampling time, t_s, must be short enough so that the full audio bandwidth fits within $1/(2t_s)$; for bandwidth of 20 kHz, choose $t_s = 25\,\mu$s. The number of taps for this example is thus:

$$N = \tau/t_s = 40$$

an easily achievable number. The closeness of g to 1 can be reduced by increasing N. For example, the reader can verify, as above, that $N = 400$ (i.e., $t_d = 10$ ms) relaxes g to 0.93, which would be much more tolerant of coefficient inaccuracy.

A final note on reverberation

The development of music was influenced by architecture. The rapid passages of Mozart would get lost in a cathedral whose echos make it suitable for slow Gregorian chant. On the other hand, a quartet of singers in a reverberant room can sound like a much larger group due to echos which come fast enough to add in with the original notes. Modern nonrecursive (chorus, flanging, etc.) and recursive delay processing is now done digitally.

A *studio-quality* digital reverb processor might have audio specifications (and corresponding digital characteristics) as follows:

Digital Effects/Reverb Processor (e.g., Lexicon 224XL)
    Frequency response: 15 kHz (sampled at 30 kHz or faster)
    Dynamic range:
        Nonreverb (nonrecursive) mode: 90 dB (= 15-bit a/d conversion)
        Reverb (recursive) mode: 80 dB (<14 bits due to accumulated error)
        Maximum delay: 6 seconds (= 180 K words of memory of 30-kHz
            samples if it were done nonrecursively)

The dynamic range in excess of 80 dB makes for hiss-free processing, at least 20 dB better than analog CCD delay lines, for example. Since the parameters and the topology are programmable, the same unit can simulate large or small concert halls, or do special effects: chorusing, echo, flanging, repeats of input sequences or resonant chords excited by input signals.

2.8.4 RECURSIVE FILTERS IN NUMERICAL ANALYSIS – TOPICS

The Integrator The true time integrator, a common circuit in analog signal processing and control, can be synthesized nonrecursively by a finite-impulse-response filter only if the number of sections approaches infinity, since its impulse response is a step function. Analogously, a simple passive RC filter can be used to make a true integrator only if used with the high gain and feedback of an ideal operational amplifier to make the current independent of the charge accumulated on the capacitor. The analog circuit block diagram and waveforms are seen in Figure 2.27.

The digital circuit architecture will depend upon the algorithm chosen. One possibility might be trapezoidal integration: the next value of the integral is the previous value plus the trapezoid area between adjacent output values.

$$y(n) = y(n-1) + \frac{1}{2}[x(n) + x(n-)] \qquad (2.85)$$

How well does this integrator behave? Evaluate the frequency response by

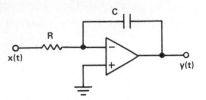

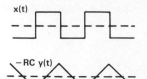

a. Analog integrator.

b. Typical time response of integrated square-wave (zero-average).

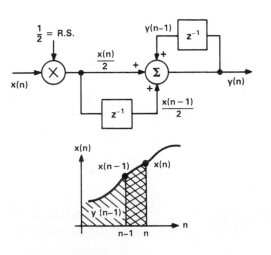

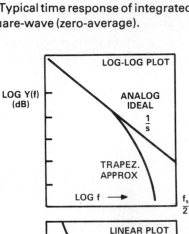

c. Digital integrator employing trapezpoidal integration algorithm.

d. Digital integrator response compared with ideal 1/s transfer function.

Figure 2.27. Integrators.

the procedure of the previous section: z-transform followed by transformation to frequency via $z = e^{st}$. The result, plotted in the Figure, is:

$$H(\omega) = j\frac{1}{2}\cot(\omega t_s/2) \qquad (2.86)$$

$$\approx 1/j\omega \quad \omega \ll \text{sampling frequency, } \omega_s$$

At low frequencies, this is the familar analog integrator (gain inversely proportional to frequency, phase shift 90°). However, though this algorithm does well at very low frequencies, it falls away as f approaches $f_s/2$. This is not surprising, since at low frequencies there are many sampling points, and the approximation to the continuous function is close; however, as frequency approaches $f_s/2$, there are fewer sampling points and the approximation deteriorates. More sophisticated algorithms improve the high frequency behavior at the expense of noise sensitivity, using power series interpolations (see Hanning (1983)).

Dead-Band Due to Finite Wordlength Arithmetic: Inadequate computational precision can end up in misleading results. Consider the first-order decay:

$$y(n) = ay(n-1) + bx(n) \qquad (2.87)$$

With infinite precision, this is a well-behaved filter. But suppose that the filter calculation is carried out in two-digit fixed-point precision. We start the filter out with an initial condition and give it a fixed input to decay to.

Example: $a = 0.96, b = 0.04, y(0) = 85, x(n) = 100$

With coefficients and data specified to only 2 digits and the result rounded to 2 digits, the filter settles to a limit, $y = 88$, as seen in Figure 2.28 (see table on next page), and never reaches the input $x = 100$. If the filter starts out higher than 100, it settles to a different limit, in this case, $y = 112$.

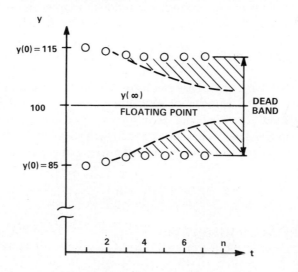

Figure 2.28. Dead-band error due to limited precision of calculation. In this case, a first-order low-pass filter is implemented recursively with only two digits of precision.

> *Exercise*: Repeat the limit-point calculation above, retaining 3 digits in the coefficients and in the individual calculated terms. What is the connection between final error and precision of coefficients or calculation?

Of course, no realistic filter calculation would be limited to 2 digits (less than 8-bit) accuracy. The example illustrates the need to be cautious about roundoff and accumulated error with recursive filter structures.

n	x	$y(n)$ calculated	$y(n)$ rounded
0	100	–	85
1	100	$4 + 81.60 = 85.60$	86
2	100	$4 + 82.56 = 86.56$	87
3	100	$4 + 83.52 = 87.52$	88
4	100	$4 + 84.48 = 88.48$	88
5	100	$4 + 84.48 = 88.48$	88
.		.	.
.		.	.
.		.	.
∞		(hung up)	88

Exercise: A putative bandpass. The filter described by Eq. 2.88, is proposed for medical electronics in the processing of EKG signals. To diagnosis cardiac anomalies, one needs to supress both low and high frequencies but bring out the essential features of the heartbeat structure. The filter in time

$$y(t) = (0.75)y(t-1) + [x(t) - x(t-4)] \qquad (2.88)$$

is intended to be a bandpass whose passband is from 7 to 35 Hz for a sampling frequency of 250 Hz. The recursive structure makes it remarkably compact; originally designed to be implemented in software, it could go into a portable low-cost instrument. How well would it accomplish its goals? Evaluate the frequency response using techniques developed earlier in this chapter.

EXERCISES FOR CHAPTER TWO

Exercise 2.1 (a) Perform the integrals for the Fourier coefficients of the square wave, Eq. 2.3(b). Show that the coefficients decay as $1/n$.

(b) A triangular wave is the integral of a square wave. Show using the Fourier series representation that the Fourier coefficients of the triangular wave decay as $1/n^2$.

Exercise 2.2 Show that the Fourier series for a square wave, Eq. 2.3(b), emerges as a special case of the Fourier series for a pulse wave, Eq. 2.4.

Exercise 2.3 Show by explicit calculation how a pulse delayed by time t_0 has the same Fourier transform as the pulse centered at $t = 0$, except for a phase factor, $\exp(-j2\pi f_0)$.

Exercise 2.4 Prove the convolution theorem, Eq. 2.15. Introduce the explicit form of the Fourier transform, change variables to bring out a single exponential, and compare the result with the definition of the Fourier inverse, Eq. 2.6.

Exercise 2.5 The simplest interference patterns are formed when light is passed through two infinitely narrow slits. Show this explicitly using the Fourier transform (which generates the diffraction pattern, as in Eq. 2.9) for two slits, represented as two delta functions at positions $+x_0$ and $-x_0$.

Exercise 2.6 Try this if you can obtain an analog multiplier. Apply as one input source a sine-wave oscillator. Use a pulse generator as the other input, with repetition rate about 1 Hz and pulse width variable over the range 0.1 ms to 0.5 s. Connect both an oscilloscope and an audio amplifier/speaker to the output. Listen to the tone as the pulse width is made narrower. How many cycles of oscillation are in the tone-burst when you can no longer clearly hear pitch information?

Exercise 2.7 Touch-tone telephones transmit information in tone bursts, with individual tones at about 1 kHz. Suppose that there are 10 tones in a band from 1 to 2 kHz. What is the upper limit to the rate at which information can be sent in this form?

Exercise 2.8 Show how the extra oscillations at the edges of the finite representation of the square wave (Figure 2.1a) may be understood precisely by the convolution theorem, given that the finite sum is the product of the full series and a finite frequency window.

Exercise 2.9 Explain why the Fourier transform of the sine and cosine must contain pairs of delta functions with the sign and real/complex parts shown in Equations 2.11 and 2.12, and Figure 2.4.

    Method (a): Do the Fourier inverse. What delta function terms must the spectrum contain?

           (b): Apply the shifting theorem. The shift from cosine to sine is by a time, $T/4$, where T is the period. Energy exists in the spectrum only at $f = \pm 1/T$.

Exercise 2.10 Verify the information theorem of Eq. 2.34 for the sine wave observed over a finite square window, Figure 2.9a.

Exercise 2.11 Suppose signals are to be processed with 16-bit resolution and real-time bandwidth of 10 kHz. How many seconds of data can fit per megabyte (16×10^6 bits) of storage medium? Repeat for 8-bit resolution and 5 kHz bandwidth.

Exercise 2.12 For the digital oscillator, first convince yourself of the explanation in the text, and then calculate the numerical harmonic-distortion amplitudes. Do they agree with the values in the Table? Hint: Expand $\sin(x)/x$ near its zeros to first-order.

Exercise 2.13 What are the spectral consequences of a sample-and-hold as in Figure 2.1?

Method: express the S&H operation as a convolution and look at the Fourier transform of the result. What is the maximum error over the meaningful range below $f_s/2$? If the S/H is used only to give the a/d converter time to encode a discrete point, and the digitized data array is only a point set, show that the error disappears.

Chapter Three

The Discrete Fourier Transform and the Fast Fourier Transform Algorithm

3.1 INTRODUCTION

Discrete Fourier transforms (DFTs) are important tools in digital signal processing; they are used to compute the Fourier transform with *discrete* frequency intervals. Straightforward evaluation of them is excessively time-consuming, but *fast Fourier transforms* (FFTs), among the most powerful general-purpose DSP algorithms, provide a means of greatly speeding up DFT computations. The FFT is not an approximation to the DFT but a family of algorithms; the results are the same, but the computation (for N points) is reduced from N^2 to about $(N/2) \log_2 (N)$ multiplications. For a 1,000-point transform, speed is typically improved by a factor of 200!

The chapter begins (3.2) with an introduction to the DFT and its relationship to the continuous FT and to the Fourier series. Considerable attention is paid to understanding the origin of the common difficulties experienced with the DFT, due to limitations in either number of samples or window size.

In 3.3, beginning with a radix-2 version, the evolution of FFT algorithms is traced, along with the data flow from input signal to output spectrum. Butterfly arithmetic and special addressing sequences—unusual features of the FFT—make special-purpose FFT processors in firmware (especially with hardware multipliers) a natural improvement over software-only implementations. While architecture will not be explored in detail until chapter 5, a foretaste is given of the kind of speed enhancement possible with VLSI architectures, which allow pipelining of the FFT butterfly steps.

In section 3.4, we show ways to avoid the pitfalls of the DFT: aliasing, leakage, and the picket fence effect. *Aliasing* is eliminated by proper filtering of the analog signal. The *only* data points which make sense are those associated with frequencies less than one-half the sampling rate ($f_s/2$). *Leakage* is due to the finite data record, or window, which effectively multiplies the signal by a rectangle and thus convolves each spectral peak with $\sin(f)/f$. The peaks are broadened, making spectral separation of nearby frequencies difficult, and the sidelobes can obscure small-amplitude spectral peaks at other frequencies. A longer signal record can lessen the leakage. However, improved *window weighting* functions, discussed in Section 3.4, can greatly enhance spectral separation. The "picket fence" effect is closely related to leakage: each point in the transform is convolved with a $\sin(f)/f$ function. This produces a ripple in the amplitude of the DFT. Longer records, smoothing windows, or even adding zeros to the record, produce a cure.

Some examples of spectral analysis are given in Section 3.4: spectral separation of common windows, the technique of *decimation* and its implications to for high-resolution frequency "zoom." The FFT, used with the convolution theorem, connects time and frequency domains to give freedom of choice for optimum-efficiency calculations, in order to increase signal-to-noise, enhance spectral features, and minimize artifacts. The chapter concludes (Section 3.6) with some FFT applications: rapid convolution and FIR filtering, deconvolution and image reconstruction, avoiding errors in inverse transforms back to the signal domain, transformations of multicomponent exponentials into a form amenable to FFT processing, and correlation and autocorrelation via the FFT.

3.2 THE DISCRETE FOURIER TRANSFORM

The discrete Fourier transform (DFT) is the Fourier transform of a sampled signal computed at equally spaced frequency intervals of size f_s/N, where N is the number of points in the sampled window. While an FT computation is always at some discrete interval, t_s/M, however fine its scale, the term DFT is usually reserved for the special case where $f_s/N = 1/(Nt_s)$, i.e., $M = N$.

DFT Special Properties

(1) The inverse DFT gives a reconstructed signal which is periodic in the interval of the window, $T_W = Nt_s$;

(2) The DFT gives as many points in the spectrum as there are in the sampled signal;

(3) The DFT is a *complete* representation of the sampled signal;

(4) Requirements for increased speed of the DFT led to an algorithm, the FFT, which revolutionized the computation of spectra.

In discussions of the Fourier transform in Chapter 2, the frequency axis was assumed continuous. What happens if the spectrum is computed only at discrete sampled frequencies? The result can be predicted by analogy with the consequences of sampling in the time domain, which forces information in the frequency domain to be periodic. The Fourier transform of a row of sampling functions is a row of sampling functions at intervals of $1/t_s$ (t_s = sampling interval), as in Figure 3.1a. Apply the idea in reverse: do a sampling of the continuous FT at intervals, f_W. The sampling in frequency is accompanied by a periodic repetition in time at intervals, $T_W = 1/f_W$, as in Figure 3.1b. According to the convolution theorem, multiplying the spectrum by the sampling-function row repeats the signal in time at intervals, T_W. A discrete FT example was seen at the bottom of Figure 2.16; notice the resulting repetition in the signal (left side). The DFT combines sampling in both time and frequency domains. For example, the pulse wave of Figure 2.14 has as its DFT a sampled version of the repeated $\sin(Mx)/\sin(x)$ function, shown in Figure 3.1c.

SIGNAL   SPECTRUM

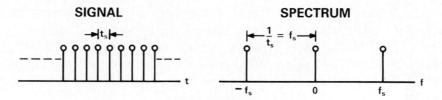

a. Sampling in time generates periodicity in frequency.

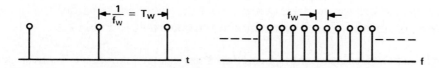

b. Sampling in frequency generates periodicity in time.

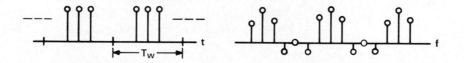

c. Sampled FT or DFT for a pulse.

Figure 3.1. Origin of the DFT periodicity.

Normally, T_W is chosen as the range of time over which data is sampled and stored in a buffer prior to computing the spectrum. Once T_W is chosen, the smallest frequency interval, f_W, of the spectrum is thereby determined. There are some subtleties about edge effects when data is discontinuous across the edges of the window; these will be considered in Section 3.5. There are also some alternative choices for time/frequency samples, which can produce a frequency *zoom* in a spectral computation (see Decimation in Spectra) or can reduce the requirements on FIR filter computations (see Decimation Filters).

The expression for the DFT is, setting $t_s = 1$ for simplicity,

$$X(k) = \sum_{n=0}^{N-1} x(n)\, e^{-j2\pi nk/N}$$

(3.1)

The inverse (DFT^{-1}) is written:

$$x(n) = \sum_{k=0}^{N-1} X(k)\, e^{+j2\pi nk/N}$$

(3.2)

DFTs are thus computed by a series of multiply-and-add steps. In each case, the finite series is a sufficient and unique representation of the sampled signal or spectrum. The *uniqueness* follows because the sampled exponential functions, $\exp(j2\pi nk/N)$, are orthogonal—even over the finite interval (see Exercises). The finite sum is *sufficient* because of the sampling: both the signal and its spectrum are repeated or aliased, hence there is no need to carry the sum beyond the edges of the window in time or the fundamental interval in frequency. A somewhat more-rigorous proof is obtainable by considering the periodicity of the DFT as expressed in Eq. 3.1 (see Exercises). As will be seen in Chapter 4, the *symmetry* between equations 3.1 and 3.2 will also be seen when they are re-expressed by the z-transform.

The Fourier transform, the Fourier series, and the discrete Fourier transform are thus members of a set of transforms which are closely connected, differing only in whether the data are continuous or discrete and whether the interval is finite or infinite. The relationships are summarized in Table 3.1.

Table 3.1 The relationships between the Fourier transform, Fourier series, and their sampled-data versions can be demonstrated by comparing the nature and range of their time or frequency variables.

	Time Domain $y(t)$ (t_{min}, t_{max})	Frequency Domain $Y(f)$ (f_{min}, f_{max})
data: range:		
Fourier Transform	continuous *infinite*	continuous *infinite*
Fourier Series	continuous *periodic*	discrete *infinite*
FT of sampled data	discrete *infinite*	continuous *periodic*
Discrete Fourier Transform	discrete *periodic*	discrete *periodic*

Because the explicit time or frequency variable is usually suppressed in writing the DFT, it is helpful to summarize and compare various units and fundamental intervals often used for discrete FT spectrum plots.

A minor point of detail: Because the point, $N/2$, is aliased to coincide with

SUMMARY OF UNITS AND FUNDAMENTAL INTERVALS IN SAMPLED SPECTRA

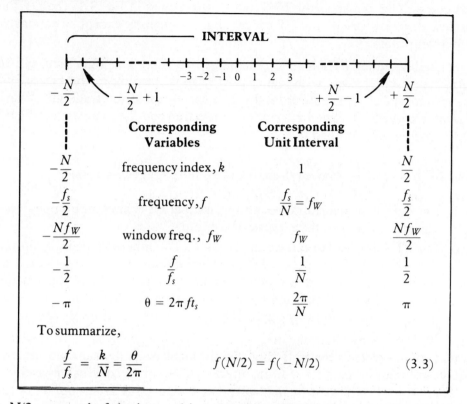

	Corresponding Variables	Corresponding Unit Interval	
$-\dfrac{N}{2}$	frequency index, k	1	$\dfrac{N}{2}$
$-\dfrac{f_s}{2}$	frequency, f	$\dfrac{f_s}{N} = f_W$	$\dfrac{f_s}{2}$
$-\dfrac{Nf_W}{2}$	window freq., f_W	f_W	$\dfrac{Nf_W}{2}$
$-\dfrac{1}{2}$	$\dfrac{f}{f_s}$	$\dfrac{1}{N}$	$\dfrac{1}{2}$
$-\pi$	$\theta = 2\pi f t_s$	$\dfrac{2\pi}{N}$	π

To summarize,

$$\frac{f}{f_s} = \frac{k}{N} = \frac{\theta}{2\pi} \qquad\qquad f(N/2) = f(-N/2) \qquad\qquad (3.3)$$

$-N/2$, one end of the interval is redundant. The open interval is more precisely expressed as $[-N/2, +(N/2 - 1)]$. We shall assume that N is even. The interval shown above, centered at $f = 0$, is equivalent to that computed in Eq. 3.1 or 3.2 with range $[0, N - 1]$, since the interval to the left of $f = 0$ maps identically over to that between $+f_s/2$ and $+f_s$.

3.3 THE FAST-FOURIER-TRANSFORM ALGORITHM

"Since I was the only one with nothing important to do, they gave me this problem to work on. With a little prodding from Garwin, I got a program out in my spare time, and I thought I would hear no more about it. The significance of the factor $N \log N$ versus N^2 was lost on me, since I had never had any use for Fourier transforms and was not aware of the extent of Fourier calculations on computers. When I realized what had happened, I told Garwin that if he had any more ideas like that I would be glad to help him out again."

(J. W. Cooley, on how the FFT algorithm came about with the encouragement of R. L. Garwin.)[1]

[1] Adapted from J. W. Cooley, *IEEE Trans. Audio & Electroacoustics* AU-17, 66 (1969).

The quotation illustrates the combination of serendipity and good judgment which characterized the rediscovery* and application of the algorithm which has come to be known as the FFT. Done directly as in Eq. 3.1, the DFT is so slow that real-time digital FT processing is unrealistic except on expensive supercomputers.

The Cooley-Tukey FFT algorithm changed all that. The algorithm will be introduced and proved with signal flow graphs, leading to the basic FFT computational element, the *butterfly*. The speed-resolution tradeoffs of various alternative FFT structures will be considered and the basic issues of FFT DSP architecture outlined.

The FFT is an algorithm for the DFT at equally spaced intervals.

1. The FFT is an algorithm, not an approximation. The computational savings are not made at the expense of accuracy.

2. The FFT is a way to carry out the DFT. The inherent accuracy limitations of the FFT are those of the corresponding DFT, no worse and no better. (However, in the actual computational process, the smaller number of steps should tend to reduce truncation errors due to limited hardware resolution—or conversely, permit the use of lower-resolution hardware.)

Attention to errors in the FFT thus should focus on understanding spectral analysis of sampled data over a finite interval. Chapter 2's conclusions apply:

There must be at least 2 (realistically 3 or 4 with noisy data) samples of the highest frequency in the signal.

The signal must first be passed through an anti-aliasing low-pass filter with cutoff frequency set well below $f_s/2$.

The consequences of measurement over a finite interval, the window, have already been touched on in Chapter 2:

A discrete peak is spread out in frequency inversely proportional to the window width in time.

An inverse transform, back in the signal domain, can contain artifacts at the window edges traceable to foldover from the adjacent periodically continued signal.

Such problems can be minimized by proper *window weighting*, which will be developed in section 3.5.

---

*The FFT algorithm was a rediscovery of an idea of Runge (1903) and Danielson and Lanczos (1942), first occurring prior to the availability of computers and calculators—when numerical calculation could take many manhours.

3.3.1 MULTIPLY-TIME BOTTLENECK OF THE CONVENTIONAL DFT

A DFT computation involves multiplication by an exponential and integration (summation) over time, as in Eq. 3.1. A separate summation is required for each frequency point in the spectrum. With N terms to the sum and N spectral points, N^2 multiplications are required. Multiply time is therefore a serious limitation. For example, suppose that the software multiply time of a microcomputer is about 100 μs. A 1,000-point transform, done directly, takes 10^{-4} s $\times 10^3 = 0.1$ s, for each point in the spectrum, and must be repeated 1,000 times for a full spectrum—about 100 s overall. This would make the *real-time* Fourier transform unrealistic. General-purpose minicomputers with *hardware* multipliers improve this by several orders of magnitude, but not even up to reasonable audio bandwidths. For example, if the multiply time for a single point is 100 ns, instead of 100 μs, the total multiply time for a 1,000-point transform is about 10^{-7} s $\times 10^{-6} = 0.1$ s.

The faster algorithm due to Cooley and Tukey (1964), and its clones and variations, have come into universal use and are generically called the FFT— abbreviation of *fast Fourier transform*. FFT algorithms reduce the N-point FT to about $(N/2)\log_2(N)$ complex multiplications.

The FFT advantage:

$$\text{FFT steps} = \frac{N}{2}\ \log_2 N$$

For $N = 2^m, \log_2(N) = m$

$$\text{FFT steps} = m\ \frac{N}{2} \tag{3.4}$$

$$\frac{\text{FFT steps}}{\text{DFT steps}} = \frac{(N/2)\log_2(N)}{N^2} = \frac{\log_2(N)}{2N}$$

For example, calculated directly, a DFT on 1,024 (i.e., 2^{10}) data points requires $N^2 = 1,024 \times 1,024 = 2^{20} = 1,048,576$ multiplications. FFT algorithms reduce this to about $(N/2)\log_2(N) = 512 \times 10 = 5,120$ multiplications, a factor-of-200 improvement.

If the software multiply time is 100 μs, the spectrum is computed in $5,000 \times 10^{-4} = 0.5$ s, a 2-Hz bandwidth compared to the 0.01-Hz bandwidth (100 seconds) prior to the FFT. Since the relative benefit increases as N increases, as shown in Figure 3.2, it is conceivable to do rather large array FFT's on quite modest computers. When carried out with fast hardware multipliers, the FFT facilitates real-time spectral analysis over wide bandwidths, as shown by the comparison of various DFT/FFT implementations for the particular example given in Table 3.2.

Table 3.2 **Time comparisons for spectrum computation between conventional DFT and FFT for 1,024 points of data. Listed are time for individual multiplications and individual solution times for DFT and FFT, assuming that multiplying time is much longer than the time for any other solution steps and thus limits the solution rate. Times are listed separately for 16-bit *fixed-point* and 32-bit *floating-point* computations. Multiply times are given for common microprocessor (8086), coprocessor (8087), and common hardware multiplier (ADSP1016A and ADSP3210) choices.**

	Multiply time μs	DFT: $N^2 T_{mpy}$ seconds	FFT : $\frac{N}{2} T_{mpy} \log_2(N)$ seconds	ms
Fixed point, 16 bits				
Software only (8086)	30	30	0.15	150
Coprocessor (8087)	4	4	0.02	20
Hardware (ADSP-1016)	0.07	0.07	0.00035	0.35
Floating point, 32 bits				
Software only (8086)	1,024	1,024	5	5,120
Coprocessor (8087)	20	20	0.1	100
Hardware (ADSP-3210)	0.1	0.1	0.0005	0.5

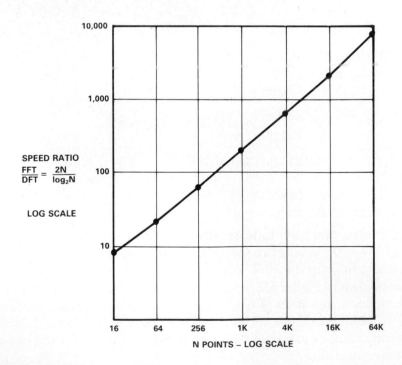

Figure 3.2.   Comparison of relative execution times for DFT and the FFT algorithm.

Once the multiply time becomes comparable to CPU instruction cycle time, the multiply bottleneck is removed. The actual bandwidth is less than Table 3.2 would imply, since overhead times for signal input, FFT program steps, and post-FFT processing all add up. But the potential is clear: high real-time bandwidth is within reach with a modest hardware investment (see Exercises). The FFT, like filtering, becomes a key real-time DSP application, as illustrated by the following example (Figure 3.3).

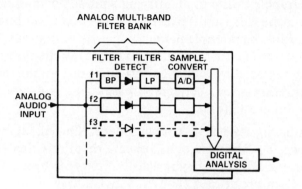

a. Block diagram of analog sonograph.

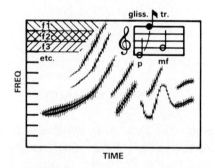

b. Frequency content as a function of time.

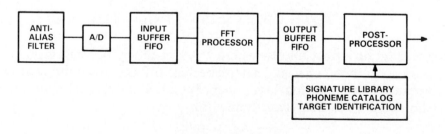

c. Digital sonography.

Figure 3.3. Sonograph block diagram and an example of a sonograph of a bird call.

3.3.2 EXAMPLE: REAL-TIME ANALYSIS OF SPEECH

A standard analog version of a sonograph, or voice-recognition instrument, passes the signal through a bank of bandpass filters (Figure 3.3a). The rectified and low-pass-filtered output is displayed as an array of points which measure the instantaneous amplitude in a given spectral channel, as in Figure 3.3b. Acquisition of the pattern is the first step to identifying the sound, be it a bird call, bat squeak, whale song, human voice-print, or sonar/radar signal. The sonograph's utility in identifying sounds having unique attributes is obvious—note the chirp in the example, as the dominant pitch slides rapidly. The analog filter-bank implementation has its limitations: the number of spectral channels (or *bins*) is limited, for example. While the analog bandpass filters could be replaced by digital equivalents, one might consider alternatively the requirements for a sonograph based on a real-time Fourier transform, shown in Figure 3.3(c).

The spectrum of Figure 3.3b is not stationary. The signal must be divided up into windows, T_W, short enough to ensure that individual features are not averaged out in the FT; all meaning is lost in the *long-term* FT of speech, for example. But T_W must be long enough to give adequate spectral resolution. What determines if the FFT will be able to keep up? The computation time per window, t_{FFT}, must be less than T_W, which is set by sampling laws:

$$T_W = t_s \times N_{pts} \geqq t_{FFT} \tag{3.5}$$

plus, to avoid losing information,

$$f_s = \frac{1}{t_s} \geqq 2 \times (\text{signal bandwidth})$$

The desired bandwidth sets the minimum sampling rate: $2 \times 8 = 16$ kHz for speech, $2 \times 20 = 40$ kHz for hi-fi audio. The window width is determined by the nature of the signal; $T_W = 50$ ms is adequate for human speech phonemes. The number of sampled points is the ratio, T_W/t_s. In the speech example,

$$N_{pts} = T_W/t_s = 0.05 \text{ ms} \times 16 \times 10^3 \text{ Hz} = 800 \text{ points} \tag{3.6}$$

If the calculation is not complete within T_W, real-time processing is not possible, and the computation would have to be done off-line. Attempting to speed up computation by reducing the number of points at the cost of reduced spectral definition does *not* help significantly. For

[FFT time per window] < [Data-acquisition time per window]

$$\frac{QN}{2} t_{mpy} \log_2 N + N t \ (\textit{overhead}) \ < \ N t_s$$

where Q specifies the number of instruction cycles necessary to carry out each FFT spectral point, and $t(overhead)$ contains other program steps, such as those relating to I/O operations. Replacing t_s by $[1/(2 \times \text{signal bandwidth})]$,

$$\text{Bandwidth} < \frac{1}{Q\, t_{mpy}\, \log_2 N + t\,(overhead)} \qquad (3.7)$$

which is not very sensitive to N (a reduction from 1,024 to 512 points reduces the $\log_2 N$ term by only 10%).

Q is strongly dependent upon how the FFT is carried out; in a properly optimized FFT processor, it can be in the neighborhood of 10. The overhead time is also a strong function of the DSP architecture; it becomes the dominant factor when multiply times become comparable to instruction cycle times. For example, if $Q = 8$ and the selected hardware multiplier has 100-ns throughput:

Number of points	Bandwidth, kHz, with $t(overhead) = 50\,\mu s$	Bandwidth, kHz, with $t(overhead) = 1\,\mu s$
64	18.2	172 kHz
256	17.7	135
1,024	17.2	111
4,096	16.8	94.3

The assumed overhead time per spectral point of 50 θs in the first case may be caused by I/O handling delays or an unoptimized FFT processor or program structure. The second case assumes an architecture optimized to keep up with the multiplier and shows a bandwidth adequate even for sonar/radar data rates.

3.3.3 FFT ALGORITHM

In this section, an FFT algorithm will be traced through signal flow graphs down to the basic operation, the *butterfly*. There are numerous closely related versions of FFT algorithms; we will not dwell on these except to compare decimation (to be defined) in time with decimation in frequency, and to mention some issues which can enhance throughput.

It is assumed that the signals and spectra are complex; each multiplication, written $A \times B = C$, means

$$C_r + jC_i = (A_r + jA_i) \times (B_r + jB_i) \qquad (3.8)$$

$$C_r = A_r B_r - A_i B_i$$

$$C_i = A_r B_i + A_i B_r$$

where the subscripts, r and i, indicate coefficients of the the real and imaginary parts of an argument.

Take an N-point transform (Eq. 3.1) and split it into two $N/2$-point transforms. This is already a computational saving, since:

$$2\left[\frac{N}{2}\right]^2 = \frac{N^2}{2} \qquad (<N^2)$$

If one split helps, why not more? So the $N/2$-point transforms are not computed at all; each is, in turn, split into two $N/4$-point transforms, and so on, until there remain $(N/2)$ two-point transforms. The process requires $\log_2 N$ splits, giving $(N/2)\log_2 N$ basic operations. As illustrated in Figure 3.4, for an 8-point example, this FFT algorithm involves first splitting the data set, $x(n)$, into odd- and even-labeled points, $y(n)$ and $z(n)$, then using periodicity of the exponential function to eliminate redundant operations—the key to why the FFT saves time. Let

$$y(n) = x(2n) \quad \text{and} \quad z(n) = x(2n + 1) \tag{3.9}$$

for $n = 0,1,2,\ldots,\frac{N}{2} - 1$. Equation (3.1) may be rewritten as

$$X(k) = \sum_{n=0}^{\frac{N}{2}-1} \left[y(n)\,e^{-j4\pi nk/N} + z(n)\,e^{-j2\pi(2n+1)k/N}\right] \tag{3.10}$$

A factor, $W^k = e^{-j2\pi k/N}$, can be divided out of the second term; it is not a function of n, hence it multiplies the $z(n)$ sum. W is a unit vector with phase angle proportional to k and has N increments of angle per 2π rotation. Thus,

$$X(k) = Y(k) + W^k Z(k) \qquad 0 \leqq k < N/2 \tag{3.11}$$

Here,

$$Y(k) = \sum_{n=0}^{\frac{N}{2}-1} y(n)\,e^{-j4\pi nk/N}$$

$$Z(k) = \sum_{n=0}^{\frac{N}{2}-1} z(n)\,e^{-j4\pi nk/N}$$

$$W = e^{-j2\pi/N} \tag{3.12}$$

Both $Y(k)$ and $Z(k)$ are periodic in the *half*-interval, since $e^{-j4\pi\theta}$ repeats over π. $X(k)$ may thus be generated for the second half of the range $N/2$ to N without further computation. For the second half of the range, with k' the excursion of k beyond $N/2$, $k = k' + N/2$,

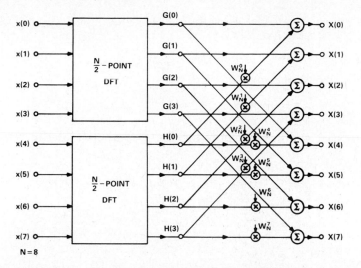

a. The 8-point transform is reduced to two 4-point transforms. The content at each summing node is a sum of arrows from the left, weighted by phase factors. Multiplication by factors W is implied by the ($\times$) symbol.

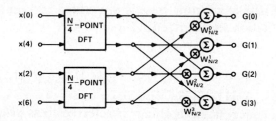

b. Splitting each 4-point transform into 2-point transforms.

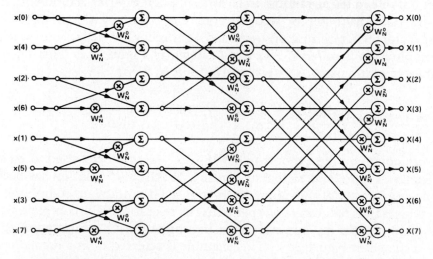

c. The final reduction represents what is actually computed in the FFT.

Figure 3.4. Signal flow graphs illustrating the combing procedure of the decimation-in-time FFT for an 8-point example.

$$X(k) = X\left(k' + \frac{N}{2}\right)$$

$$= Y\left(k' + \frac{N}{2}\right) + W^{\left(k' + \frac{N}{2}\right)} Z\left(k' + \frac{N}{2}\right)$$

$$= Y(k) - W^k Z(k) \tag{3.13}$$

The minus sign on the W factor for the second half of the range occurs because:

$$W^k = e^{-j2\pi\left(k' + \frac{N}{2}\right)/N} = e^{-j2\pi k'/N} e^{-j2\pi N/(2N)}$$

$$= e^{-j\pi} e^{-j2\pi k'/N} = e^{-j2\pi k'/N}$$

$$= -W^{k'} \tag{3.14}$$

Thus, the work of computing an N-point transform has been reduced to computing two $N/2$-point transforms with multiplicative phase factors, W^k and $-W^k$. Further splitting by 2 is illustrated in Figure 3.4b, with the final FFT computation summarized in Figure 3.4c.

3.3.4 SIGNAL FLOW GRAPHS, DECIMATION-IN-TIME FFT, BUTTERFLIES, AND BIT REVERSAL

The procedure of successive combings is referred to as *decimation*, from the Latin (in the time of the Roman Empire, to "decimate a cohort" meant to select out one in ten at random to be killed). Each Fourier coefficient, $X(k)$, may be traced back step by step as a sum of two terms with a multiplicative phase factor, W^k, reaching the original data after $\log_2 N$ steps. The basic unit of computation is called the *butterfly*, shown in Figure 3.5. This butterfly combines two inputs (A,B) to give two outputs (C,D) as

$$
\begin{aligned}
C &= A + W^k B &&\text{Decimation-in-time} \\
D &= A - W^k B &&\text{butterfly calculation}
\end{aligned}
\tag{3.15}
$$

The periodicity of the exponential, $W^{(N/2)} = -1$, has been used to go from the form of Figure 3.5a to that of Figure 3.5b; the 180° rotation has become a minus sign on one of the summing junctions. The number of complex multiplications is thus reduced to one per butterfly operation. The complete signal-flow graph for the 8-point example using the butterfly computation of Eq. 3.10 is shown in Figure 3.5c. This example is referred to as a *decimation-in-time* FFT, since alternate *time* samples are combed out by the process; note the difference in labeling at the left of Figures 3.4c (decimation in time) and 3.6a (decimation in frequency).

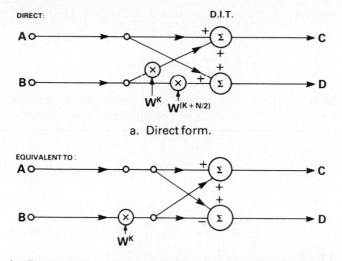

a. Direct form.

b. Reduced form with one multiply operation per butterfly.

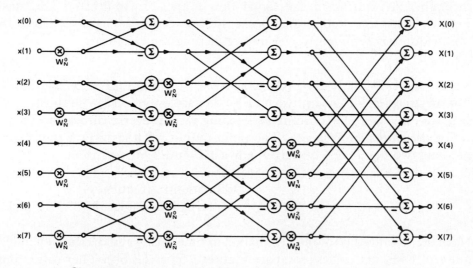

c. Signal-flow graph for the 8-point decimation-in-time example.

Figure 3.5. The butterfly computation for the decimation-in-time FFT.

The-signal flow graph shows that a given piece of data is used in only one butterfly per step (or "pass"). The results (C,D) of a given butterfly become the input (A,B) for the next pass. Since the input data to each butterfly may be *replaced* by the computed points, which become the inputs to the next stage, only $2N$ rather than $4N$ memory locations need be set aside—2 for each (complex) sample. This class of FFTs is called *computation in place*.

The FFT lends itself to rapid pipelining, since the form of the butterfly is identical for each step, although the addresses are non-sequential. To get an ordered set of $X(k)$ requires an unusual ordering of the input data in Figure 3.3c.

$$x(0), x(4), x(2), x(6), x(1), x(5), x(3), x(7)$$

This is called *bit-reversed order*, since the sequence is just what one gets if the numbers 0-7 are written in binary and the bit-order is reversed:

Decimal numbers	0	1	2	3	4	5	6	7
Binary equivalents	000,	001,	010,	011,	100,	101,	110,	111

Bit-reversed binary	000,	100,	010,	110,	001,	101,	011,	111
Decimal equivalents	0	4	2	6	1	5	3	7

Bit-reversed FFT program sequences can be reordered in a few extra program steps. Also, there exist versions of the FFT which avoid the bit-reversal through restructuring of the signal flow graphs. In optimized DSP fast-Fourier-transform processors, special address-generator hardware does the offset addressing efficiently with minimum program overhead.

3.3.5 DECIMATION-IN-FREQUENCY FFT AND THE INVERSE FFT

For the inverse FFT, an equivalent FFT signal-flow graph can be constructed (Figure 3.6) with the combing proceeding in the reverse direction. This is called *decimation in frequency*. The input-output relationship for the corresponding two-point DFT butterfly shown in Figure 3.6b is written:

$$\begin{aligned} C &= A + B & \text{Decimation-in-frequency} \\ D &= (A - B)\,W^k & \text{butterfly calculation} \end{aligned} \qquad (3.16)$$

Note the resemblance of the decimation in time (DIT) and decimation in frequency (DIF) butterflies (compare Figures 3.5b and 3.6b). They differ only in whether the multiplication by W_k comes before (DIT) or after (DIF) the summation step. The two are identical topologically if the direction of signal flow is reversed and the labels (A,B) and (C,D) are exchanged. This topological identity can lead to an *inverse* FT computation within the same program or hardware structure. Take the decimation-in-time butterfly equations (3.15) and solve for (A,B) in terms of (C,D). The result is:

$$A = \frac{1}{2}(C + D)$$

$$B = \frac{1}{2}(C - D)\,W^{-k} \qquad (3.17)$$

D.I.F.

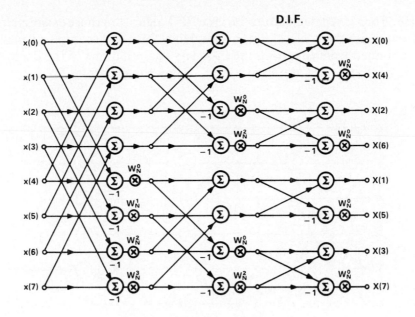

Figure 3.6a. Signal flow graph for an 8-point example, beginning at the left with an ordered data set. This is called decimation in frequency.

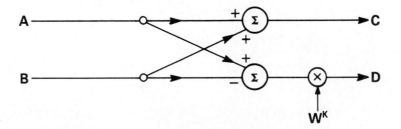

Figure 3.6b. Butterfly computation for decimation in frequency.

This is identical in form to the decimation in frequency equations except for exchange of labels, plus, in addition, the extra factor of ($\frac{1}{2}$) and the replacement of W^k by W^{-k}. Here's why: for a DFT of the form

$$X(k) = \sum x(n)\, W^n(k)$$

the inverse DFT is just

$$x(n) = \frac{1}{N}\, \sum X(k)\, W^{-n}(k)$$

i.e., reverse the labels, x and X, replace W^k by W^{-k}, and divide by N (which is 2 for the 2-point butterfly FT).

This greatly simplifies computation of the inverse DFT. The same program can be used for both, merely replacing W^k by W^{-k} and dividing by N for the inverse. For a given decimation-in-time FFT algorithm there exists an inverse DFT which is of the decimation-in-frequency form. An example of an inverse FFT obtained from inverting the butterfly computations of the decimation-in-time FFT (Eq. 3.15 and Figure 3.5c) is shown in Figure 3.7.

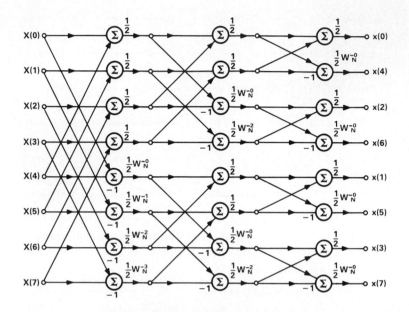

Figure 3.7. Inverse FFT from inverting the butterfly computations of the decimation-in-time FFT (Eq. 3.15 and Figure 3.5c). The multiplier symbol has been suppressed but is implied in each W^k step.

3.3.6 BUTTERFLIES FOR COMPLEX SIGNALS AND SPECTRA

In general, both the signal and the spectrum are complex. Even if the signal is real, the spectrum has both real and imaginary parts in order to represent the phase. Computations are done with real numbers, re-assembled into complex components. Tabulated below is a summary of the form of the coefficients C_r, C_i, D_r, D_i (r = real, i = imaginary) for the butterfly equations, 3.15 and 3.16. For decimation in time, let $\theta = 2\pi k/N$; since $W^k = e^{-j2\pi k/N}$, $W^k = \cos\theta - j\sin\theta$,

$$
\boxed{
\begin{aligned}
&\text{DIT, Complex Coefficients:} \\
&\quad C_r = A_r + B_r\cos\theta + B_i\sin\theta \\
&\quad C_i = A_i + B_i\cos\theta - B_r\sin\theta \\
&\quad D_r = A_r - B_r\cos\theta - B_i\sin\theta \\
&\quad D_i = A_i - B_i\cos\theta + B_r\sin\theta
\end{aligned}
}
$$

(3.18)

For decimation in frequency,

DIF, complex coefficients:

$$C_r = A_r + B_r$$
$$C_i = A_i + B_i$$
$$D_r = (A_r - B_r) \cos \theta. - (A_i - B_i) \sin \theta$$
$$D_i = (A_r - B_r) \sin \theta + (A_i - B_i) \cos \theta$$

(3.19)

Four real multiplications are required for a given butterfly computation in either form.

3.3.7 REAL-VALUED SIGNALS

If the input signal is real, the butterfly calculations are simplified.

Decimation in time for real data:

$$C_r = A + B \cos \theta$$
$$C_i = -B \sin \theta$$
$$D_r = A - B \cos \theta$$
$$D_i = B \sin \theta$$

(3.20)

and

Decimation in frequency for real data:

$$C_r = A + B$$
$$C_i = 0$$
$$D_r = (A - B) \cos \theta$$
$$D_i = (A - B) \sin \theta$$

(3.21)

With phase information represented, the spectrum is complex, even if the signal is real.

FFT Algorithms for Real-Valued Data: It may appear that something is "wasted" in reserving space not occupied by the imaginary part when the signal is real; the multiplications by zero are in effect performed anyhow, since the length of the computation has not been reduced when half of the inputs are set to zero. Algorithms exist to make use of the $2N$ pieces of information which emerge from an N-point FFT (see Brigham, 1974, for more details).

One can, for example, transform *two* real signals, $h(n)$ and $g(n)$, simultaneously, and the linear nature of the DFT will keep their transforms unmixed: the two signals are combined to form a complex signal, $y(n)$

$$y(n) = g(n) + jh(n) \tag{3.22}$$

Its FFT,

$$Y(k) = R(k) + jI(k) \tag{3.23}$$

can readily be separated to give the individual signal FFTs, $G(k)$ and $H(k)$, using FT symmetry theorems in Table 2.1.:

(a) The DFT of a real function (i.e., the real part, $g(n)$) has an even real part and an odd imaginary part;

(b) The DFT of an imaginary function (i.e., the component $jh(n)$) has an odd real-part and an even imaginary part. The separated spectra are:

$$G(k) = R_e(k) + jI_o(k)$$

$$H(k) = R_o(k) + jI_e(k) \tag{3.24}$$

Here, the even (e) and odd (o) parts are constructed (see Exercises) by taking sums and differences of $R(k)$ and $I(k)$ reflected about the origin, or—with the frequency interval defined as $[0, N-1]$ rather than $[-(N/2), N/2-1]$—reflected about the upper end point. This method has especial utility in real-time spectral analysis, with two buffers of signal accumulated continuously in order to avoid losing any points.

A variation of this technique is used to transform a signal window having $2N$ points in the time it takes for an N-point transform, by decomposing even and odd signal points into the real and imaginary components (each occupying N points) of a new N-point signal. The reconstruction is incorporated into the computational algorithm using DFT symmetry properties (see Exercises). However, the spectral resolution is that of an N-point, rather than a $2N$-point, transform.

What has been gained? With the signal real, the real part of the FT is even and the imaginary part is odd. That means there was really no need to compute the redundant halves in the first place. The N-point FFT on $2N$ signal points eliminates the redundant computations; all spectral points are meaningful, and it is done in half the time. The interval $[0, N-1]$ is *all* meaningful, while half of the interval $[0, 2N-1]$ in the direct computation would contain useless repeated information (see Exercises).

3.3.8 FFT PROGRAM EXAMPLE

The most convenient size for a data array, N, is a power of 2. In the general case, N can be written as a *composite number* comprising powers of 2. But since the program becomes more complicated—and therefore slower, there is a clear advantage when computing FFTs by DSP techniques to arrange for N to be a power of 2, even if it is necessary to pad the signal with zeros at the window edges.

In the period following the Cooley-Tukey rediscovery, a flurry of activity produced a number of algorithms and programs. Much of this information is summarized in the Bibliography. These will not be discussed here, since the $(N/2)\log_2 (N)$ improvement provided by the basic power-of-two, or *radix-two*, algorithm is sufficient for our present purposes. An example in FORTRAN is shown here to illustrate the butterfly computation steps, loop structure, and bit reversal.

Analysis of the Simple FFT Program: This example illustrates how to turn the FFT idea into a working program. The signal $DATA(\mathcal{J})$ is assumed complex, with real and imaginary parts stored in adjacent memory locations. The data length is assumed to be a power of 2, with the quantity $NN = \log_2 (\text{length})$. The program carries out the decimation-in-time algorithm. The reordering (bit reversal) is done at the beginning (DO 5 loop), as can be verified by example (see Exercises). Transform values are returned in the array DATA, replacing the input signal. The spectral points will appear in normal order. The FFT is carried out in the lower half of the program. The nested DO loops carry out a given pass of $N/2$ complex (or N real) butterfly operations. The butterflies combine samples initially close together and increasingly far apart, following the DIT signal flow graph of Figure 3.5c. The GO TO 6 progresses through $\log_2 (\text{length}) = NN$ passes, recalculating W^k for the next pass and repeating the butterfly computation sequence.

The phase rotations, W^k, often referred to as "twiddle" (twirl or rotate) factors, are carried out between labels 8 and 9, using an approximation to the true rotation specified below label 7. Labeling the real and imaginary parts of W as WR and WI,

$$WR^{n+1} = WR^n \cos (\Delta\theta) - WI^n \sin (\Delta\theta)$$

$$WI^{n+1} = WI^n \cos (\Delta\theta) + WR^n \sin (\Delta\theta) \tag{3.25}$$

But since the angle increment $\Delta\theta$ is small, the cosine is approximated as $1 - (\frac{1}{2}) \sin^2 \theta$. Only a sine-table lookup is ever needed, and only once per pass.

Though not an optimized program, this example has a clear correspondence with the original algorithm; the multiplication steps $(*)$, which potentially limit the rate of computation when performed in software, are easy to see.

Table 3.3 A simple FFT program in FORTRAN. The program carries out the FFT by decimation in time. The signal input, as array DATA, is put by the subroutine into bit-reversed order. The spectrum replaces the signal. The program can compute either the forward or the reverse transform by setting ISIGN to +1 or −1.

```
                  SUBROUTINE FFT(NN,ISIGN,DATA)
C PASS PARAMETERS:
C     LENGTH = 2**NN = (POWER OF 2)
C     ISIGN = +1 FORWARD TRANSFORM
C             -1 REVERSE TRANSFORM
C DATA POINTS COMPLEX
C     DATA(J) = COMPLEX DATA
C             REAL PART INDEX J=EVEN
C             IMAGINARY PART INDEX J+1
C SIZE OF DATA ARRAY SPECIFIED IN DIMENSION [GREATER THAN] 2**NN
                  DIMENSION DATA(1024)
                  N = 2*NN
                  J=1
C PUT SIGNAL INPUT INTO BIT-REVERSED ORDER
                  DO 5 I=1,N,2
                  IF(I-J) 1,2,2
        1         TEMPR = DATA(J+1)
                  TEMPI = DATA(J+1)
                  DATA(J) = DATA(I)
                  DATA(J+1) = DATA(I+1)
                  DATA(I) = TEMPR
                  DATA(I+1) = TEMPI
        2         M=N/2
        3         IF(J-M) 5,4,4
        4         J = J-M
                  M = M/2
                  IF(M-2) 5,3,3
        5         J = J+M
                  MMAX = 2
C NN PASSES
        6         IF(MMAX-N) 7,10,10
        7         ISTEP = 2*MMAX
C CALCULATE TWIDDLE FACTORS FOR THIS PASS
                  THETA = 6.2831853/FLOAT(ISIGN*MMAX)
                  SINTH-SIN(THETA/2)
                  WSTPR = -2.*SINTH*SINTH
                  WSTPI = SIN(THETA)
                  WR = 1.
                  WI = 0
C N/2 BUTTERFLIES PER PASS
                  DO 9 M=1,MMAX,2
                  DO 8 I=M,N,ISTEP
                  J = I+MMAX
C BUTTERFLY CALCULATION
                  TEMPR = WR*DATA(J) - WI*DATA(J+1)
                  TEMPI = WR*DATA(J+1) + WI*DATA(J)
                  DATA(J) = DATA(I) - TEMPR
                  DATA(J+1) = DATA(I+1) - TEMPR
                  DATA(I) = DATA(I) + TEMPR
        8         DATA(I+1) = DATA(I+1) + TEMPR
                  TEMPR = WR
C TWIDDLE FACTOR INCREMENT FOR NEXT PASS
                  WR = WR*WSTPR - WI*WSTPI + WR
        9         WI = WI*WSTPR + TEMPR*WSTPI + WI
                  MMAX = ISTEP
                  GO TO 6
       10         RETURN
                  END
```

3.4 WINDOWS AND WINDOW WEIGHTING

3.4.1 WHY WINDOW WEIGHTING?

A signal observed for a finite interval of time (window) may have distorted spectral information in the Fourier transform due to the ringing of the $\sin(f)/f$ spectral peaks of the rectangular window. To avoid or minimize this distortion, a signal is multiplied by a window-weighting function before the DFT is performed. Window choice is crucial for separation of spectral components which are near one another in frequency or where one component is much smaller than another. The choice of window function is an art; it depends upon the designer's skill at manipulating the tradeoffs between the various constraints, depending on what one wants to get out of the spectrum or its inverse. A group of selected popular windows, whose properties will be discussed in this chapter, are shown—with their transforms—in Figure 3.8. The mathematical relationships are summarized in Table 3.4, which follows Figure 3.8.

The spectrum of a sine wave of infinite duration peaks at a single frequency. But if the sine wave is observed over a finite interval, the single spectral peak is spread into a range of frequencies. This is called *spectral leakage*. Spectral leakage can be understood by noting that the finite interval in effect multiplies the sine wave by a pulse of magnitude 1 during the observation window and magnitude 0 elsewhere. In the frequency domain, the broadened spectrum is the convolution of the delta-function peak with the $\sin(x)/x$ spectrum of the pulse.

The broadening can, alternatively, be traced to the discontinuities introduced at the edges of the finite window, either by bringing the signal to zero in the FT or by the periodic extension of the signal (Figure 3.9) in the DFT. If the signal is *commensurate* with the window (integer number of periods fits within the window, or $f = Mf_W$), the discrete DFT frequencies all fall on zeros of $\sin(x)/x$, except at Mf_W, and there will be a single peak in the DFT, shown in Figure 3.9a as the commensurate, or "integral-multiple" case. If the signal frequency falls midway between sampled frequency values, Mf_W ("non-integral-multiple" case), the sampled DFT spectrum gives a very different picture—a broad peak whose frequency is poorly determined and whose amplitude is inaccurate, since no point in the discrete spectrum falls on the peak of the $\sin(x)/x$ function.

The *frequency* is poorly determined because the sampled spectral points fall at the peaks of the sidelobes. The sidelobe amplitudes decay slowly, at $-6\,dB$ per octave, since the $\sin(x)/x$ peaks decay as $1/f$.

The *amplitude* is inaccurate because two sine waves of identical amplitude can have quite different DFT sampled spectral peak-height, depending on where the samples occur.

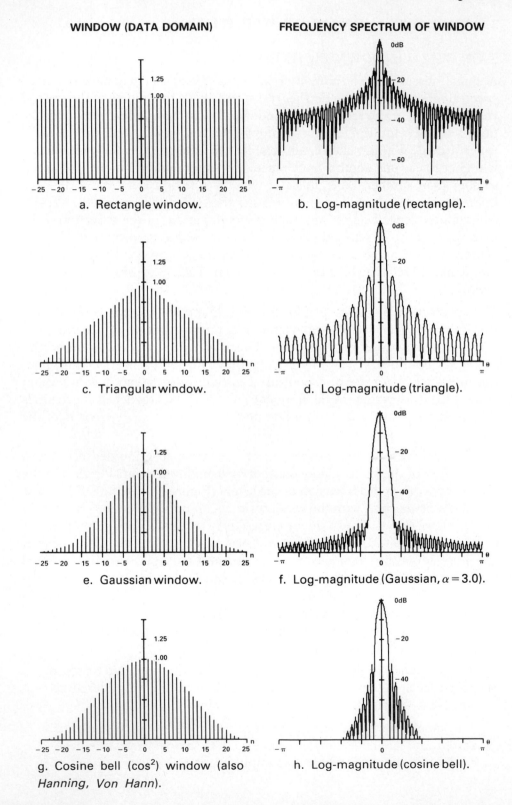

WINDOW (DATA DOMAIN)

FREQUENCY SPECTRUM OF WINDOW

a. Rectangle window.

b. Log-magnitude (rectangle).

c. Triangular window.

d. Log-magnitude (triangle).

e. Gaussian window.

f. Log-magnitude (Gaussian, $\alpha = 3.0$).

g. Cosine bell ($\cos^2$) window (also *Hanning, Von Hann*).

h. Log-magnitude (cosine bell).

WINDOW

SPECTRUM

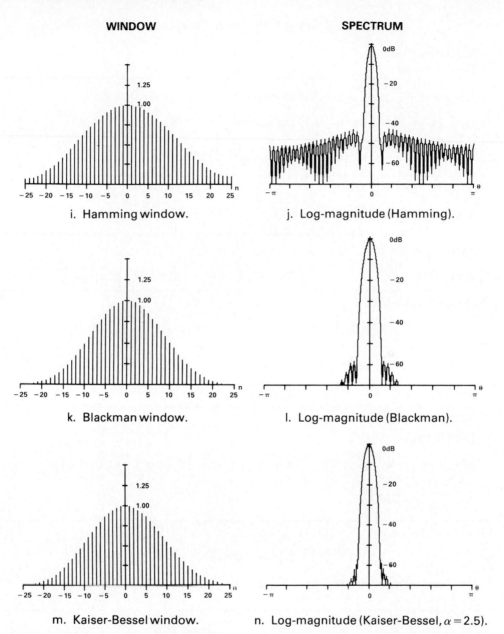

i. Hamming window.

j. Log-magnitude (Hamming).

k. Blackman window.

l. Log-magnitude (Blackman).

m. Kaiser-Bessel window.

n. Log-magnitude (Kaiser-Bessel, $\alpha = 2.5$).

Figure 3.8. Window shopping? Here are some popular window-weighting functions (left) and their Fourier transforms (right). The window functions are shown as N-point samples with $N = 50$, which sets the scale for the frequency axis $\theta = 2\pi k/N$. The comb-like appearance of the transforms comes from the pattern of zeros, occurring at multiples of the window frequency f_s/N. The range, $-\pi < \theta < \pi$, corresponds to the Nyquist range, $-f_s/2 < f < f_s/2$, or the dimensionless frequency range $-N/2 < k < N/2$. (Adapted from Harris, ©1978 IEEE—see Bibliography).

Table 3.4 Window Catalog

Window Function: "Time" Domain	DFT of Window Function: Frequency Domain
Range $n = -\dfrac{N}{2}$ to $\dfrac{N}{2} - 1$	Range $k = 0$ to $N - 1$ Define $\theta = \dfrac{2\pi k}{N}$ $D(\theta) = \exp\left(\dfrac{j\theta}{2}\right) \dfrac{\sin(N\theta/2)}{\sin(\theta/2)}$

Rectangle (Figure 3.8a)

$$w(n) = 1.0 \qquad\qquad W(\theta) = \exp\left(-j\,\frac{N-1}{2}\right)\frac{\sin(N\theta/2)}{\sin(\theta/2)}$$

Triangle (Figure 3.8c)

$$w(n) = 1.0 - \frac{|n|}{N} \qquad W(\theta) = \frac{2}{N}\,\exp\left(-j\left(\frac{N}{2}-1\right)\theta\right)\left[\frac{\sin N\theta/4}{\sin\theta/2}\right]^2$$

Gaussian (Figure 3.8e)

$$w(n) = \exp\left[-\frac{1}{2}\left(\frac{\alpha n}{N/2}\right)^2\right] \qquad W(\theta) = \frac{1}{2}\frac{\sqrt{2\pi}}{\alpha}\,\exp\left[-\frac{1}{2}\left(\frac{\theta}{\alpha}\right)^2\right] \star D(\theta)$$

$$\simeq \frac{N}{2}\frac{\sqrt{2\pi}}{\alpha}\,\exp\left[-\frac{1}{2}\left(\frac{\theta}{\alpha}\right)^2\right] \qquad \begin{array}{l}\alpha > 2.5 \\ \theta \text{ small}\end{array}$$

Cosine Bell, $\cos^2$, von Hann
(Figure 3.8g)

$$w(n) = \cos^2\left[\frac{\pi n}{N}\right] \qquad W(\theta) = \frac{1}{2}\left\{D(\theta) + \frac{1}{2}\left[D\left(\theta - \frac{2\pi}{N}\right) + D\left(\theta + \frac{2\pi}{N}\right)\right]\right\}$$

$$= \frac{1}{2}\left(1 + \cos\frac{2\pi n}{N}\right)$$

Hamming (Figure 3.8i)

$$w(n) = \alpha + (1 - \alpha)\cos\left(\frac{2\pi n}{N}\right) \qquad W(\theta) = \alpha D(\theta) +$$

$$\text{commonly, } \alpha = 0.54 \qquad\qquad +\frac{1}{2}(1-\alpha)\left[D\left(\theta - \frac{2\pi}{N}\right) + D\left(\theta + \frac{2\pi}{N}\right)\right]$$

Blackman (Figure 3.8k)

$$w(n) = 0.42 + 0.50\cos\left(\frac{2\pi n}{N}\right) \qquad W(\theta) = \sum (-1)^m \frac{\alpha_m}{2}\left[D\left(\theta - \frac{2\pi m}{N}\right) + D\left(\theta + \frac{2\pi m}{N}\right)\right]$$

$$+ 0.08\cos\left(\frac{4\pi n}{N}\right) \qquad \text{commonly, } \alpha_0 = 0.42, \alpha_1 = 0.50, \alpha_2 = 0.08$$

Kaiser (Figure 3.8m)

$$w(n) = \frac{I_o\left[\pi\alpha\sqrt{1 - \left(\frac{n}{N/2}\right)^2}\right]}{I_o(\pi\alpha)} \qquad W(\theta) \simeq \frac{N}{I_o(\alpha\pi)}\frac{\sinh\sqrt{\alpha^2\pi^2 - (N\theta/2)^2}}{\sqrt{\alpha^2\pi^2 - (N\theta/2)^2}}$$

The off-channel/on-channel comparison (error E in Figure 3.9b) is a worst case. This "picket fence" effect is measured by the ratio of the gain for a tone located ½ bin from the DSP sample to the gain if the tone is located at the DSP sample point, i.e.,

$$\frac{W\left(\dfrac{2\pi f_s}{2N}\right)}{W(0)}$$

The discontinuities in the time domain (Figure 3.9a) result in leakage in the frequency domain (b), because many spectral terms are needed to fit the discontinuity. As the frequency moves from off-channel to on-channel, the discontinuity vanishes. This can be used to advantage for signals whose components are harmonically related; the window can be adjusted during the sampling or after the fact to embrace an integral number of cycles, so that all spectral peaks are essentially delta functions.

Since this is not usually possible, window weighting functions (often referred to as just "windows") of shapes other than rectangular are chosen to multiply the data so as to minimize the effect of the discontinuity, by bringing the signal (and as many derivatives as possible) to zero at the edges of the window. But if the windowing causes too much of the signal to be lost, the loss of information produces a spectrum with broader peaks and less definition. Window selection requires a compromise between these effects.

Consider the *triangular* window function (Figure 3.8c); its value goes linearly from unity, at $n = 0$, to zero at the endpoints. Its spectrum is shown in Figure 3.8d. The logarithmic amplitude scale used in Figure 3.8 for the frequency plots is the most convenient way to compare window functions, since the decay range of the envelope, which may cover a wide range of amplitude, is more readily apparent. The improvement over the rectangular window in definition of the peaks is apparent in the comparison of DFTs in Figure 3.10. Spectral leakage is reduced: the sidelobes fall off at 12 dB/octave rather than 6 for the rectangle, since the FT of the triangle is proportional to $[\sin(f)/f]^2$. On the other hand, the central peak is twice as wide as that of the rectangle, so spectral resolution has deteriorated (see Exercise 3.9).

The peak is wider because the triangular window function throws away information by attenuating the signal towards the time-window boundaries. More-advanced window designs attempt to optimize the results of the tradeoff between enhanced sidelobe suppression and central peak width, which determines how well two spectral terms can be separated.

Spectral resolution is always better enhanced by taking more cycles within a window (when possible) than by optimizing window functions. For example, the sum of two nearby frequencies has a beat pattern (variable envelope), as

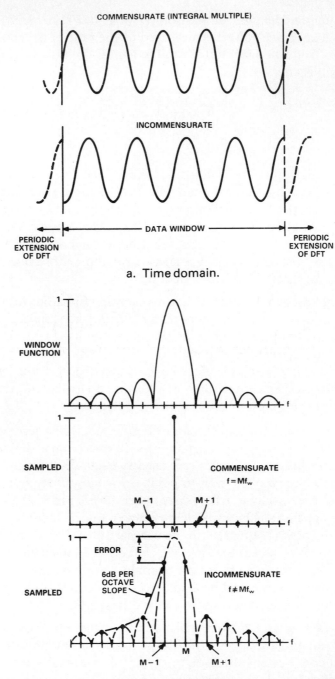

COMMENSURATE (INTEGRAL MULTIPLE)

INCOMMENSURATE

PERIODIC
EXTENSION
OF DFT

DATA WINDOW

PERIODIC
EXTENSION
OF DFT

a. Time domain.

1

WINDOW
FUNCTION

f

1

SAMPLED

COMMENSURATE
$f = Mf_w$

M − 1   M + 1

M   f

1

ERROR   E

6dB PER
OCTAVE
SLOPE

INCOMMENSURATE
$f \neq Mf_w$

SAMPLED

M − 1   M   M + 1   f

b. Frequency domain.

Figure 3.9. A signal commensurate with the window or "on-channel" has in its spectrum only one non-zero component, while the non-commensurate or "off-channel" signal DFT spectrum falls near the peaks of the $\sin(x)/x$ which decay slowly. The window is transparent to commensurate signals but distorts incommensurate signals.

in Figure 3.11a. The two peaks in the spectrum (Figure 3.11b) are resolved only if the observation window ($T_W = 1/f_{W2}$) includes a full cycle of the beat (see Exercise 3.10), no matter how clever the choice of window weighting.

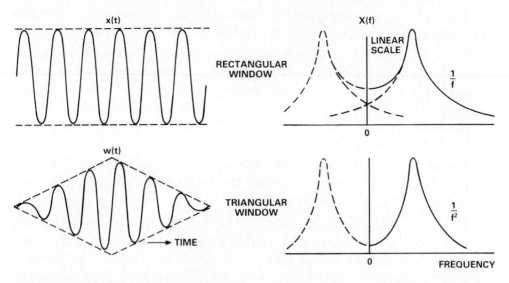

Figure 3.10. Spectrum of a single sine wave weighted first by a rectangle and then by a triangle. Note reduction of spectral leakage across $f = 0$.

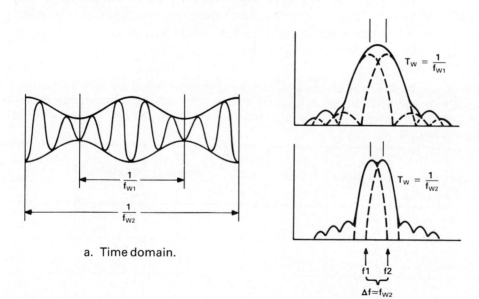

a. Time domain.

b. Frequency domain.

Figure 3.11. Spectral resolution is illustrated by the beat from the sum of two sine waves. The two frequencies can be resolved in the FT only if the data contains a full cycle of the beat.

Window Function Parameters

Window features may be compared by a few parameters, which are summarized in Table 3.5 and discussed below.

The *width of the central peak* is measured by the distance from the origin to either the first zeros or a specified dB reduction from the peak.

The *6-dB point* is a measure of the closest two frequencies which can be resolved, as in the example of Figure 3.11. If two spectral peaks are so close together than their amplitude has not fallen to half the peak value when they cross, there will be only one peak. The key value is -6 dB rather than -3 dB, the half-power point, since it is the amplitudes of the window functions which are added coherently.

The *highest sidelobe* and *rate of sidelobe falloff* measure how well a weak spectral peak will emerge amidst the background of sidelobes from stronger terms in the spectrum.

The *equivalent noise bandwidth* (ENBW) is a better measure of spectral sharpness than the width of the central peak alone. A window with a very narrow central peak but high-amplitude slowly descending sidelobes will give poor spectral separation of weak spectral components from sidelobes. The equivalent noise bandwidth is the width of a rectangular peak whose area is equal to the integral of the window-function magnitude. The ENBW specifies how well the window function *concentrates* spectral information, since slowly descending sidelobes will contribute to ENBW on an equal footing with central peak width.

Table 3.5 Numerical characteristics of common window functions. Here, width characteristics are measured in *bins*, the smallest frequency interval of the DFT, with 1 bin = 1/*N* (dimensionless), which corresponds to $f_W = \Delta f = f_s/N$.

Name:	Rectangle	Triangle	Cosine Bell	Hamming	Blackman	Kaiser $\alpha = 3.0$
6-dB point (bins)	1.21	1.78	2.00	1.81	2.35	2.39
Full width (bins) of central peak	2.0	4.0	4.0	5.0(zero)	7.0(zero) 6.0(-40dB)	7.0(-70dB)
Highest side lobe (dB)	-13	-27	-32	-43	-58	-69
Rate of side-lobe fall-off (dB/octave)	-6	-12	-18	-6	-18	-6
Equivalent noise bandwidth (bins)	1.00	1.33	1.50	1.36	1.73	1.80

3.4.2 GAUSSIAN WINDOW

A Gaussian window function (Figure 3.8e and f) removes sidebands completely, in principle, since the FT of a Gaussian is another Gaussian. The Gaussian illustrates the fundamental limits of information, since we cannot simultaneously sharpen our knowledge of both the signal and its FT, as illustrated in Figure 3.12. A Gaussian chosen to zero out the endpoint discontinuity of the signal (curve 1) is so broad in frequency that spectral resolution is very poor. Broadening the window-function width (curve 3) sharpens the peak, but brings back the ringing sidelobes since the window function is no longer zero at the endpoints.

The Gaussian is in principle the minimum-uncertainty window, since only it satisfies the uncertainty principle (Chapter 2)

$$\Delta T \, \Delta f \geq 4\pi$$

with an *equality*. However, other functions are faster computationally and bring out spectral information better.

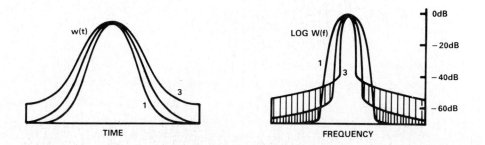

Figure 3.12. The inverse relationship between the width of a Gaussian window and its FT.

3.4.3 COSINE BELL WINDOW

The cosine bell function, defined in Table 3.4 and shown in Figure 3.8g, is also known as the cosine-squared, raised-cosine, and von Hann* window. This is the $n = 2$ member of a series of (cosine)^n window functions, which bring increasingly higher-order derivatives to zero at the endpoints as n is increased (see Exercise 3.11).

The sidelobes of the cosine bell (Figure 3.8h) fall off faster than those of the triangle, because both the function and its first derivative go to zero at the window edge. Since the first discontinuity shows up only in the second derivative, the lobes fall off as f^{-3}, or -18 dB/octave.

---

*We avoid the term Hanning, sometimes encountered, since it leads to irretrievable confusion with a different window invented by Hamming.

An alternative explanation for the fast falloff of sidelobes suggests a simplified approach to window design. The cosine bell can be viewed as the sum of a rectangular window and a single-cycle cosine of frequency, f_W, since

$$\cos^2\left(\pi \frac{n}{N}\right) = \left(\frac{1}{2}\right)\left[1 + \cos\left(2\pi \frac{n}{N}\right)\right] \qquad (3.26)$$

The Fourier transform of this can thus be written as the sum of the two corresponding FTs:

$$W(f) = \frac{1}{2}\frac{\sin(f)}{f} + \frac{1}{4}\left[\frac{\sin(f - f_W)}{f - f_W} + \frac{\sin(f + f_W)}{f + f_W}\right] \qquad (3.27)$$

where $f_W = 2\pi/N$ for N samples in the window.

The translated $\sin(f)/f$ functions are half the size of the central $\sin(f)/f$, and are located at its first zero. Their sidelobes are exactly out of phase with those of the central $\sin(f)/f$, so the sidelobe structure tends to cancel out, as seen in Figure 3.13.

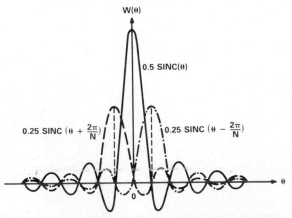

Figure 3.13. The raised cosine or Von Hann window can be viewed as the sum of three sin(f)/f functions.

The von Hann window leads to economy in computation if the window weighting is performed as a convolution in the frequency domain rather than as a multiplication in the time domain. While convolution is normally much more time consuming than multiplication, the fact that the cosine bell's spectrum can be written as in Eq. 3.27 makes the computation very compact (see Exercises). The window-weighting is carried out as

$$Y'(f_n) = Y(f_n) \star W(f_n)$$

$$= \frac{1}{2}\left\{Y(f_n) - \left(\frac{1}{2}\right)\left[Y(f_{n-1}) + Y(f_{n+1})\right]\right\} \qquad (3.28)$$

where $Y(f_n)$ is the unweighted DFT of the data. The advantages are apparent; because the factors are binary fractions, the multiplications in the convolution are just simple right shifts (division by 2) rather than full multiplications.

Thus, cosine bell windowing in the frequency domain takes only $2N$ binary shifts-plus-adds for N data values. Done in the time domain, it would take N real multiplications.

The origin of the negative sign in Eq. 3.28 is not apparent from the window function of Eq. 3.27; it comes from a phase shift, $\exp(-j(N/2)f_W/f_s)$, caused by the $N/2$ shift in the data domain to bring both window and data to cover the same range (see PHASE CORRECTION below).

PHASE CORRECTION IN WINDOW CONVOLUTION

Data samples are conveniently written from 0 (e.g., time zero) to $N - 1$, while window functions are usually plotted between $-N/2$ and $+(N/2 - 1)$ to exploit the computational simplification of the even symmetry. To bring both data and window function to a common origin, the window is shifted to the $[0,N]$ range. A shift in the signal domain by $N/2$ translates to a phase factor $e^{jN\theta/2}$ in the FT. For the cosine bell, the phase shift applies negative signs to two of the $\sin(x)/x$ functions. This phase shift is often neglected, erroneously. The correction scheme defined here is simple and general.

3.4.4 HAMMING AND BLACKMAN WINDOWS

The cosine bell window nearly cancels sidelobes in its transform. Is it worthwhile to consider sidelobe cancellation further? Are there other simple functions—combinations of rectangle and cosines—that can cancel sidelobes better? Since the first sidelobe is the biggest, why not cancel it exactly? The Hamming window seeks to do this (Figure 3.8i), using the form

$$w(n) = \alpha + (1 - \alpha)\cos\left(2\pi\frac{n}{N}\right) \tag{3.29}$$

Its associated transform (Figure 3.8j) is listed in Table 3.4. The value of the coefficient, α, that will exactly cancel the first sidelobe of the transform is about 0.543478261, not an easy degree of precision to attain computationally. Fortunately, it's not that necessary; with $\alpha = 0.54$, a new zero is forced at $\tau = 2.6(2\pi/N)$, with 50-dB cancellation of the first sidelobe. Because of the non-zero value of the window at the edges, the FT displays slow falloff at 6 dB/octave of higher-order sidebands, though at levels at least 40 dB below the peak.

One can generate windows of arbitrary shape, adding higher order terms in the Fourier series for the window function.

$$w(n) = \sum_{0}^{\frac{N}{2}} a_m \cos(2\pi mn/N) \tag{3.30}$$

The computational cost is not high, since the cosines are stored in the DFT lookup tables. However, the more terms included for narrowing the peak in the signal domain, the wider will be the spectral peak (information theory wins again!)

A marked improvement with modest computational increase is afforded by the Blackman window (Figure 3.8k and l); it adds one more cosine term to cancel the 3rd and 4th sidelobes (though not the first ones) and cancels the window function to zero at the edges:

$$a_0 = 0.426590 \simeq 0.43$$
$$a_1 = 0.496560 \simeq 0.50$$
$$a_2 = 0.076848 \simeq 0.08 \qquad (3.31)$$

The Blackman window, with the approximate coefficients above, provides excellent attenuation of sidelobes (all down at least 60 dB) with only a modest increase in computation over that required by the Hamming window (the extra cosine term requires no new computation, just table lookups at different places); the main cost is a central peak which is about twice as wide as that for the rectangle window.

3.4.5 KAISER-BESSEL WINDOW

This window function is designed to keep most of its energy concentrated in the central peak. The computation is more complicated than for the above examples but avoids table lookups; helpful design rules have evolved from this window function's popularity in digital filter design.

The window function (Table 3.4) is a Bessel function like the fundamental mode of a drumhead oscillation. Its FT is a hyperbolic sine, which has virtually zero sidelobes if α is chosen so as to bring $w(n)$ to zero inside the window edges. The Bessel function is pre-computed by the indicated power series and stored as a lookup table. As Table 3.5 demonstrates, this window has higher ENBW than the others listed, but at the cost of a relatively wide central peak.

Design rules computed by approximations from the exact expressions help select the parameters and explore the tradeoffs between record length, spectral resolution, and leakage (see Kaiser and Schafer 1980):

$$\alpha = 0.1\,R/\pi \qquad \text{(within 10\% for } 40 < R < 100\,\text{dB)}$$

$$\frac{\Delta f}{f} = \frac{6(R + 12)}{155}$$

$$N = \frac{12(R + 12)}{155\,\Delta fT} \qquad (3.32)$$

Here, R is the relative amplitude of the first sidelobe in dB, and α sets the scale-factor on the window function. Δf is the full-width of the spectral peak, defined by the point at which the central lobe falls to the value of the first sidelobe; since the function is very steep in this region, other definitions of Δf do not significantly change the above expressions. To design a window for prescribed values of R and Δf, use the first expression above to get α. The bandwidth is set by the second expression once R is known. N, the number of sampled points which must fit in the window to meet this spectral criterion, is computed from the third expression.

3.4.6 WINDOW WEIGHTING SUMMARY

Many other windows have been worked out; an extensive tabulation is given in Harris (1978). The selection can be summarized as shown in the following table.

PLUSES AND MINUSES OF COMMON WINDOW FUNCTIONS

Triangle:
- \+   Computation simple: no sin/cos lookup table
- \+   Narrow spectral peak
- −   Large sidelobes, only moderate falloff

Cosine Bell:
- \+   Reasonable sidelobe rejection;
  - First lobe down 32 dB, with − 18 dB/octave falloff
- \+   Central peak as narrow as triangle
- \+   Computationally *very* efficient:
  - 3 terms in frequency domain

Hamming
- \+   Moderately sharp central peak
- −   Poor sidelobes;
  - First lobe only 40 dB down, with 6 dB/octave falloff

Blackman
- \+   Good sidelobe rejection:
  - First lobe 60 dB down
- 0   Moderate computation: 2 cosines
- −   Broad central peak
- \+ + Excellent ENBW for spectral separation

Kaiser
- \+ + Best sidelobe rejection, − 70 dB or better
- 0   Computation unlike others above: non-cosine
- \+   Design rules well established from filters
- \+ + Best ENBW for spectral separation

3.5 SPECTRAL ANALYSIS EXAMPLE

SPECTRAL SEPARATION FOR THE COMMON WINDOWS

The importance of windowing is well illustrated by the problem of Fourier-transforming the sum of two nearby sine waves of widely differing amplitudes. The spectral separation problem is made difficult by chosing one test signal 40 dB less than the other. This example shows how sidelobes pop up with a rectangular window when one of the signal frequencies is not an

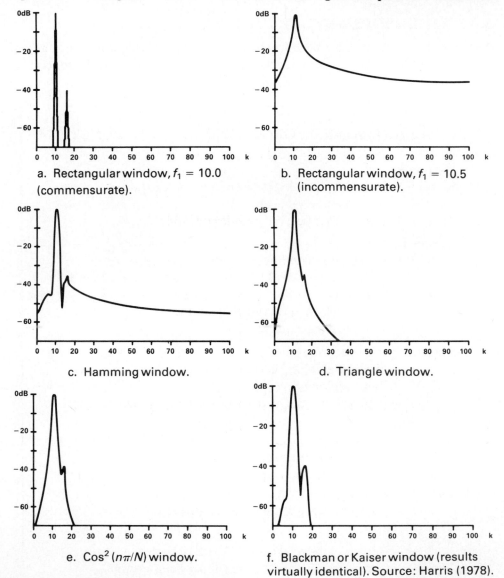

a. Rectangular window, $f_1 = 10.0$ (commensurate).

b. Rectangular window, $f_1 = 10.5$ (incommensurate).

c. Hamming window.

d. Triangle window.

e. $\cos^2 (n\pi/N)$ window.

f. Blackman or Kaiser window (results virtually identical). Source: Harris (1978).

Figure 3.14. Spectral separation example, using the windows of Figure 3.8 (envelopes of transforms). The data are: $f_1 = 10.5$ (except in a), $f_2 = 16.0$, $A_1 = 1.00$, $A_2 = 0.01$. The x-axis label, k, specifies the frequency values at which the transform is computed. (Harris, ©1978 IEEE—see Bibliography)

integral number of bins, then the degree of success attained with various windows in suppressing spectral leakage and recovering the lost frequency.

If both terms are commensurate with the window (Figure 3.14a, two peaks are well resolved, even with a rectangular window. More typically (Figure 3.14b) the second peak is not even visible with rectangle weighting. Other window choices display the tradeoff between sharpness of the peaks and decay of sidelobes. For this example, only the Blackman and Kaiser windows (Figure 3.13f) bring out the weaker term as a well-resolved peak. Simulations such as this are strongly advised for the more demanding spectral-separation problems.

3.6 FFT APPLICATIONS: DECIMATION, CONVOLUTION, DECONVOLUTION

The chapter concludes with some applications that illustrate the virtues of the FFT as a signal processing tool: (1) Decimation for higher-resolution FFT's, equivalent to a zoom lens; (2) speeded-up convolution using the FFT; and (3) deconvolution for image restoration, or deblurring. The section also sets the stage for digital filtering, the subject of the next chapter.

3.6.1 SPECTRAL CONSEQUENCES OF DECIMATION AND INTERPOLATION

Decimation (also called *downsampling*) and *interpolation* (also called *upsampling*) are closely related techniques (for an in-depth tutorial, see Crochiere and Rabiner, 1981). Their opposing effects in the data domain and the spectral domain are summarized in Figure 3.15.

Decimation-by-M selects one out of each M samples of the input signal. A freeze-frame movie is an example: the growth of a flower can be observed in a viewing interval of seconds if shutter "samples" were taken slowly compared to the playback rate. The effect on the spectrum (Figure 3.15) is to broaden the spectral information: since the time scale of the data has been speeded up, the frequency range of the spectrum has been expanded by the decimation index (or ratio of total samples to samples kept), M. This will have a useful application in a frequency "zoom" as described below. The signal to be decimated must first be low-pass filtered to prevent aliases in the result. The lowpass cutoff is set to eliminate frequencies above $f_s/(2M)$, to keep the decimated spectrum from leaking across the edges of the rescaled frequency interval. Decimation has many applications: coherent detection, multiplexing in time or frequency, data compression, and computationally compact filtering, which will be discussed in Chapter 8.

Interpolation is the familiar smoothing operation for filling in gaps in a set of data. Available points are stretched out and the gaps filled by numerical analysis "best fits," which are equivalent to passing the data through a digital low-pass filter. Since the interpolated data is stretched out in time, the spectral information is compressed compared to the original scale (Figure 3.15). Interpolation can be applied to curve fitting and real-time fast function generation, to be discussed in Chapter 8.

Decimation "Zoom" for High-Resolution FFTs

Since the DFT (and, as a result, the FFT) by definition compute N equally spaced frequency points, a high-resolution look at a selected portion of the

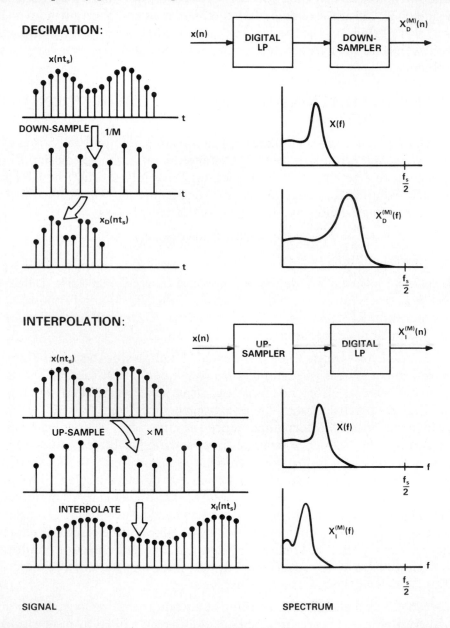

Figure 3.15. Decimation and Interpolation: the sampling process, the time-domain consequences, and the frequency-domain consequences.

spectrum cannot be done in the same way as a continuous FT, where the neighborhood of any frequency can be selected and magnified for inspection. However, decimation prior to an FFT makes it possible to view any desired spectral region with high resolution and only modest computation.

Consider the spectrum of Figure 3.16a. Suppose the fine-structure near f_0 needs to be determined with precision, yet it is computationally prohibitive to do a full FFT with the required resolution. The zoom method shifts the band of interest (b) to center at $f = 0$, then expands that region to magnify the fine-structure (c).

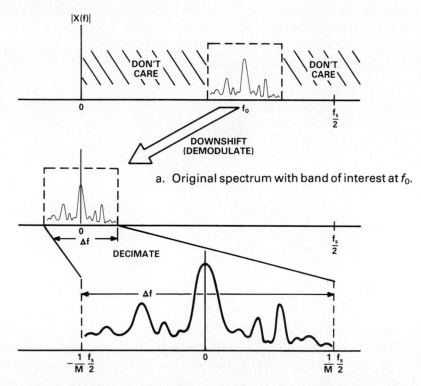

a. Original spectrum with band of interest at f_0.

b. Band shifted to the frequency origin.   c. Decimation by M expands the scale.

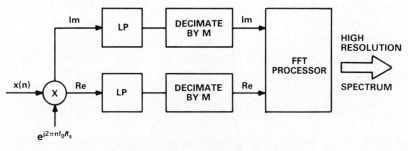

d. Decimation mechanism.

Figure 3.16. Decimation in spectral analysis.

The mechanism is shown in Figure 3.16d. The shift in band origin is accomplished, according to the shifting theorem, by multiplying the signal by an exponential, $\exp(j2\pi f_0/f_s)$, producing in effect *heterodyning*. Since substantially fewer samples are needed for the narrower bandwidth, decimation selects one out of M samples and throws away the rest. This compresses the signal (as in Figure 3.15), and therefore expands the frequency domain proportionately. Decimation-by-M gives a factor-of-M zoom in the spectrum. This signal is then passed to an FFT processor. The benefit: a small FFT can be used, since the number of points is compressed by M.

Example: In hi-fi audio, take $f_s = 65,536$ Hz, or spectral bandwidth 32,768 Hz. To get 1-Hz resolution directly with the FFT would take a time-consuming 32,768-point transform. Using decimation, a spectral region can be examined with 1-Hz resolution using only a 128-point FFT! Here's how:

Design Criteria:

$$f_s/2 = 32\,\text{kHz}$$
$$N = 128$$
$$\triangle f = 1\,\text{Hz} = f_W$$

Calculate the decimation index, M. Since Nf_W is the new spectral bandwidth after decimation by M,

$$\frac{f_s}{2M} = N f_W = 128 f_W$$
$$M = \frac{65,536}{256} = 256$$

The equivalent in resolution of a 32K-point FFT is created at an arbitrary region of the spectrum with only a 128-point FFT, by decimation-by-M with $M = 256$. For spectral analysis, the window in frequency can be swept or "panned", or one can do a "frame advance" to other spectral regions of interest, simply by changing the value of the parameter, f_0.

The decimation zoom does *not* circumvent information theory, which limits frequency resolution to the inverse of the data-window interval. The same data window must be sampled as would be required if the full 32K-point FFT were being computed (intuitively, one would expect to need at least 1 second to obtain 1-Hz resolution):

$$T_W = \frac{1}{f_W} = 1\,\text{second}$$

What has been accomplished is that a large amount of unneeded computation has been eliminated.

3.6.2 FFTs FOR RAPID CONVOLUTION

The convolution of two functions, $x(t)$ and $y(t)$, can often be performed more rapidly as a complex multiplication of their Fourier transforms, $X(f)$ and $Y(f)$, followed by an inverse FFT.

$$x(t) \star y(t) = \text{FFT}^{-1}[X(f) \, Y(f)] \tag{3.33}$$

The speed of the FFT makes the procedure attractive. Even though 3 FFT's are required, the log reduction in number of multiplications still wins over doing it directly for records longer than a few hundred points. The actual comparison requires looking at the periodic properties of the DFT "circular" convolution.

Circular Convolution
To implement the rapid convolution requires taking care that the record length does not overlap due to the wraparound of the DFT. The FFT interval must be expanded to length P, such that

$$P \geqq M + N - 1 \tag{3.34}$$

where M and N are the number of data points in arrays $x(t)$ and $y(t)$, respectively. In an actual computation, both $x(t)$ and $y(t)$ are extended to length P by filling with zeros prior to performing the P-point FT, to avoid erroneous overlaps with periodic extensions or garbage values in adjacent memory.

Any convolution has width equal to the sum of widths of its two inputs, as demonstrated by the convolution of two rectangular pulses, $x(n)$ and $h(n)$, in Figure 3.17a. In circular convolution, the periodicity brought about by the DFT effectively wraps both functions, $x(n)$ and $h(n)$, around a cylinder, as seen earlier in Figure 2.15. To avoid spillover across the window edges in the convolution, the circumference of the cylinder is extended to the value P above. If either of the functions has nonzero value over only a portion of its range, the corresponding N or M can be reduced to the nonzero width, reducing the convolution interval P accordingly.

The speed comparison between direct convolutions in time and convolution by transformation and frequency-domain multiplication depends upon the algorithm and the nature of the functions: real or complex, nonzero domain, and any symmetries. The extended range, P, makes the breakeven point occur at larger N values than for the direct–FT/FFT comparison. Assuming the signals are real, the P-point FFTs can be reduced to $(P/2)$–point FFT's, as discussed earlier. Here is an approximate comparison:

Direct (Time Domain) Convolution
For x nonzero over range N, and y nonzero over range M, with $N > M$, there

are two regions of output indicated in Figure 3.17(a):
 (1) Total overlap, range $N - M$, with M multiplies per convolution point;
 (2) Partial overlap, two regions of range M, with $M/2$ non-zero multiplies in each (regions where either $h(m)$ or $x(n)$ are zero need not be computed).

The number of multiplications is thus

$$(N - M)M + 2M(M/2) = NM$$

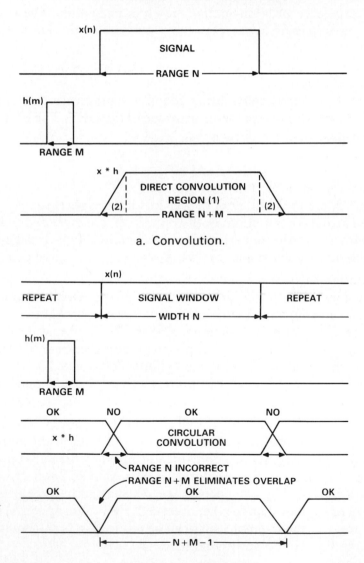

a. Convolution.

b. Circular convolution.

Figure 3.17. Convolution and circular convolution. The extended range of the convolution output must be accounted for in the DFT circular convolution.

Circular (Frequency-Domain) Convolution
Assume real functions, and $P/2$-point FFTs.

Step (1)	FFT of $x(t)$ and of $y(t)$ $\left(\dfrac{P}{2}\ \text{complex outputs}\right)$	$2P\log_2\left(\dfrac{P}{2}\right)$
Step (2)	multiply $X(f)$ by $Y(f)$	$4\left(\dfrac{P}{2}\right)=2P$
Step (3)	FFT^{-1} of step (2) output	$P\log_2\left(\dfrac{P}{2}\right)$

$$\text{Real multiplication steps (total):}\qquad 3P\log_2\left(\frac{P}{2}\right)+2P$$

taking into account the 4 real multiplications per complex butterfly. Here are two numerical comparisons.

(1) *Two real functions of equal length.* For $N = M = P/2 = 512$, the direct convolution takes about $NM = 256{,}000$ multiplications, while the FFT version takes

$$3{,}064\log_2(512)+2{,}048 = 29{,}624,$$

a factor-of-9 difference.

(2) *One long (signal) record of $N = 1{,}024$ points convolved with a shorter (filter) record of $M = 64$ points.* Here, $M + N - 1 = 1{,}024 + 64 - 1 = 1{,}087$; therefore the next higher power of 2 gives $P = 2{,}048$ (per Eq. 3.34). The direct convolution takes $NM = 64$ K multiplies, while the circular convolution takes about

$$3P\log\left(\frac{P}{2}\right)+2P = 6{,}144\log_2(1{,}024)+4{,}096$$
$$= 65{,}536$$

Thus, there is no improvement in this case, since P had to be extended to 2,048, the nearest power of 2.

However, for very long or continuous signals being filtered by fixed-coefficient filters, the FFT convolution is *always* a saving over a direct convolution, since the number of points, N, in the signal window can be adjusted to make $N + M$ a power of 2, and the Fourier transform of the filter coefficients need be done only once and stored. In this case, the comparison of multiply operations is:

Direct Convolution Filter:	NM	
FFT Convolution Filter:	$2N\log_2(N/2)$	$N \gg M$

assuming that the filter coefficient FFT has already been done. Whenever $2 \log_2 (N/2) < M$, there is a net computation-time saving. For example, for a 50-tap filter, $M = 50$,

N	$2 \log_2 (N/2)$	Ratio: $\dfrac{\text{Direct Convolution}}{\text{FFT Convolution}}$
2^{10}	18	3

The benefit increases as higher filter-performance demands push M upwards. The precise improvement depends upon the time for other steps and details of the algorithm for either alternative. For $N = M$, the breakeven point comes near $N = 256$, above which 3-step circular convolution is faster.

Circular Convolution of Long Records

If the signal goes on indefinitely or continues for longer than a practical FFT window, it must be broken up into segments of length N for segment-by-segment convolution with the second signal (e.g., the filter of length Q). Circular wraparound as discussed above complicates the summing up of each segment to reconstruct the full convolved record. The two main methods to correctly add up output convolution segments are listed below; the reader is referred to standard DSP references, (e.g., Oppenheim and Shafer, 1975, Rabiner and Gold, 1975) for details.

(1) Overlap-save method: Do the DFT convolution of length N but throw away the initial Q values which are wraparounds before connecting the segments. Make up for the loss by overlapping the signal segments by an amount Q.

(2) Overlap-add method: Do the DFT convolution on non-overlapping segments of length, $N - Q$, and add them up directly. The spillovers when one output segment is winding down, as in the triangular tails of Figure 3.17, are just made up for by the other end of the next segment which is just winding up. The reader may best understand these remarks by sketching an example in analogy with Figure 3.17. If the second method sounds dubious, try it for a row of pulses (see Exercises).

3.6.3 DECONVOLUTION FOR IMAGE ENHANCEMENT

When a signal has suffered deterioration (finite record, motional blurring, defocussing, etc.), deconvolution provides a way to improve the image. Image enhancement works best when the distortion is known or can be estimated from the improvement in the deconvoluted image, as in the Caruso example (Figure 1.21), where distortion is created by the recording horn—and in pinhole distortion (Figure 2.6), which depends on the hole geometry. Writing the distortion or *point-spread function* as $p(t)$, the observed image is expressable as a convolution of the true image with $p(t)$:

$$y'(t) = \int y(\tau) p(t - \tau) dt \qquad (3.35)$$

To improve the picture by *deconvolution*, first turn the convolution into a multiplication: take the FFT of the above equation to get the observed spectrum, $Y'(f)$; then divide through by the FFT of the blurring to recover the FT of the true image $Y(f)$:

$$Y(f) = \frac{Y'(f)}{P(f)}$$

Provided that this procedure does not diverge (i.e., $P(f)$ must not cross zero) or become too noisy, the true image can be recovered by an inverse FFT:

$$y(t) = \mathrm{FFT}^{-1}\left[\frac{Y'(f)}{P(f)}\right] \qquad (3.36)$$

which itself can be written as a convolution:

$$y(t) = y'(t) \star \mathrm{FFT}^{-1}\left(\frac{1}{P(f)}\right) \qquad (3.37)$$

The procedure has obvious appeal, but noise or artifacts can be introduced in the division step.

(1) If $P(f)$ has any zeros, the deconvolution diverges. For example, *motional blurring* in photography with relative motion of camera and object is representable as convolution of the signal with a pulse of width T_b, the motion-per-frame. The DFT of the blurring pulse is $\sin(\pi T_b f)/\pi T_b f$. If the blur is severe enough that the zeros of the sinc occur at frequencies less than the Nyquist frequency, the deconvolution procedure fails. Data which has passed through a sharp aperture (the rectangular window in 1-dimension) suffers from the same problem in deconvolution.

(2) Deconvolution tends to enhance noise, since the point-spread function usually becomes small somewhere, for example at high frequencies in the case of low-pass blurring. This is analogous to the noise which may be introduced when using a differentiator-type filter to compensate for lag in a control signal.

Improved deconvolution methods modify the function $P(f)$ to reduce noise.

(1) Put a floor (lower bound) on the point-spread function to keep it from crossing zero:

$$P''(f) = P(f) + \epsilon \qquad (3.38)$$

While this adds an error, the consequences are negligible if ϵ is down in the noise level of the signal.

(2) *Power-spectrum equalization* (PSE) calculates an estimate $Y''(f)$ of the deblurred frequency domain image from the measured $Y'(f)$ as

$$Y''(f) = \frac{Y'(f)}{P''(f)} \qquad (3.39)$$

with

$$P''(f) = \sqrt{P^2(f) + \frac{\phi_n}{\phi_{sig}}} \qquad (3.40)$$

Here, ϕ_n and ϕ_{sig} are the power spectra of the noise and of the signal, respectively. Here is why this helps: In low-noise regions of the spectrum, ϕ_n approaches zero, and the PSE approach becomes the direct deconvolution of Eq. 3.25. In spectral regions where signal power is small, ϕ_{sig} approaches zero and $P''(f)$ becomes large, so the signal estimate, $Y''(f)$, is small, even if the point spread function has zeros. Signal estimation is a rich subject. Consult Hunt (1978), Andrews and Hunt (1977), Pratt (1977), or Hunt (1972) for further information.

3.6.4 FOURIER FILTERING AND INVERSE-FFT FIDELITY

Suppose that important information in $y(t)$ lies in the amplitude envelope, $a(t)$, of a certain frequency component, f', which is obscured by other spectral components and noise (Figure 3.18a). The shape of the envelope is reflected in the shape of the corresponding peak in the spectrum (Figure 3.18b), since

$$\text{FT}\,[a(t)\sin(2\pi f't)] = A(f) \star \delta(f - f') \qquad (3.41)$$

where $A(f)$ is the FT of the envelope $a(t)$. A common example is a decaying sine wave whose decay time is the key signal parameter to be determined. Since the FT of an exponential is a Lorentzian, the spectral peak will have Lorentzian shape.

$$\text{FT}\left[e^{-\frac{t}{\tau}} \sin(2\pi f't) \right] = \frac{1}{1 + (f\tau)^2} \star \delta(f - f') \qquad (3.42)$$

The peak in the spectrum thus contains the envelope information of the signal. For exponential ringing, τ is the decay time that is sought; it is inversely proportional to the linewidth (bandwidth) of the peak.

Suppose the envelope form is not known in advance; for example, the decay of a nonlinear impulse response, which is nonlinear and therefore not exponential. "Fourier filtering" can isolate the spectral peak and then do an inverse transform to recover $a(t)$. Multiply the spectrum by a bandpass filter function, $H(f - f')$, centered at f' and with a form carefully chosen so as not to distort the spectral peak, as in Figure 3.18b. Now take the inverse transform. The result, shown in Figure 3.18c, is disappointing. The reconstructed signal droops at the left edge and shows too large an amplitude at the right edge (compare the original signal; what is happening?)

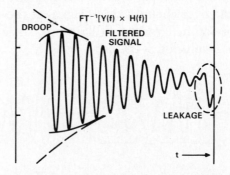

a. Signal with spectral term whose envelope is of interest.

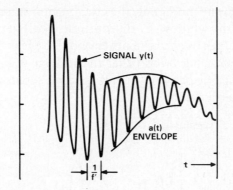

b. The signal's FT. Shown superimposed on the corresponding peak is a band-pass filtered spectrum.

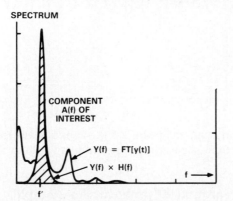

c. The FT^{-1} does not display the envelope faithfully.

Figure 3.18. Digital filtering in the frequency domain.

Since the filter function, $H(f)$, multiplies the spectrum, the reconstructed signal will be a convolution of the signal with the FT of the filter. A filter, $H(f)$, slightly wider than the spectral signal peak, will have a time-domain representation, $h(t)$, whose width is a little shorter than the decay time of the signal. (In the example shown, a Gaussian filter in frequency was used to avoid ringing in time.) When the inverse DFT is computed to recover the envelope, $h(t)$ gets convolved with the signal, as in Figure 3.19. But, since the DFT^{-1} periodically repeats the signal, the convolution smears information across the boundaries of the window. A discontinuity will get smoothed; a large signal at one end will leak over to the other end. The leakage can be minimized, for example by narrowing $h(t)$ so that the leakage is limited in range. Unfortunately, this just broadens $H(f)$ and spectral separation suffers: the bonds of information theory again.

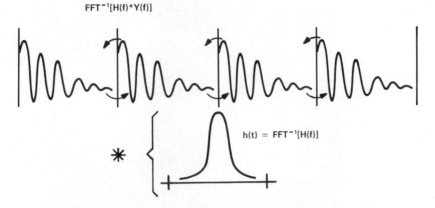

Figure 3.19. Explanation for the problem of Figure 3.18. A periodically repeated reconstructed signal (FFT^{-1} (FT(s(t)))) has leakage due to width in the time of $h(t)$ (the FFT of the filter function, $H(f)$.

Artful procedures exist to get around the problem. For example,

  (1) Append zeros to the original signal (Figure 3.20), as in the circular convolution, so a big signal at one end of the window does not spill over and overwhelm a small signal at the other end. However, the envelope at the large end still droops, since those zeros will leak over in the convolution of the DFT^{-1}.

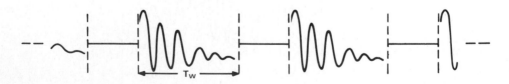

Figure 3.20. One solution for the problem of Figure 3.18. Append zeros so that the large signal at one end does not leak over into the other end of the reconstructed signal.

(2) Since we know that there is no true signal in the region of appended
zeros, find a function to make that happen in the inverse DFT. This
correction function, $d(t)$ is obtained by going through the entire double FT process on an invented signal which is everywhere 1 (Figure
3.21a). The result (Figure 3.21b) shows leakage where there should
be 0's and droop where there should be 1's. This is just the function,
$d(t)$, needed to correct the problem. Multiply the final filtered signal
by $d(t)^{-1}$ (Figure 3.21c) to obtain a corrected envelope. Careful: do
it only over the data window, since $d(t)$ goes to zero and $d(t)^{-1}$ diverges in the region of appended zeros.

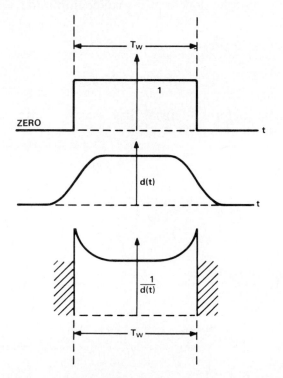

Figure 3.21. Droop in the envelope towards the endpoints can be corrected by a function evaluated by performing the FFT^{-1} (FFT) sequence on a data-set everywhere
equal to 1.

3.6.5 TRANSFORMATIONS INTO THE SPECTRAL DOMAIN:
MULTICOMPONENT EXPONENTIAL EXAMPLE*

It can be fruitful to transform complicated signals which do not look like conventional spectra into a form where key signal parameters can be extracted
from spectral peak-positions and heights.

*D. N. Swingler, *IEEE Transactions on Biomedical Engineering*, vol BME-24, no. 4, p. 408 (1977). The author
is indebted to D. Pons for bringing this method to his attention, and to both Pons and R. Cordeau for discussions
of its application to deep-level-transient spectroscopy in solid-state physics. Though it is not uncommon that
a DSP method moves between disciplines as widely separated as biochemistry and applied physics, it is also
not uncommon that a clever method lies underutilized because diverse disciplines share no common literature.

Consider a multicomponent exponential decay:

$$y(t) = \sum_{i=1}^{M} A_i\, e^{-\alpha_i t} \tag{3.43}$$

Signals of this form occur, for example, in chemical reactions where the α_i are the rate constants and the A_i are the concentrations. The key signal processing problem in *transient spectroscopy* is separating the amplitudes and rate constants of the various components in order to model the system. When plotted on a log scale, the components are sometimes evident (Figure 3.22a). In this two-component example, the fast-decaying term has large-enough amplitude, A_1, to be identified, but the slow term of smaller amplitude, A_2, is soon buried in noise.

With exponentials in the presence of noise,

$$S = A\, e^{-\alpha t} + \text{Noise} \tag{3.44}$$

the time available before a signal has decayed into the noise (i.e., $A\, e^{-\alpha t} = \text{Noise}$) is only

$$t = \frac{\ln\left(\dfrac{A}{\text{Noise}}\right)}{\alpha} \tag{3.45}$$

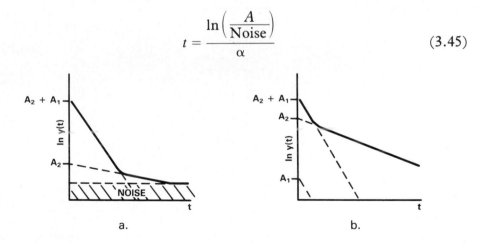

a.   b.

Figure 3.22. A sum of two exponential decays with (a) comparable amplitudes and (b) fast-decay of relatively small amplitude.

Noise and baseline uncertainty limit the simple graphical analysis of Figure 3.22. If the fast-decay term is the smaller in amplitude, it is difficult to recover, even with good signal-to-noise (Figure 3.22b), since it soon disappears into the slow term.

The multicomponent exponential signal (Eq. 3.43) can be made to look like a spectrum by a series of transformations, shown graphically in Figure 3.23 for a single term. The time variable is first plotted on a logarithmic *time* scale, so that the decay makes a rapid transition in the vicinity of the time constant.

The derivative of this function looks like a spectral peak. The peak position gives the rate constant, and the form can be unfolded, using convolution, to give the amplitude. For multicomponent decay, the technique transforms a complicated decay waveform into separated peaks, which facilitates the extraction of individual decay parameters.

First, transform the time variable:

$$t = e^x \qquad x = \ln(t)$$

The transformed signal becomes

$$y(e^x) = \sum_{i=1}^{M} A_i \, exp(-\alpha_i e^x) \tag{3.46}$$

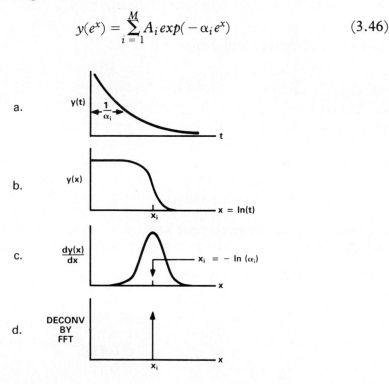

Figure 3.23. Transformation of variables (a) and (b), and differentiation (c), reshapes an exponential decay to a form similar to (d), a spectral term.

Each component has a sharp dropoff after one time constant in x-space, since $exp(-exp(\alpha x))$ varies very rapidly, as shown in Figure 3.23b. Now take the derivative of Eq. 3.46.

$$\frac{dy(e^x)}{dx} = -\sum A_i e^{x-x_i} \, exp(-e^{x-x_i})$$

$$x_i = -\ln(\alpha_i) \tag{3.47}$$

This second transformation makes the signal look like a peak, as in Figure 3.23c; the position, x_i, locates the decay rate, α_i. A multicomponent spectrum will show a series of such peaks. Spectral resolution can be enhanced by recognizing that each term is being convolved with the same function; since it is *known*, it can be deconvolved.

$$\frac{dy(e^x)}{dx} = [-e^x \exp(-e^x)] \star \sum_i A_i \delta(x - x_i) \tag{3.48}$$

The first term on the right is a function whose FT can be computed; the second term is our objective, a series of peaks (Figure 3.23d) whose location and amplitude determine all properties of the signal.

The DFT deconvolution method, carried out by an FFT, provides the mechanism to obtain the spectrum.

$$\sum_i A_i \delta(x - x_i) = \text{FFT}^{-1}\left[\frac{\text{FFT}\left[\frac{dy(e^x)}{dx}\right]}{\text{FFT}[e^x \exp(-e^x)]} \right] \tag{3.49}$$

The spectrum is obtained by an inverse FFT of the quotient of two terms, the FFT of the differentiated signal, which may in practice be taken as a first difference of adjacent samples, divided* by the FFT of the derivative of $\exp(-e^x)$. The quotient is multiplied by a window function before inversion in order to reduce sidelobes in the spectrum.

Consider this challenging example with amplitudes, A_i, differing by an order of magnitude, and time constants spanning more than 3 decades.

Amplitudes A_i	Decay Rates α_i
12.6	0.1433
2.2	0.0055
0.9	0.00042
13.3	0.000042

The results will be compared with those given by an older classic method using the same change of variables but obtaining spectrum-like properties by a multiplication by $\exp(x)$ rather than a derivative. A similar convolution can be recognized in the result:

$$e^x y(e^x) = g(x) \star \left[\sum_i \left(\frac{A_i}{\alpha_i}\right) \delta(x - x_i) \right] \tag{3.50}$$

where

$$g(x) = [-e^x \exp(-e^x)] \tag{3.51}$$

*Give the function a floor to avoid amplifying computational noise, as described in the deconvolution section above.

Here, $g(x)$ is the same function that appears in the derivative method. It has similar spectral form to Eq. 3.46, except that the spectral amplitudes are proportional to A_i/α_i, rather than A_i. The results are compared in Figure 3.24.

The derivative method shows well-separated peaks (Figure 3.24a), whose amplitudes and time constants can readily be identified as corresponding to the above table. The multiplication method shows inferior performance (Figure 3.24b), because the coefficients, (A_i/α_i), must span a very wide dynamic range, 80 dB in this example. In fact, a Blackman window had to be used to keep sidelobes down over this dynamic range, with consequent sacrifice of center peak width; for the derivative method, a sharper-peaked Hamming window could be used.

In the presence of noise, the differences are even more pronounced. A small amount of noise added to the data makes little difference in derivative processing (Figure 3.24c) but makes the e^x-weighted method almost useless (Figure 3.24d). The noise is amplified because of multiplication by a growing exponential.

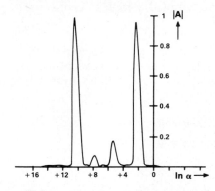

a. Derivative method, Hamming window.

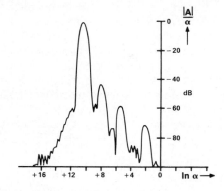

b. Classical method, Blackman window.

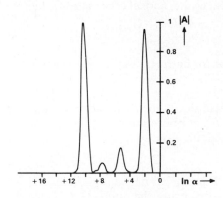

c. Derivative method in the presence of noise.

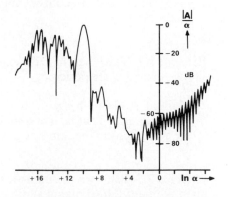

d. Classic method in the presence of noise.

Figure 3.24. Examples of results of the derivative transformation and spectral analysis of multicomponent exponentials. (Swingler, ©1977 IEEE—see Bibliography)

The reader may have noted that the comparison is plotted on a linear amplitude scale for the derivative method but a log scale for the e^x method, which automatically makes peaks look better relative to sidelobes and noise. In the e^x method, two signals of equal amplitude but widely differing time constants (e.g. first and fourth terms in the example) differ in amplitude by many orders of magnitude (about 70 dB) because the peaks go as A/α. These demands upon dynamic range make the e^x method fragile in the presence of noise. The derivative method is more robust. It also eliminates any baseline (often difficult to separate from a slow decay term) since the derivative automatically eliminates dc-offsets in the signal.

EXERCISES FOR CHAPTER 3

Exercise 3.1 Prove the orthogonality of the sampled complex exponential function over the finite interval.

Method: Multiply two exponentials together and sum, like taking the DFT of an exponential. The sum can be shown to be zero except when the two (sampled) frequencies are equal. The proof follows from the geometric series r_n, which has the sum

$$\sum_0^N r^n = \frac{1 - r^N}{1 - r} = \begin{cases} 0, n \neq 0 \\ N, n = 0 \end{cases}$$

with $r = e^{j2\pi n/N}$.

To see why the series sum is as shown, recall that this series was shown in Chapter 2 to be equal to the repeating $\sin(Mx)/\sin x$ function; the sampled points come out nonzero only on the peaks, and all other samples fall on the zeros.

Exercise 3.2 Show that the DFT expansion for both $x(t)$ and $X(f)$, via equations 3.1 and 3.2, are periodic.

Method: Add f_s to the frequency interval in Eq. 3.1 or T_W to the time interval in Eq. 3.2. That is, n becomes $n + N$, or r becomes $r + N$. Then use the periodicity of the complex exponential function.

Exercise 3.3 Estimate the number of FFT points which can be calculated maintaining full hi-fi digital audio bandwidth, using the numbers in Table 3.2 as an example. Follow the method of Eq. 3.7. What would the same hardware have accomplished before the FFT? First calculate the rate-limiting multiply steps, and then add an overhead estimate for signal I/O and nonmultiplication process steps.

Exercise 3.4 Write the explicit computation for $G(k)$ and $H(k)$ in the FFT computation of two real signals simultaneously, following the method outlined in the text.

Method: The only tricky step is the formation of the even and odd parts. Review how to do this with functions centered at 0, and translate to the conventional FFT summation over $[0, N - 1]$ by recalling that the interval $[N/2, N - 1]$ contains the same information as the negative half-interval, due to aliasing. That is, $[f_s/2, f_s]$ maps over into $[-f_s/2, 0]$.

Exercise 3.5 Write the explicit computation for $G(k)$ and $H(k)$ in the N-point FFT computation of a $2N$-point real signal, following the method suggested in the text.

Method: The splitting of even and odd points of the original $2N$-point signal is precisely equivalent to the initial combing of the decimation-in-time FFT in Section 3.3.3. Reproduce those steps, then pull out the real and imaginary parts using the same symmetry arguments outlined in the text for the two-signals-simultaneously algorithm. That is,

$$x(\text{even}) \rightarrow g(n)$$
$$x(\text{odd}) \rightarrow h(n)$$

whose Fourier transforms, $G(k)$ and $H(k)$, have the symmetry properties of purely real and purely imaginary signals, respectively. Combine *their* real and imaginary parts to generate $X_r(k)$ and $X_i(k)$.

Exercise 3.6 Convince yourself that the N-point FFT on $2N$ real data points does not lose spectral resolution but merely throws away redundant portions of the spectrum. Sketch the form of a $2N$-point spectrum for real data. The coefficients between $(2N)/2$ and $(2N) - 1$ are equivalent to the "negative-frequency" spectral components. Your sketch should reflect the symmetries in the spectrum: real-part even and imaginary-part odd. Now show on this sketch what is computed in an N-point FFT of this same signal, when even and odd points are combed into real and imaginary portions of an N-point signal.

Exercise 3.7 Verify the operation of the FFT program given in Table 3.3 for 4 complex (8 real) data points. Set up a table labeled D0 – D7 and index parameters N, J, I, and M. Go through the (DO 5) loop and verify that the pairs get properly bit-reversed, so pairs $(0,1,2,3)$ come out as $(0,2,1,3)$. That is,
   Inputs: D0 D1 D2 D3 D4 D5 D6 D7
   Outputs: D0 D1 D2 D5 D6 D3 D4 D7
where D0,D1 is the first pair, etc.

Exercise 3.8 Carry out the FFT computation of Table 3.3 symbolically for 4 complex points, beginning at label 6 (data already bit-reversed). Verify the correspondence with the signal-flow graph.

Exercise 3.9 For the FT of the triangular window, Figure 3.8d, the central peak is twice as wide as that of the rectangle, and, in addition, there are only half as many zeros per unit of frequency. Explain.

Method: Show that the triangle is related to the integral of a rectangular window whose width is only half that of the observation window. Differentiate the triangle. To get a recognizable rectangle window, use the shifting theorem (Table 2.1), which inserts a phase factor and adds a constant (both unimportant) to the FFT.

Exercise 3.10 Explain the statement: When two nearby frequencies sum to give a beat pattern (Figure 3.11a), two resolved peaks are seen in the spectrum (Figure (3.11b) only if the observation window includes a full cycle of the beat.

Method: relate the position of the first zero of $\sin(x)/x$ to the beat frequency, $(f_2 - f_1)$.

Exercise 3.11 Show that the family of $(\text{cosine})^n$ window functions reduces endpoint discontinuities by bringing to zero derivatives of order up to n.

Method: expand the cosine in a Taylor's series and look at the derivatives of the leading terms at the end points.

Exercise 3.12 A window function with n derivatives going to zero at the edges of the window will have sidelobes diminishing at $(-12 - 6n)$ dB/octave. Explain.

Method: How many times must the window function be integrated before a discontinuity appears? Look up the integration theorem of Fourier transforms, Table 2.1

Exercise 3.13 Window-weighting by the von Hann window, calculated as a convolution in the frequency domain, reduces to simple differences of three FT values as in Eq. 3.28. Show why.

Method: The window function is expressible as a sum of three $\sin(x)/x$ functions, one at $f = 0$ and the others shifted as in Eq. 3.27. Use the fact that the FT of unweighted data is convolved with $\sin(x)/x$. What is the effect in the time domain of shifting the $\sin(x)$ pair away from $f = 0$?

Exercise 3.14 Try the overlap-add method of DFT convolution for a row of arbitrary non-repeating pulses. Let the second function, $h(t)$, be a pulse of length Q. Divide the signal into windows of length labeled $N - Q$, and perform the convolution on each, first sliding the window back to the origin. Glue the results together, but now at intervals of length, $(N - Q) + Q = N$. The wind-up of one segment should compensate for the wind-down of the next, giving a result equivalent to the continuous convolution.

Chapter Four

Digital Filters

4.1 DIGITAL FILTER OVERVIEW

4.1.1 DIGITAL FILTERS: WHEN, WHY, WHAT, HOW

This chapter brings together the sampling theory of Chapter 2 and the spectral concepts of Chapter 3 to develop the design tools for both FIR and IIR digital filters. First, a brief recapitulation of the advantages of filtering digitally. Digital filters are:

Stable: no drift with time or temperature

Repeatable: component values replaced by digital parameters

Adaptable: parameters are programmable and reconfigurable

High-performance: meet demanding frequency response without phase errors

Predictable: design testable by simulation in software

Where to use a digital filter? Anywhere that signals need enhancing relative to noise. Here are a few specific applications:

Digital Audio (as listed in Chapter 1): digital filters go at every stage from recording to reproduction.

Sound generators: shape the signal for music or voice synthesis. Adaptive modeling capability of digital filters is especially valuable here.

Telecommunications: increase communication speed with smart noise-cancellation or with data compression algorithms.

Frequency synthesizers: enhance spectral purity.

Harmonic cancellers: locate zeros to cancel unwanted line or noise interference terms.

Overload detectors: sense electrical or physical overloads by changes in the pattern of signal harmonics; adjust programmable gains automatically to eliminate the problem.

Phase-sensitive detectors: absolute noise rejection (coherent filter) of very high Q. Related to heterodyning, decimation.

Correlators: correlate events, or correct adaptively for phase shifts between several detectors.

To set the stage for digital filters in depth, here are two examples which illustrate problems often encountered, the antialiasing filter and the reconstruction filter. Consider the sequence of steps in Figure 4.1: the analog signal, $v(t)$, is passed through an antiliasing filter with output $x(t)$, digitized to get $x(n)$, put through a digital process which yields $y(n)$, and brought back to the analog world as $y(t)$.

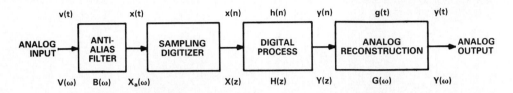

Figure 4.1. Key components of a digital signal processing system, with labels in time-, frequency-, and z-domains for this chapter.

Example: Reconstruction filter

The reconstruction filter on the output side in Figure 4.1 is a necessary bridge between the discrete set of points, $y(n)$, and the continuous output signal, $y(t)$. This particular reconstruction filter, the sample-and-hold, illustrates linear phase but with a less-than ideal frequency response below f_s. The first-order sample-and-hold also passes some distortion above its nominal cutoff at f_s. Both effects are calculable and can be eliminated by a smoothing filter as part of the analog reconstruction.

The sample-and-hold, illustrated in Figure 4.2, converts discrete points into a stairstep. What is its frequency response, $G(\omega)$? Ideally, $G(\omega)$ would have the form:

$$G(\omega) = 1, \qquad |\omega| \leq \frac{\pi}{t_s} \tag{4.1}$$

$$G(\omega) = 0, \qquad |\omega| > \frac{\pi}{t_s}$$

(where $t_s = 1/f_s$ and $\omega = 2\pi f$) since the reconstruction filter's purpose is to eliminate the high-frequency structure of the repeating spectrum (Figure 4.2a).

How closely does the sample-and-hold achieve this aim? The sample-and-hold has time response:

$$g(t) = 1, \qquad 0 < t < t_s \tag{4.2}$$
$$0, \qquad t > t_s$$

Calculate the frequency response, $G(\omega)$, which corresponds to this behavior in time (see Exercises).

$$G(\omega) = \frac{\sin\left(\dfrac{\omega t_s}{2}\right)}{\omega \dfrac{t_s}{2}}\, e^{-j\omega t_s/2} \tag{4.3}$$

This filter has so-called *linear* phase, equivalent to a constant delay of duration $t_s/2$ (see Exercises). The amplitude response (magnitude) is plotted in Figure 4.2b and c for comparison with the ideal. The sample-and-hold attenuates the signal undesirably in the pass-band, and passes some high-frequency information in the stop-band. Aliasing of the repeating (digital) spectrum leaks through as high-frequency noise. This chapter discusses techniques that will help to improve upon such performance.

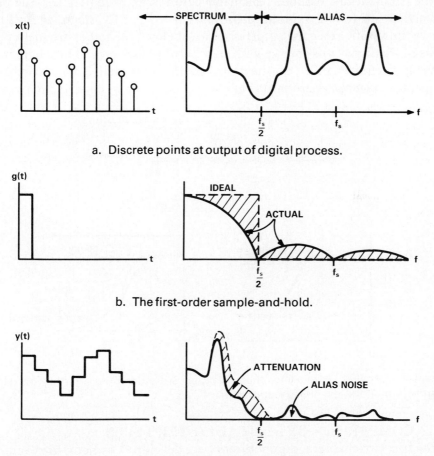

a. Discrete points at output of digital process.

b. The first-order sample-and-hold.

c. Output of (b) prior to smoothing.

Figure 4.2. Sample-and-hold reconstruction filter in time (left) and frequency (right) domains.

Example: Antialias Filter and Phase Distortion

It is useful to compare the criteria for the reconstruction filter with the anti-alias filter on the input side of Figure 4.1. The antialias filter has superior *amplitude* spectral response but usually introduces phase distortion, which is correctable in the digital domain if we choose.

For example, the 3rd-order Butterworth filter has a transfer function whose magnitude squared is:

$$|B(\omega)|^2 = \frac{1}{1 + \left(\dfrac{\omega}{\omega_c}\right)^6} \qquad (4.4)$$

A simple analog circuit useful for antialiasing is shown in Figure 4.3. Behavior of such circuits (pole locations as a function of component values, etc.) will be discussed in Section 4.3. While the Butterworth filter is monotonically flat in the passband and exhibits reasonable cutoff of high frequencies, the phase distortion is severe: maximum phase shift of $\pi/2$ per RC stage, or $3\pi/2$ total above the cutoff. Since the cutoff is set well below $f_s/2$, higher-frequency portions of the signal will emerge with significant phase error. This chapter will develop filters which do not have this problem, as well as correction techniques for use when phase shifts have been introduced by circuits such as this.

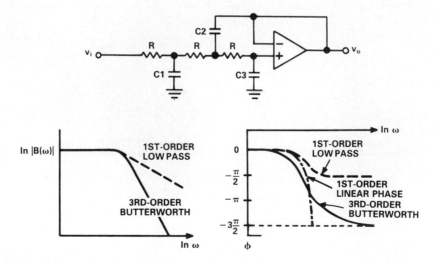

Figure 4.3. Analog antialiasing filter example: 3rd-order Butterworth. Top: circuit; bottom: amplitude and phase response.

4.1.2 COMPARISON OF DIGITAL FILTER TYPES

In the first two chapters, digital filters were developed by heuristic examples: the moving average and the first-order low-pass. Two points of view evolved:

> *Filtering as convolution.* This point of view is manifested in filter design in the data (time) domain, for example, the moving average.

> *Filter design as the shaping of transfer functions.* Here, the design is done in the frequency domain.

This chapter will link these points of view, in the context of practical digital filter design. A few limitations in scope will keep the treatment finite in length:

> The FIR filter is often preferred since it is unconditionally stable and usually has a linear phase response.

> The IIR filter, however, because of its simple architecture, also has its place, given appropriate recipes for design and guidelines about stability and accuracy.

> The low-pass (LP) will be the FIR prototype, with transformations available to go from the LP to the high-pass (HP), band-pass (BP), or band-reject (BR) cases.

> Analysis of IIR filters will emphasize the 2nd-order section, since complicated IIR filters can be reduced to a cascade of 2nd-order sections (plus one first-order section, for filters having odd numbers of sections).

> Since computer-aided design (CAD) is stressed, analytic mappings and transformations are de-emphasized; see the formal books on digital filters for that. Our approach is to work from the desired response curves rather than predict them analytically.

The three most common digital filter topologies are summarized in Figure 4.4. Here are some of their key features.

Nonrecursive or FIR Filters

The finite impulse-response (FIR) filter shown in Figure 4.4a has no feedback terms, only feedforward. The output values are a function of a finite number of previous input samples. FIR filters have advantages:

> *Stability*: FIR filters are stable with respect to oscillation, since there are no poles.

> *Accuracy*: FIR filters do not tend to accumulate errors, because of their finite memory of past events. 12 to 16 bits is generally adequate for FIR filters. An IIR filter may require 16 to 24 bits (to handle truncation and roundoff errors) for comparable performance.

> *Design Ease*: FIR filters are easy to understand and design, even for the non-specialist. While FIR filters do not correspond to familiar analog filters, they are the digital offspring of an analog predecessor, the delay line filter (see CCD filter, Figure 2.21).

> *Easy for CAD*: The filter coefficients are computed from a best-fit approximation to the transfer function, specified to desired limits.

On the down side, FIR filters can require many coefficients to achieve high performance, resulting in more multiplications and lower bandwidth limits in those cases where an IIR filter could do the same job.

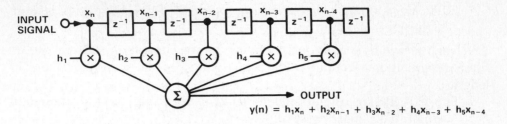

$$y(n) = h_1 x_n + h_2 x_{n-1} + h_3 x_{n-2} + h_4 x_{n-3} + h_5 x_{n-4}$$

a.   FIR or finite impulse response.

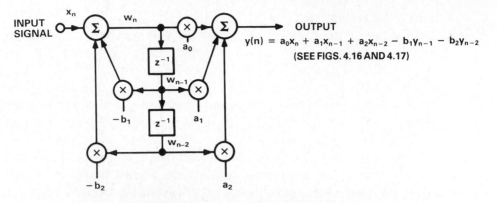

$$y(n) = a_0 x_n + a_1 x_{n-1} + a_2 x_{n-2} - b_1 y_{n-1} - b_2 y_{n-2}$$
(SEE FIGS. 4.16 AND 4.17)

b.   Basic IIR or infinite impulse response.

Figure 4.4.   Block diagrams of the three most common digital filters.

Recursive or IIR Filters

The infinite impulse-response (IIR) filter, shown in Figure 4.4b, includes feedback of previous outputs as well as inputs. The designer manipulates the poles and zeros in the complex plane, as with analog filters. Advantages of the IIR filter include:

High efficiency: IIR filters have simpler architectures that require fewer coefficients and a smaller number of multiply operations, giving higher throughput while requiring less memory than FIR. They also lend themselves to multiplexing.

Analog relatives: IIR filters retain close contact with analog filter design and its familiar terminology (Butterworth, Chebyshev...).

Analytic Control: If you need to force a zero at a particular place, the IIR will do it easily. If you need to model a particular process analytically, that model translates directly to an IIR filter.

CAD design ease: IIR filters also can be designed with CAD tools, via programs that optimize the coefficients of standard filter forms to meet stated transfer-function criteria.

But IIR filters have their quirks: stability must be carefully considered in the design. IIR filters can suffer from overflow, noise, and quantization errors, because of their long "memory" of past data.

Lattice Filters
Lattice filters are a type of IIR filter whose performance lies between FIR and IIR, promising greater stability than IIR with fewer coefficients to compute than with FIR. The lattice filter is a preferred building block for adaptive prediction, such as in speech analysis and speech synthesis, since the small number of coefficients allows an adequately large number of "formants" to be modeled in real time (see Friedlander, 1982).

When IIR and when FIR?
If $H(f)$ can be specified as a set of pass- and stop-bands, use an FIR filter. CAD techniques provide an interactive cookbook approach. Though IIR may be faster (fewer coefficients), FIR speeds in selected cases can be improved by the use of tricks (see Decimation). The guaranteed stability and linear phase give FIR the advantage where the transfer-function shape calls for FIR and computation time permits it.

If the analytic form of $H(f)$ matters, use IIR. For example, formant filters for speech (e.g, the Vocoder) and music are naturally modeled by poles and zeros, and implemented by IIR or lattice filters. Although IIR filters do make contact with analog classics (Butterworth, Bessel, Chebyshev, Elliptic...), this is not a real advantage, since better methods appear when one begins to think digitally.

The box below lists the principal design methods to be covered in this chapter.

DIGITAL FILTER DESIGN METHODS
FIR filter design methods (Section 4.2)
  a. The window method or Fourier-series approach.
  b. Computer-assisted design with the Remez exchange algorithm.

IIR filter design methods (Section 4.3)
  a. From analog counterpart (impulse invariant method); from second-order bandpass to elliptic filter;
  b. CAD design, such as via the Fletcher-Powell optimization developed by Steiglitz.

Examples of algorithms for filter computation are given in the box below. They are written as they might appear in a computer program, except that the computer multiply symbol ($*$) is left out to avoid confusion with convolution. For example, the FIR equation, written conventionally, would read:

$$y(n) = \sum_{k=0}^{L-1} h(k) \, x(n-k)$$

FILTER ALGORITHMS

FIR:

$$Y(N) = [S\,K = 0, L - 1]\,H(K)\,X(N - K)$$

IIR (Canonical direct form):

$$W(N) = X(N) + [S\,K = 0, L]\,A(K)\,W(N - K)$$

$$Y(N) = [S\,K = 0, L]\,B(K)\,W(N - K)$$

BIQUAD (Cascade):

$$Y(N) = [S\,R = 0, 2]\,B(R)\,X(N - R) + [S\,K = 0, 2]\,A(K)\,Y(N - K)$$

4.1.3 SUMMARY OF KEY DIGITAL FILTER RELATIONSHIPS

While digital filter computations are performed in the *time* domain, time-domain operations have a transfer function in the frequency domain. The connection is the Fourier transform, modified for sampled data and sampled frequencies. A summary is given below of key relationships between the filter function in time, $h(n)$, and its transfer function, $H(z)$, via the difference equation, which models the filter function.

KEY FILTER DIFFERENCE EQUATIONS

Sampled linear system: The output and input are coupled by finite difference equations.

$$\sum_{r=0}^{N} -b_r\,y(n-r) = \sum_{k=0}^{M} a_k\,x(n-k) \tag{4.5}$$

Causal system: Changes in the output do not precede changes in the input. The output is computed from previous inputs and outputs as

$$y(n) = \sum_{r=1}^{N} b_r\,y(n-r) + \sum_{k=0}^{M} a_k\,x(n-k) \tag{4.6}$$

IIR filter: Contains feedback from previous outputs (the b_r terms). The filter set in motion may continue to respond forever (*infinite* pulse response), though usually at diminishing amplitude.

FIR filter: Contains only feedforward terms (the b_r are zero). The system response will be of finite duration (*finite* impulse response). An FIR system can be described as:

(continued on next page)

$$y(n) = \sum_{k=1}^{M} a_k x(n-k) \tag{4.7}$$

The response disappears after M samples of zero input.

Filtering as Convolution in Time: The filter output, $y(n)$, is the convolution of previous inputs, $x(n)$, with the filter coefficient set, $h(k)$.

$$y(n) = \sum_{k=0}^{\infty} h(k) x(n-k) \tag{4.8}$$

Here, $h(k)$ is the impulse response of the filter (compare the example of Section 1.7.6). The causal assumption sets the lower limit at zero. For an FIR filter, the set, $h(k)$, terminates the sum at finite M.

Frequency Response: The z-transform of the difference equation (Eq. 4.6) gives:

$$H(z) = \frac{Y(z)}{X(z)} = \frac{\sum_{k=0}^{M} a_k z^{-k}}{1 - \sum_{r=1}^{N} b_r z^{-r}} \tag{4.9}$$

where $Y(z)$ and $X(z)$ are the z-transforms of $y(n)$ and $x(n)$, and $H(z)$ is the z-transform of the impulse response, $h(n)$.

The frequency response is evaluated on the unit circle, $z = e^{j\omega t_s}$

$$H(e^{j\omega t_s}) = \frac{\sum_{k=0}^{M} a_k e^{jk\omega t_s}}{1 - \sum_{r=1}^{N} b_r e^{jr\omega t_s}} \tag{4.10}$$

This is a Fourier expansion, the basis for much of the FIR filter design that follows. For IIR filter design, other prescriptions may connect digital and analog frequency, but the common link in drawing filter block diagrams is:

$$z^{-1} = \text{UNIT DELAY}$$

In the next section, for simplicity, the symbol $H(f)$ is used as the frequency response of a filter designed by Fourier expansion. The more correct but cumbersome designation, $H(e^{j\omega t_s})$, is used when an explicit connection with $H(z)$ above is to be drawn, since the frequency response is obtained by evaluating $H(z)$ on the unit circle, $z = e^{j\omega t_s}$.

4.2 FINITE IMPULSE-RESPONSE FILTERS

This section introduces powerful design methods based on shaping the transfer function, $H(f)$; its Fourier coefficients, $h(n)$, are the FIR filter coefficients. Two methods are introduced: (1) the window method works for arbitrary forms of $H(n)$ and minimizes the consequences of truncating $h(n)$ at finite N; (2) a CAD optimization procedure is best for standard filter shapes (LP, BP, BR, HP); it generates filter coefficients from given input parameters and tolerances.

4.2.1 FIR FILTER CONCEPTS AND PROPERTIES

The FIR filter is a generalization of the moving window on the flow of data, summarized schematically in Figure 4.5a. The aperture covers a finite win-

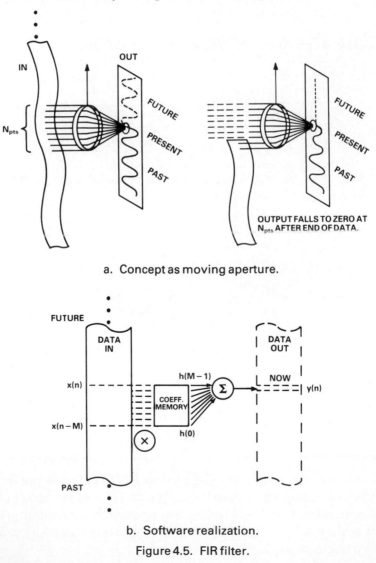

a. Concept as moving aperture.

b. Software realization.

Figure 4.5. FIR filter.

dow of width N_{pts} at each point along the data stream. Because the aperture is of finite width, the output must fall to zero after N_{pts} successive samples of zero input.

How can this idea be carried out digitally? As shown in Figure 4.5b, the data is stored as a buffer in read-write memory (RAM). The coefficients are stored in a separate buffer (RAM or ROM). A program loop executes the convolution of Eq. 4.8, looping through the range M ($=N_{pts}$) once for each output sample $y(n)$. The output buffer is shown dashed since in real-time applications there may be no need to store the outputs.

The sections which follow show how to *design* the FIR filter, i.e., how to select the coefficient set, $h(n)$.

Linear Phase Shift: FIR filters with a symmetric set of coefficients are called *linear-phase*, because the phase shift goes linearly with frequency (see discussion and Eq. 4.13). This is advantageous in avoiding waveform distortion. Phase information is often neglected, yet the magnitude is only half the information. Although phase is unimportant in perception of much *monaural* (e.g., telephone) audio information, it is crucial in the perception of "position" in stereophonic audio, for example—as well as position measurement in radar, sonar, and any image processing.

For example, if a square wave is represented by a DFT and the signs of the coefficients are neglected, the waveshape of the inverse DFT emerges totally changed, as shown in Figure 4.6. The symmetric expansion of the square wave

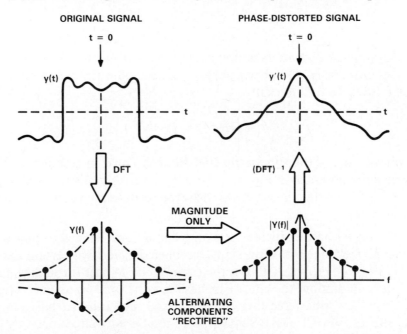

Figure 4.6. When phase information is ignored, waveshape is not preserved, as in this example of a square wave represented by a Fourier series.

has alternating signs. If the sign (phase) information is neglected, the waveform does not look at all square. If the digital signal processor's job were to filter noise from a pulse and show where a sharp transition occurred, the result of neglecting phase would surely be disappointing.

The problem is analogous to that of pulse propagation down a transmission line. Even if the *attenuation* is independent of frequency, *dispersion* (frequency-dependent propagation velocity) can cause a clean input pulse to emerge with severe waveform distortion.

Phase distortion can be avoided in DSP in two ways:

(1) FIR filters, with coefficients that are either symmetric (e.g., the usual LP) or purely antisymmetric (e.g, HP and derivative filters), do not introduce waveform distortion. More precisely, the phase shift is linear in frequency. To see this, write $H(f)$ as a Fourier series.

$$H(f) = \sum_{k=0}^{N-1} h(n) e^{j2\pi nf} \qquad (4.11)$$

The $h(n)$ are the filter coefficients. Center the sum at $n = 0$ by making a change of variables, $n \rightarrow n - N/2$, which brings out a phase shift, $e^{j\pi Nf}$, and express the exponential by its sine and cosine components.

$$H(f) = \underset{\substack{\text{fixed} \\ \text{time} \\ \text{shift}}}{e^{j\pi Nf}} \sum_{n=0}^{N/2-1} [h(n)\cos(2\pi nf) + jh(n)\sin(2\pi nf)] \qquad (4.12)$$

If $h(n)$ is *symmetric* about its midpoint, $h(-n) = h(n)$, the sine series will vanish by symmetry. $H(f)$ is then a purely real series multiplied by a phase factor which is linear in f. As a result,

$$\phi(H(f)) = -\pi Nf \qquad (4.13)$$

The phase shift of the symmetric FIR filter is linear in frequency. If $h(n)$ is *antisymmetric* about its midpoint, the cosine series vanishes, and the phase shift of $Y(f)$ remains linear; it has the same form as above, with an extra $\pi/2$ (see Exercises).

Why is linear phase sufficient to avoid distortion? Isn't zero phase better? "Linear phase" means that the signal has been delayed by a certain fixed interval, the slope of Eq. 4.13. This is the time it took to pass through the N-stage FIR filter calculation; the linear FIR has the same time lag as an N-stage delay line. To get *zero* phase requires adding a fixed time *lead*, such as a unity gain phase-shifter (see All-pass). Signal phases can be calculated correctly without this, since the delay is known and fixed.

4.2.2 THE FOURIER-SERIES APPROACH TO FIR FILTERS

Both the window method and CAD optimization methods approximate the filter transfer function, $H(z)$, as a finite Fourier series. Because of the relationship between the shape of the transfer function in the frequency domain and the number of coefficients, $h(n)$, in the time domain, it is a design challenge to get a transfer function close enough to the desired shape and at the same time truncate the Fourier series to a realistic number of coefficients (see BOX on the next page). The connection is:

$$h(n) = z\text{-transform of } H(z) \tag{4.14}$$

$$H(e^{j\omega t_s}) = H(z), \text{ evaluated on the unit circle, } z = e^{j\omega t_s}$$

Because $H(e^{j\omega t_s})$ is periodic, the filter transfer function is connected with the coefficient-set, $h(n)$, by a Fourier series:

$$H(e^{j\theta}) = \sum_{k=0}^{N-1} h(n) e^{-jn\theta} \tag{4.15}$$

where

$$h(n) = \int_{-\pi}^{+\pi} H(e^{j\theta}) e^{j\theta} d\theta \tag{4.16}$$

and $\theta = 2\pi f/f_s$. It is convenient to shift the range from 0-to-N to $\pm M$, where $M = (N - 1)/2$, to bring out the symmetries. Consider a low-pass symmetric filter, $h(n) = h(-n)$.

$$H(e^{j\theta}) = \sum_{k=-M}^{M} h(n) e^{-jn\theta}$$

$$= h(0) + \sum_{k=1}^{M} h(n) \cos(n\theta) \tag{4.17}$$

The result of shifting the range is to eliminate (for the *calculation only*) the linear phase factor of Eq. 4.12. The redefined limit gives $N = 2M + 1$. As a result, *there are $(N-1)/2$ independent Fourier components, $h(n)$, in a symmetric N-tap filter whose impulse response lasts Nt_s seconds.* A brief graphical excursion into the consequences of the above equation will bring out the principles (the "facts" in the BOX) underlying both the window method and CAD optimization methods of FIR filter design. The low-pass will be the prototype, transformed as needed to other types.

FACTS UNDERLYING FIR FILTER DESIGN METHODS

A particular filter coefficient, $h(q)$, is the numerical value of the amplitude of the qth Fourier coefficient of the frequency response, $H(f)$.

The smaller the low-pass band width of $H(f)$, the larger the number of time samples over in the central peak of $h(n)$.

The sharper the cutoff demanded in $H(f)$, the more coefficients must be included in $h(n)$.

An ideally sharp cutoff of $H(f)$ is usually not desirable, since the step response in time will ring.

Truncation of the set, $h(n)$, to limit the computational burden forces Gibbs oscillations (see below) into $H(f)$.

Windowing softens the corners of $H(f)$, lowering the number of significant Fourier coefficients and consequently the number of significant terms in $h(n)$. CAD optimization methods accomplish the same aim, keeping Gibbs oscillations within stated bounds by easing transition-bandwidth requirements.

Filter coefficients in time are Fourier coefficients of the transfer function in frequency.

FIR filter design is an art that calls for specifying a finite set of N coefficients in $h(n)$ to approximate an idealized filter form. How does N affect FIR filter performance? For the low-pass, recall the Fourier series for a pulse in frequency, which resembles the ideal low-pass response, redrawn in Figure 4.7a. The $\sin(x)/x$, which sets the amplitudes of the Fourier coefficients, is the result of applying Eq. 4.16 to the ideal low-pass. Thus each filter coefficient, $h(q)$, is the amplitude of the qth Fourier coefficient of $H(f)$. This can be seen another way, in Figure 4.7b. A given cosine in frequency corresponds to a single point in time; the higher the frequency, the further out in time will be its Fourier coefficient. Since it is the higher harmonics which add up in phase to make a steep edge, the sharper the cutoff demanded in $H(f)$, the more taps must be included in $h(n)$.

In fact, an ideally sharp cutoff in $H(n)$ is usually a mistake. Consider the step response of the "ideal" brick wall LP, Figure 4.7c. Since the filter output is the convolution of the step input with the impulse response, the output will overshoot and ring; not desirable in most cases. But since the time window in which the ringing persists and its amplitude are predictable from the form of $H(f)$ and its associated impulse response, $h(n)$, step response problems can be anticipated and incorporated into design criteria (see Exercises).

Since the filter is a tapped delay line, the outputs are delayed appropriately. The step response of an N-tap filter has its biggest change about $N/2$ sampling intervals after the step input, as shown. Were it not for the delay, the response would appear to have instantaneous step response with anticipatory ripples—a violation of real-world causality.

More insidious is the effect of truncating $h(n)$. If one could just draw an ideal $H(f)$ and take its Fourier transform to obtain whatever number of filter coefficients, $h(n)$, are needed, FIR design would be trivial. But the number of taps, N, must be restricted to attain adequate throughput with minimal computational delay on real-time signals. Truncating $h(n)$ at finite N is in effect multiplying by a pulse of width Nt_s, as in Figure 4.7d. The ideal transfer function, $H(f)$, is thereby convolved with a $\sin(x)/x$ of width $1/(Nt_s)$, or, in units of f_s, of width $1/N$. This puts ringing into the ideal $H(f)$. The width of the ringing can be compressed by increasing N, but this does not reduce the amplitude of the ringing, which stays constant. This is called the Gibbs★ phenomenon. For a step discontinuity fit by a finite Fourier series, the Gibbs overshoot amounts to 9% of the discontinuity.

The window method "rounds the corners" of idealized transfer functions, softening sudden discontinuities. This lessens the number of filter taps required. To see this graphically from Figure 4.7b, note that rounding the corners is equivalent to attenuating higher harmonics in the Fourier expansion of $H(f)$, thus decreasing the weight of samples in $h(n)$ further out in time. Truncating $h(n)$ where its coefficients are negligible lowers the magnitude of the Gibbs oscillations in $H(f)$ to a negligible amplitude. In fact, it is no longer abrupt truncation, since convolution by a gentle window in frequency is multiplication by a correspondingly gentle window in time.

Designing Bandpass and High-Pass Filters from Low-Pass Filters
The passband width of the LP is connected to the width in time over which $h(n)$ is largest. The smaller the passband width of $H(f)$, the larger the number of time samples in the central peak of $h(n)$. For the ideal low-pass of Figure 4.7a, the bandwidth, β, expressed as a fraction of f_s, sets the number of coefficients which will lie within the central peak of $h(n)$.

$$H(f): \text{pass-band width}, f_c = \beta \frac{f_s}{2} \tag{4.18}$$

$$h(n): \text{width of central peak} = \frac{1}{f_s}, \text{containing}$$

$$N \text{ coefficients} = \frac{f_s}{2f_c} = \frac{1}{\beta}, \text{in the central peak of } h(n).$$

These relationships between $H(f)$ and $h(n)$ show how to make an FIR bandpass, high-pass, and band-stop filter, by transformation of a low-pass design.

★After the same J. W. Gibbs who also developed much of thermodynamics.

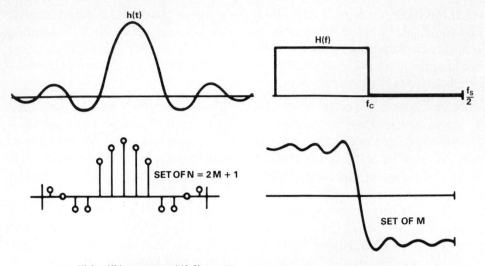

a. "Ideal" low-pass, H(f), and its corresponding time function, h(n).

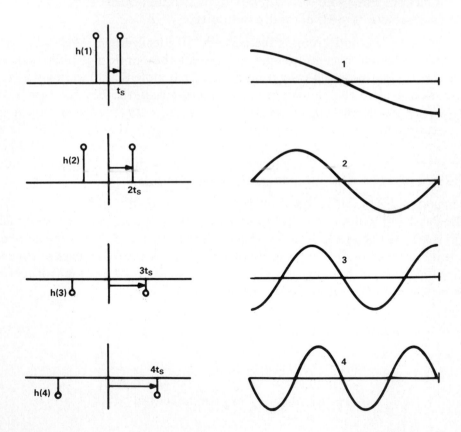

b. When H(f) is Fourier-expanded, the rapidly varying terms which give the steep cutoff give terms in h(n) out at longer time.

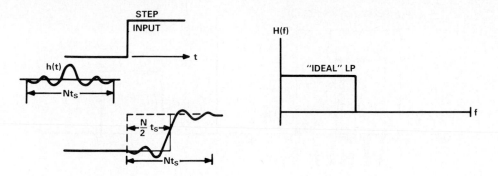

c. An "ideal brick-wall" LP will ring given a step input; note the convolution of $h(n)$ with the step on the left.

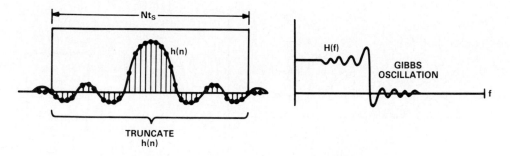

d. Truncating $h(n)$ at finite N results in Gibbs oscillations in $H(f)$.

Figure 4.7. Some basic digital filter facts.

Bandpass: What does $h(n)$ look like for an ideal bandpass? The bandpass is specified by its center frequency, f_0, and bandwidth, δf, which is the result of the convolution of an impulse at f_0 with a pulse of width δf. Since convolution in frequency corresponds to multiplication in time, the form of $h(n)$, shown in Figure 4.8a, is therefore dominated by an oscillation in time at frequency f_0, with an envelope set by the $\sin(x)/x$ form of the FT of the pulse. The number of taps of $h(n)$ per sine-wave cycle is f_s/f_0. The width in time of the central envelope peak is $1/\delta f$, and it contains $f_s/\delta f$ samples. The narrower the pass-band, the larger N will have to be in the FIR bandpass filter.

How does this notion of the bandpass relate to the analog "stagger-tuned" approach, which approximates a flat-top passband by a sequence of second-order sections at nearby frequencies? Each section has a peak in the frequency domain at f_i. Each contribution to $h(n)$ is therefore a sine wave at f_i, some at frequencies above f_0, some below f_0. The sum will therefore display interference, giving $h(n)$ which oscillates but decays in amplitude when the furthest frequency contributions have gotten out of phase by π. The envelope of $h(n)$ can be calculated readily (see Exercise 4.5) and gives the same result as this heuristic argument.

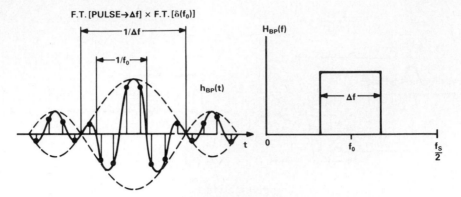

Figure 4.8a. What the set of "ideal" bandpass coefficients looks like in time.

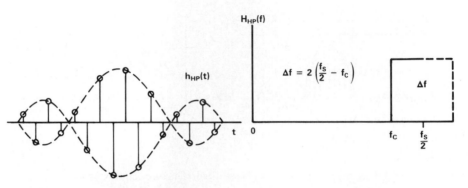

Figure 4.8b. What the high-pass looks like in time (see Exercises).

High-Pass: What do the high-pass coefficients look like in time, and how is the pattern related to the low-pass? The high-pass (Figure 4.8b) is the convolution of a peak at $f_s/2$ with a pulse of half-width equal to the pass-band width. The FIR coefficients $h(n)$ will therefore have rapid oscillation at $f_s/2$ with a $\sin(x)/x$ envelope, which decays in a time of the order $1/(\text{passband})$ (see Exercises).

Alternatively, the high-pass can be obtained from the low-pass by the transformation of f to $(f_s/2 - f)$ applied to the low-pass Fourier series. The effect (see Exercises) is to multiply each term in the series by $(-1)^k$, which changes the sign of every other term in $h(n)$—just the result already shown.

Band-Stop: The band-stop is simply expressed as the sum of a low-pass and a high-pass; it needs no special explanation.

4.2.3 THE WINDOW METHOD OF FIR-FILTER DESIGN

The window method of FIR filter design begins with a desired transfer function, $H(f)$; it is fit by a Fourier series, then any rapid transitions are smoothed to avoid the consequences of truncation of $h(n)$. We follow the exposition of Kaiser (1966, summarized in Hanning 1983) and Kaiser and Reed (1977,

1978). The window method can do $H(f)$ of *any* shape, while the CAD optimization techniques (coming in Section 4.2.4) are best for filters like LP, BP, and HP (passband magnitude 1 and stopband magnitude 0).

As an example, consider a differentiator, Figure 4.9a. Its frequency response within the window, $-f_s/2$ to $+f_s/2$, is ideally a ramp (constant slope), with constant $+90°$ phase shift. This filter, a basis for trend-detection in curves and edge-detection in images, is challenging to design, since the discontinuity at $f_s/2$ has to be smoothed without losing usable signal bandwidth. The exercise at the end of Section 2.8.2 showed that the simple first-difference filter, $y(n) = x(n) - x(n-1)$, makes a differentiator of poor frequency response,

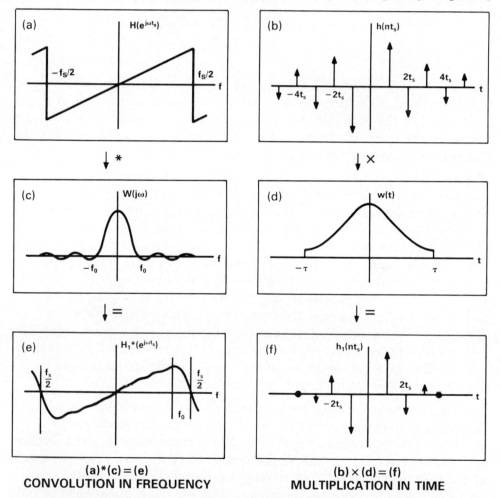

(a)*(c) = (e)
CONVOLUTION IN FREQUENCY

(b) × (d) = (f)
MULTIPLICATION IN TIME

Figure 4.9. The window or Fourier-series method of FIR filter design. (a) is the desired $H(f)$: the ideal differentiator; (b) is its Fourier series, $h(t)$. If $h(t)$ is truncated to limit computations, the truncated series represents $H(f)$ poorly. If (a) is convolved with (c) a window function, $W(f)$, to reduce endpoint discontinuities as in (e), the Fourier series goes to zero naturally (f) without truncation, being multiplied by (d) a finite-width $w(t)$. (from Kaiser, in Rabiner and Rader ©1972 IEEE—see Bibliography.

except at very low frequencies. For improvement in frequency response, the window method can be used to add higher-order differences.

Since a derivative of $e^{j\omega t}$ multiplies the argument by $j\omega$, the ideal differentiator transfer function is odd and linear in ω.

$$H(\omega) = j\omega \qquad \text{(ideal differentiator)} \qquad (4.19)$$

The differentiator's transfer function, when plotted in (repeating) frequency, as in Figure 4.9a, is a sawtooth waveform with Fourier coefficients, $h(n)$, in time (Figure 4.9b):

$$b_n = \frac{2j}{\pi} \frac{\sin\left(n\frac{\omega_s}{2}\right)}{n^2} - \frac{\frac{\omega_s}{2}\cos\left(n\frac{\omega_s}{2}\right)}{n} \qquad (4.20)$$

where $\omega_s = 2\pi f_s$. The coefficients, b_n, are the values of $h(n)$ in time. We note that $h(n)$ is odd, like the high-pass, and has the $1/n^2$ behavior similar to the coefficients of the triangular wave. The second term comes from the extra unit-step that must be added to generate the endpoint discontinuity in the ramp joining $f_s/2$ and $-f_s/2$ (see Exercise 4.6).

The number of Fourier filter coefficients will be truncated to keep the $h(n)$ computation (i.e., filter length) manageable. To avoid ringing due to truncation, as discussed above, relax the filter requirements so no true discontinuity remains. Convolve the ideal $H(f)$ with a broadening function $W(f)$ which leaves most of the spectral range unchanged but smears the discontinuity, as in Figure 4.9c and e. For a smearing function of width $2f_0$, a width approximately f_0 is lost off the transition region of $H(f)$. Convolving $W(f)$ with $H(f)$ multiplies the filter coefficients, $h(t)$, by $w(t)$ (the FT of $W(f)$, also shown). The sooner $w(t)$ goes to zero, the fewer filter coefficients, $w(t)h(nt_s)$, will be significant. The tradeoff of information theory appears again: since the widths of $W(f)$ and $w(nt_s)$ are inversely related, narrowing $W(f)$ to lose the least usable signal bandwidth broadens $w(nt_s)$, calling for more filter coefficients to be included in the filter computation.

Window functions were discussed extensively in Chapter 3; similar selection criteria apply to windowing in filter design. Here are some guidelines:

   What sets the transition bandwidth? For a given window, the transition bandwidth over which the discontinuity is smeared decreases as N increases. The transition bandwidth is not a strong function of window choice.

   What sets the stopband attenuation? The stopband attenuation is only slowly dependent on N. The stopband attenuation *is* a strong function of window choice.

For example,

Window	Transition Width	Max Stopband Attenuation (dB)
Rectangular	$\dfrac{4\pi}{N}$	20
Cosine Bell	$\dfrac{8\pi}{N}$	50
Hamming	$\dfrac{8\pi}{N}$	45
Blackman	$\dfrac{12\pi}{N}$	80

These values for the stopband attenuation are typical for N values of about 50; they vary by a few dB for other N, since this specification depends upon the rate of sidelobe falloff in Table 3.5

For discussion here, it is convenient to adopt the Kaiser window, from Chapter 3; because it has performance comparable to the Blackman window, it allows a tradeoff between transition width and stopband attenuation, and precise empirical design rules have been worked out. The window, $W(f)$, and its Fourier transform, $w(n)$, are

$$W(f) = \frac{\sinh f}{f} \tag{4.21}$$

$$w(n) = I_0(n)$$

where $I_0(n)$ is a Bessel function of order zero (see Table 3.4 for the precise form). As long as $w(n)$ reaches zero at both ends of the window, the Kaiser window hyperbolic sine has no sidelobes in the frequency domain. If that gives too broad a smearing, the parameter α (Table 3.4) may be decreased. The time-domain window widens proportionately, and may reach the ends of the filter window, in which case some ringing appears in $W(f)$, though at very small amplitude.

A computed example is shown in Figure 4.10b: a 19-tap windowed differentiator. For comparison, the transfer function of the first 9 terms of a Fourier series (i.e., a 9-tap unwindowed truncated filter) is also shown. The superiority of the window design in terms of smooth linear-in-f response up to the cutoff is evident.

How many taps are required for a given usable bandwidth? In the case of the differentiator, it depends upon how much of the upper end of the signal bandwidth has been traded away to gain a given degree of smooth behavior, the region f_d in Figure 4.10 (bottom). For the Kaiser window, the relationship between number of taps, N, and the fraction of signal bandwidth lost, $f_d/(f_s/2)$ is (see Kaiser, 1966)

$$N = A \frac{\frac{f_s}{2}}{f_d} \qquad (4.22)$$

where A ranges from 8, for 3% overshoot, to 11.5, for 0.01% overshoot. For example, the filter of Figure 4.10b loses about 20% of the usable bandwidth; to keep overshoot to 3% would take $8/(0.2) = 40$ coefficients. (The overshoot cannot be seen in the plot, but it would be evident on a log scale.)

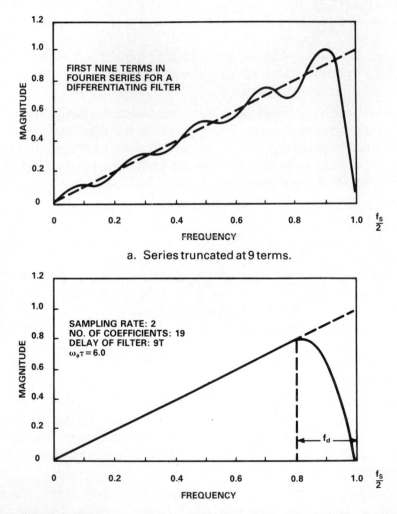

a. Series truncated at 9 terms.

b. Windowed differentiation carried out by a 19-tap windowed truncation.

Figure 4.10. Examples of Kaiser window differentiator filters. (*Kaiser and Reed, ©1977, reprinted with permission*)

Computer-Aided Filter Design by the Window Method

Kaiser (1977, 1978) has provided programs for standard filter functions. The design approach is graphical, through specifying the tolerances of a number of filter parameters. While it is preferable in practice to use a different CAD optimization approach (next section) for standard filter functions (LP, BP, HP), a brief review of the Kaiser results for a low-pass illustrates the tradeoffs between number of taps and signal-to-noise enhancement.

The designer specifies

e_p, the allowed ripple or deviation from unity in the passband, in dB
e_s, the allowed ripple in the stopband, in dB
b, the cutoff frequency, as a fraction of $f_s/2$
d, the transition band width, as a fraction of $f_s/2$.

There are several approaches to fit different needs. They include:

NER (nearly equal ripple)
MPERS (monotone passband, equiripple stop-band)
MAXFLAT (monotone passband and stopband)

The NER approximation spreads the truncation error over the whole of the pass and stop bands; $e_p = e_s = e$. The number of taps is related to the design parameters by the (empirical) relation:

$$N = \frac{-e(\text{dB}) - 8}{29\,d} \qquad (4.23)$$

For example, for $d = 0.2$ and $e = -34$ dB (ripple limited to $20 \log(0.02)$),

$$N = \frac{(34 - 8)}{(29 \times 0.2)} = 4.5$$

Since there are two bands in which the error must be equally controlled, the number of taps must be doubled (and usually increased to the nearest odd number, in order to maintain linear phase). In this case, 9 taps will barely suffice, and 11 taps will be quite adequate.

Computer programs for designing Kaiser filters are given in the references cited above.

It is illuminating to compare Kaiser's NER design and MAXFLAT methods; MAXFLAT demands smooth behavior: $H(f)$ nonincreasing in either passband or stop-band. The filter is forced to tangency at $f = 0$ with a given number of zero derivatives, and also at $f = f_s$. For 5% error (0.45 dB in passband and $-26\,\text{dB}$ in stopband), the required number of taps goes approximately as

$$N - 1 = \frac{2}{d^2} \qquad (4.24)$$

where d is the transition frequency fraction between passband edge (magnitude 95%) and stopband edge (5% magnitude). The number of coefficients goes as the inverse *square* of the passband fraction rather than just inversely as for the NER. The MAXFLAT is thus computationally more demanding then the NER.

A data example comparing NER and MAXFLAT is shown in Figure 4.11. The test signal is a single cycle of a cosine, chosen because its spectrum occupies a large portion of the frequency range, hence the slow falloff of the peak at f_0. Wideband noise has been added; its magnitude is comparable to that of the signal. Examples of filters to reduce the noise so as to avoid aliasing are:

$$\text{NER}, e = -80\,\text{dB}, b = 0.325, d = 0.5$$
$$\text{so } N_p = 5, \text{ or } 10 \text{ taps}$$
$$\text{NER, same except } d = 0.15$$
$$N_p = 17, \text{ or } 34 \text{ taps}$$
$$\text{MAXFLAT}, b = 0.1, d = 0.2$$
$$50 \text{ taps}$$

The 10-tap NER (b) leaves mid-band noise. The 34-tap NER (c) makes the data more noise-free, but the fidelity to the signal spectrum is not perfect, since it fails to get the shape right above $f_u = 0.15$, the usable bandwidth. By contrast, the MAXFLAT (d) faithfully follows the signal spectrum up to about $f_u = 0.4(f_s/2)$ of the potential bandwidth, at the cost of increased taps.

As these examples suggest, filter design is an art, best accomplished with a specific signal situation in mind.

4.2.4 COMPUTER-AIDED OPTIMIZATION FOR FIR-FILTER DESIGN

The window method and the CAD optimization method both use Fourier series to expand the filter's transfer function. The window method rounds off the consequences of truncation at finite N with known window functions. The CAD optimization method takes a different approach: recognize from the start that N will be finite, Fourier-expand $H(f)$ over finite N, and seek a set of Fourier coefficients which will minimize the stated error criteria—passband ripple, stopband attenuation, and transition bandwidth—at given

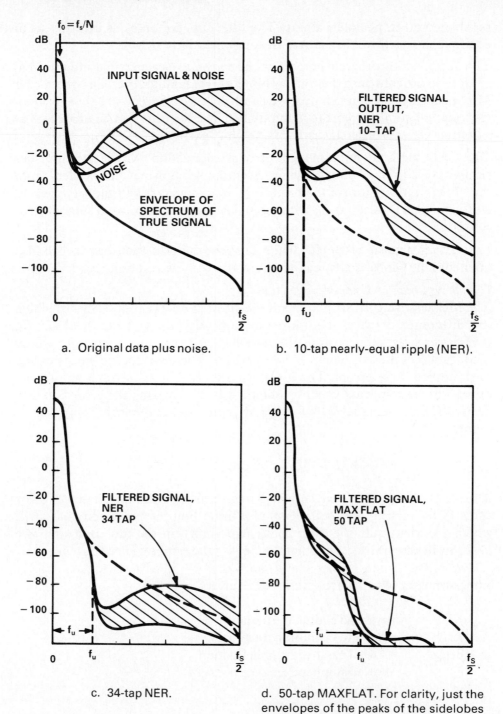

a. Original data plus noise.

b. 10-tap nearly-equal ripple (NER).

c. 34-tap NER.

d. 50-tap MAXFLAT. For clarity, just the envelopes of the peaks of the sidelobes in the spectrum have been drawn.

Figure 4.11. Spectrum examples before and after passing through a low-pass filter designed by the Kaiser window method. (*Kaiser and Read, ©1977, reprinted with permission*)

passband and stopband locations. The filter function, $h(n)$, is just this set of coefficients that gives the the best-fit approximation to $H(f)$.

The optimization program iterates on the extrema (local maxima and minima) of $H(f)$ in order to push them all within the error bounds, increasing the order M as needed to accomplish this. For a filter of M pairs ($N = 2M + 1$ taps), there are about M extrema to manipulate, since the filter can be expressed as a Fourier series of order M (try it for a small-N example).

The CAD methods do numerical error minimization, which accomplishes similar goals to those of windowing, even though this may not be clear from the "black-box" nature of the procedure. Windowing is a formula approach, while the CAD method is a direct tailor-made numerical trial-and-error solution.

Restricting the size of ripples while relaxing the transition bandwidth also quenches the Gibbs oscillations of a set of $h(n)$ truncated at finite N.

How to Optimize an Unknown Function

The McClellan algorithm plays with the *alternations*, the frequencies at which the difference between the approximation, $H_d(f)$, and the ideal, $H(f)$, crosses zero. Between a pair of alternations will be a region of error for which the location of the extremum, f_i, is estimated. The set of coefficients is varied to minimize these errors. The low-pass (Figure 4.12) illustrates the procedure. The best approximation is obtained by minimizing the error between desired H and the actual H_d (d for digital), computed at the set of "worst case" points, f_i:

$$W(f_i)[H_d(f_i) - H(f_i)], = -(-1)^i \epsilon \qquad (4.25)$$

The set of points, $H_d(f_i)$, are adjusted to give the smallest simultaneous error at all f_i, by varying the coefficients of a finite Fourier expansion of $H_d(f_i)$, which are identically the filter coefficients, $h(n)$, in the time domain. The $(-1)^i$ term takes into account that the signs of the error regions alternate.

The parameters available to specify the fit are:

f_p: passband ending frequency
f_c: stopband beginning frequency
δ_1:   allowed passband ripple fraction
δ_2:   allowed stopband leakage fraction
ϵ: error bound

The transition bandwidth, $f_c - f_p$, between passband ending and stopband beginning is the relaxation needed to keep N finite without the truncation consequences of Figure 4.7, serving the same purpose as transition width in

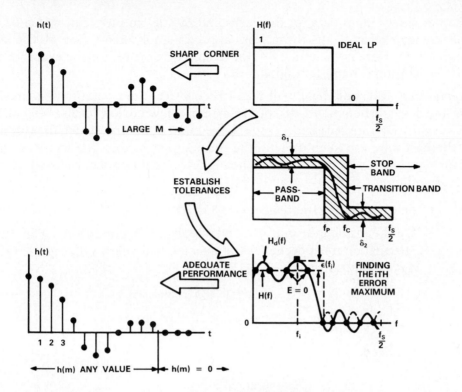

Figure 4.12. Filter design criteria for CAD optimization; ideal LP example. The filter transfer function, $H(f)$, is approximated by a finite sum of sine waves whose amplitudes form the filter coefficients, $h(n)$.

the window method. $W(f_i)$ here is a *weighting* function (*not* a window), typically set equal to 1 in the stopband and equal to the ratio of stopband to passband ripple in the passband.

$$W(f_i) = 1 \qquad\qquad \text{(stopband)}$$

$$W(f_i) = \frac{\delta_2}{\delta_1} \qquad\qquad \text{(passband)} \qquad\qquad (4.26)$$

One worst-case point lies between each pair of alternations (where H_d crosses H). Although the f_i are not known exactly, their locations can be approximated by computing a starting guess midway between a pair of points where the error goes to zero (see Figure 4.12). The number of points in the set, f_i, is established by the number of taps. Since there are M cycles of the highest Fourier component of $H(f)$ in the 2π interval, there are at most $M - 1$ local maxima and minima in the $[0,\pi]$ interval (see Exercises).

CAD FIR Design by Parks and McClellan Program
Given this concept of minimizing maximum errors, the rest is done by "turning the crank," using readily available CAD programs. The Appendix to this

chapter shows input data for the Parks and McClellan program, which uses the Remez exchange algorithm from approximation theory (see McClellan *et al*, 1973). Here is a summary of the design procedure, for the example of a low-pass filter suitable for digital audio.

Application: full audio bandwidth (25 kHz), in real time, want to pass music with little distortion ($<$ 0.5 dB) but bring noise down to low levels ($-$ 80 dB). How many taps are required? Tradoff: the number of taps, which determine if the hardware can keep up with the throughput, against the width of the transition band, which determines the highest signal frequency passed without attenuation.

(1) Select sampling frequency: $f_s = 50\,\text{kHz}$

(2) Define passband end, f_p, and stopband beginning f_c. The full attenuation must occur for frequencies well below $f_s/2$, say $0.8\,f_s/2$. Try a transition band:

$$f_p = 0.6\,f_s/2 = 15\,\text{kHz}$$
$$f_c = 0.8\,f_s/2 = 20\,\text{kHz}$$

Normalized, $$\frac{f_p}{f_s} = 0.30$$

$$\frac{f_c}{f_s} = 0.40$$

(3) Define weighting function.

Passband ripple $<$ 0.5 dB

Stopband attenuation $<$ $-$ 80 dB

This will put more weight on stopband attenuation than on passband ripple.

Stopband deviation $\delta_1 = 3 \times 10^{-4}$
$20\log\delta_1 = -70\,\text{dB}$

Passband deviation $\delta_2 = 0.06$
$20\log(1 + \delta_2) = \pm 0.5\,\text{dB}$

Ripple ratio $= 0.06/0.0003 = 200{:}1$

(4) Run the design program. Among the outputs when the actual deviations are specified is the number of taps. A second mode allows the number of taps as well as the ripple *ratio* to be specified, and the actual ripple values emerge as outputs. An example of input parameters and output filter specifications for a problem that doesn't differ greatly is given in Table 4.1.★

★An example of a program for FIR filter design will be found in the Appendix to this Chapter.

(5) Can the filter hardware keep up? The answer will depend upon the design; it must satisfy

$$N \times t_{mac} < \frac{1}{f_s} \qquad (4.27)$$

where t_{mac} is the time-per-tap to execute one multiply-accumulate step (i.e., one term in the convolution sum, Eq. 4.8). Assuming that multiplication time is the limiting factor on computing speed, t_{mac} is as short as the hardware MAC time if done in hardware, or as long as the multiply cycle time for a microprocessor or microprocessor-accelerator.

Table 4.1 FIR filter design input criteria and output coefficients via the Remez Exchange Algorithm. (Ref: IEEE 1979; see also Peled and Liu)

```
    RUN FIRGRAPH

INPUT DATA

17,1,2,0,16            ; SPECIFY FILTER LENGTH AND TYPE
0,0.15,0.3,0.5         ; SPECIFY LOWER AND UPPER BAND-EDGES
1,0                    ; SPECIFY DESIRED FUNCTION VALUE IN EACH BAND
1,29                   ; SPECIFY WEIGHT FUNCTION IN EACH BAND


                       FINITE IMPULSE RESPONSE (FIR)
                       LINEAR PHASE DIGITAL FILTER DESIGN
                       REMEZ EXCHANGE ALGORITHM

                       BANDPASS FILTER

              FILTER LENGTH =    17

    ***** IMPULSE RESPONSE *****
              H(  1) =   0.45984692E-03 = H(  17)
              H(  2) =   0.10255111E-01 = H(  16)
              H(  3) =   0.23305153E-01 = H(  15)
              H(  4) =   0.62355087E-02 = H(  14)
              H(  5) =  -0.48847619E-01 = H(  13)
              H(  6) =  -0.59384681E-01 = H(  12)
              H(  7) =   0.74297398E-01 = H(  11)
              H(  8) =   0.30020890E+00 = H(  10)
              H(  9) =   0.41522482E+00 = H(   9)

                       BAND  1        BAND  2         BAND
    LOWER BAND EDGE    0.000000000    0.300000012
    UPPER BAND EDGE    0.150000006    0.500000000
    DESIRED VALUE      1.000000000    0.000000000
    WEIGHTING          1.000000000   29.000000000
    DEVIATION          0.028284101    0.000975314
    DEVIATION IN DB    0.491475552  -60.217113485

    EXTREMAL FREQUENCIES
       0.0000000   0.0659722   0.1215278   0.1500000   0.3000000
       0.3138888   0.3486109   0.3937497   0.4458328   0.5000000
```

(6) If the answer to (5) is YES, the design is done. Go ahead with a simulation of performance (next section), then build it. If NO, and the hardware is unchangeable, relax the transition band requirement to reduce the number of taps, and iterate the design program. If software multiplication is used, consider hardware solutions; if hardware multipliers are used, it is conceivable to seek a faster multiplier, or to perform several multiplications in parallel using more than one multiplier—both somewhat more costly solutions.

Computational Speed Limits on Number of Taps
The upper limit on number of taps is set by the computation time per tap, essentially the multiply-accumulate time t_{mac} of one step in Eq. 4.8. Rewriting Eq. 4.27, at the upper limit,

$$N(\text{taps}) = \frac{1}{f_s \times t_{mac}}$$

The comparison shown below is revealing. For a rather demanding digital audio filter with $f_s = 50\,\text{kHz}$,

	$t_{mac}\,(\mu s)$	Max N(taps)	Comment
Hardware mpy ADSP-1010A	85 ns	235	Extra time for other operations significant
Coprocessor – 8087	4 µs	5	Won't handle the filter of Table 4.1
Microprocessor 8086 (5 MHz)	30 µs	< 1	Won't handle any filter at 50 kHz
68000 (12.5 MHz)	5.5 µs	3	Won't handle the filter of Table 4.1

The microprocessor or coprocessor solution is limited to quite low-order filters, inadequate for FIR—except for low-audio-frequency signals—but possible for simple IIR filters at reduced sampling rates (see Section 4.3).

4.2.5 FILTER SIMULATION AND WORD-LENGTH EFFECTS

While the bulk of numerical issues will be left for Chapter 5, one key principle and one example will be given here in connection with FIR filter simulation. The principle deals with how much of the noise present in the signal will get through an FIR filter of ideally high resolution, i.e., one that does not add noise due to computation errors. The example then asks, given noise-free data, how much noise is added by computation with finite-word-length arithmetic?

Noise Propagation Through FIR Filters
What is the noise amplification of an FIR filter? Even though the filter may be trying to eliminate the noise, noise overload in a computation can saturate the filter and make it useless. It is possible to estimate this effect so that noise overload can be eliminated by adaptive gain design.

Consider first the output, $y(n)$, of an N-coefficient filter whose only input is noise, $\epsilon(n)$

$$y(n) = \sum_{m=0}^{N-1} h(m)\,\epsilon(n-m) \tag{4.28}$$

The mean-square noise of the *output* is obtained by squaring both sides of Eq. 4.28 and taking the statistical mean:

$$\overline{y^2} = \sum_{n=0}^{N-1}\sum_{m=0}^{N-1}\sum_{q=0}^{N-1} h(m)h(q)\mathrm{P}[\epsilon(n-m),\epsilon(n-q)]\,\epsilon(n-m)\,\epsilon(n-q) \tag{4.29}$$

where $P(\epsilon)$ is the probability of a noise value, ϵ. Suppose the noise has zero mean, a mean-square value, σ^2, and is uncorrelated:

$$P[\epsilon(m),\epsilon(q)] = P[\epsilon(m)]\,P[\epsilon(q)] \tag{4.30}$$

Then all the sums over m and q with unequal indices, $m \neq q$, vanish; the term with $m = q$ is just equal to the mean-square noise, σ^2, leaving

$$\overline{y^2} = \sigma^2 \sum_{n=0}^{N-1} h^2(n) \tag{4.31}$$

Thus, the sum of the squares of the filter coefficients measures the noise amplification of the ideal infinite-resolution FIR filter. This subject will be developed for practical (i.e., non-ideal) finite-word-length filters in Chapter 5. An illustrative example is given below.

Filter Simulation: Introduction to Finite Wordlength
An example of a filter simulation is shown in Figure 4.13. The filter was designed using the program described above, with the following specifications:

Filter specifications:

passband ripple $<0.5\%$
stopband gain $< -70\text{dB}$

$$f(\text{pass}) = 0.2\frac{f_s}{2}$$

$$f(\text{stop}) = 0.3\frac{f_s}{2}$$

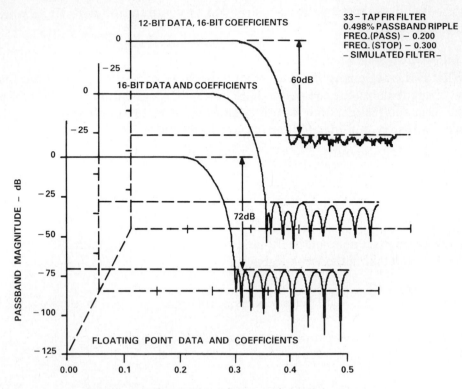

Figure 4.13. Low-pass filter: 33-tap FIR transfer function. A floating-point calculation is compared with simulations using 16-bit coefficients and 12- and 16-bit data.
Source: Analog Devices

The effects of finite word length arithmetic can be estimated by carrying out the filter calculation with variable word length in both data and coefficients. The filter calculation (the convolution of Eq. 4.8) is carried out with broadband input data, i.e., a cosine wave of unit amplitude for each value f in the filter spectrum. The output amplitude *is* the frequency response, $H(f)$. In this example, floating-point calculation is compared with two others:

Data	Coefficients
floating point	floating point
16 bit	16 bit
12 bit	16 bit

Although small errors can be seen in the 16-bit simulation, they are at the −70-dB level. When the data has only 12-bit resolution, the quality of the zeros deteriorates, but still the stop-band output is below 60 dB. The passband ripple and pass/stop frequency range specifications are still met.

Thus, 12 bit data with 16-bit coefficients is adequate even for a relatively stringent filter.

Exercise: Why does it make sense that the filter specification which deteriorates first is the stopband attenuation? Why is the observed attenuation to 60 dB reasonable for 12-bit data?

Method: The filter is trying to force zeros in the stopband. Just how close to zero can you make a finite wordlength convolution? An approximate zero is similar to approximating an orthogonal integral such as

$$\int_{-\pi}^{+\pi} \cos\theta \sin\theta \, d\theta = 0$$

Let the integral become a sum over the $M = 33$ taps, and let the data be an ideal cosine plus a 1 out of 12 bit (1 part in 4,096) quantization uncertainty. The integral is non-zero by an amount at most

$$\text{Error} = (33 \text{ taps}) \times (1 \text{ in } 2^{-12} \text{ error}) = 2^{-7} = -42 \text{ dB}$$

This is worst case; the actual error of -60 dB in Figure 4.13 includes some random error cancellation (see Exercises).

Later chapters will explore rounding methods in depth: Biased/unbiased, randomized, and truncation vs. rounding. See also Hanning and Peled & Liu.

4.2.6 HARDWARE IMPLEMENTATION OF FIR FILTERS

Hardware concepts can be introduced as shown in Figure 4.14. Compare the memory map with the conceptual scheme of Figure 4.5b. Although the memory map could be of a software-only filter, the hardware organization of Figure 4.14 shows how architecture tailored to a class of filter structures speeds execution.

Two separate memory banks are used. One is dedicated to the coefficients, fixed in value during a calculation. This bank could be ROM for single-purpose filters or RAM for programmable filters. The other memory bank is set aside for the data, with counter A specifying the base address and counter B incrementing the pointer within a moving window. The use of two memory banks halves the number of data fetches and data writes. The data memory acts as a circular buffer. Counter A points to the most recent data (right side of Figure 4.14b (moving window)), while counter B points to successively less-recent samples, moving downwards and (wrapping around) to the oldest data $(N-1)$ points earlier (where N is the number of taps in the filter).

Data accessed at address $(A + B)$ is brought out, together with the coefficient pointed to by counter C. Both are loaded into the multiplier/accumulator (MAC) and the product-sum accumulates as B and C go through the N-point range of a single filter output point. When the count reaches N, counter C is reset, counter A is incremented, a new data point replaces the oldest point in the buffer (note the R/W line and the multiplexing of the data memory write address from counter A), counter B is preset with A's incremented contents. On the next CLK cycle, the process begins again.

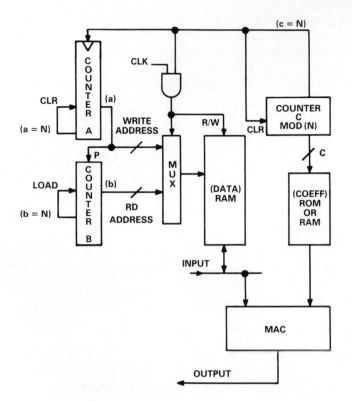

a. Hardware implementation of an FIR filter.

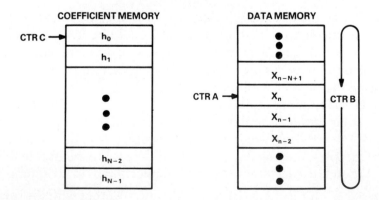

b. FIR filter memory organization for coefficients and data.

Figure 4.14. FIR filter organization.

A speedup of nearly a factor of two results if the FIR filter coefficients are symmetric, as they are in a linear-phase low-pass filter (for example, see Table 4.1). Data which are to be multiplied by equal-valued coefficients can be added before the multiplication. For example, in a 4-point filter, $h(0) = h(3)$, and $h(1) = h(2)$, so

$$y(n) = \sum_{m=0}^{3} h(m)\,x(m-n) \tag{4.32}$$

$$y(17) = h(0)[y(17) + y(14)] + h(1)[y(16) + y(15)]$$

For a 32 tap-filter, assuming a 165-ns MAC and neglecting time for other operations, the direct approach yields a 5.3-μs computation per output point, for a bandwidth of 190 kHz. The pipelined add-then-multiply approach drops this to about 3.7μs or 270 kHz, assuming a 150-ns multiplication followed by an 80-ns addition.

When general-purpose FIR filters are implemented in a microcomputer system with hardware-multiply capability, the two-memory architecture of Figure 4.12 is retained, except that, instead of counters, the micro ALU does the address generation and handles the multiplexing. However, the multiplier's speed can be a problem in the opposite sense. If $N_{taps} < 16$, the DSP filter is too fast for an NMOS micro such as the 8086! Handshaking must slow the throughput to keep up with the necessary data manipulations in software.

4.3 INFINITE IMPULSE-RESPONSE FILTERS

Recursive filters (in analog hardware) are familiar territory to analog filter designers* but less so to readers with mostly digital background. This section provides an introduction that will help to close knowledge gaps for both types. The 2nd-order section (also called the *biquad*), covered in (4.3.1), serves as the IIR prototype, and introduces signal flow graphs for digital filter block diagrams. Special cases of the biquad reviewed in (4.3.2) perform the LP, BP, HP triad and model the behavior of commonly found systems (e.g., formants, in speech). Higher-order filters (4.3.3) may be synthesized from series or parallel cascades of biquads. Hardware implementation and performance will be illustrated in (4.3.4) with a biquad prototype.

Several alternative mappings (4.3.5) permit "old (analog) wine to be put into new (digital) bottles," though by thinking digitally from the beginning, mappings can largely be avoided. Ready to push around poles and to quench errant limit cycles, we briefly review (4.3.6) high-performance analog filters (Butterworth, Chebyshev, and elliptic), then turn (4.3.7) to a black-box CAD

---

*Although we often think of recursiveness as a property of feedback circuits, even the simplest passive one-resistor-one-capacitor filter is characterized by a pair of recursive equations (see Figure 1.17a). It could be modeled by an IIR filter with one step of delay.

approach for high-performance IIR filters synthesized from such 2nd-order sections.

In earlier sections it was sufficient to think of $H(f)$ as the filter function, but IIR filters require the more precise examination of $H(z)$ and the frequency response evaluated on the unit circle, $H(e^{j\omega t_s})$. Figure 4.15 summarizes a road-map of the steps in design and simulation: Enter the digital design either with a block-diagram of the digital circuit or with a difference equation; each leads to the other. Evaluate $H(z)$ by the z-transform (review Section 2.7 if necessary), then plot the frequency response $H(e^{j\omega t_s})$. Simulation and adjustment of the design end the brief tour. Further stages of design sophistication, shown in the dashed box in Figure 4.15, may be attacked as needed, either to translate analog lore to digital realization or to manipulate z-plane poles directly. Although it is helpful to understanding, most DSP users will not need this level of sophistication, once the terminology is clear, since the CAD approach incorporates much of it and allows a cookbook approach.

4.3.1 THE SECOND-ORDER SECTION AS A PROTOTYPE

Here is developed the correspondence between common IIR difference equations, the corresponding block diagram for filter implementation, and key properties as defined by the poles and zeros in the z-plane. The analog designer should feel quite at home but will encounter some new elements, because of the inherent wraparound of the z-plane and the nature of finite word-length arithmetic. The biquad is a general-purpose module from which either more complicated higher-order filters or multiplexed multi-band filters can be constructed.

The Biquad: Direct and Canonical Forms
Write the recursive filter difference equation for $n = 2$. This defines the *biquad*, a quadratic in *both numerator and denominator*.

$$y(n) = \sum_{k=0}^{2} a_k x(n-k) + \sum_{k=1}^{2} b_k y(n-k) \qquad (4.33)$$

$$= a_0 x(n) + a_1 x(n-1) + a_2 x(n-2) + b_1 y(n-1) + b_2 y(n-2)$$

Take the z-transform of both sides

$$Y(z) = \sum_{k=0}^{2} a_k z^{-k} X(z) + \sum_{k=1}^{2} b_k z^{-k} Y(z) \qquad (4.34)$$

$$= (a_0 + a_1 z^{-1} + a_2 z^{-2}) X(z) + (b_1 z^{-1} + b_2 z^{-2}) Y(z)$$

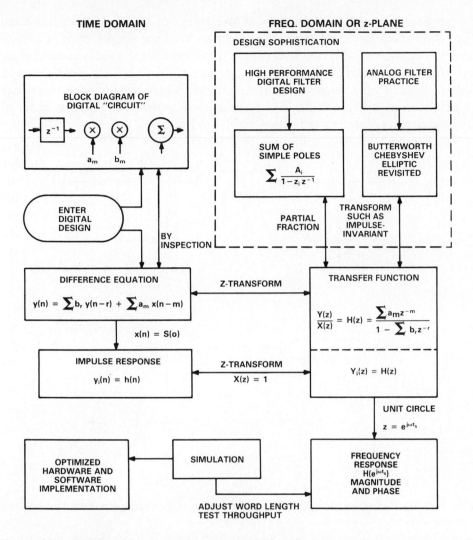

Figure 4.15. Roadmap for digital filter design. The dashed box contains more-sophisticated design tools which can be avoided on first acquaintance.

Both $Y(z)$ and $X(z)$ can be factored out of the sum. Solve for $H(z)$:

$$H(z) = \frac{Y(z)}{X(z)} = \frac{\displaystyle\sum_{k=0}^{2} a_k z^{-k}}{1 - \displaystyle\sum_{k=1}^{2} b_k z^{-k}} \qquad (4.35)$$

$$= \frac{a_0 + a_1 z^{-1} + a_2 z^{-2}}{1 - b_1 z^{-1} - b_2 z^{-2}}$$

By implementing the summation of the terms of the difference equation (Eq. 4.33) as a block diagram, the direct form of the digital biquad is constructed as shown in Figure 4.16. The left half realizes the zeros and the right half realizes the poles (Eq. 4.35). Though direct, this is not the most common biquad form. By breaking the central summation into two parts, the topology can be seen as a direct cascade of two filters in series. Reverse the order of the two circuits in series to get the structure of Figure 4.17.

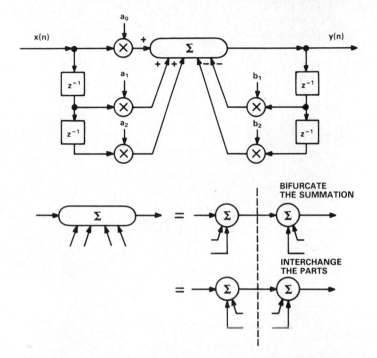

Figure 4.16. Biquad (recursive) filter, direct form and first steps towards deriving the canonical form.

The operations in the center of Figure 4.17a are equivalent and can be coalesced into one. This gives the biquad usually seen, Figure 4.17b, called *canonical* because, in minimizing the number of delays, it has become the commonly *prescribed* imple   entation; a canonical form of an nth order filter has only n delay stages, each written as z^{-1}.

The recursive nature of the calculation is evident. The difference equation for filter calculations can be obtained by inspection of the block diagram, the reverse of the procedure for the *direct-form* diagram. Label the newly created intermediate variable $w(n)$

$$y(n) = a_0 w(n) + a_1 w(n-1) + a_2 w(n-2) \qquad (4.36)$$

$$w(n) = x(n) - b_1 w(n-1) - b_2 w(n-2)$$

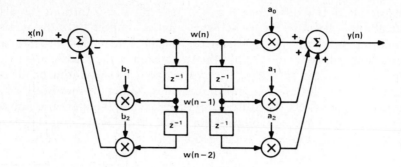

a. Reverse cascade of the filter in Figure 4.16.

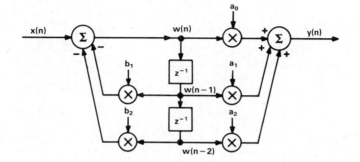

b. Equivalent operations coalesced.

Figure 4.17. Biquad, canonical form.

The design can go from block diagram to difference equation or the reverse. This introduction to the elegant subject of transformations by signal flow graphs will suffice for our needs; see Oppenheim (1975), etc., for more examples.

Exercise: Synthesize an integrator. The integrator was introduced in Chapter 2 as a filter which can *only* be made recursively. Here we will explore digital integrator design further. Roundoff accumulation will play the role that dc drift did in an analog integrator, so the ideal integrator by itself is not a common filter. Integration is favored over differentiation in closed-loop *analog* feedback systems. The dc drift of analog op-amps is generally minimized in closed loop, and also the integrator circuit does not suffer from the high-frequency noise of analog differentiators. Because of this analog origin, integration is often encountered in translating the differential equations of closed-loop control systems into digital form.

The form of the digital integrator circuit depends upon the algorithm used. Consider the trapezoid rule for the integral, $y(n)$, of the previous $x(n)$:

$$y(n) = y(n-1) + \frac{1}{2}[x(n) + x(n-1)] \qquad (4.37)$$

Inspection of this difference equation gives the direct form of the digital integrator—shown in Figure 4.18a; the initial multiply by 1/2 requires no multiplier, since a 1-bit right-shift (RS) will do. Need it take two delay stages for a first-order operation? Split the summation in the middle and reverse the order of the cascade (18b). The two delays are now operating on the same data and can be coalesced to give the canonical (minimum number of delays) form of Figure 4.18b. The difference equation, and therefore the computation sequence, for the canonical form of the recursion relation is obtained, as in the previous example, by inspection of the block diagram (do it).

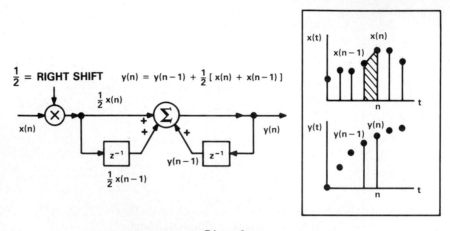

a. Direct form.

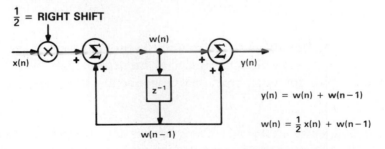

b. Reverse cascade or canonical form.

Figure 4.18. Integrator using trapezoid rule.

Connection Between Filter Coefficients and z-Plane Poles and Zeros

How are the a's and b's of the filter related to the poles and zeros of $H(z)$? For the biquad, $H(z)$ has at most two zeros in the numerator and two poles in the denominator. A polynomial can be factored; when this is done for Eq. 4.35, the biquad can be expressed in the form:

$$H(z) = G \frac{(z - r_0 e^{-j\theta_0})(z - r_0 e^{+j\theta_0})}{(z - r_p e^{-j\theta_p})(z - r_p e^{+j\theta_p})}$$

(4.38)

where G sets the gain and the r's and θ's locate the zeros and poles in the z-plane, as in Figure 4.19. (We pick in this case the subset of biquads where poles and zeros occur in pairs.) It is straightforward to show that the relationship between the a's and b's of $y(n)$ and the r's and θ's of $H(z)$ is as shown in the BOX.

RELATING NORMALIZED BIQUAD FILTER COEFFICIENTS TO PAIRED POLE AND ZERO LOCATIONS

$$a_0 = 1$$
$$a_1 = -2r_0\cos\theta_0$$
$$a_2 = +r_0^2$$

$$\text{(4.39)}$$

$$b_0 = 1$$
$$b_1 = +2r_p\cos\theta_p$$
$$b_2 = -r_p^2$$

SECOND-ORDER FILTERS IN STANDARD FORM

	s-PLANE, H(s)	**z-PLANE, H(z)**
LP	$\dfrac{a_0}{b_2s^2 + b_1s + b_0}$	$\dfrac{Gz^2}{z^2 - (2r_p\cos\theta_p)z + r_p^2}$
BP	$\dfrac{a_1s}{b_2s^2 + b_1s + b_0}$	$\dfrac{Gz(z-1)}{z^2 - (2r_p\cos\theta_p)z + r_p^2}$
HP	$\dfrac{a_2s^2}{b_2s^2 + b_1s + b_0}$	$\dfrac{G(z-1)^2}{z^2 - (2r_p\cos\theta_p)z + r_p^2}$

Practically any manipulation of poles and zeros which was useful in analog circuits has a z-plane digital counterpart. This is especially useful in building high-order multiple-stage filters (Section 4.3.3) from analog filter theory, using the biquad as the basic element. As another example, one can even make an oscillator with proper choice of parameters in a second-order section (see Exercises, also Chapter 8); in practice, though, IIR quantization errors make a lookup table method more common for oscillators.

4.3.2 BIQUADS FOR SPECIAL PURPOSES

Although there are several distinct forms of the biquad, a few are encountered more frequently, because their poles and zeros model especially useful physical systems. An example gives a useful geometrical interpretation. Consider a biquad with a single zero along the positive *real* z-axis and a single pole-pair.[*] The frequency response of any filter is obtained by evaluating $H(z)$ around the unit circle, $z = e^{j\omega t_s}$; hence, in this case,

$$|H(e^{j\omega t_s})| = \frac{R}{P_2 P_1} \tag{4.40}$$

where R, P_2, and P_1 are radii from the frequency in question to the zero and the poles, as shown in Figure 4.19b. For the zero at $z = q$ and the pole-pair at $z = R\,e^{\pm j\omega_0 t_s}$, the general biquad (Eq. 4.38) becomes

$$H(z) = \frac{1 - qz^{-1}}{(1 - re^{-j\theta}z^{-1})(1 - re^{+j\theta}z^{-1})}$$

$$= \frac{1 - qz^{-1}}{1 - 2\,r\cos(\omega_0 t_s)z^{-1} + r^2 z^{-2}}$$

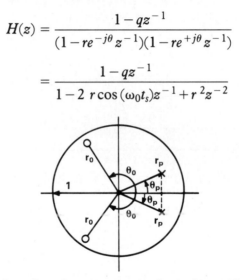

Figure 4.19a. Location of zeros and poles for the biquad filter of Eq. 4.35.

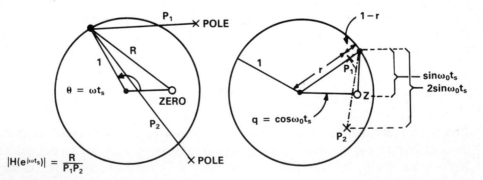

Figure 4.19b. Geometric interpretation of biquad frequency response for general example (left) and constant gain at resonance (right).

[*]These examples are adapted from the early classic, Rader & Gold (1967).

To evaluate the frequency response, let $z = e^{j\omega t}$. A few steps of calculation lead to the magnitude,

$$\left|H\!\left(e^{j\omega t_s}\right)\right| = \sqrt{\frac{1+q^2-2q\cos(\omega t_s)}{(1+r^2-2r\,\cos(\omega-\omega_0)t_s)(1+r^2-2r\,\cos(\omega+\omega_0)t_s)}} \qquad (4.42)$$

A few cases are especially significant:

Case 1. Constant Gain at Resonance The gain at resonance, $\omega = \omega_0$, can be made independent of resonant frequency by selecting $q = \cos(\omega_0 t)$ and letting r approach 1 (high-Q resonator). This choice can make a compact bank of equal-gain resonators, for example the front-end of a voice-recognition circuit. A single hardware array can be multiplexed with a series of natural frequencies, changing the ω_0 parameters in software in real time. This choice of q makes $|H(\omega_0)|$ independent of ω_0, as shown geometrically in Figure 4.19b. At resonance, $\omega = \omega_0$, Eq. 4.42 takes on the form

$$R^2 = 1 - \cos^2(\omega_0 t_s)$$

$$P_2{}^2 = 1 + r^2 - 2r = (1-r)^2$$

$$P_1{}^2 = 1 + r^2 - 2r\cos(2\omega_0 t_s)$$

$$\cong 2[1 - \cos(2\omega_0 t_s)] \qquad (r \cong 1)$$

$$= 4[1 - \cos^2(\omega_0 t_s)] = 4R^2$$

$$|H(e^{j\omega t_s})| = \frac{r}{P_2 P_1} \cong \frac{1}{2(1-r)} \qquad (4.43)$$

As ω crosses ω_0 (black dot in Figure 4.19b), the bandpass amplitude is determined by the distance, $(1-r)$, of the pole from the unit circle; the ratio, R/P_1, of radii to the zero and the other pole is (nearly) independent of the choice of ω_0.

Case 2: Resonator with fewer multiplications. Choose $q = r\cos(\omega_0 t_s)$. Recall how the coefficients of powers of z^{-1} in the numerator and denominator of $H(z)$ (Eq. 4.9) are related to the x and y terms of the filter difference equation (Eq. 4.6); then equation 4.41 for this special case of q corresponds to

$$y(n) = x(n) - r\cos(\omega_0 t_s)\,x(n-1) + 2r\cos(\omega_0 t_s)y(n-1) - r^2 y(n-2)$$

Regrouping terms, the output is computed to be

$$y(n) = r\cos(\omega_0 t_s)[2y(n-1) - x(n-1)] - r^2 y(n-2) + x(n)$$

With two pre-computed coefficients, $x \cos \omega_0 t_s$ and r^2, this biquad requires only two multiplications, rather than 5 (general biquad) or 3 (general case of Eq. 4.41). The reduced number of multiply operations makes for highest possible biquad throughput.

Case 3: True zero forced at dc. Choose $q = 1$. This resonator has absolute dc blocking. Why is there a zero at dc? The dc limit corresponds to $z = 1$ for $z = e^{j\omega t_s}$.

Case 4: Formant vocoder prototype. Choose $q = 0$. This class of resonators has no s-plane zeros and non-zero dc gain, like the spring-mass or vocal-cord resonance. With coefficients chosen to overdamp the resonance, this is also the biquad lowpass of Section 2.7.3 (Figure 2.19). The z^{-1} term in the numerator may be understood by the impulse invariant transformation between z-plane and s-plane (see Section 4.3.5 and Exercises).

4.3.3 HIGHER-ORDER FILTER SYNTHESIS

Higher-order filters are usually synthesized as a series or parallel cascade of biquad modules. High-performance filters (Butterworth, Chebyshev, elliptic, Bessel...) are constructed from second-order sections with coefficients that differ according to a prescription based on type of filter and number of sections. This approach has an advantage which tends to minimize accumulated error: by grouping poles in pairs, a given recursive calculation is limited to a second-order section, which can be well-behaved. Also, errors in the *ratio* of two polynomials of comparable order tend to cancel, as long as a pole is not approached too closely.

Series Cascade

This is the familiar analog method; for example, a sequence of three second-order sections form a 6th-order filter, as in Figure 4.20. For the series circuit, the polynomial is factored in both numerator and denominator to bring out factors which combine into second-order sections.

$$H(z) = \frac{\displaystyle\sum_{k=0}^{M} a_k z^{-k}}{\displaystyle\sum_{r=0}^{N} b_r z^{-r}} = B_1(2) \times B_2(2) \times B_3(2) \times \dots \quad (4.44)$$

where $B_i(2)$ stands for a polynomial of at most second order in numerator and denominator. The factoring is done by seeking terms of the form $(z - z_0)$ in numerator and $(z - z_p)$ in the denominator.

Parallel Summation

Rather than a series cascade, break the high performance filter up into a set of second-order sections whose outputs can be *summed* in parallel, as in Figure 4.20b. Although less familiar in an analog counterpart, digital accumulation of parallel sections poses no problem. The error accumulation is equivalent in series and parallel cases (see Exercises).

To build the parallel form, expand $H(z)$ of Eq. 4.35 as a partial fraction* and group the poles in pairs. The result has the form:

$$H(z) = \sum_{r=1}^{\frac{N}{2}} \frac{a_{0r}' + a_{1r}'z^{-1}}{1 - b_{1r}'z^{-1} - b_{2r}'z^{-2}} + \sum_{i=0}^{M-N} c_i'z^{-i} \qquad (4.45)$$

where a_r', b_r', and c_i' are expressed in terms of the original a,b,c but are not the same. The second sum is necessary only if $M > N$ in Eq. 4.44.

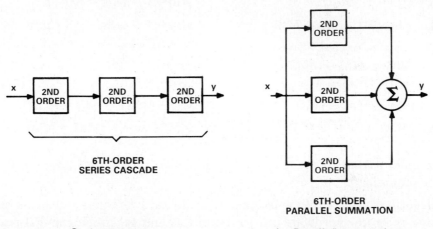

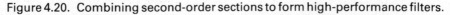

a. Series cascade.   b. Parallel summation.

Figure 4.20. Combining second-order sections to form high-performance filters.

4.3.4 HARDWARE IMPLEMENTATION OF THE BIQUAD

The biquad lends itself to multiplexing so that *one* set of high-speed hardware can perform in such applications as a high-order filter, a multi-band equalizer, or a tracking multi-harmonic bandpass (frequency comb) filter. For example, memory is assigned to the various sets of coefficients and intermediate data to be fed to a multiplier-accumulator.

*With partial fractions, an arbitrary ratio of two polynomials (like Eq. 4.9 for $M < N$) can be expanded as a series of first-order poles in the form:

$$H(z) = \sum \frac{A_k}{(z_p^{-1} - z^{-1})} \qquad (4.46)$$

where A_k is evaluated as the *residue* of the pole, z_p^{-1}, i.e., the value of $(z_p^{-1} - z^{-1}) H(z)$ as z approaches z_p. These first-order terms are grouped in twos and combined to make the second-order sections of Eq. 4.45.

Consider a single-stage, the biquad computation for the canonical form of Figure 4.17. The filter calculation for each (Nth) biquad is written by inspection:

$$w(n) = x(n) - b_1 w(n-1) - b_2 w(n-2)$$

$$y(n) = w(n) + a_1 w(n-1) + a_2 w(n-2)$$

(4.47)

The canonical form has an intermediate data stage which, in computation, corresponds to temporary storage locations $w(n)$, $w(n-1)$, and $w(n-2)$. Here is the series of computation steps which must be carried out (* = multiplication):

	OPERATION	ACCUMULATOR CONTENTS
(1)	w(N − 1)*(− b1) → TEMP AND → ACC	w(N − 1)*(− b1)
(2)	x(n)*c1 → ACC	x(n)*c1 + w(N − 1)*(− b1)
(3)	w(n − 2)*(− b2) → ACC	x(n)*c1 + w(N − 1)*(− b1) + x(n)*c1
(4)	ACC = w(n) → MEM w(n − 1)*a1 → ACC	w(n − 1)*a1
(5)	w(n − 2)*a2 → ACC	w(n − 1)*a1 + w(N − 2)*a2
(6)	MEM = w(n) → ACC	w(n) + w(n − 1)*a1 + w(n − 2)*a2

The quantity left in the accumulator after the last step is just $y(N)$, by the above equations.

This form of program flow assumes a certain hardware pipelining capability: in one cycle, the hardware can both multiply and accumulate and also make a memory write or read. With this assumed architecture, the 2nd-order section takes only 6 cycles. Assuming a cycle time of about 200 ns, the single biquad computation takes $1.2\,\mu s$, for an impressive data throughput rate of 800 kHz.

A multiplexed 6th-order filter from three 2nd order sections is shown in Figure 4.21. Three direct-form biquads are chained together, with equivalent delay stages coalesced analogously to Figure 4.17. The intermediate storage locations, $w(n)$, and hence the computation sequence, become identical with that of the *canonical* form just discussed. The memory allocation for the above steps requires 5 coefficient locations per section, and 4 data locations: one for $w(n) = $ TEMP, one each for $w(n-1)$ and $w(n-2)$, and one for a stage output which is also the next stage input. Only a single MAC is required for the 3-stage filter shown. In operation, the common hardware sequentially loads in a new set of coefficients and then runs through the above sequence for each filter section.

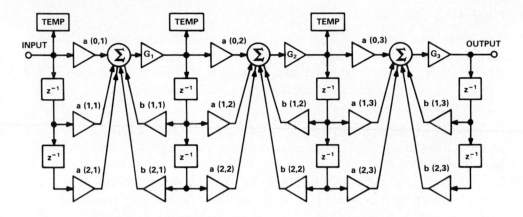

SECTION LOCATION	RAM			ROM	
	SECTION 1	SECTION 2	3	LOCATION	VALUE
1	X(1,1)			1	a (0,1)
2	X(2,1)			2	a (1,1)
3	Y(1,1)	X(1,2)		3	a (2,1)
4	Y(2,1)	X(2,2)		4	b (1,1)
5		Y(1,2)	X(1,3)	5	b (2,1)
6		Y(2,2)	X(2,3)	6	a (0,2)
7			Y(1,3)	7	a (1,2)
8			Y(2,3)	8	a (2,2)
9	TEMP	TEMP	TEMP	9	b (1,2)
				10	b (2,2)
				11	a (0,3)
				12	a (1,3)
				13	a (2,3)
				14	b (1,3)
				15	b (2,3)

a. A 6th-order IIR filter is constructed as a series of 3 second-order sections. A memory map illustrates that, to do so, the filter requires only nine RAM registers and 15 ROM locations.

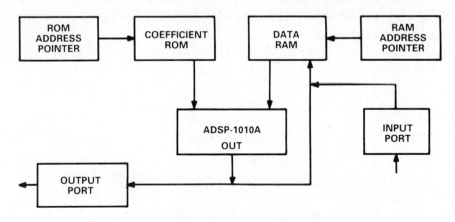

b. Hardware block diagram.

Figure 4.21. Biquad digital hardware implementation.

Comparison with Microprocessor Software IIR Filter

The filter just described illustrates a hardware-intensive solution: though very fast, it is single-purpose and potentially complex to debug. As an alternative strategy, a microcomputer can help with the filter calculation. This imparts flexibility, since the computer can readily load coefficients and alter configurations.

As another hardware-software comparison, here is how a *software-only* implementation of the 2nd order section performs. Nagle and Nelson (1981) have given a complete exposition of digital filter calculations on 16-bit microprocessors, including program listings for specific microprocessors in common use. For a fourth-order filter, their results for the 8086 μP were:

> Performance, microprocessor software IIR filter:
> execution time: 610 μs
> maximum sampling rate: 1,640 Hz
> lag between input and output: 207 μs.

This is dominated by the 8086's 30.6-μs multiply time in software (4-MHz clock rate). If the computation time were solely given by $N_{taps} \times t_{mac}$, the execution would have been only $8 \times 30 = 240 \mu s$, rather than 610 μs, so the software approach of even an optimized program is bottlenecked by architecture. The comparison is revealing: the same 4th order filter would take (two 2nd-order sections) $\times$ (1.2 μs per section) $= 2.4 \mu s$ in *hardware*, more than 100 times faster. The best of both worlds is a marriage of hardware speed and microprocessor flexibility: the *DSP microprocessor* (Chapter 6).

4.3.5 MAPPING ANALOG DESIGN TO DIGITAL: IMPULSE-INVARIANT AND OTHER COMMON TRANSFORMATIONS
The extensive collection of analog filter lore can be implemented digitally by mapping between the *s*-plane of analog design and the *z*-plane of digital filters. In their favor, mappings
- show the route between an analog circuit diagram and a set of digital filter coefficients,
- introduce via classic analog designs key performance choices one has to make in CAD design specification, and they
- relate the (unfamiliar to most) *z*-plane to the familiar *s*-plane view of filters, showing how a specified analog $H_a(s)$ is connected to a digital $H(e^{j\omega t_s})$ or coefficient set, $h(n)$.

However, an extensive discussion of mappings is unlikely to be helpful to today's designers, since:
- many readers find conformal mapping confusing
- several mappings that are of quite different form are in common use
- the form of the *digital H(z)* obtained may bear little resemblance to the analog $H_a(s)$, and—most important—

- direct digital design using computer-aided-design (CAD) tools works very well (thus, an intermediate analog design is not really necessary).

For these reasons, we limit the discussion in the following section to two common mappings:

- The *impulse-invariant transformation*, which synthesizes the digital filter whose impulse response, $h(n)$, matches that of the analog filter, $h_a(t)$, at sampled points, $t = nt_s$, and
- *the bilinear transformation*, which synthesizes the digital filter whose performance mimics that of the analog system's differential equation, integrated at finite sampling intervals.

Section 4.3.6 then reviews analog classics often implemented by the tools of this section, and section 4.3.7 introduces CAD design tools based on best-fits to desired performance, subject to specified limit criteria.

Impulse invariance requires that the digital filter's impulse response, $h(n)$, be equal to that of an analog filter at discretely sampled points in time:

$$h(nt_s) = h_a(t) \text{ for } t = 0, \ t_s, \ 2t_s, \ 3t_s, \ldots \qquad (4.48)$$

Here, $h_a(t)$ is the inverse Laplace transform of the analog filter transfer function, $H_a(s)$. The general procedure is best illustrated by an example, summarized in the box below.

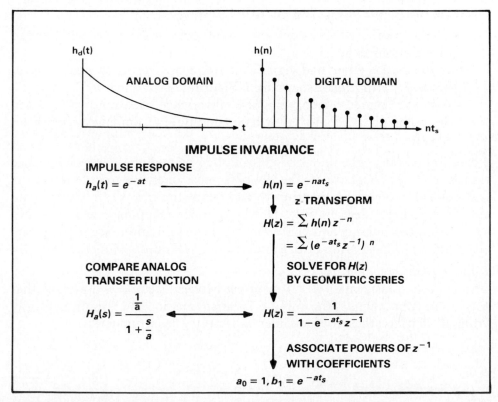

To generalize this example to a design procedure based on any analog filter, the analog $H_a(s)$ is reduced to a partial-fraction expansion.

$$H_a(s) = \sum_{i=1}^{M} \frac{A_i}{s + s_i} \tag{4.49}$$

By analogy with the above first-order example, the requirement that the impulse responses of the two designs match at sampled points forces $H(z)$ to have the form:

$$H(z) = \sum_{i=1}^{M} \frac{A_i}{1 - e^{-s_i t_s} z^{-1}} \tag{4.50}$$

The strengths and limitations of the impulse-invariant procedure are related to the domains thereby mapped from s-plane to z-plane, as summarized in Figure 4.22a. The unit circle in the z-plane maps a *strip* in the s-plane bounded by $\pm f_s/2$. The strong points of the procedure are:

- The frequency response is mapped from analog to digital domains without distortion, since the frequency along the imaginary s-axis is linearly mapped to the circumference of the circle in the z-plane.
- A stable analog design will map to a stable digital design, since the left-half s-plane maps to the interior of the z-plane unit circle.

However, there are some limitations to be aware of:

- The gain of $H(e^{j\omega t_s})$, as designed by impulse invariance, will depend on the sampling rate.
- If the analog filter had significant frequency response extending to beyond $f_s/2$, the digital mapping of Figure 4.22a will alias it.
- For the same reason, impulse invariance doesn't work when designing digital high-pass or band-stop filters from analog high-pass filters.

These limitations, once understood, can often be tolerated, especially since transformations exist to relate the low-pass (most amenable to impulse-invariant design) to high-pass, band-pass, and band-stop filters (see Oppenheim and Shafer, 1975, Section 5.2). However, other mappings, which do not suffer from this aliasing and consequent leakage limitation, are more generally useful. The most common of these is the *bilinear transformation*.

Bilinear Transformation
The bilinear transformation is a mapping between s-plane and z-plane that gets around limitations of the impulse-invariant method and is more-naturally "digital" in its treatment of fundamental sampling consequences:

$$s = \frac{2}{t_s} \frac{z-1}{z+1} \tag{4.51}$$

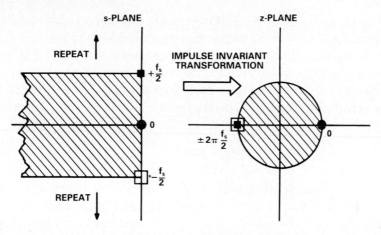

a. Mapping of domains in the s-plane to the z-plane by the impulse-invariant method.

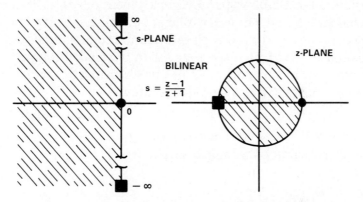

b. Mappings of domains in the bilinear transformation.

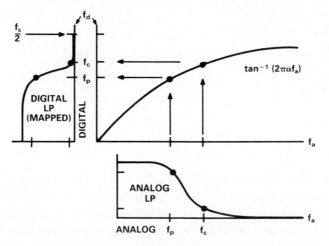

c. Frequency warping from digital to analog frequency domain with the bilinear transformation.

Figure 4.22. Impulse-invariant and bilinear-transformation mappings.

As shown in Figure 4.22b, the *bilinear* transformation takes the *entire* left half of the s-plane (rather than just a horizontal strip) onto the interior of the unit circle in the z-plane. The advantage is a one-to-one mapping. The bilinear is often chosen for mapping analog LP filters into digital form, since the analog filter performance does not deteriorate from aliasing. However, the relationship between analog frequency and its digital equivalent is no longer linear:

$$2\pi f_a = \tan\left(\frac{2\pi f_d t_s}{2}\right) \tag{4.52}$$

This is not a serious disadvantage, since desired filter behavior can be "prewarped", as in Figure 4.22c. A related bilinear with numerator and denominator reversed takes an analog low-pass design into a digital high-pass, with similar analog-digital warping (see Exercises).

In addition to the useful properties already mentioned, the bilinear transformation has an intuitive physical explanation, which is proved in the standard DSP-theory texts. The bilinear transformation results if the system differential equation is solved by integration (as in the analog computer simulation of a system), and then the integrals are approximated by recursive differences to arrive at a "digital" (i.e., a sampled) solution.

Example: Translate the first-order analog low-pass into a digital filter via the bilinear transformation, Eq. 4.51.

Solution: By straightforward substitution,

$$H(z) = \frac{2}{t_s}\left(1 + \frac{at_s}{2}\right)\left(\frac{1 + z^{-1}}{1 - \dfrac{1 - a\dfrac{t_s}{2}}{1 + a\dfrac{t_s}{2}}}\right) \tag{4.53}$$

The filter coefficients are obtained from powers of z^{-1} in numerator and denominator:

$$a_0 = 1, \, a_1 = 1, \, b_1 = \frac{1 - a\dfrac{t_s}{2}}{1 + a\dfrac{t_s}{2}} \tag{4.54}$$

Interpretation: The bilinear transformation introduces a zero not present in the analog first-order low-pass. This is common; while poles map one-for-one, terms in the numerator tend to differ from their analog counterparts. The pole is located within the unit circle, since $b_1 < 1$ for positive decay-constant, a.

This concludes our excursion into mappings. The analog devotee might wish to translate standard analog biquads in the s-plane into digital equivalents by the two transformations introduced here (see Exercises).

4.3.6 REVIEW OF HIGH-PERFORMANCE ANALOG FILTERS

Although it is not our intention to delve into the lore of analog filter design, it is useful to summarize some terminology of high-performance filters. There are several common analog prototypes often synthesized digitally. Their names and key characteristics are summarized in the BOX, for a low-pass prototype. Given below are the *analog* forms, $H(s)$, for $s = j\omega$, along the imaginary axis of the s-plane. The transformation to digital filters can be accomplished by a variety of mappings, with the bilinear being the most common.

REVIEW OF COMMON ANALOG FILTERS
$\quad H(j\omega) = H(s)$, for $s = j\omega$

BUTTERWORTH
Designator: Max-flat
Frequency response: Nth order polynomial

$$|H(j\omega)|^2 = \frac{1}{1 + \left(\dfrac{\omega}{\omega_c}\right)^{2N}} \tag{4.55}$$

Features: keeps frequency response flat until nearly $\omega = \omega_c$. Monotonic passband and monotonically decreasing stopband.
$\quad$ Poles: poles of $|H(s)|^2$ equally spaced around the unit circle in the s-plane.
$\quad$ Zeros: none in the s-plane, except at ∞.

CHEBYSHEV
Designators: Equiripple Passband (type 1), Equiripple Stopband (type 2)
Frequency response (type 1):

$$|H(j\omega)|^2 = \frac{1}{1 + \epsilon^2 C_N^2\left(\dfrac{\omega}{\omega_c}\right)} \tag{4.56}$$

where C_N are Chebyshev polynomials.
Features: minimizes the maximum error with fewer taps than Butterworth.

	Passband	Stopband
Type 1	Equiripple	Monotonically decreasing
Type 2	Monotonic	Equiripple

Poles: on an ellipse in the s-plane
Zeros: none in the s-plane, except at ∞ (type 1);
$\quad$ type 2 is smooth in the passband and has zeros in the stopband.

(Continued on next page)

ELLIPTIC (CAUER)
Designator: Nearly equal-ripple
Frequency response:

$$|H(j\omega)|^2 = \frac{1}{1 + \epsilon^2 E_N{}^2\left(\dfrac{\omega}{\omega_c}\right)} \qquad (4.57)$$

where E_N are elliptic functions.

Features: spreads the error uniformly through both pass and stop bands. Elliptic filters have both poles and zeros.

Elliptic filters are optimal in the sense of yielding the sharpest cutoff between passband and stopband for a given order N. The errors between ideal and real filter are distributed evenly across passband and stopband. By contrast, the *Butterworth*, in meeting a desired transition bandwidth, overperforms in smoothness low in the passband, since $(1 = x^n)$ is increasingly (and perhaps unnecessarily) close to 1 as x gets smaller. The *Chebyshev* trades away this un- necessary smoothness, replacing it with equiripple in the passband (type 1) and a monotonically decreasing error in the stopband. The elliptic filter takes the final step; it is equiripple in both pass and stopbands. As a result, the order, N, needed for comparable performance is significantly reduced. For example, a design with specified passband smoothness, transition bandwidth, and stopband attenuation might be met by the following three alternatives:

Butterworth	6th order
Chebyshev	4th order
Elliptic	3rd order

The elliptic filter, requiring half as many computations as the Butterworth, will therefore have a factor-of-2 speed advantage, for these criteria. While elliptic functions are not something one memorizes, they are in the hand- books, and—better yet—in the common CAD programs.

Analog filters, such as the Butterworth, can be designed with nearly linear phase response in the passband, improving as the order is increased. This property is largely preserved in the digital transformation; but for truly linear phase response, the FIR type is used instead of an IIR filter.

4.3.7 CAD APPROACH TO HIGH-PERFORMANCE IIR FILTERS

As with FIR filters, a tailored-to-fit approach is possible for IIR. A readily available program, due to K. Steiglitz, which adjusts the parameters of a series cascade of second-order sections to fit specified frequency response character- istics, can be found in Peled and Liu (1976) or the IEEE digital filter collection (1979). The optimization minimizes the mean-square error between ideal and actual transfer functions, the same goal as in the FIR version of Eq. 4.25, but with biquad sections for $H(z)$ replacing the Fourier series.

The optimization is by the Fletcher-Powell algorithm. The required number of biquad stages and the coefficients for each stage emerge as outputs. An example of an input-output dialogue is given in Table 4.2.

The reader will find an excellent discussion in the original Steiglitz (1970) paper; it is also discussed in the standard digital filter texts (see Oppenheim and Shafer or Peled and Liu). The strategy observes constraints for stability and minimum phase. Since the program uses the canonic form, cascades are easy, and the resulting filters can be realized accurately and simply. Steiglitz demonstrates as design examples a low-pass, a wide-band differentiator, a linear discriminator, and vowel formant filters. The strategy pursued gives minimum-phase designs which are not far from linear over the passband,

Table 4.2 Input/output dialogue for an IIR filter design. (From Peled and Liu, ©1976 Wiley. Reprinted with permission).

```
                INFINITE IMPULSE RESPONSE (IIR)
                FLETCHER-POWELL OPTIMIZATION ALGORITHM
                RECURSIVE FILTER OF ORDER  12
     BAND EDGES ARE  0.0        0.10      DESIRED VALUE = 1.00
     BAND EDGES ARE  0.20       0.30      DESIRED VALUE = 0.0
     BAND EDGES ARE  0.40       0.60      DESIRED VALUE = 1.00
     BAND EDGES ARE  0.80       1.00      DESIRED VALUE = 0.0
     FINAL ERROR FUNCTION VALUE = 0.14498175D+01
     MAXIMUM ABSOLUTE DEVIATION IN PASSBAND =    2.49276 DB

     MINIMUM STOPBAND ATTENUATION = -12.05523 DB

     CONSTANT MULTIPLIER C*= 0.25244430D+00
     COEFFICIENTS FOR CASCADE DECOMPOSITION

     A(1,1) =  0.53264255D+00    A(2,1) = -0.46317444D+00
     B(1,1) = -0.44271071D+00    B(2,1) =  0.59043832D+00
     A(1,2) =  0.53264255D+00    A(2,2) = -0.46317444D+00
     B(1,2) = -0.25154394D+00    B(2,2) = -0.22687963D+00
     A(1,3) =  0.53264255D+00    A(2,3) = -0.46317444D+00
     B(1,3) = -0.25154393D+00    B(2,3) = -0.22687963D+00
     A(1,4) =  0.53264255D+00    A(2,4) = -0.46317444D+00
     B(1,4) = -0.25154393D+00    B(2,4) = -0.22687963D+00
     A(1,5) = -0.11724331D+01    A(2,5) =  0.66611937D+00
     B(1,5) = -0.61800777D+00    B(2,5) =  0.21600159D+00
     A(1,6) = -0.10132787D+01    A(2,6) =  0.72480545D+00
     B(1,6) =  0.48195038D+00    B(2,6) = -0.38382065D+00
     ROOTS
          REAL          IMAG          REAL          IMAG
       0.46450118D+00                -0.99714373D+00
       0.22135535D+00 0.73582615D+00  0.22135535D+00-0.73582615D+00
       0.46450118D+00                -0.99714373D+00
       0.61841607D+00                -0.36687214D+00
       0.46450118D+00                -0.99714373D+00
       0.61841607D+00                -0.36687214D+00
       0.46450118D+00                -0.99714373D+00
       0.61841607D+00                -0.36687214D+00
       0.58621654D+00 0.56786401D+00  0.58621654D+00-0.56786401D+00
       0.30900389D+00 0.34715730D+00  0.30900389D+00-0.34715730D+00
       0.50663934D+00 0.68419443D+00  0.50663934D+00-0.68419443D+00
       0.42377265D+00                -0.90572303D+00
```

using all-pass delay functions as needed to compensate for the changes in phase shift when passing near a pole. This minimum-phase trick recognizes that the magnitude of $H(e^{j\omega t_s})$ is unchanged when $H(z)$ is multiplied by

$$\frac{1 - az}{z - a} = \frac{z^{-1} - a}{1 - az^{-1}} \qquad \text{All-pass delay} \qquad (4.58)$$

which has constant magnitude, a, when z is on the unit circle. This is the digital *all-pass delay*, a familiar analog phase shifter (see Exercises).

EXERCISES FOR CHAPTER 4

Exercise 4.1 Calculate the frequency response of the 1st-order sample and hold, Eq. 4.3. If you translate $g(t)$ to be symmetric, the origin of the linear phase shift will become clear.

Exercise 4.2 Use the convolution theorem to show why $h(n)$ for the high-pass looks as shown in Figure 4.8b.

Exercise 4.3 Show that the transformation, $f \rightarrow (f_s/2 - f)$, applied to the low-pass Fourier series, multiplies each term in the series by $(-1)^k$, making for the rapid oscillation in $h(n)$ shown in Figure 4.8b.

Exercise 4.4 Explore the connection between time for step response ringing to decay and limits, N, of $h(n)$. Develop a qualitative connection with transition bandwidth.

Method: Transition bandwidth is set by the highest "significant" Fourier component in $H(f)$, whose frequency in turn sets the time for "significant" ringing in time to die out. This is only a qualitative connection since a convolution is required to do it right.

Exercise 4.5 Explore the phase-cancellation consequences of the stagger-tuned bandpass filter discussed in Section 4.2.2. Show that the width of the central peak of $h(n)$ when the contributions are π out of phase is just

$$\Delta t = \frac{1}{\Delta f}$$

where Δf is the bandwidth.

Exercise 4.6 Obtain the Fourier coefficients, b_n, for the ideal differentiator, Eq. 4.20. First, do it by brute force, from the integrals which define the Fourier coefficients. Then, show how the first and second terms come from writing the repeating ramp as the sum of two functions.

Method for (b): Draw the repeating pattern of the ramp, redraw that as the sum of a sloping line and a stairstep. Take the derivative of each to get a constant plus a row of unit impulses. Fourier-transform each of these. The Fourier transform of the ramp is then obtained by going the other way, dividing by n (equivalent to the integral, the inverse of the derivative).

Exercise 4.7 The "linear phase" argument is developed in the text only for low-pass filters, whose phase acts as if the signal had been passed through a simple N-tap delay line. Does linear phase apply to other FIR filters? Try a. Differentiator, first ideal, then windowed. b. High-pass. c. Band-pass. Hints: a. recall the factor $+j$ in the ideal differentiator is a phase *lead*. How can anything which is being delayed lead? b. Try the consequences of the transformation $f \rightarrow (f_s/2 - f)$ on the linear phase argument. c. Look in the time domain at the convolution of a signal frequency just below, and then just above, f_0.

Exercise 4.8 Why is there a second term in the Fourier-expansion filter coefficients of the differentiator, Eq. 4.20? Isn't the Fourier series just $\sin(k\omega_s/2)/k^2$, falling off as $1/k^2$, in the same way as for a triangular wave?

Method: The triangular wave (terms in $1/k^2$) is the derivative of the square wave (terms in $1/k$). The ramp wave is the derivative of the sum of what two functions? Fourier-expand the two functions and then integrate once to get the expansion of the ramp. Do not be afraid to Fourier-expand a row of delta functions.

Exercise 4.9 Why are there at most $M-1$ local maxima and minima in the $[0,\pi]$ interval of $H(e^{j\theta})$, expressed as an M-term Fourier series, as in Eq. 4.15?

Method: Sketch the behavior of the M-term Fourier series. There are M cycles of the highest Fourier component of $H(e^{j\omega t_s})$ in the 2π interval.

Exercise 4.10 The estimate of roundoff error in filter simulation in Section 4.2.5 gave a worst-case stopband ripple of -42 dB, while the actual error from the ideal was -60 dB. Show that if the errors do not add up uniformly but randomly, the error estimate is much closer.

Method: Random errors add up as $\sqrt{N}$ rather than as N. Compute the ratio $\sqrt{N}/N$ for $N = 33$ taps. Converted to dB, this is the difference between the worst-case estimate and the actual random error. Ans: $\sqrt{(1/33)}$ gives a 15 dB difference; 42 (worst case estimate) + 15 = 57, reasonably close to 60 (observed).

Exercise 4.11 In the chapter introduction, the first-order sample-and-hold was introduced as a reconstruction filter whose performance can be improved upon. Suggest how to do it.

Methods: For example, eliminate the high-frequency terms by analog low-pass post-filtering, and put back what is lost in the pass-band by peaking up the digital filter's amplitude response where needed.

Exercise 4.12 Examine the statement in Section 4.3.1 comparing integrators and differentiators in the digital context. Does roundoff accumulation cancel in a closed-loop feedback system with digital integrators in the same sense that it does for the closed-loop analog counterpart? Also, is there any reason to avoid differentiators in digital simulation and control systems, as one avoided analog differentiators? Hint: this question is intended to help those with a strong analog background think digitally. Those without closed-loop feedback background are advised to skip it.

Exercise 4.13 Write the canonical form of the difference equation for the digital integrator calculation by inspection of the block diagram, Figure 4.18b.

Exercise 4.14 The second order section may remind those familiar with op amp circuits of the state-variable analog filter, which, with proper choice of coefficients, can make an oscillator. Try it for digital systems. By analogy with what it takes in the pole-zero plot to make an oscillator, extract a second-order filter and draw the digital circuit. Compare it to the analog state-variable filter. Hint: Only for the analog-experienced.

Exercise 4.15 What is wrong with the following statement? Error or quantization noise is added to a signal each step of a P-term series cascade, so the last stage filters an input which has P-times the error. With a parallel arrangement, each stage sees a "fresh" signal, so the error is P times less for the parallel form.

Method: consider the accumulated error of the parallel cascade. Recall a theorem in network theory on how parallel and series networks are related.

Exercise 4.16 Why are there no z^{-2} terms in the numerators of the partial fraction expansion, Eq. 4.45?

Method: The partial fraction expansion gives terms of the form $A/(z-z_p)$. When two of these are grouped to form a 2nd order denominator, what does the numerator look like?

Exercise 4.17 In Chapter 1 the digital filter was introduced heuristically as the substitution of a difference equation for an analog system differential equation. Show that this is not the impulse-invariant transformation $z-1 = e^{st_s}$ but rather a different mapping: $s = (1-z^{-1})/t_s$

Those already tired of mappings may take comfort that this one corresponds to expanding e^{-st} in a power series to 1st order, and thus corresponds to an *approximation* of the impulse invariant mapping in the limit of small sampling times.

Method: As in Section 4.2, take a look at a system differential equation, approximate a derivative by

$$\frac{dy}{dt} = \frac{y(n) - y(n-1)}{t_s}$$

take the z-transform of the result, solve for $H(z)$, and compare with the analog form of $H(s)$ involving the same coefficients of the original differential equation.

Exercise 4.18 Design a digital audio multi-band equalizer. Given the single-stage throughput listed above, how many 6th-order sections could be multiplexed with one set of hardware and maintain adequate bandwidth for hi-fi audio (15 kHz)? Answer: (1 MHz/3)/(2 × 15 kHz) = 10 sixth-order sections.

Exercise 4.19 Write the program steps for the *direct*-form 2nd-order section calculation of Eq. 4.33. Show memory organization for a multi-stage cascade of N biquads. Show that data memory required is $3(N + 1)$ for an N-stage filter. Discuss address generation. Procedure: follow the procedure used in the text for the *canonical* form. With the direct form, the three output data elements of one section become the three input data for the next section.

Exercise 4.20 Verify the translation between the s-plane analog biquads of the Box associated with Eq. 4.39 and their z-plane equivalents.

Exercise 4.21 For the bilinear transformation of Eq. 4.51, but with numerator and denominator reversed, why is it that an analog lowpass design transforms into a digital highpass? Hint: Sketch where on the unit circle in the z-plane the points that can be found that correspond to $s = 0$ and $s = \infty$.

Exercise 4.22 The all-pass filter, a useful analog phase-adjuster, plays the
same role in its digital equivalent:

$$H(z) = \frac{z^{-1} - a}{1 - az^{-1}} \quad \text{All-pass delay}$$

a. Locate the pole and zero in the z-plane. Answer: both at same angle, one
inside and one outside the unit circle.
b. Translate $H(z)$ back to the s-plane and evaluate the transfer function there.
Answer: The result should be equivalent to the analog circuit of Figure 2.26.
c. Translate $H(z)$ back to $h(n)$ and construct the digital filter block diagram.
Answer: Figure 2.26.

APPENDIX TO CHAPTER 4

*A Computer Program for FIR Filter Design**

This appendix contains the instruction comments of a computer program for
FIR linear phase filter design. The program with only minor modifications is
after McClellan et al.

```
C     PROGRAM FOR THE DESIGN OF LINEAR PHASE FINITE IMPULSE
C     RESPONSE (FIR) FILTERS USING THE REMEZ EXCHANGE ALGORITHM
C     JIM MCCLELLAN, RICE UNIVERSITY, APRIL 13, 1973
C     THREE TYPES OF FILTERS ARE INCLUDED--BANDPASS FILTERS
C     DIFFERENTIATORS, AND HILBERT TRANSFORM FILTERS
C
C     THE INPUT DATA CONSISTS OF 5 CARDS
C
C     CARD 1--FILTER LENGTH, TYPE OF FILTER.  1-MULTIPLE
C     PASSBAND/STOPBAND, 2-DIFFERENTIATOR, 3-HILBERT TRANSFORM
C     FILTER.  NUMBER OF BANDS, CARD PUNCH DESIRED, AND GRID
C     DENSITY.
C
C     CARD 2--BANDEDGES, LOWER AND UPPER EDGES FOR EACH BAND
C     WITH A MAXIMUM OF 10 BANDS.
C
C     CARD 3--DESIRED FUNCTION (OR DESIRED SLOPE IF A
C     DIFFERENTIATOR) FOR EACH BAND
C
C     CARD 4--WEIGHT FUNCTION IN EACH BAND.  FOR A
C     DIFFERENTIATOR, THE WEIGHT FUNCTION IS INVERSELY
C     PROPORTIONAL TO F.
C
C     CARD 5--RIPPLE AND ATTENUATION IN PASSBAND AND STOPBAND.
C     THIS CARD IS OPTIONAL AND CAN BE USED TO SPECIFY LOWPASS
C     FILTERS DIRECTLY IN TERMS OF PASSBAND RIPPLE AND STOPBAND
C     ATTENUATION IN DB. THE FILTER LENGTH IS DETERMINED FROM
C     THE APPROXIMATION RELATIONSHIPS GIVEN IN :
C     L. R. RABINER, APPROXIMATE DESIGN RELATIONSHIPS FOR
C     LOWPASS DIGITAL FILTERS, IEEE TRANS. ON AUDIO AND
C     ELECTROACUSTICS, VOL. AU-21, NO. 5, OCTOBER 73.
```

```
C      **** WHEN THIS OPTION IS USED THE FILTER LENGTH ON
C      CARD 1 SHOULD BE SET TO 0. ******
C
C      THE FOLLOWING INPUT DATA SPECIFIES A LENGTH 32 BANDPASS
C      FILTER WITH STOPBANDS 0 TO 0.1 AND 0.425 TO 0.5, AND
C      PASSBAND FROM 0.2 TO 0.35 WITH WEIGHTING OF 10 IN THE
C      STOPBANDS AND 1 IN THE PASSBAND.  THE IMPULSE RESPONSE
C      WILL BE PUNCHED AND THE GRID DENSITY IS 32.
C      SAMPLE INPUT DATA SETUP
C      32,1,3,1,32
C      0,0.1,0.2,0.35,0.425,0.5
C      0,1,0
C      10,1,10
C
C      THE FOLLOWING INPUT DATA SPECIFIES A LENGTH 32 WIDEBAND
C      DIFFERENTIATOR WITH SLOPE 1 AND WEIGHTING OF 1/H.  THE
C      IMPULSE RESPONSE WILL NOT BE PUNCHED AND THE GRID
C      DENSITY IS ASSUMED TO BE 16.
C      32,2,1,0,0
C      0,0.5
C      1.0
C      1.0
C
C      THE FOLLOWING INPUT DATA SPECIFIES A LOWPASS FILTER
C      BY GIVING DIRECTLY THE RIPPLE AND ATTENUATION.
C      0,1,2,0,32
C      0,0.2,0.3,0.5
C      1,0
C      1,1
C      0.2,40
C
       COMMON DES,WT,ALPHA,IEXT,NFCNS,NGRID,PI2,AD,DEV,X,Y,GRID
       DIMENSION IEXT(66),AD(66),ALPHA(66),X(66),Y(66)
       DIMENSION H(66)
       DIMENSION DES(1045),GRID(1045),WT(1045)
       DIMENSION EDGE(20),FX(10),WTX(10),DEVIAT(10)
       DIMENSION OMEGA(50),RESPA(50)
       DOUBLE PRECISION OMEGA,SUMAR,SUMAC,RESPA,ATTN,RIPPLE,SVRIP,DELTAF
       DOUBLE PRECISION PI2,PI
       DOUBLE PRECISION AD,DEV,X,Y
       PI2=6.283185307179586
       PI=3.1415926535899793
C
C      THE PROGRAM IS SET UP FOR A MAXIMUM LENGTH OF 128, BUT
C      THIS UPPER LIMIT CAN BE CHANGED BY REDIMENSIONING THE
C      ARRAYS IEXT, AD, ALPHA, X, Y, H TO BE NFMAX/2 + 2.
C      THE ARRAYS DES, GRID, AND WT MUST DIMENSIONED
C      16(NFMAX/2 + 2).
C
       NFMAX=128
  100 CONTINUE
```

Chapter Five

The Bridge to VLSI

5.1 CHAPTER OVERVIEW AND ISSUES

5.1.1 INTRODUCTION

The reader who has followed this text through the first four chapters should now be well-acquainted with the fundamentals of signal processing. What issues come to the foreground when these ideas are implemented with IC chips? Going beyond its title, this chapter involves not only building bridges but also the use and design of silicon networks, as in Figure 5.1, which are not unlike highway networks. The network must intelligently lead data where it needs to go as directly and rapidly as possible. The pathways must be wide enough to allow for accuracy in data flow, and sufficient in number and strategic location to minimize bottlenecks. The network must be usable, through a software or a microcode roadmap, which picks optimum paths without requiring the user to keep track of needless detail.

As one moves from theory into practice, tradeoffs must be made between:
Speed of execution
Accuracy
Cost (including both hardware and programming)

Architectural decisions influence the silicon real estate. For example, an on-chip multiplier (shown in the Figure) is an important feature for DSP speeds, though it takes up more space for one function than the rest of the arithmetic functions combined. Tradeoffs between floating point and fixed point are another decision point. A full floating-point solution will bring the highest accuracy, at about double the chip cost of fixed point. Block floating point

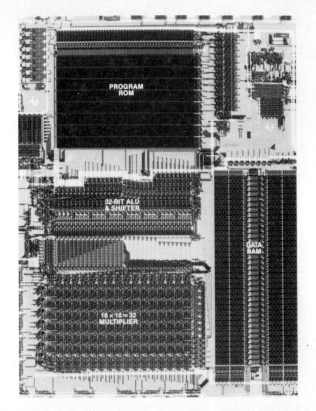

Figure 5.1. Microphotograph of a first-generation VLSI single-chip digital signal-processor, TMS32010. (Courtesy Texas Instruments Semiconductor Group)

provides an in-between compromise. If one also requires highest speed, costs will escalate by factors of ten, as in the fully parallel array processor.

This chapter discusses solutions to such issues through improved architecture and careful error analysis, largely in fixed point. Chapter 6 will extend this to accurate transportable floating-point computational standards and floating-point chips.

Structural ideas such as parallelism and pipelining (Figure 5.2a), which improve DSP architecture for higher throughput, are discussed in Section 5.3. Pipelining removes bottlenecks when operations can be broken into subunits, as in the auto assembly line.

Parallelism helps when more identical operations, performed at the same time, could speed progress, as in supermarket checkout lines. Quantization error, familiar in the dot matrix of Figure 5.2b, determines how wordlength and formats (fixed and floating point) affect the fidelity of DSP computations.

Quantization error will be discussed in Section 5.4 from the standpoint of numerical analysis and in Section 5.5 from the point of view of I/O and signals. In Section 5.4, two key applications will be emphasized: (1) The fast Fourier transform (FFT) illustrates both throughput limitations and error growth sensitivity of an array operation on an entire data set. (2) Digital filters exhibit error growth when high-performance is demanded; and there are differences in error sensitivity between nonrecursive FIR and recursive IIR topologies. Section 5.5 gives a brief overview of analog input/output (I/O) requirements and methods, highlighting needs for dynamic range and throughput in several application areas.

DSP ISSUES AND SOLUTIONS

Throughput of
array operations   →   Parallelism and pipelining alternatives.

Quantization error   →   Effect of fixed-point register length and
                                                    floating-point alternatives.

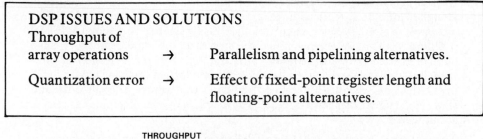

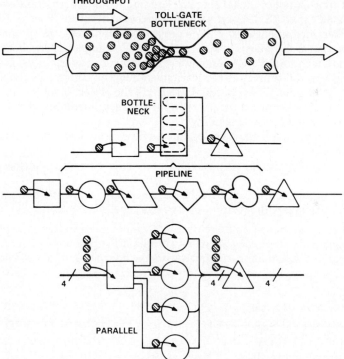

a. A bottleneck may be removed by pipelining a stream of sub-operations, or paralleling identical operations.

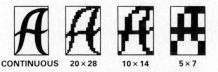

CONTINUOUS   20 × 28   10 × 14   5 × 7

b. Quantization error, seen in the familiar dot matrix.

Figure 5.2. Architecture bottlenecks and quantization error determine DSP speed and accuracy.

How do DSP computer chips resemble general purpose microcomputer chips? How are they different? What application guidelines set the design criteria? A key difference: background computing can take its time, but real-time signal processing demands immediate response, since the throughput must keep up with the signal.

General-purpose vs. DSP computer chips

Douglas Hofstadter, artificial-intelligence specialist and author of *Gödel, Escher, and Bach*, in his PhD thesis, calculated an elegantly symmetric diagram of fractal complexity: the more one magnified the structure, the more complexity came out. He referred to the small computer that did the calculation as Rapunzel, after the fairy-tale heroine who toiled tirelessly all night long to spin the "golden hair" upon which she eventually escaped from her tower prison. Though small and slow, Hofstadter's computer could be left to toil away hour after hour as results accumulated.

The real-time nature of digital signal processing requires a different emphasis. For example, the U.S. Department of Defense Advanced Projects Research Agency identified among its key VLSI projects the intelligent "pilot's assistant," an expert system that assists combat pilots by offloading lower-level chores and performing monitoring and communications functions so the pilot is freer to focus intellectual resources on tactical and strategic objectives. As conceived in the DARPA "Strategic Computing Initiative", the project involves "intelligent systems" machines from artificial intelligence projects interacting with DSP-systems for real-world signals, including

    Speech recognition in the noisy cockpit environment, with a natural language interface;

    Monitoring all basic flight systems and integrating them into a data base in real time;

    Speech output to alert the pilot to the existence of an incipient problem;

    Communications with satellite-based navigational and surveillance systems.

The system as envisioned would require intelligent interpretation of 20,000 rules and an aggregate processing speed of 10 billion instructions per second contained, not in a huge mainframe, but within the crowded confines of a small airframe.

This chapter sets the stage for the following chapters that cover alternative approaches to digital signal processing using VLSI chips: *The single-chip DSP microprocessor*, the *Word-Slice^{TM} chip sets*, and *floating-point standards* and chips that implement them. These approaches will be defined in Section 5.2.3. A *DSP microprocessor*'s architecture is compared with that of a well-known *general-purpose* microprocessor, the Intel 8086/80186 family. While the rapid progress in development of *dedicated DSP* microcomputer chips makes an in-depth critique of specific products of little lasting value, design evolution will be demonstrated with T.I.'s TMS32010, an important pioneering device.

To set the stage for a philosophical digression on VLSI design opportunities in Section 5.2, we review performance of familiar systems, from calculators to supercomputers, with respect to speed (Section 5.1.2) and accuracy (5.2.2).

5.1.2 COMPUTING-SPEED PERFORMANCE EVALUATION

Benchmark Speed Criteria

Figures of merit for speed are measured in MIPS and MFLOPS:

- *MIPS: Millions of Instructions Per Second.* MIPS measures instruction execution rate but is not very useful, since the mix of instructions must be specified for any useful benchmark example.
- *MFLOPS: Millions of Floating-Point Operations Per Second.* MFLOPS is a widespread figure of merit in comparing floating-point processors. It also depends upon the mix of instructions and their relative speed.

A simple benchmark in compiled BASIC, a slight variant of one which was tried out as an Exercise in Chapter 1, illustrates the range of speeds in various implementations of a multiply-accumulate (MAC) operation.

```
100  REM MAC SPEED EXAMPLE
200  INPUT A, B, W
300  C = 0
400  PRINT "START TIMING"
500  FOR I = 1 TO 10000
600  C = A*B + W
700  NEXT I
800  PRINT "DONE"
```

(a) Compiled with *software* floating-point routines, the arithmetic line could take as many shifts and adds as bits in the word, giving a throughput of 0.01 to 0.1 MFLOPS on typical micro/minicomputer systems.

(b) In floating point *hardware*, if the multiply and the add each take a 200-ns instruction cycle, the step runs at 2.5 MFLOPS, for a throughput of 2.5 MFLOPS.

(c) If the multiply and the add operations of version (b) can both be done in one cycle, the speed goes up to 5 MFLOPS.

(d) If 10,000 identical processors of type (c) could be connected in parallel, the throughput would be 50,000 MFLOPS!

While version (d) is not very feasible at this writing, multi-megaflops speeds are now realizable at modest cost.

Computation speed is not solely a function of processor speed or memory access time, or address generation time, though these are all important. Algorithms and architecture are even more important, and the two interact strongly in DSP systems. We illustrate below some examples from general-purpose computing.

The software specialist with a good understanding of machine architecture can design an efficient compiler which, while nearly transparent to the application programmer, takes full advantage of speedup potential of the hardware. The Cray series vector computers are pioneering examples, operating on a dimensioned array in an efficient pipeline so as to speed the execution of loops. The rewards are appearing in many fields. Cryptography, for example, has in recent years been in a race to keep ahead of the properly programmed supercomputer's ability to factor increasingly large numbers in search of primes, the basis for data encryption.

With the availability of many necessary functions in VLSI, the hardware designer benefits from incorporating new architectural advances, which speed up specific classes of end-use algorithms (e.g., vector processing), while avoiding over-complex structures (e.g., truly independent "systolic" parallel processors) which may be very difficult to program efficiently.

The examples below compare performance of common microcomputer and mainframe systems. The balance of the chapter will amplify the discussion in the context of digital signal processing.

(1) Computations made on a *small* computer that *rounds* the results of floating-point operations can be more accurate than computations on a *mainframe* that *truncates* them (as does the IBM 4331/4341). Statistical errors will be smaller on the computer that rounds, since the total error in an integration, for example, increases proportionally to N with truncation, but only $\sqrt{N}$ with rounding.

(2) Computations made on a small computer that uses the IEEE floating-point standard may have a wider *dynamic range* than those made on a VAX using the DEC floating-point standard, since the DEC standard ("D-floating") defaults to an 8-bit exponent, even in double precision, while the IEEE standard extends the exponent to 11 bits. The difference is significant: 10^{38} as opposed to 10^{99}. On the other hand, for the same word length, bits given to the exponent take *precision* away from the mantissa.

(3) Raw instruction cycle time is a poor measure of benchmark execution time when comparing computers. For example, consider the following FORTRAN program, an exercise in number-crunching speed. The benchmark program exercises arithmetic (including multiply) and function (square root) calculation 100,000 times in a loop:

```
Implicit Real *8
        DO 10           I = 1, 100000
                        A = 1
                        B = A*A
                        C = B + 2.0
                        D = C/2.0
                        E = D - 1.0
                        F = DSQRT(E)
    10  CONTINUE
        STOP
        END
```

The comparison shown in Table 5.1 was obtained:

Table 5.1. Execution-time comparison of common microcomputer and mainframe systems.

Computer	Time (seconds)
Z80 system	17,760 (1,730 single precision)
IBM PC	1,222
IBM PC with 8087	194
VAX 11/780 supermini	7.25
APS 164 array processor	0.90

(Ref: A. E. DePriston and S. T. Elbert in *Physics Today*, July 1984, p. 100.)

In this example, *all* CPUs operate at nearly the same clock speed, 4-5 MHz. The more than four orders of magnitude difference between microcomputer and array processor is due to differences in wordlength, number of operations per clock-cycle, and implementation of parallelism, e.g., instruction pre-fetching and decoding.

(4) High speed can offset a higher initial cost, especially when the computer is used primarily for number-crunching—delegating time-wasting input/output operations with displays, printers, etc., to an auxiliary processor. The comparison of Table 5.2 is revealing.

Table 5.2. Relative cost/speed comparison in megabucks (millions of U.S. dollars) per MFLOPS (million floating point operations per second) of common microcomputers and mainframes. All computations in the example were carried out in FORTRAN double precision.

Computer	Cost $	Speed, MFLOPS	Relative Cost 10^6/MFLOPS
IBM PC	4 k	0.00037	10.8
IBM PC with 8087	5 k	0.0073	0.68
VAX 11/750	80 k	0.057	1.4
Cray-1S	11 M	18	0.61
Cray XMP	15 M	53	0.28

Ref: Jack J. Dongarra, Argonne National Laboratory Report ANL-MCS TM-23 (December, 1983).

While the comparison will differ in detail depending upon the benchmark, and also depends upon selecting software (in this case a FORTRAN compiler able to fully utilize the hardware), the cost in "megabucks per megaFLOPS" gives an advantage to hardware optimized for number crunching. While the supercomputer leads the pack here, the PC-plus-coprocessor is surprisingly cost-effective, at least by this criterion (and with plugin accelerators even more so). However, a comparison of raw processing power (speed in MFLOPS column in Table 5.2) reminds us that several hundred unaugmented PCs working day and night still do not keep up with a mainframe number-cruncher, nor does a PC normally support the more sophisticated operating system and multi-user programming environment of a major mainframe.

Table 5.2 is necessarily a snapshot performance *comparison* at one particular time. However, generations succeeding those cited by Dongarra *et al*—at all levels from PC (with 16 bits yielding to 32-bit 80386 μPs) to supercomputer (with Cray-3 based on sub-100-picosecond GaAs logic)—appear to retain the *relative* raw speed and cost/performance comparisons, even though performance at all levels continues to improve.

5.1.3 FIRST WORDS ABOUT ACCURACY

Accuracy issues will be addressed in detail in Section 5.4. At this point, however, the reader may wish to consider the following preliminary ideas and examples.

Fixed-point and floating-point formats
The *fixed-point* representation is the usual way of writing a decimal number: 12.3456, for example. The format F $(b.c)$ for unsigned numbers, specifies b, the total number of digits, and c, the (smaller) number of digits which lie to the right of the decimal point. Fixed-point numbers suffer from dynamic range limits: The largest number that can be written is limited by the number of digits available. For example, if a register has 8-digit capacity, the largest and smallest decimal fractions that can be stored directly in BCD (binary-coded decimal) are 0.99999999 and 0.00000001.

The *floating-point* representation, sometimes called scientific notation, represents a base-10 number, y, as

$$y = m \cdot 10^E \tag{5.1}$$

where E is the exponent and m is the mantissa (for example, $1.23456 \cdot 10^1$, also written as 1.23456E1). Hand calculators often have a built-in function to switch over to floating point when the result of a computation originally entered in fixed point exceeds the size of the registers (see Exercises).

Accumulator Size to Avoid Overflow
To optimize dynamic range, data arrays are kept scaled so that the largest values nearly fill all bits of the data word. The two 8-bit calculations below, two alternative ways to add numbers that would round to 2 + 1,

Pre-rounded	**Pre-scaled**
00000010.	010.01010
+ 00000001.	+ 001.01111
00000011.	011.11001

$$2 + 1 = 3 \qquad\qquad 2.3125 + 1.46875 = 3.78125 \quad (2 + 1 = 4??)$$

provide different results when rounded before computing; given that 8 digits are available—the one on the right, prescaled to more completely fill the available range, provides the higher resolution.

Word length must also *increase* during arithmetic if accuracy is not to be lost by overflow or underflow. The sum of two numbers added in fixed point can overflow unless they were pre-scaled to each be less than one-half of "full scale". The rule is:

For N additions, allow $\log_2 N$ extra bits to accumulate (worst-case) to guarantee no overflow.

Multiplication poses more-severe requirements, since the product of two N-bit numbers will require a register $2N$-bits long if all the bits of the product are to be kept. Of course, the result can be truncated or rounded, but if the computation is of an integral or sum of products, severe errors can accumulate unnecessarily if the truncation is performed at each step (see Exercises).

Word Growth vs. Size in an Integral

Suppose an integral of nearly full-scale numbers must be evaluated without overflow. How many bits must be added to the word length? The answer is a function of the number, N, of points being summed and of the signal itself.

(a) Worst case assumption: all the numbers have the same large value, so no cancellation occurs. The integral can be up to N times as large as any input number, or $\log_2 N$ bits larger.

(b) White noise inputs, as in random walk or Brownian motion: the integral would go up as $\sqrt{N}$. Only half as many bits would have to be added as were required in case (a), since $\log_2\sqrt{N} = \frac{1}{2}\log_2 N$.

(c) Sine wave of amplitude 0.99 full-scale. The integral could go on essentially indefinitely without overflow (except for a very slow drift due to a net buildup of truncation or rounding errors).

A computation comparing these rules illustrates that the optimum scaling algorithm depends upon the anticipated signal-to-noise ratio, since scaling an output register of given resolution to prevent an anticipated overflow loses bits of precision.

Number of input data points	Number of bits added to output accumulator		
	Data = const.	Data = noise	Data = sine
16	4	2	0
64	6	3	0
256	8	4	0
1,024	10	5	0
4,096	12	6	0

In the absence of a basis for more-intelligent checks (see *block-floating point*, Section 5.4.2), the worst-case assumption is chosen. The combination of signal wordlength and number of points being computed is sufficient to determine the safe width of an accumulator. For example:

(1) For 1-dimensional data array of up to 256 points coming from an 8-bit a/d converter, 16-bit fixed point arithmetic (8 + 8) is adequate.

(2) If the inputs are 12-bit numbers and the array has 1,024 points, the array calculations must be performed to (12 + 10 = 22) bits. A 24-bit accumulator would be adequate.

These requirements are typical for one-dimensional array processes, such as spectral analysis. For digital filters, where not all points need to be within the calculation at once, the requirements can be relaxed proportionately (see Exercises).

Precision vs. Dynamic Range
Floating point numbers stored in binary format set aside some of the bits for exponent and some for the mantissa; the exponent precedes the mantissa. For example,

$$(\text{exponent: 4 bits})(\text{mantissa: 12 bits})$$

might be a typical choice for a 16-bit word length. Floating point trades off some precision to gain dynamic range, since the number of bits in the mantissa sets the precision of the word. However, the gain in dynamic range is significant, and grows with word size, since dynamic range (in dB) grows linearly with word size in fixed point but exponentially in floating point.

This can be shown by an example. The exponential format for decimal numbers, for example, $3.1415926 \times 10^6 (= 3{,}141{,}592.6)$, has a binary equivalent,

$$y = m\,2^E$$

and the exponent, E, when expressed as an N-bit binary number, has a maximum value equal to

$$E_{max} = 2^N - 1$$

A binary floating-point number may thus have N bits assigned to the exponent and M bits assigned to the mantissa; it is written as a word with $(N + M)$ bits (exponent first). For example, a 16-bit number formatted (or "blocked"), with 4 bits assigned to the exponent and 12 bits assigned to the mantissa, is referred to as being in a $(4E,12M)$ floating-point format. In principle, such a number could span the dynamic range:

	Mantissa	Exponent		
Largest	1111 1111 1111	1111	=	$(2^{12} - 1) \times 2^{15}$
Smallest	0000 0000 0001	0000	=	1×2^0

The *dynamic range is set by the number of bits assigned to the exponent*, and the *precision available for computation is set by the number of bits assigned to the mantissa.*

The dynamic range and precision vary greatly with wordlength. For example,

WORD LENGTH:	16	32	64
FIXED POINT			
Dynamic range	6.6×10^4	4.3×10^9	1.8×10^{19}
Maximum precision	>4 digits	>9 digits	>19 digits
FLOATING POINT	$(4E,12M)$	$(8E,24M)$	$(11E,51M)$
Exponent dynamic range	3.3×10^4	5.8×10^{76}	$10^{1,419}$
Mantissa precision	>3 digits	>7 digits	>15 digits

Thus, a change from 32-bit fixed point to $(8E,24M)$ floating point loses somewhat more than 1 digit of precision, but gains 10^{67} of dynamic range.

The tradeoff between dynamic range and precision depends upon how the binary word is blocked into mantissa and exponent. For a 32-bit word, the range could be, among other possibilities,

Format	Dynamic Range	Precision
$(8E,24M)$	$2^{255} = 5 \times 10^{76}$	>7 digits
$(11E,21M)$	$2^{2,047} = 10^{1,419}$	>6 digits

Note the spectacular change in dynamic range when the exponent is expanded from 8 to 11 bits. However, while 6 or 7 digits of precision may seem adequate, two numbers which differ from one another by 2 parts in 10^7 are represented by the *same* bit pattern in a 24-bit mantissa representation, and roundoff errors can easily accumulate in a repetitive calculation to give final answers of much lower accuracy. When this is a concern, one should consider expanding from 32 to 64 bits by performing double-precision calculations to retain much better accuracy, albeit at the cost of slower processing.

Many calculators and computer languages have a flexible format arrangement that does computations in a fixed-point format until register overflow approaches, and then switches into exponential format. An underflow or overflow test program can determine the internal register-width of a processor whose architecture is unknown, since going out of range in fixed point is a statement of dynamic range (see Exercises).

Floating point standards have been adopted; they will be discussed in Chapter 6. They differ considerably in detail from the simple conceptual examples introduced here; a sign bit is included; the mantissa is scaled to a standard fraction format, and the exponent has an offset to permit both positive and negative exponent values.

5.2 SOME VLSI-DSP DESIGN PHILOSOPHY

While VLSI design *per se* is not within our scope here, the low cost and ready availability of DSP chips compared to larger mainframe systems offer a testing ground for VLSI concepts and computer architecture in general.

The bottleneck of Von Neumann is uncorked in Section 5.2.1, along with a brief introduction to concurrent, or parallel, processor interconnection.

Desirable features for DSP processors are stated in Section 5.2.2, and the rationale for them is made clear by an examination of some shortcomings of conventional microprocessors. DSP chip families fill various niches, which are defined in Section 5.2.3 and elaborated in later chapters.

5.2.1 A PHILOSOPHICAL FRAMEWORK FOR VLSI-DSP

DSP is a subfield of general purpose computing in which architectural improvements such as pipelining and parallelism are *required* to get the speed needed to handle signals in real time. Accompanying that should be a software environment such that program development proceeds rapidly and logically, and the results on one system are transportable to another. For example, a floating-point *standard* leads to more portable programs, less concern about bit-width tradeoffs, and less worry about error accumulation. On the other hand, advanced architectures with parallelism execute code at a speed which depends upon how well the program and/or the compiler utilizes the architecture.

Architecture Improvements Through VLSI
Conventional microprocessor speed is bus-limited in speed due to the historical adoption of a single shared memory for both instructions and data (Von-Neumann architecture). This may have been a rational choice in earlier days; instructions and data are both binary numbers, and the association of an instruction with the address of its operand lessened the control problem (for the primitive electronics of the time) of keeping separate instruction and data sequences in synchronism. However, the shared memory and single instruction/data bus is now a significant bottleneck (see Figure 5.3).

FIRST-GENERATION MICROPROCESSOR

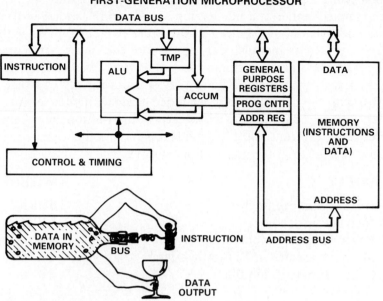

Figure 5.3. Bottleneck of a shared instruction/data bus, as seen in a 1st generation microprocessor.

CONCURRENT MICROPROCESSOR

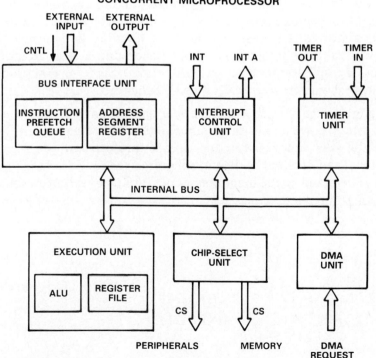

Figure 5.4. Architecture of the 80186, an example of independent subunits operating concurrently.

Instructions being decoded, followed by data being operated on, both pass through the same channel. It is like having to cook from a recipe written with ingredients and instructions written in the same paragraph, rather then set aside separately. The sequential operation is like the step-by-step operation of a checkers or chess game, a single piece at a time, even though there is a board full of pieces which would be in play simultaneously in a real battle; compare this with the parallel action of a video game with many events taking place at once. In a Von-Neumann architecture, instructions may take many clock cycles, and different kinds of operations have to be assigned separate time slots to avoid getting intermixed. For example:

	WHAT	**WHERE**
T0	Instruction (pre)fetch	data bus
T1	Instruction decode	CPU
T2	Processor puts out data address	address bus
T3	Memory selected in T2 supplies data	data bus
T4	Data settles to valid value	data bus

While not all of these cycles are utilized for all operations, an operation which must leave the CPU registers and access memory pays a speed penalty of a factor of two or three.

The next generation of microprocessors attempts to circumvent part of this limitation. For example, in the 80186, the bus interface unit (Figure 5.4) acts as an arbitrator to the outside world bus, queuing both instructions and address calls onto an internal bus. Each inside unit acts as an independent processor or controller for a subset of functions. Since several processes can be operating at once (*concurrently*), speed is enhanced.

Even though no true parallelism (identical replication) exists here, a speedup results from the reduction in number of separate phases, T1-T4 above, because the next instructions are already prefeteched (during previous instructions) and decoded by the time they are needed. For example, compare the 8088, 8086, and 80186 in number of operations to execute a simple fetch/execute for 16-bit data:

	8088	8086	80186
Data Bus:	8-bit	16-bit	16-bit
			(PREFETCH (Interleaved))
	FETCH L	FETCH	EXECUTE
	FETCH R	EXECUTE	
	EXECUTE (L,R)		
	OUT L		
	OUT R		
Cycles	5	2	1

Architecture, accuracy, and speed interact. The Intel 8088 brought 16-bit computations over an 8-bit data bus structure. Although the half-width bus brought compatibility with 8-bit components, with the economy of low pin count, its speed was less than half that of the 8086, since 8-bit data had to be demultiplexed for 16-bit computations. Thus, 8088 systems (e.g., the original IBM PC) run at half the speed of those using a true 16-bit 8086, which in turn is beaten out (due to an instruction prefetch pipeline) by its twice-as-fast 80186 descendant.

This example illustrates *general-purpose* microprocessor evolution. Let us now compare idealized architectures and see how they help solve specific classes of problems, in particular the computation-intensive class of *digital signal processors*.

Computer Architectural Topologies
Several idealized topologies illustrated in Figure 5.5 show how to improve architectures for DSP problems.

• *Trees* (Figure 5.5a) reduce the number of separate elements which must be considered at one time, as in the processing of computer vision from pixels to more global constructs. Examples include the frog vision discussion of Chapter 1 (see also Chapter 8) and speech-recognition processors that extract from multipoint signals a few key parameters as templates to match against template dictionaries in memory.

• *Dataflow architecture* (Figure 5.5b) sets up a number of independent processors (each with memory and arithmetic capability) and dynamically assigns the flow of new data to be processed to a sub-element which is not busy. This works well in straightforward problems organizable like supermarket checkout lines, but poorly when the results of one computation have an effect on the system as a whole. For example, 5 bright programmers working as a team on a single problem do not solve the problem in 1/5 the time of a single bright programmer.

Key issues in parallel architectures are: how to most efficiently interconnect hardware resources, and how to avoid programming difficulties if instructions to concurrent processors must be interleaved.

• *Structured parallel networks* (Figure 5.5c) are capable of the highest speed since no interconnects need be changed during a given computation. Fixed parallel networks lend themselves only to highly structured computations, for example,

> Stress/strain and vibrations of a plate;
> Diffusion;
> Electromagnetic field distributions;
> Magnetism;
> Image processing

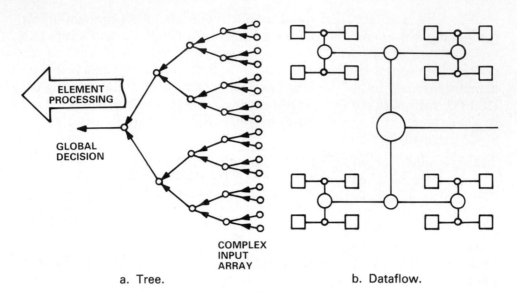

a. Tree.   b. Dataflow.

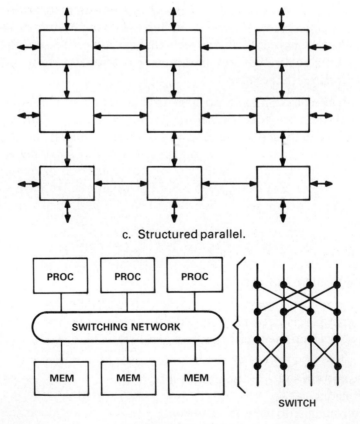

c. Structured parallel.

d. Flexibly switched.

Figure 5.5. Topologies of computer architecture.

These problems can be effectively solved by a fixed parallel system, since each cell need interact with only a few neighbors. The final state is the solution of the problem, and the evolution gives the dynamics. The "game of life" is a familiar computer-graphics example: a cell on a graphics screen gets eaten or replicated, depending on the contents of neighboring cells. In IC design, an array multiplier has a fully parallel solution. Among program structures, FFT spectral analysis has sufficient symmetry for parallelism without significant processor switching difficulty.

• *Flexible parallel processing* (Figure 5.5d) demands a switching network capable of changing the interconnects between processors. For example, the butterfly switch shown can channel information from any processor to any other processor with only as many bits of switch selection as the number of processors being interconnected.

True parallelism of a *single instruction* acting on *multiple data* (SIMD) may be simulated with less replication of computation power by arithmetic *pipelining*. For example, in a DO loop with identical operations on array elements, the entire array may loaded into a pipeline queue and passed through the arithmetic with no need for further instruction decoding or loop overhead. Vector processors such as the Cray-1 utilize this speedup. Even without a vector processor, the overhead in nested looping can be reduced considerably (see Exercises).

5.2.2 DESIRABLE FEATURES OF A DSP PROCESSOR
Listed here are some system features by which to judge DSP chips. While any such list is bound to be incomplete, they include some important ones for DSP speed and accuracy. How to achieve speed and accuracy will be the subject of the balance of this chapter.

• *Bus-width*: Wordlength matched to the application, with double-precision capability. 16 bit words are typical for fixed-point digitized signals, while 32 or more bits are required to represent floating-point numbers.

• *Expanded arithmetic*: Wide enough to minimize roundoff error, i.e., multiplier outputs twice the data-bus width, with selectable single-word rounding or double-precision output operations. An accumulator or multiplier/ accumulator (MAC) must have enough additional guard bits to not overflow in the longest loop anticipated.

• *Data-address generation*: It is desirable for data addresses to be generated automatically for commonly used program loops without making demands upon the ALU. In a limited class of cases, a counter will provide a simple solution to the bulk of address-generation requirements when addresses are sequential. When simple logic elements are insufficient, it is useful to employ an *address generator* unit, which is itself a computing microprocessor. For example, the nonsequential data addresses of the FFT may be computed by an address generator.

- *Instruction sequencing*: Instruction sequences should not place overhead on the main computing units. An *instruction sequencer* unit should be programmable, to execute common DSP instruction sequences with no overhead cost for jumps and loops.

- *Scaling*: Wide dynamic range should be allowed for without requiring excessive scaling computations. A *barrel shifter* is a commonly used element to reduce a wide output to the data-bus width in a single cycle without making demands on the main processor, or to do multi-bit, single-cycle shifts and rotations, or to normalize/denormalize, for floating-point and block floating-point operations.

- *Parallelism in architecture*: The subunits described above should be able to operate independently of one another, with enough pipelining registers to queue the results of one operation onto the shared bus while the next operands are being computed.

5.2.3 DSP CHIP FAMILIES DEFINED

DSP Processor systems are built up by a number of methods: Bit-Slice, Word-Slice™, all-in-one microcomputers, and microprocessor-plus-memory systems.

Bit-Slice systems are assembled from small but fast subunits arranged in parallel to build the required word-length. Bit-Slice offers flexibility and high performance, at the cost of high parts count, large power consumption, and hardware-intensive development. While Bit-Slice was a key stage in assembling supercomputers when the only superfast chips were MSI, this approach is being eclipsed by the VLSI Word-Slice approach.

Word-Slice® assemblies stack larger subunits than Bit-Slice systems, and are a natural VLSI replacement for Bit-Slice, with comparable performance and far fewer components. The basic modular components of Word-Slice systems are microcode memory, address generators, and program sequencers. With these, a whole gamut of systems can be built by using them to control in parallel any number of fixed- or floating-point multipliers or multiplier/accumulators, ALUs, barrel shifters, memories, etc.

The chips can be assembled in diverse ways to make a total system with high performance, comparable to array processors. A typical Word-Slice system, shown in Figure 5.6, uses microcode program memory to exercise control over the chip functions in parallel. Word-slice solutions to system-design problems are discussed in Chapter 7.

Single-chip DSP Microcomputers, such as the TMS32010 or the NEC-7720, form self-contained systems at moderate cost having moderate performance. On-chip program memory (RAM or ROM) and a modest amount of on-chip

® Registered trademark of Analog Devices, Inc.

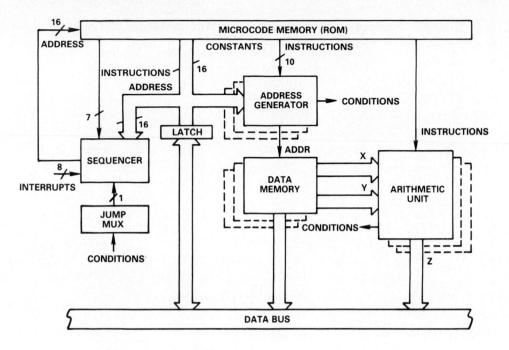

Figure 5.6. Microcoded Word-Slice™ components.

data-memory are coupled with an on-chip multiplier to bring full-audio bandwith performance.

DSP Microprocessors have system architectures that lie between the single-chip microcomputer and the Word-Slice family. The design of DSP micro-processors and microcomputers follows the Harvard architecture (a historical alternative to the Von Neumann architecture, also defined in the 1940's) with separate data and instruction buses. An example is shown in Figure 5.7. A multiple-bus architecture with separate instruction and data busses permits two data operands to be loaded in one cycle, and allows a single-cycle multiply-accumulate. With data memory off-chip, large data segments can be addressed, for high-performance digital filters and spectrum analyzers. On-chip components typically include fast multiplier-accumulator, ALU, barrel shifter, data-address generator, and instruction cache memory. Internal pipe-lining—for example, via the input and output buffer registers shown on com-putational units—and a highly parallel instruction set permit up to three com-putational operations to be performed in a single cycle.

Qualitative cost-perfomance and level-of-integration comparisons for these families are summarized in Table 5.3 and graphed in Figure 5.8. "Perform-ance" is not a one-dimensional measure but a combination of speed and accuracy, to be discussed later in the chapter.

Table 5.3 Performance-cost comparison for common DSP systems.

METHOD	Speed Performance	Level of Integration (typ. chips per system)	Relative Cost	Typical Application
Bit Slice	Ultrafast	Low (70)	Very high	Supercomputer
Word Slice	Very fast	Med (30)	High	Array processor
DSP microprocessors	Fast	Very high (10)	Medium	Accelerator Professional instruments
DSP microcomputers	Medium	High (10)	Low to Medium	Telecommunications Consumer Low-end professional

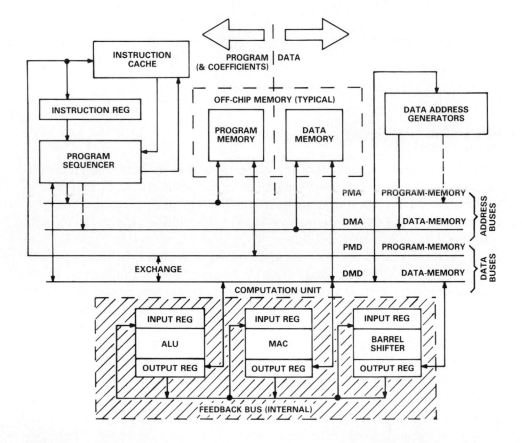

Figure 5.7. DSP microprocessor following the Harvard architecture with separate pro-gram-memory and data-memory. Separate busses for each function facilitate single-cycle execution of any instruction. (Source: Analog Devices)

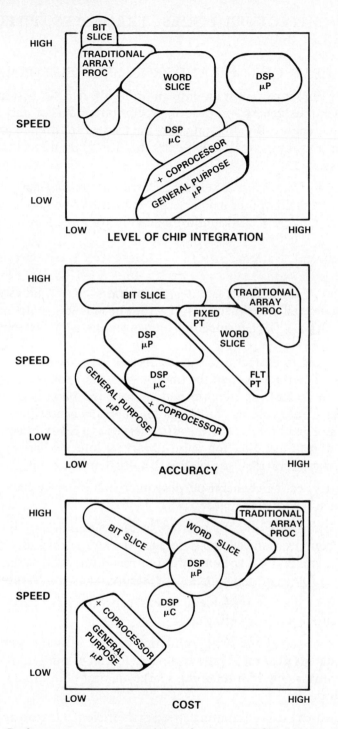

Figure 5.8. Performance-cost comparison of common DSP chip families. The traditional array processor is included as a performance/cost calibration.

5.3 DSP ARCHITECTURE ISSUES: TRADEOFFS, PIPELINING, AND PARALLELISM

5.3.1 ON-CHIP VS. OFF-CHIP AND PROCESSOR PERFORMANCE

The choice of functions to put on-chip and off-chip in DSP systems echoes similar problems in general-purpose microcomputer design and computer architecture in general. We illustrate how the tradeoff of immediate on-chip data availability vs. chip area available to other functions affects throughput by large factors.

Example: Memory Hierarchies in a General-Purpose Microcomputer
Fetching data from computer memory can be likened to book borrowing, to illustrate the advantage of partitioning memory into a speed hierarchy:

Local cache	Small size	Local	Fast retrieval
to	to	to	to
Archives	Large size	Central	Slow retrieval
(Office bookshelf	(Almanac vs.	(Suburban vs.	(Open shelf vs.
vs. research lib.)	encyclopedia)	metropolitan)	attended stacks)

The time required to extract information from a book is short if the book is on one's desk. The time to go to the library, find the book, etc. may greatly exceed the time to look up the information once the book is in hand. Yet, libraries make a large volume of information centrally accessible *concurrently* to many users. The photocopy in the office serves as a ready reference, while a departmental reading collection (mini-library) is intermediate in speed and the range of information that is accessible to more than one reader.

The same issues occur in general-purpose microcomputer systems as well as in DSP systems, as shown in Figure 5.9.

Data words copied into on-chip registers or cache memory are immediately available for computation, while a reference to external memory make take several cycles; archival storage on a different medium, such as disk, (even if directly addressable as in virtual-memory systems) takes much longer.

External memory may be available to several processors in a parallel system, speeding execution while lowering costs.

The regular architecture and large production volume of memory chips causes RAM chip costs to fall at rapid rates compared to that of computational chips, so a cost premium is paid for retaining a large amount of internal memory on a processor chip.

A memory *hierarchy* aids computing speed and efficiency. Mead and Conway (1980) estimate that access time grows as $\sqrt{M}$ for memory size M, based on the lengths of the interconnects on a chip. For two different systems having memory sizes, M_1 and M_2, the estimated access times are related as

$$\frac{t_1}{t_2} = \sqrt{\frac{M_1}{M_2}} \qquad (5.2)$$

A specific example in Table 5.4 illustrates the hierarchy of speeds in a typical microcomputer system.

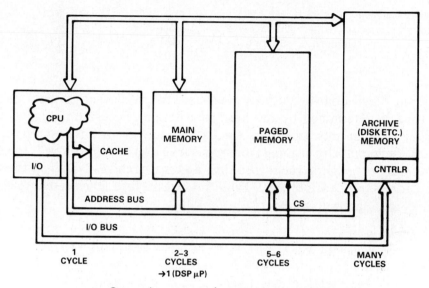

a. General purpose microprocessor system.

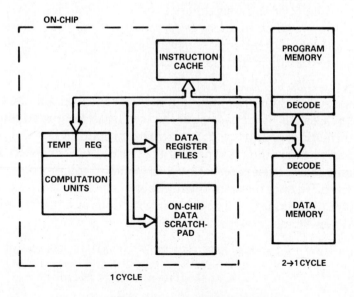

b. DSP microprocessor system.

Figure 5.9. A hierarchic memory structure balances speed of immediate access against cost and size of available storage.

Table 5.4. Hierarchy of speed, size, and access frequency. Example is a typical microcomputer system at the time of the Mead-Conway rule.

Type of Instruction	Access Time (cycles)	Typical Size (bytes)	Relative Access Frequency (% of accesses)
Register reference (cache)	1	32	50
Memory reference	3	64K	35
Paged memory reference	5 or 6	1M	14
Disk memory	100	10M	1
			100

While the Mead-Conway estimate was based on physical arguments, the actual time is determined by the hardware design. Access time grows with memory size in this example, since the number of instruction cycles increases substantially in going from register reference ("desk-top") to memory reference ("reading room") to disk access ("archives"). Given the estimated access frequencies (f_i), of Table 5.4, the *average* time needed to access memory in the above example is

$$\frac{\sum t_i f_i}{\sum f_i} = (0.50 \times 1) + (0.35 \times 3) + (0.14 \times 6) + (0.01 + 100)$$

$$= 0.50 + 1.05 + 0.84 + 1 = 3.4 \text{ cycles} \qquad (5.3)$$

The average time to access megabytes is only 3 or 4 times slower than the register access time when memory is partitioned as shown. The division of access time and access frequency is nearly optimal in this example, since all contributions to the sum in Eq. 5.3 are nearly equal.

The Memory On-Chip/Off-Chip Tradeoff
Determination of the amount of on-chip memory involves a delicate balance. Chip area ("real estate") is already squeezed by the area required for expanded on-chip arithmetic capability, e.g., an array multiplier.* But the penalty for going off-chip to access data is typically a factor of 2 or 3 in speed, due to the cycles added to address an external bus. In standard microprocessors, this is evident in the difference between the number of clock cycles to execute register-to-register or arithmetic-on-register instructions and memory-to-register or arithmetic-on-memory instructions.

Speed for Various Types of Instruction (80186 Microprocessor)

Type	X = Register	X = Memory
Move Register to X	2	9 to 12 clock cycles
Arithmetic on data in X	3	10

*Total chip area is restricted and expands very slowly with the years since designers find the *yield* of chips that work after fabrication to be a strongly decreasing function of chip area, due to substrate material defects.

Instruction execution is already facilitated in modern microprocessors such as this by an on-chip *instruction queue register*, a FIFO (first-in-first-out memory) which prefetches and decodes instructions. In older microprocessors, such as the 8080/8085—without instruction prefetch, all instructions take several more cycles, so the ratio between off-chip memory instructions and on-chip register appears less severe: a factor of 2.5, rather than 3 to 5.

DSP Architecture Example: TMS-32010 DSP Microcomputer
Modern DSP microprocessors attempt to reduce all instructions to (ideally) single-cycle execution, for enhanced speed and easier signal-throughput estimation. Rather than implementing the instruction set in microcode, whose execution takes several clock cycles (as in the 80186 example above), the architecture is translated directly to a logic layout with pathways adequate to carry out any given instruction directly. The architecture of the TMS32010, shown in Figure 5.10, partially achieves this goal.

The TMS32010 features separate data and instruction buses (the *Harvard architecture*, introduced in Section 5.2.3). The following observations are not intended to derogate this important and evolving chip family, but rather to serve as an example of how architecture and throughput interact.

This pioneering chip made great inroads into DSP by putting a full multiplier on-chip. But the on-chip/off-chip memory tradeoff imposed by its architecture is quite apparent: any operation requiring access to external data memory is slowed by a factor of 2. In spite of the separate program and data buses, the first-generation TMS32010 was bus-limited, partly due to internal bus widths and partly due to pinouts. The architecture is optimized only for data contained in the internal scratchpad memory, which is limited in size: 144 16-bit words in the TMS32010.

On-chip memory limitations are removed in later generations to lessen or eliminate the speed penalty for going off-chip; only small register files and perhaps a program cache memory are retained on-chip.

- The (internal) data-address shares the program bus, so data addressing and instruction fetching must be in different phases. Program addresses and program contents also share the same bus.
- External data must be brought in via an input/output (I/O) call, which requires 2 cycles, one to put the port address lines out onto the A/PA bus, and a second to read in the data.
- Only data which fits within the 144-word internal scratchpad (labeled DATA RAM) can be directly addressed.
- Computational units connect only to the data bus. The two internal buses talk to one another through a bus interchange module. A table lookup of coefficients stored in program memory can get to the computational portion only by taking 3 cycles, using the *table read* and *table write* instructions.

•All I/O, *including* access to external memory, must pass through a single 16-bit pathway, the DO bus. External memory is addressed via the same 12-bits that access I/O ports; external memory references are therefore a two-cycle operation, rather than the single cycle for internal memory references.

•A penalty is paid for looping, since each branch instruction adds two cycles: one to test the condition and a second to execute the jump.

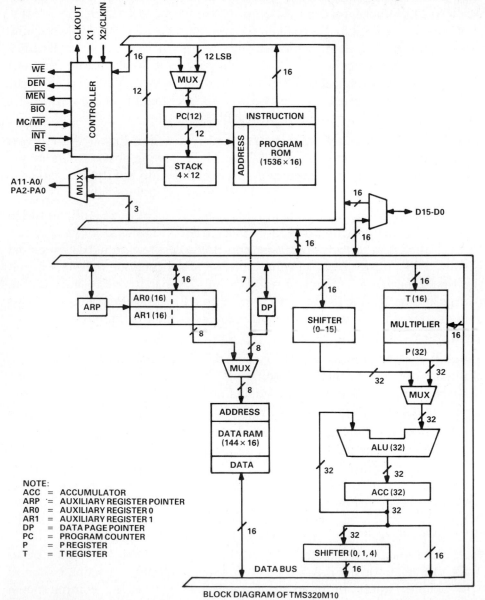

Figure 5.10. Architecture of the TMS32010, a pioneering single-chip DSP microcomputer. Compare with the chip layout photo of Figure 5.1. (Courtesy Texas Instruments Semiconductor Group)

•The lack of parallelism in computation and of pipelining in temporary storage registers results in a multiply-accumulate, the basic DSP primitive, taking 2 cycles.

On-Chip Cache Memory in the DSP Microcomputer

On-chip scratchpad *cache memory* or *register files* play a key role in maintaining highest throughput. In architectures which slow when external memory reference is made, a speed benchmark will suffer a sudden slowing when the data-array size passes the on-chip cache boundary. In the TMS32010, for example, a factor-of-two penalty is paid for data buffers in excess of 64 points, as the FFT example of Figure 5.11 shows.

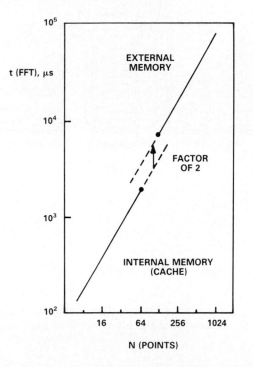

Figure 5.11. FFT speed vs number of data points, for the TMS32010 DSP microcomputer. Note the factor-of-two jump when data size passes the internal cache-memory boundary.

5.3.2 DATA-FLOW BOTTLENECKS AND THEIR RESOLUTION

Data-flow bottlenecks are as familiar as traffic slowdowns on a highway under repair or the progress of shoes through a shoe factory (Figure 5.12). The throughput in a process consisting of a series of steps is limited by the slowest step, as in the tollgate on a highway. When one operation in shoemaking takes 10 times longer than either of the other two operations, the time to make a *single* pair of shoes (latency) is $12T$, decreasing only to $10T$ in *continuous* production (with handshaking lines between adjacent operations).

With *pipelining*, the slow step is broken up into a sequence of separate operations (make heel, make body,... put on laces, trim scraps), each of length T, as shown in Figure 5.12b. A new shoe can be put into the pipeline each T seconds. The bottleneck is broken into a sequence of steps whose steady-state flow matches that of the rest of the system. Pipelining is the

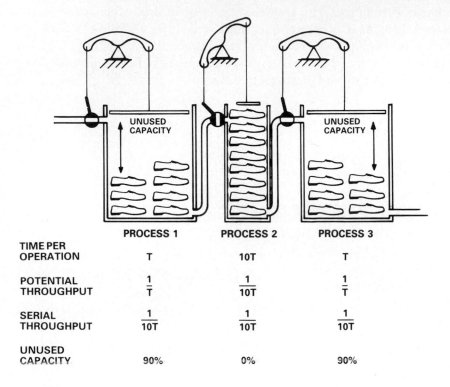

	PROCESS 1	PROCESS 2	PROCESS 3
TIME PER OPERATION	T	$10T$	T
POTENTIAL THROUGHPUT	$\frac{1}{T}$	$\frac{1}{10T}$	$\frac{1}{T}$
SERIAL THROUGHPUT	$\frac{1}{10T}$	$\frac{1}{10T}$	$\frac{1}{10T}$
UNUSED CAPACITY	90%	0%	90%

a. Data-flow bottleneck; a step ten times longer than the others limits the throughput rate. Mechanical feedback shown closes the input valve to a given process when its capacity is full.

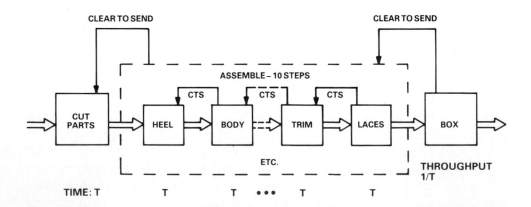

b. Pipelining (partitioning) of the slow step speeds the steady-state throughput.

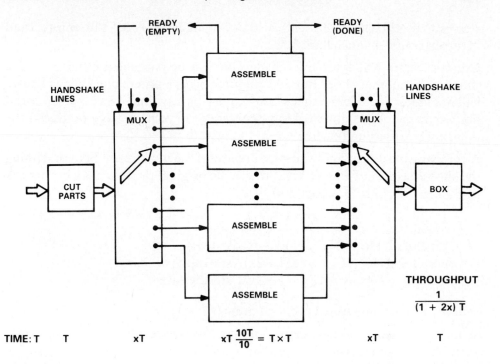

c. Paralleling a number of identical processors increases the speed, even for a single product to pass through.

Figure 5.12. Shoe factory throughput.

classic auto assembly line procedure. While human workers tire of putting the same nut on the same bolt all day long, computing and DSP circuits don't get bored.

While not all processes can be divided into chunks of optimal size, the addition of a few extra storage registers can often speed up a process significantly. For example, in a conventional microcomputer, a multiply operation requires that input X be kept in a holding pattern until Y arrives. Separate data buses for X and Y speed the multiplication by a factor of two. The pipelined biquad filter of Chapter 4 is an example in hardware.

On the negative side, when a process many levels deep is pipelined, the startup or termination is slow while the pipe fills or empties. Several levels of pipelining are justified only when loops of relatively long length are anticipated, or when the same instruction is to be performed repetitively on a sequence of data points (vector operation).

Paralleling the slow step with a number of identical processors is an alternative speedup method. The difficulty of partitioning a process into subunits for pipelining is avoided, as is the startup transient; but the cost is that of replicating the hardware. A distribution network is required at input and output, to schedule data into and out of available processors. The distribution may be scanned (if all tasks take identical time) or flagged with "I'm empty" and "Output ready" handshaking lines.

Only processes which are interchangeable can be paralleled efficiently; one shoe is like another shoe. But 10 computer programmers working as a team will not crack a problem 5 times faster than a tightly linked two-person team, nor would two Einsteins working (separately) in parallel have produced the theory of relativity any earlier.

A comparative throughput estimate illustrates the improvement in pipelining the shoe assembly of Figure 5.12, where the original process has a bottleneck that takes $10T$, plus $1T$ at each end:

Serial:	$12T$ to $10T$
Pipelined, ideally:	$1T$ (steady state)
Paralleled (10):	$(3 + 2x)T$ startup to
	$(1 + 2x)T$ steady state

where xT is the time added in distribution.

This subject is a good deal more complex than space allows in this introduction. Other points could be made: parallel channels may be costly (for example, a whole new assembly line); and there may be quality and synchronization problems if channels differ. Utilization of available channels is statistical, and "phase transitions" occur, as when a small accident in one lane of a highway slows all lanes of traffic for miles when traffic density is above a critical level. Protocols in high-speed parallel processor systems must be highly evolved to anticipate such problems (Ref: Hockney (1981)).

Example: FFT with Pipelining and Parallel Butterflies
The FFT provides a good example of the use of alternative signal-processing architectures to improve throughput. A discussion of FFT computation can be found in Section 3.3.8. Here, we consider the implementation in VLSI of alternative FFT architectures. The key comparison is that of butterfly time, t_B, and the time, $(N/2)T \log_2 N$, to cycle through all butterflies of an FFT. The interval, T, includes the butterfly computation time and any overhead in address generation or looping. Realistic alternatives to consider are:

Serial (direct)
Pipeline $\log_2 N$ stages deep, with $N/2$ steps
Parallel $N/2$ butterfly processors, iterate $\log_2 N$ times

Typical computation flow in each of the three alternatives is shown in Figure 5.13. In the *serial* approach, there is a single processor; each butterfly is computed, one step at a time. In the *pipelined* arrangement, there are $\log_2 N$ butterfly processors, corresponding to the number of passes (3 in the case of 8 data points—B1, B2, B3); each is used to compute the butterflies pertinent to its pass in series; as each pass is computed, the processors are ready to accept

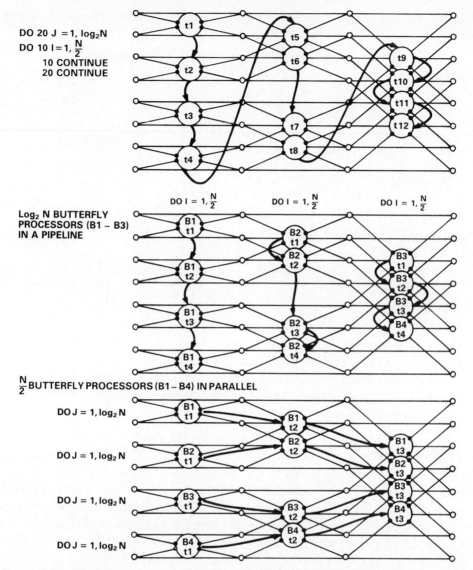

Figure 5.13. The FFT is an example of a computation which can benefit from either pipelining or paralleling. B1, etc., are labels for separate butterfly processors, while t1, etc., label process cycles. The arrows designate the sequence of operations performed by a given butterfly processor hardware unit with (top) a single processor (middle) pipelined processors, and (bottom) parallel processor's.

a pair of inputs for the next pass, and when the pipeline is full (steady state), a set of outputs will be produced by each pass ($N/2$ computations). In a *parallel* computation, there is one processor for each of the $N/2$ steps per pass; all butterflies for that pass are computed at the same time; as soon as one pass is completed, all are ready for the next pass; in the steady state, there will be an output for every computation cycle. The times and hardware costs, assuming that t_B dominates computation time, are summarized in Table 5.5.

Table 5.5 FFT speed enhancement by pipelining and by paralleling butterfly processor units.

Architecture	Computation time, t in units of t_B	Number of Butterfly Processors
Serial	$\dfrac{N}{2}\log_2 N$	1
Pipeline	$\dfrac{N}{2}$	$\log_2 N$
Parallel	$\log_2 N$	$\dfrac{N}{2}$

For example, consider a 1,024-point FFT

Serial	5,120	1
Pipeline	512	10
Parallel	10	512

The cost/performance ratio (the megabuck/megaFLOPS of Table 5.2) is *unchanged* by these partitioning alternatives, since $(t/t_B) \times B = (N/2)\log(N)$ in all cases. However, when the spectral analysis must keep up with the required data rate, whatever the price, the throughput comparison above is the correct performance measure.

Parallel processing is the fastest but also uses the most hardware. Speeding a highly regular fixed-purpose computation like the FFT is not necessarily an overexpensive solution when the processing element (in this case the butterfly) is simple and can be replicated in VLSI. In highly parallel architectures, the toughest problem becomes not the arithmetic but the data organization and address generation. In the FFT example, fully parallel access of $N/2$ data points would be required each butterfly cycle to keep up with the arithmetic. Since ordinary memory access can pass but one data point per cycle over the bus, a special resident memory buffer would have to be set up to supply the data needed at the required rates. The need to compute the characteristic FFT data-address sequences would also slow the overall computation—unless special address-generator chips (Chapter 6) are used.

The FFT lends itself ideally to partitioning of identical tasks (the butterfly) in either pipelined sequential or in simultaneous parallel form, because it is so symmetric. But pipelining is not a speedup for one-shot (nonrepetitive) operations. The pipelined throughput rate is misleading for a *single* operation. A single multiply in a processor pipelined four-levels deep will take (*latency time*) four times the quoted throughput interval; the first three are wasted waiting for the pipe to empty.

Pipelining Summary

The proper place of pipelining is in steady-state operations, not in one-shot operations. The digestive system is a steady-state system with a typical throughput of 3 meals/day, plus snacks, though an average time in the pipeline of about 1 day is needed to complete the extraction of useful products from a given meal or snack.

On the other hand, human reproduction is in the one-shot category. Cells multiply and divide in seconds, yet the cycle time to produce a baby is nine months, since the pipeline, once primed, cannot accept new inputs. Two additional examples below illustrate when pipelining can be of little or of large benefit.

Example of *minimal gains*: The speedup of one portion of a process must be balanced not only against the cost but also against other portions unchanged by the improvement. If 50% of the process time is spent in number crunching and the other 50% in moving data, improvement in number-crunching without improvement in I/O is of little benefit, as illustrated in Figure 5.14

Importance of Using Fewer Clock Cycles: Improvements in raw clock speed happen slowly, as logic families shrink in size or as new families develop. While shrinking the scale of feature size increases integrated-circuit switching speed, the 1-micrometer mark in photo-lithography was a difficult boundary to cross reliably; beyond it, the speedups of further scale shrinking battle with higher development costs. Clock speeds have increased by only about a factor of ten in the last decade. New families are slow to develop; GaAs logic, for example, has been under development for several years but the factor of 3 to 10 speed enhancement it brings will always be at much higher cost per chip.

The architectural improvements discussed above take a different approach: *pipelining and parallelism speed processing by doing more in fewer clock cycles.* This favors chips which do a multiply-accumulate step (the most fundamental primitive operation in numerical computation) in a *single* clock cycle. Three architectural features produce an eightfold increase in speed despite limitations in the clock cycle.

(1) A multiply-and-accumulate step pipelined within the chip doubles the throughput speed, since no intermediate memory reference is required.

(2) If the instruction fetch-execute can also be executed in one cycle, a further doubling results.

(3) A further doubling comes if both the x operand and the y operand are separately bussed to the inputs, as double-memory permits. The data (x) and coefficients (y) of an FIR filter are loaded and execute in a single cycle. So do the data and twiddle-factor multipliers of an FFT.

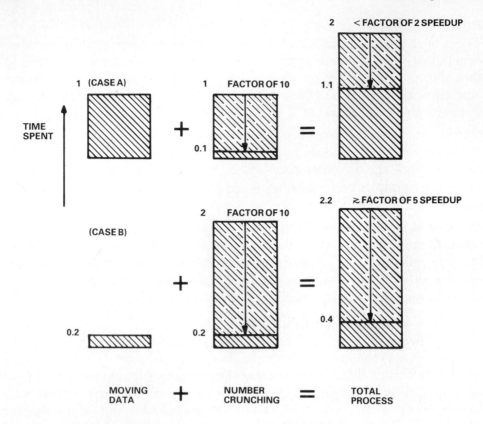

Figure 5.14. An example of minimal gains. If (Case A) moving the data occupies comparable time to number crunching, a factor-of-10 speedup in the latter improves total throughput by only a factor of about 2. But if a large disparity is reduced (Case B), the speedup is greatly enhanced.

Algorithms Tailored to Architectural Strengths—and Weaknesses

(1) *Array speedup by choosing the order of operations.* Now that multiply operations need take no longer than any other full-word operation, a considerable speedup can be accomplished by choosing *when* in an array operation to do the multiplys, the adds, the shifts. Although the order of operations (ideally) does not matter in arithmetic, it *does* influence computational speed considerably. For example,

$$(\text{Add, then multiply}) = (\text{Multiply, then add})$$

$$(A + B) \times (C + D + E) = AC + AD + AE + BC + BD + BE \qquad (5.4)$$

Both sides are identical mathematically, but the left half takes 3 adds and 1 multiply while the right half takes 5 adds and 6 multiplies: a significant speed difference!

A similar speedup option exists in a double-precision multiply-accumulate sequence or in a double-precision dot product. Writing the double-precision, $2N$-bit numbers as $(MS\ LS)$, where MS stands for more-significant half and LS stands for less-significant half, a sum of products has the form:

$$(MS\ LS)_1 \times (MS\ LS)_2$$
$$+ (MS\ LS)_1 \times (MS\ LS)_2$$
$$+ (MS\ LS)_1 \times (MS\ LS)_2$$

$$\cdot \qquad \cdot$$
$$\cdot \qquad \cdot$$

$$+ (MS\ LS)_1 \times (MS\ LS)_2$$

MS = more-significant half
LS = less-significant half

Each product contains four partial products, each a $2N$-bit number, which must be shifted to its correct place in the $4N$-bit field:

$$(MS\ LS)_1 \times (MS\ LS)_2 = (2^N MS_1 + LS_1) \times (2^N MS_2 + LS_2)$$

$$= 2^{2N} MS_1 MS_2 + 2^N MS_1 LS_2 + 2^N LS_1 MS_2 + LS_1 LS_2$$

For example, when each portion of MS or LS is an 8-bit number, the product, $(MS\ LS)_1 \times (MS\ LS)_2$ is the sum of partial products

$$(2^8 MS_1 + LS_1) \times (2^8 MS_2 + LS_2)$$

$$= 2^{16} MS_1 MS_2 + 2^8 MS_1 LS_2 + 2^8 LS_1 MS_2 + LS_1 LS_2$$

Recognizing that multiplication by 2^{16} means left-shifting $MS_1\ MS_2$ by 16 places and filling with zeros, and that multiplication by 2^8 means left-shifting $MS_1\ LS_2$ and $MS_2\ LS_1$ by 8 places and filling with zeros, the sum may be written out:

$$(\text{---}MS_1 \times MS_2 \text{--})0000000000000000$$
$$+ 00000000(\text{---}MS_1 \times LS_2 \text{--})00000000$$
$$+ 00000000(\text{---}MS_2 \times LS_1 \text{--})00000000$$
$$+ 0000000000000000(\text{---}LS_1 \times LS_2 \text{--})$$

The partial products are shifted left in the same way that the digits in decimal partial products are shifted. But here's the time-saving trick: For double-precision product sum or double-precision dot-product operation, rather than shifting as each product is computed, all of the partial products of the same weight are computed, accumulated (using three paralleled accumulators), and *then* shifted to the correct position. The savings depend on the length of the array; for long arrays, they approach 33% if MACs and N-bit shifts take the same amount of time.

(2) *The Winograd Algorithm.* There are many other such algorithmic tricks. For example, the Winograd algorithm [Winograd (1972)] speeds either the complex number manipulations in the FFT or double-precision multiplies by a simple transformation of variables. The operation

$$R = X \cdot Y$$

where

$$X = a + bq \qquad Y = c + dq$$

can be either the computation of an FFT butterfly (q being the twiddle factor) or a double-precision product (q being the scale factor for the least-significant half). The direct product is the sum of terms:

$$(a + bq)(c + dq) = ac + adq + bcq + bdq^2$$

Accumulating M such products would be accomplished by calculating the running sums of the terms, ac, ad, bc, bd, separately and then adding the results, requiring $4M$ MACs and then 5 adds (3 direct and 2 carry-related) at the end.

The Winograd algorithm redefines a new set of variables

$$
\begin{aligned}
X_1 &= a - b & Y_1 &= c - d \\
X_2 &= a + b & Y_2 &= c + d \\
X_3 &= b & Y_3 &= d
\end{aligned}
$$

with which the equivalent product may be written

$$\mathrm{R} = \frac{R_1 + R_2}{2} + qR_3 + q^2 R_4$$

where

$$
\begin{aligned}
R_1 &= X_1 \cdot Y_1 - X_3 \cdot Y_3 \\
R_2 &= X_2 \cdot Y_2 - X_3 \cdot Y_3 \\
R_3 &= (R_2 - R_1)/2 \\
R_4 &= X_3 \cdot Y_3
\end{aligned}
$$

A running product of M inputs requires only $3M$ MACs (the running sums of $X_1 \cdot Y_1$, $X_2 \cdot Y_2$, $X_3 \cdot Y_3$) followed by six adds and two shifts at the end. The savings are significant, especially for large M:

Direct: $4M$ MACs and 5 adds at the end
Winograd: $3M$ MACs, followed by six adds and two shifts

M multiplies are saved at the cost of only one add and two shifts.

(3) *Straight-line vs. Looped Code.* The execution of loops is standard in almost any computation. However, it is quite common that a branch test takes more time than other operations. For example,

INSTRUCTION CLASS – TIME IN CYCLES

PROCESSOR	MOVE	ADD	MAC	BRANCH
80186 general-purpose μP	2	3-4	>20	13
TMS32010 pioneer DSP μC	1	1	2	2
ADSP-2100 DSP microprocessor	1	1	1	1

DSP microprocessors reduce the number of cycles per operation compared to generic microprocessors (80186). The last entry, the ADSP-2100, discussed in Chapter 6, executes all classes of instructions in 1 cycle. The extra cycle penalty paid for each loop test may be avoided in the case of the TMS32010 by writing the code in straight line, repeating over and over again the instructions which would ordinarily appear in a loop. The time savings are significant, though practitioners would agree that the technique is inelegant.

But, as Figure 5.15 demonstrates, memory demands make straight-line code worthless for loops of significant size (N); the relative savings shrink to zero as N increases.

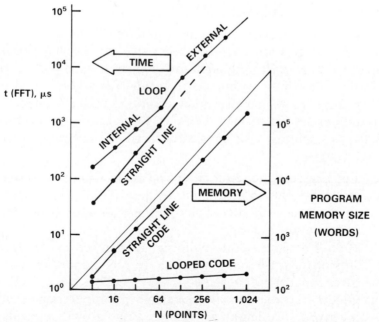

Figure 5.15. Straight-line code reduces execution time of smaller loops in the FFT example, using chips like the TMS32010 with branching overhead—at the cost of much larger program memory. (Adapted from data presented by S. Magar et al, *Electronic Design*, Aug. 15, 1982, pp. 149-154.)

5.4 FINITE-WORDLENGTH ARITHMETIC

5.4.1 INTRODUCTION

Finite word length leads to noise growth during any computation, either fixed- or floating-point. An extensive literature is available on the topic; for a summary, see Oppenheim and Schafer (1975) or Rabiner and Gold (1975). This discussion will be limited to a few fixed-point examples, which illustrate how to estimate the accuracy of:

> Digital filters (FIR and IIR)
> Spectrum analysis (FFT)

Typical fixed-point quantization errors will be demonstrated by three examples:

Rounding issues: how to minimize noise growth with finite-length coefficients; fixed point vs. floating point vs. block floating-point.

Noise growth in a filter computation: how does noise vary as a function of the wordlength, the number of stages, and the structure (FIR vs. IIR). The IIR *limit cycle* will be discussed briefly: in a finite-wordlength feedback loop, small output fluctuations are not fed back as input corrections, so the output cannot go to zero after the input has gone to zero.

The FFT: To avoid worst-case error growth in a butterfly computation, rules must be observed for scaling to avoid overflow; block floating point is shown to be a useful approach.

These accuracy estimates will be limited to *fixed-point* calculations. *Floating-point* DSP calculations are less sensitive to error accumulation, but doing everything in floating point rather than in fixed point is not a miraculous solution to finite-arithmetic problems; for a given total word length, floating point trades some accuracy off in the mantissa, also floating point computations are not without their own finite-arithmetic problems. Finally, a fixed-point solution *does not* always sacrifice performance when used for real-world signals:

(a) For example, noise on an analog phone-line may restrict telephone speech signals to a dynamic range of 50-60 dB. Quantization steps smaller than -72 dB (12 bits) will provide a better representation only of the noise!

(b) A *wide* dynamic range audio signal (hi-fi) covers 80 dB and takes 14 bits of quantization. Therefore, 16 bits is more than adequate for a *given* input or output value.

(c) On the other hand, *cumulative* errors in computation usually determine if a given word-length will suffice. Error growth depends upon the algorithm that is being performed. While error accumulation is small for modest digital filters, extension bits (also called *guard bits*) are required for common DSP vector or matrix operations, such as convolution, correlation, and spectral analysis. A summary of the principal rules of thumb is developed in Section 5.4.4.

The resolution of the a/d converter is also a DSP accuracy issue. How many bits should there be, and how should this be matched to word-length in the digital process that follows? There is no definitive answer to the question. Two examples illustrate why:

APPLICATION	ADC	DSP OPERATION
CAT scanner	8 bits	floating point (32 or more)
Telecommunications	10-12 bits	8 to 10 bits with compression

Because many data points are transformed as long arrays (matrix rotation, zoom, reconstruction...), the CAT scan requires high-resolution computations, usually in floating point, even though each data point is restricted to modest resolution (8 bits) because of limitations in the transduction process and/or the high volume of data acquisition. On the other hand, in telecommunications, a continuous throughput of data puts a premium on low-resolution computations, storage, and transmission, even though the input signal's dynamic range may require 10-12 bits per point. *Data-compression* algorithms are thus a must in telecommunications.

Issues in input quantization are discussed in Section 5.5; data compression is introduced in Chapter 8.

5.4.2 ARITHMETIC ERROR SENSITIVITY

Fixed- vs. Floating Point

Fixed- and floating-point representations require somewhat different handling for common arithmetic operations, and the rules to ensure adequate word-growth without overflow are somewhat different. In *fixed point*, addition/subtraction rules are straightforward. An overflow bit must be anticipated in addition, and an underflow flag in subtraction. Multiplications require an error budget for wordlength growth, since the product of two unsigned *b*-bit numbers can be $2b$ bits wide.

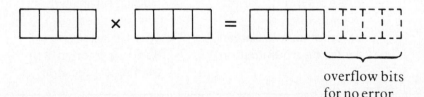

overflow bits
for no error

In *floating point*, word growth in multiplication is less likely to cause overflow than in fixed point.

$$(2^{E_1} M_1) \cdot (2^{E_2} M_2) = 2^{E_1 + E_2}(M_1 \cdot M_2) \tag{5.5}$$

The floating point representation has a wider dynamic range than fixed point, so *multiplications* can go on with less concern for word-growth. As long as the mantissas (M) are expressed as fractions and the exponent (E) sum does not exceed the number of exponent bits available, overflow will not occur. On the other hand, there is always roundoff noise in the mantissa in single-precision calculations, since the double-length product will be rounded to single word-length. Because the exponent is like a gain factor, *absolute* errors can grow rapidly in floating point.

Floating point *addition*, however, requires a prior denormalization of smaller numbers to make all exponents the same, just as in decimal floating-point addition:

NORMALIZED	DENORMALIZE (Common Exponent)
9.8765×10^{-10}	0.0987×10^{-8}
$+ 7.0345 \times 10^{-12}$	$+ 0.0007 \times 10^{-8}$
$+ 4.3210 \times 10^{-8}$	$+ 4.3210 \times 10^{-8}$
	4.4204×10^{-8}

For a given finite accumulator width, denormalizing the numbers with smaller exponents means that precision is lost in the mantissa, since the number with the larger exponent dominates.

The box shows a brief summary of rules-of-thumb.

Arithmetic Error Sensitivity: Fixed vs. Floating Point		
	Fixed Point	**Floating Point**
Add:	Straightforward; No error for single add of normalized binary fraction (with overflow bit).	Denormalize smaller number. Roundoff error possible.
Mpy:	For b-bit inputs, $2b$-bit outputs must be accommodated. Care in rounding accumulation.	Product bit-growth slow (exponential representation). Mantissa error growth (for single-precision).

Block Floating Point

The bits assigned to the exponent are bits of precision given up by the mantissa. *Block floating point* is an attractive compromise when word-length is limited. As shown in Figure 5.16, binary numbers in a block

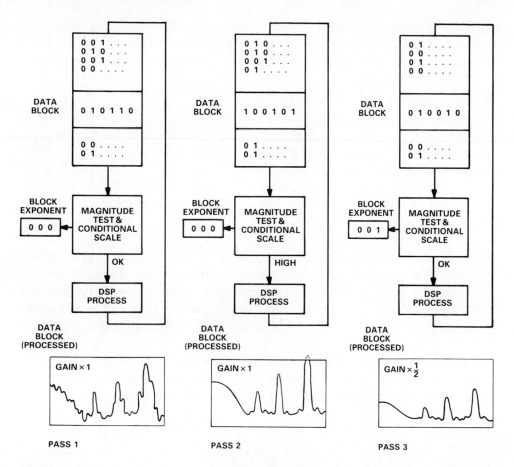

Figure 5.16. Block floating point example for unsigned data. The overflow condition of any point is detected, the entire array is rescaled downwards on the next pass, and the scaling is stored in the block exponent register.

undergoing a computation are scaled to occupy close to the full word-length, for minimum roundoff error, and a *single* exponent register is assigned to the *block exponent* of the whole block. Full-length mantissas are preserved—at the cost of only one extra register and some scaling logic. As the computation proceeds, some point in the array could approach full-scale (in Pass 2 of the example sketched). The magnitude test logic detects this condition for *each* member of the array as it passes through the process, and scales the *entire* array down (note scale change in Pass 3) to avoid overflow.

5.4.3 OVERFLOW, UNDERFLOW, AND ROUNDING

The results of an arithmetic operation often extend beyond the wordlength of the data bus. Guidelines developed in Section 5.4.4 give estimates of word-length growth. The following section examines how best to throw away bits when necessary. *Double precision* keeps all $2b$ bits of information resulting from a b-bit multiply, while *extended precision* adds some extra bits to an accumulator, thereby providing overflow guard bits without the slowdown of double precision. *Rounding* and *truncation* pare down the result to fit register or bus-width requirements.

Extended Precision and Double Precision
When two b-bit words are multiplied, maximum word-length doubles. If the limited available bus width rules out keeping full double precision, but accuracy demands more than the single word-length result, an *extended-precision* compromise defined below is often adequate.

$(b \text{ bits}) \times (b \text{ bits}) = (2b \text{ bits})$ Double precision
$\rightarrow (b + g \text{ bits})$ Extended precision

where g is the number of guard bits—in an accumulator, for example. While the result of a computation sequence will eventually be rounded to fit the data bus, keeping extended-precision intermediate values allows error-free accumulation of 2^g additions—and even more if $+$ and $-$ signed numbers result in partial cancellation. Likewise, in multiplying 16-bit numbers, keeping a 24-bit extended precision result ($g = 8$) may often suffice. Extended precision throws away the portion of the result containing the least information. To see this, write each b-bit number as $[MS \, LS]$, as if it were a double-precision $(b/2)$-bit number, and track what happens to the partial products:

$$[MS \, LS]_1 \times [MS \, LS]_2 = \left(2^{\frac{b}{2}} MS_1 + LS_1\right)\left(2^{\frac{b}{2}} MS_2 + LS_2\right)$$

$$= 2^b (MS_1 {\cdot} MS_2) + 2^{\frac{b}{2}}(MS_1 {\cdot} LS_2 + MS_2 {\cdot} LS_1) + (LS_1 {\cdot} LS_2)$$

$$= 2^b (MS_1 {\cdot} MS_2) \quad [\text{16-bit left-shift, KEEP}] \qquad (5.6)$$

$$+ 2^{\frac{b}{2}}(MS_1 {\cdot} LS_2 + MS_2 {\cdot} LS_1) \quad [\text{8-bit left-shift, KEEP}]$$

$$+ (LS_1 {\cdot} LS_2) \quad [\text{unshifted, IGNORE}]$$

Noise and signal fluctuation buildup, as a string of multiplies is accumulated (as in a convolution or FFT), makes the least significant part, $LS_1 \cdot LS_2$, relatively unimportant. In practice, that partial product can often be ignored.

Even a true *double-precision* DSP operation may compromise at truncating the multiplication result to allow for growth in the accumulation step. For example, an algorithm for a double-precision FIR filter has the approximate form:

DOUBLE-PRECISION FIR

$$Y(N) = \sum_{K=0}^{L-1} \{H_M(K)X_M(N-K) + [H_M(K)X_L(N-K) + H_L(K)X_M(N-K)]2^{-16}\} \quad (5.7)$$

for 32-bit fractional coefficients written in 16-bit halves as $[H_M H_L]$ and 32-bit data as $[X_M X_L]$. The least-significant term, $[H_L(K)X_L(N-K)2^{-32}$, has been truncated away to keep each product output at 48 bits in order to allow free accumulator growth to 64 bits without overflow.

Truncation, Rounding, Unbiased Rounding
As indicated above, the result of a multiplication typically has to be reduced in width, with the more significant bytes, MSB, kept and the less-significant bytes, LSB, eliminated. For example,

(MSB LSB)	→	(MSB)	
0010 1001	→	0010	truncation
0010 1001	→	0011	rounding

Three alternatives are commonly encountered: *truncation*, *rounding*, and *unbiased rounding*. Here is a decimal example:

NUMBER	TRUNCATE	ROUND	UNBIASED ROUND
1.44	1.4	1.4	
1.47	1.4	1.5	
1.49	1.4	1.5	
1.X5	1.X	1.(X + 1)	1.X or 1.(X + 1) (50-50 chance)

└──────── Randomly even or odd digit

NUMBER	TRUNCATE	ROUND	UNBIASED ROUND
1.111145	1.11114	1.11115	1.11114
1.111155	1.11115	1.11116	1.11116
1.111165	1.11116	1.11117	1.11116
1.111175	1.11117	1.11118	1.11118

Truncation of the LSB is to be avoided, since it systematically underestimates the values: 1.49 becomes 1.4 when truncated from 3 to 2 digits.

Rounding pushes a result downwards if the bits to be deleted are less than half of the full-scale LSB, and upward if more than half-scale. For example, 1.44 becomes 1.4 and 1.47 becomes 1.5 when rounded to two digits.

Unbiased Rounding: When a number falls at the half-way point, round it up half the time and down the other half. This is usually done by rounding towards an even number ($1.35 \to 1.4$, $1.45 \to 1.4$) and relying on the random distribution of even and odd numbers. The result is a slight skewing of the even-odd distribution, but an averaging of the roundoff errors.

A comparison of truncation and rounding summarizes why rounding is preferred, unless you happen to be the recipient of the truncated "spare change" in an interest calculation.

NUMBER 4 digits	TRUNCATION to 3 digits	ROUNDING to 3 digits
0.047	0.04	0.05
0.044	0.04	0.04
0.049	0.04	0.05
0.050	0.05	0.05
0.043	0.04	0.04
0.048	0.04	0.05
0.051	0.05	0.05
0.049	0.04	0.05
0.046	0.04	0.05
SUM: 0.427	0.38	0.43

The example below demonstrates the action of unbiased rounding for a string of decimal numbers which happen to all fall at the halfway point between digits:

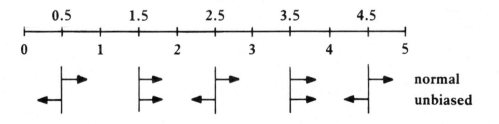

The *normal* round at the halfway point is up (rounding towards $+\infty$); the *unbiased* round tends to go down half the time.

Saturation Arithmetic: A Desirable Option.
When a number passes full-scale magnitude and overflows into the register extension, a substantial error can occur in computation, since the largest positive number suddenly starts counting upwards from zero (unsigned arithmetic) or wraps around to count upwards from minus 1 (twos complement):

$$(11...11) + 1 = (00...00) + \text{ovfl} \quad \text{(unsigned)}$$
$$(011..11) + 1 = (111..11) \quad \text{(twos complement)}$$

A *saturation arithmetic* option in computational devices simulates the "pegged" or full-scale condition of analog electronic circuits: the bits are kept at all 1's whenever bits are found in the overflow register. Here's why:

Suppose a signal level or noise spike larger than full scale occasionally comes through, as shown in Figure 5.17 (top). It may not be feasible to reduce the gain; signal dynamic range would then be reduced, or perhaps data flow cannot be interrupted.* While occasional in-scale noise can readily be

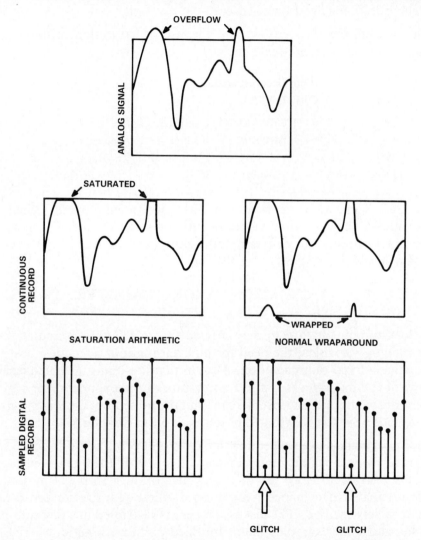

Figure 5.17. Consequences of overflow in saturation arithmetic (left) as compared with normal wraparound (right). The glitch shown in the sampled data, while accurately representing the wraparound values, is erroneous and hard to filter. By contrast, saturation arithmetic has no glitches.

---

*Successive-approximation and flash a/d converters are inherently saturating, but counter types (for example, dual-slope) can overflow; if they do, the overrange flag can be used to cause the processor to saturate the data.

recognized and filtered, wraparound of overflows in the *sampled digitized* data creates sharp opposite-direction transitions (at right) which have broadband harmonic content (perhaps introducing aliases) and are difficult to filter; if later converted to analog form, they will create noise spikes far worse than the original noise.

Saturation to avoid wraparound lessens the error; there are no "wrong-way" glitches on the left. The same option is useful in DSP elements, such as MACs, where computational overflow can occur.

Here is an example of overflow, and the use of saturation arithmetic or rescaling, for 8-bit data (full-scale = 255):

Decimal Value	Binary Representation OVF DATA	
293	0 0 1 00100101	Overflowing datum
37	00100101	if 8-bit byte alone output
255	11111111	if saturation arithmetic
293/2 = 146	10010010	if overflow rescaled (right shift)

Clearly 255 (saturation) is closer to 293 (overflowing result) than 37 (wraparound)! The saturation value, as with a pinned voltmeter, can signal overflow to the process downstream, for a reduction in gain at an opportunity when data flow will not be lost.

5.4.4 FILTER QUANTIZATION-ERROR TRADEOFFS IN FIXED-POINT ARITHMETIC

While floating-point arithmetic may provide a more-accurate computation for widely ranging variables, the need for high throughput at modest cost will often dictate a fixed-point solution. Also, in floating-point, as noted earlier, accuracy of the mantissa is lost in trading bits to the exponent, for a given wordlength constraint. The lost bits can limit the noise floor rapidly in certain computations. (See the Newton-Raphson example in Chapter 8.)

Filters, in particular, are well suited to fixed-point computations, since real-world signals have dynamic ranges that are often within the fixed-point word-width, the computed arrays are of modest length, and 16-bit computations provide adequate performance in many cases. Filter quantization errors have been extensively studied. The discussion here is restricted to a few summary guidelines and some examples, *mostly* for fixed-point arithmetic, where error accumulation is most likely. For further discussion, see Rabiner and Gold (1975) or Peled and Liu (1976).

(1) IIR error is greater than FIR error for a given transfer function.
Compare two prototypes, the 1st-order IIR low-pass and an FIR low-pass of comparable performance (Figure 5.18).

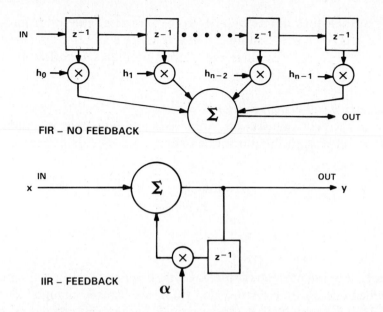

Figure 5.18. The FIR filter does not feed back its output signal; but even the simplest (first order low-pass) IIR filter returns a sample of the output back to the input indefinitely.

For the recursive IIR system, the number of bits required to represent an exact result grows with time, since each iteration around the loop returns to the summing point a feedback term $\alpha y(n-1)$, which is b bits wider (if unrounded) than the last, for b-bit coefficient and data width.

n	$x(n)$	$y(n)$
0	0	0
1	$x(1)$	$x(1)$
2	$x(2)$	$x(2) + \alpha x(1)$
3	$x(3)$	$x(3) + \alpha x(2) + \alpha\alpha x(1)$
4	$x(4)$	$x(4) + \alpha x(3) + \alpha\alpha x(2) + \alpha\alpha\alpha x(1)$

.
.
.

This wordlength growth is not possible in practice, of course. Rounding at the multiplier output keeps the number of bits in $y(n)$ from increasing without limit. But a high-performance IIR filter may have a long memory: if coefficients α lie close to 1, the higher-order products are not small, and rounding-error accumulation is therefore significant.

A related issue, the *limit cycle*, arises in IIR filters, for example, the biquad bandpass. The finite-wordlength feedback loop has a deadband within which output fluctuations are not fed back as input corrections. Therefore, in an IIR filter, the output does not necessarily go to zero after the input has gone to zero. More precisely, if rounding of a nearly-unity coefficient places it on the unit circle, a *limit cycle* results: the output oscillates indefinitely at small amplitude.

(2) Rounding is better than truncation.
As shown in the previous section, unbiased rounding, that is, choosing the *b*-bit number closest to the unrounded result, gives error of zero mean and width 2^{-b}.

$$\epsilon = \frac{y(\text{rounded}) - y(\text{exact})}{y(\text{exact})} \qquad (5.8)$$

$$\frac{1}{2}2^{-b} < \epsilon < +\frac{1}{2}2^{-b}$$

Truncation, lopping off less-significant bits, gives an error of the same width but the mean is shifted downscale, since all "small change" is thrown away indiscriminately. The quantization noise probability density is summarized in Figure 5.19. Only positive numbers are considered in this analysis; with a sign bit, the error mean and width depends upon the choice of representation.

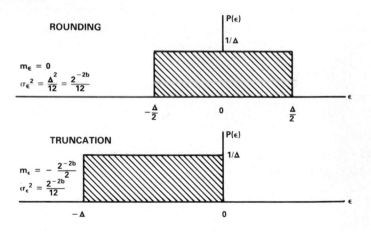

Figure 5.19. Error distribution in truncation and rounding.

(3) Error can arise from quantization of input or filter coefficients, or from arithmetic roundoff.

(a) Input or output quantization. Accurate computation cannot readily make up for too-coarse input or output quantization, as in the displayed letter "A" of Figure 5.2b. Input quantization, when an analog signal is digitized by an a/d converter, acts like noise equal to the difference between ideal (continuous analog) and quantized value. While the error for an ideal *b*-bit a/d converter, shown in Figure 5.20, is *bounded* by half the quantization step size (above),

the likely error range for most input signals is like that of white noise of zero mean and of variance $\sigma_e{}^2$:

$$<\sigma_e>^2 = \frac{2^{-2b}}{12} \qquad (5.9)$$

Here, the subscript, "e", stands for error and the 12 comes from the variance calculation (see Exercises). Input quantization error is the benchmark against which coefficient and arithmetic roundoff errors are to be compared. As soon as output error grows above the value in Eq. 5.9, available information is being lost, since noise has been added by the DSP computation.

The *noise-to-signal ratio*, NSR, in dB ($= -$ SNR, the familiar signal-to-noise ratio) is a better measure of information preserved than noise growth alone, since signal scale factors are also included in NSR. To obtain a rule-of-thumb for NSR, scale the signal to be as large as possible without significant chance of clipping (overflow). For a broadband signal, such as audio, a gain, G, of

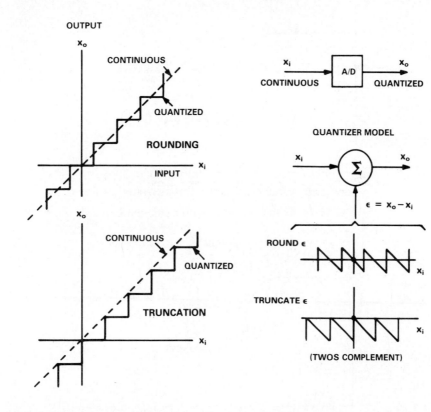

a. Output/input relationship for ideal ADC.   b. Noise source model and its error bounds for rounding (typical ADC) and truncation (1/2-LSB bias).

Figure 5.20. Quantization noise model.

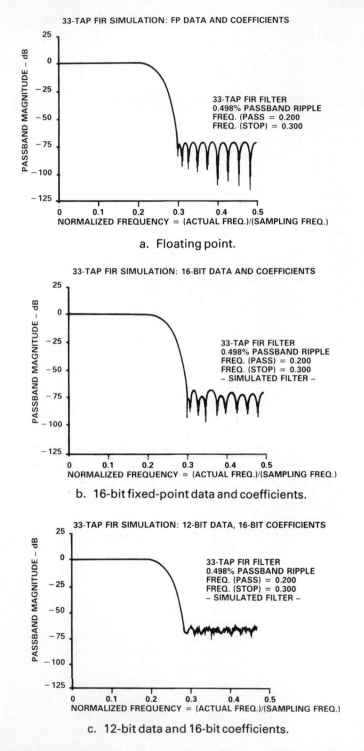

a. Floating point.

b. 16-bit fixed-point data and coefficients.

c. 12-bit data and 16-bit coefficients.

Figure 5.21. Comparison of floating-point and fixed-point 33-tap filter transfer function with finite-wordlength coefficients. (Source: Analog Devices)

roughly one quarter the rms signal level, σ_x, (or $G\sigma_x = 1/4$) stands little chance of clipping. The NSR is then

$$\text{NSR} = 10 \log \left(\frac{\sigma_e}{G\sigma_x}\right)^2$$

$$= 10 \log \left[\frac{4}{3} \cdot 2^{-2b}\right] \qquad (5.10)$$

The noise-to-signal ratio for input quantization is thus★, (approximately)

$$\text{NSR (dB)} = -6\,b + 1.24 \qquad \text{[INPUT QUANTIZATION]} \qquad (5.11)$$

for b-bit information. Thus, an NSR of -80 dB or less (SNR of 80 dB or more) requires quantization to at least $b = 14$ bits.

(b) Coefficient quantization. When finite word-length coefficients are used to calculate filter outputs (product-sums of the form N/D, the poles $(D = 0)$ and zeros $(N = 0)$ are shifted from their ideal positions. This is ordinarily not very serious, except when IIR poles lie close to the unit circle. Zeros may no longer reach zero, but this is of less significance since there is only slight degradation, even to the stopband attenuation. An FIR example is depicted in Figure 5.21.

Disturbance of a pole position is more serious, especially for high-performance IIR filters with poles close to the unit circle. Stability criteria have been evaluated for numerous cases but will not be discussed here, since performance will have been degraded so as to be unacceptable before an actual instability is encountered. IIR filters may be computed in several equivalent ways, corresponding to parallel, cascade, or direct architectures (Chapter 4). The parallel and cascade forms are less susceptible to coefficient errors than the direct form and are therefore preferred. While moving a pole onto the unit circle causes both deadband and limit-cycle errors, the same phenomenon can be useful: an oscillator of stable amplitude and frequency can be made from a fixed-point IIR limit cycle (Chapter 8).

(c) Computational Roundoff Error. The effect of roundoff is modeled in Figure 5.22 as an independent white noise source, ϵ. It injects noise at each multiplication (fixed point) or at each multiplication and accumulation (floating point). Fixed-point computations are also susceptible to *overflow* at any accumulation step. This limitation of dynamic range by overflow and rounding affects the noise-to-signal ratio and makes careful scaling necessary. Floating point is relatively insensitive to dynamic range limitations; however, it is susceptible to noise at both accumulation and multiplication, since, prior to the add, the mantissa of the smaller number is shifted right to make the exponents identical. The numerical example in Figure 5.22 illustrates how formatting the same pairs of numbers as a binary fraction (implied decimal point at

★Oppenheim and Schafer (1975), p, 417. Note that for *sine waves*, signal-to-noise ratio is $6b + 1.76$dB.

the left) leaves the product unchanged but pushes word growth off to the right automatically. The size of the noise source, ϵ, depends upon what roundoff is performed; $\epsilon = 1/128$ in the numerical example if all bits are kept, and $\frac{1}{32}$ ($= \text{LSB}/2$) if the result is rounded to the 4-bit input width.

A result of random variable theory, Parseval's theorem (see 2.3.4) can be used to *predict* the noise added to a signal by a computation:

The variance at the output of a system is the autocorrelation of the system's impulse response.

$$\sigma_0{}^2 = \frac{2^{-2b}}{12} \sum_{n=0}^{\infty} h^2(n) \qquad (5.12)$$

where $h(n)$ is the filter's response to a unit impulse introduced at the noise-generating location. Equation 5.12, proven by probability theory in Section 4.2.5, has an intuitive physical explanation: The $2^{-2b}/12$ term is the noise set by the quantization step-size of the filter computation, as in Eq. 5.9. But any noise input, even a quantization noise impulse, causes a filter to ring, and ringing response is governed by the impulse response, $h(n)$ (Chapter 2). The net rms output noise, $\sigma_0{}^2$, is, by the convolution theorem, the integral of the quantization noise level times the [impulse response]^2 over all past times—precisely what Eq. 5.12 states.

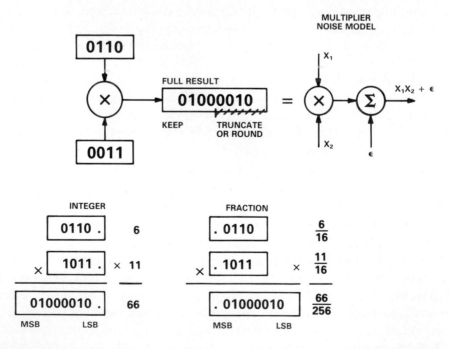

Figure 5.22. Modeling filter roundoff error as a noise source. The binary fraction format (at right) keeps the most-significant half of the output in the bit position of the inputs.

The noise added by a filter will therefore grow in proportion to its impulse response time, or (equivalently) to the sum of the squares of the filter coefficients.

Error accumulation as time goes on is, alternatively, predictable from the behavior of the filter transfer function in the frequency domain, since the square of the impulse response summed over all time is, by Parseval's theorem, the integral of the transfer function's magnitude round the unit circle.

$$\sigma_0^2 = \frac{2^{-2b}}{12} \int_{-\pi}^{+\pi} |H(e^{j\omega t_s})|^2 \, d\omega \qquad (5.13)$$

As with input quantization, translation of computation noise rules, such as Eqs. 5.12 and 5.13, into a useful rule-of-thumb requires input scaling and the computation of a noise-to-signal ratio. The rule that follows is over-conservative but allows a comparison of several common filter types. Fixed-point scaling ordinarily calls for binary-fraction format, as shown in the example of Figure 5.22.

Signal Scaling Rules: To avoid overflow, keep the output magnitude, $|y(n)| < 1$. Since $y(n)$ is the convolution of filter impulse response with the input, $x(n)$, this requires:

$$1 > |y(n)| \geq x_{\max} \sum_{r=0}^{\infty} |h(r)| \qquad (5.14)$$

To guarantee no overflow in the filter computation, the input must thus be scaled so that

$$x_{\max} < \frac{1}{\sum |h(r)|} \qquad (5.15)$$

The correct scaling of input signal *amplitudes* is thus related to actual *filter coefficient* values.

Noise to Signal Ratios for Common Filters: A filter's rms noise depends on the spectrum of the signal because allowable filter scaling depends on the signal spectrum. Two limiting cases, white noise and a pure sine wave, demonstrate limiting cases of broadband and narrowband signals. Inputs are scaled to their maximal value, given by Eq. 5.15, e.g., to $(1 - \alpha)$ for white noise input to the 1st-order low-pass, $y(n) = x(n) - \alpha y(n - 1)$. The results of calculations for these two cases are summarized in Table 5.6 for a first-order low-pass and a second-order biquad with one complex pole pair.

Comparing this to input quantization, NSR $\cong 2^{-2b}$, of Eq. 5.9, it is clear that, for $\delta \rightarrow 0$, noise is greatly increased relative to the signal. Physically, as a pole approaches the unit circle, the impulse response decays more slowly, and noise accumulation increases, since noise is proportional to the integral of the square of the impulse response in Eq. 5.12.

Table 5.6 Noise-to-signal ratios for white noise and pure tone b-bit inputs for 1st-order lowpass and 2nd-order biquad with a single complex pole pair. In both cases, δ measures the distance of the pole from the unit circle. Following the notation of Section 4.4, for 1st-order LP, $\delta = (1-\alpha)$; α is the feedback coefficient, and $\delta = (1 - r)$ for the biquad, where r measures the radius of the pole in the denominator. The maximal input signal amplitude is set by the guideline of Eq. 5.15. For more details, see Oppenheim and Weinstein (1972).

	1st-order lowpass	2nd-order biquad
Noise-to-signal for:		
White noise	$\dfrac{2^{-2b}}{4\delta^2}$	$\dfrac{2^{-2b}}{2\delta^2\sin^2\theta}$
Sine wave	$\dfrac{2^{-2b}}{48\delta}$	$\dfrac{2^{-2b}}{2\delta\sin^2\theta}$

The key result in Table 5.6 in helping the user design for accuracy is the dependence upon the power of δ. A summary guideline is given in Table 5.7:

Table 5.7 Number of bits which must be added to avoid error as the pole is brought closer and closer to the unit circle to make a more selective filter.

Signal	Bits added to keep same NSR
Broadband (white noise)	One bit added as δ is halved
Narrowband (sine wave)	One bit added as δ is reduced to $\delta/4$

The above rules are illustrative of the trend and, while safe, give an overly pessimistic estimate of actual error magnitudes. The input scaling bound is over-protective, since an equality in Eq. 5.15 is unlikely. For example, *saturation arithmetic* (above) allows for occasional upper bounds to be reached and clamped at a saturation value without wraparound or other glitches.

Noise Estimation Rules for FIR vs. IIR Filters

The above example is specific to two particular IIR filters. A general guideline can be written that can be evaluated in principle for any filter. For IIR types, noise estimation requires evaluation of a complex integral around the unit circle; the output is sensitive to the poles in the transfer function. By contrast, for FIR types the estimate is simple and general: noise is proportional to number of taps and *independent* of coefficient values. Noise estimation and control is therefore simpler for FIR filters.

Since the variances of independent noise sources add, the quantization noise in a multiple-tap filter is proportional to the number of taps. Basic filter theory shows that the net noise is:

GENERAL EXPRESSION FOR NOISE ESTIMATION

$$\sigma^2 = (N + M + 1)\left(\frac{2^{-2b}}{12}\right)\frac{1}{2\pi}\int_0^{2\pi}\frac{d\omega}{D^2(e^{j\omega t_s})} \tag{5.16}$$

where $D(e^{j\omega t_s})$ is the denominator of the overall transfer function; $(N + 1)$ is the number of (nonrecursive) coefficients in the numerator; and M is the number of (recursive) coefficients in the denominator. This result follows from application of Parseval's relation to noise variance, Eq. 5.13 above.

Here is the basic explanation of Eq. 5.16: In computing $y(n)$, $(N + M + 1)$ products are formed. Each product computation has its own computation error, which may be assumed independent and white. Thus, the equivalent noise has a variance proportional to the number of multiplications performed. This noise is passed recursively around the filter, with a transfer function proportional to the *denominator*, $D(e^{j\omega t_s})$, of the overall transfer function. Since terms in the numerator are not recursive, they contribute no feedback.

Thus, there are two classes of filter noise:

IIR Filter Noise: The evaluation of the integral in Eq. 5.16 above is complicated and specific to the filter. Guidelines for first- and second-order IIR prototypes already summarized in above in Table 5.6 will be sufficient for our purposes.

FIR Filter Noise: For the FIR, the denominator in Eq. 5.16 is 1, so the noise takes on the much simpler form:

$$\sigma^2 = (N + 1)\left(\frac{2^{-2b}}{12}\right) \tag{5.17}$$

The noise is *independent* of filter coefficient *values*, since (referring to Figure 5.23), the noise passes directly to the output summation without going

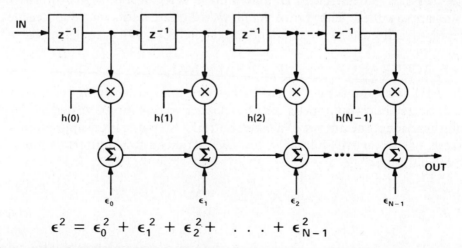

$$\epsilon^2 = \epsilon_0^2 + \epsilon_1^2 + \epsilon_2^2 + \ \cdots\ + \epsilon_{N-1}^2$$

Figure 5.23. Equivalent noise model for a digital FIR filter. Noise sources, ϵ, replace the quantization error of finite arithmetic.

through the coefficients. The noise increases linearly with the *number* of taps N, so doubling the number of taps requires one more bit to retain the same accuracy. This is the noise growth, or NSR, for *properly scaled* inputs, since the dependence of σ upon $h(n)$ (Eq. 5.12) drops out in the scaling (Eq. 5.15) to prevent signal overload.

As with the specific 1st and 2nd order examples above, the relevant quantity is not just noise but noise-to-signal ratio. An input gain factor, A, is chosen as large as possible to minimize quantization error but not so large as to cause overflow. To estimate noise-to-signal, two input scaling rules for A are useful limiting cases.

$$GENERAL\ INPUT\ SCALING\ RULE: \tag{5.18}$$

Broadband signal $\qquad A \;<\; \dfrac{1}{x_{max} \displaystyle\sum_{n=0}^{N-1} |h(n)|}$

Narrowband signal $\qquad A \;<\; \dfrac{1}{x_{max} \max\,[\,|H(e^{j\omega t_s})|\,]}$

To estimate the NSR, plug in filter coefficients in time (for broadband signal) or transfer function in frequency (for narrowband signals), together with anticipated maximum input signal amplitude, x_{max}; then compute the appropriate gain, A, of Eq. 5.18; then compute the NSR from the ratio σ^2/A, using the noise, σ^2, computed from Eq. 5.17.

Avoiding Noise Overload in Filter Cascades Scaling of filter gains and the ordering of filter sections is a key to minimizing overall noise and avoiding overload due to overflow of amplified noise when several signal-processing stages are cascaded. The approach follows the quantization noise source ideas of the above discussion. See Oppenheim and Schafer (1975), for example.

5.4.5 ACCURACY IN FFT SPECTRAL ANALYSIS

The FFT is a prototype of a long array-computation, with $\log_2 N$ passes of $(N/2)$ butterfly computations for each pass. For real inputs scaled as fixed-point fractions, the output of a given butterfly step can grow approximately a factor of two for each butterfly. For decimation-in-time butterfly inputs X, Y, and outputs X', Y'

$$X' = X + WY$$
$$Y' = X - WY \tag{5.19}$$

where W is the twiddle factor, $e^{j\theta}$. Since the sum of two independent fractions is at most 2, the simplest approach for real data would give the data-set a

1-right shift after each pass. Some comments about the consequences of this strategy:

(a) Scaling down a bit each pass is *much* better than pre-attenuating the input by a large factor (the total anticipated growth) and doing no scaling during the butterfly computations. The computation noise level will be low, since noise introduced at early stages of FFT computation will be attenuated by the downscaling that takes place at later stages. By contrast, input preattenuation by a large factor and no downscaling at later stages amplifies the computation noise added at the earlier passes of the FFT.

(b) The output with bit-per-pass downscaling is decreased by $1/N$ from the DFT output defined in Chapter 3, since $\log_2 N$ downscalings of ½ each equals N. This extra $1/N$ scale factor is often ignored erroneously.

(c) If available, *block floating point*, as defined in the previous section, is preferable to blind downscaling, since the growth towards overflow of any spectral point depends upon the "peakiness" in the spectrum. *Parseval's theorem* (Section 2.3.4) tells us that, for signals scaled to unit rms amplitude over the observation window, the integrated area under the signal's spectrum is a constant (see Figure 2.8). As the DFT sum is performed, the single peak associated with a pure tone will therefore grow more rapidly than any point in a broadband noise spectrum, since both have the same integrated area. A block floating-point algorithm checks output magnitudes after each pass and tests for values greater than ½. Should any value be out of range, the entire array is renormalized before the next pass.

For an FFT on *complex* signals, the guidelines are a little different, since the butterfly computations are sums of real and imaginary components. To estimate the growth, write the FFT decimation-in-time (DIT) butterfly for complex input data, $(X_0 + jY_0), (X_1 + jY_1)$, as:

$$X_0' = X_0 + (X_1 C - Y_1 S)$$

$$X_1' = X_0 - (X_1 C - Y_1 S)$$

$$= X_0 - (X_0' - X_0) = 2X_0 - X_0'$$

$$Y_0' = Y_0 + (X_1 S + Y_1 C)$$

$$Y_1' = Y_0 - (X_1 S + Y_1 C)$$

$$= 2Y_0 - Y_0' \tag{5.20}$$

where S and C stand for the sine- and cosine components of W. Factor-of-two per-pass scaling may not be adequate, as indicated in the following guideline:

Complex FFT worst-case growth per butterfly pass $\qquad\qquad$ (5.21)

$1 + 2^{1/2}$ $\qquad\qquad$ Decimation in time

$2 \cdot 2^{1/2}$ $\qquad\qquad$ Decimation in frequency

$-\tfrac{1}{4} < X < \tfrac{1}{4}$ $\qquad$ Optimum (blind) prescaling:

To see why, first consider decimation in *time*. Note in Figure 5.24 that output X_0', for example, is the sum of X_0 plus a rotated vector having components X_1 and Y_1. While the angle could take on any value during the computation, the worst case shown adds $\sqrt{2}$ to the magnitude of X_0, giving worst-case growth by a factor of $1 + \sqrt{2}$. In the case of decimation in *frequency*, the butterfly pattern takes the twiddle factor as a postmultiply. The worst case is the lower term in the Figure:

$$X_0' + jY_0' = (C + jS)(X_0 + jY_0 - X_1 - jY_1)$$

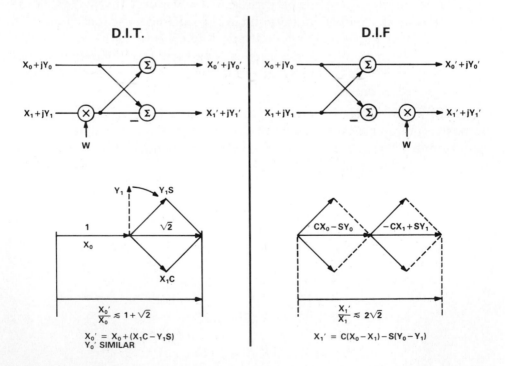

Figure 5.24. Geometric explanation of worst-case butterfly output growth in the decimation-in-time and decimation-in-frequency FFTs.

and the component, X_1, for example, is the sum of *two* vectors

$$X_1' = (CX_0 - SY_0) + (-CX_1 + SY_1)$$

which in worst case could each add a factor of $\sqrt{2}$ growth.

As a result, prenormalization of initial data inputs to magnitude less than ¼ (rather than ½) is guaranteed not to overflow. In practice, block floating point is preferred when possible, and leads to considerably less noise/signal growth than blind 2-bit downscaling per butterfly (See Weinstein references quoted in Oppenheim and Schafer Chapter 9).

5.5 ANALOG I/O METHODS: DYNAMIC RANGE AND THROUGHPUT TRADEOFFS

5.5.1 OVERVIEW

Analog I/O is essential for nearly every DSP system. For the reader unfamiliar with analog signals, the basic requirements will be briefly explained; refer to D. Sheingold, ed., *Analog-Digital Conversion Handbook* (1986) or to Higgins (1983) for a more thorough discussion of analog-digital issues and methods. Analog I/O for digital signal processing requires resolution and throughput that is adequate not only for the signal's dynamic range and bandwidth but also for the digital computations to be performed. The previous section discussed quantization error from the numerical analysis point of view. The present section reviews specific real-world I/O requirements and the capabilities and limitations of specific methods from the signal's point of view.

The key components of a typical system are shown in Figure 5.25a. The input is anti-alias filtered with a low-pass designed to reject $f_s/2$ and all higher frequencies, while passing the highest permitted signal frequencies. Modern antialias filters often use *switched-capacitor* techniques and are available as inexpensive self-contained ICs, which embody analog transfer functions with high speed, parametric stability, and programmability, and with flat passband and sharp cutoff. The signal is then sampled with the *sample-and-hold* (S/H) circuit (shown in its simplest form to exemplify the principle)—a capacitor's charge follows the input voltage until a switch is opened at the sampling instant, then the capacitor holds the charge until the sample is digitized, the switch is closed, and a new cycle starts, repeating at a rate, f_s.

The *analog-to-digital converter* (ADC) converts the analog signal to binary format. The most common ADC for DSP uses the successive-approximation method, the digital equivalent of a two-pan weighing scale. The sampled analog input is compared against a binary approximation generated by the *digital-to-analog converter* (DAC), which gets its input from a successive-approximation register (SAR). The logic exercises each bit in turn, starting

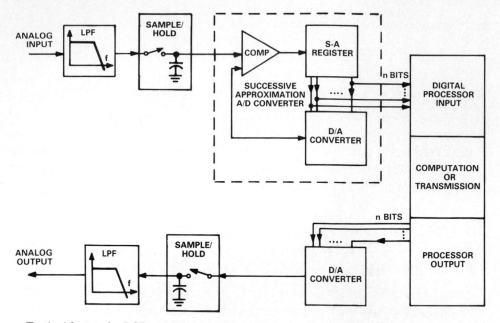

a. Typical form of a DSP analog input (showing successive-approximation hardware) and output system.

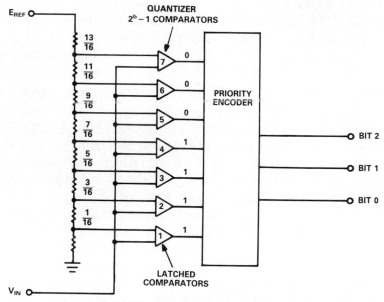

QUANTUM LEVEL TO BINARY ENCODING

LEVEL	0	1	2	3	4	5	6	7
BIT 2	0	0	0	0	1	1	1	1
BIT 1	0	0	1	1	0	0	1	1
BIT 0	0	1	0	1	0	1	0	1

b. High-speed "flash" converter; comparator output states shown for $V_{IN} = 8/16$.

Figure 5.25. Conversion system and converter.

with the most-significant bit, thus trying approximations to the input at successively lower powers of two; each added "weight" is either kept (1) or eliminated (0) from the SAR, depending upon the sign of the comparator output. The binary output, after all b bits have been tested and set, is passed to the digital processor. The throughput of this ADC method scales inversely with resolution: the conversion time is approximately b clock cycles for a b-bit word.

Successive-approximation converters are generally limited to sampling rates somewhat less than 1 MHz. For acquiring digital samples at faster rates, parallel "flash" converters are used, in single or multiple stages (Figures 5.25b and 5.27a). A "flash" converter consists of a priority encoder and $2^n - 1$ comparators in a string, biased 1 LSB apart, starting at $\frac{1}{2}$ LSB. For 0 input, all comparators are off. As the input increases, it causes an increasing number of comparators to switch state in sequence, rising, like the liquid in a classic thermometer. The outputs of the comparators are applied to a *priority encoder*, which provides at the output a code that depends on the number of comparators that have been consecutively turned on.

The evident advantage of this approach is that conversion occurs in parallel, with speed limited only by the settling time of the input circuitry and the switching time of the comparators and encoder. As the input changes, the output code changes. Thus, this is the fastest kind of conversion circuit. In practice, the comparators are latched simultaneously in order to establish a precise sampling time and minimize the effects of small timing differences between the comparators.

Analog output to drive display devices or generate control voltages is provided through a *digital-to-analog converter* (DAC), usually a resistive ladder network with currents switched in binary weights to develop an output voltage proportional to the value of the binary word. A sample/hold circuit on the DAC output lessens switching transients (or DAC *glitches*) and frees the processor output register for other uses; a smoothing filter suppresses the stairstep changes in the S/H output to create a smooth analog waveform. Both the ADC and the DAC may be multiplexed to bring several signals in or out of a given processor port. It is also feasible to use additional a/d and d/a channels in parallel to maximize throughput.

5.5.2 I/O DYNAMIC RANGE AND THROUGHPUT REQUIREMENTS

Since computational accuracy is closely related to the dynamic range and accuracy available in the input signal and required of the output signal, the resolution, accuracy, and speed of the ADC and DAC must be matched to both signal bandwidth and digital-process requirements. A few common examples are tabulated in Table 5.8. The requirements for two common applications—laboratory measurements and digital audio—are discussed briefly below.

Table 5.8 Analog I/O Dynamic Range and Binary Resolution.

Bits	Resolution	Dynamic Range, dB	Application Example	ADC Method for Fastest
8	.4%	48.2	Transient Recorders	Flash
10	.1%	60.2	Radar	
12	.02%	72.2	Telecommunications	Subranging
14	.006%	84.3	Lab Measurements	Successive approx.
16	.0015%	96.3	Digital audio	
18	.00038%	108.4	Digital multimeter	Analog Integration
20	.000095%	120.4	Vision limits	
22	.000024%	132.4	Precision DVM	

Laboratory Instrumentation I/O Requirements and Methods
Measurement resolution is no better than the system signal-to-noise. Noise in analog signal conditioning is kept to a minimum by careful grounding, shielding, and using the narrowest bandwidth required. For signals at the ADC scaled to 10 volts and ADC input drifts of 100 μV, 16-bit resolution is a typical upper limit for broadband signals. Higher resolution is met with a digital-voltmeter-type converter. Speed and resolution are traded off, with 6½-digit (20-bit +) measurements limited to <100Hz in a precision DVM, while the *system DVM* is capable of audio range at 16-18 bits. The successive-approximation ADC covers a range from about 16 bits, 100 kHz to 12 bits, 1-MHz, while faster measurements use the simultaneous, or *flash*, ADC, which can reach video rates (>200 MHz) at reduced (typically 6 to 8-bit) resolution. Measurement speed can go no faster than the cycle time of the ADC logic.

The above ADC families differ in the number of clock cycles required to make a measurement. The flash ADC applies the signal to a resistive ladder and multi-comparator network; the speed is limited only by the network and comparator settling time plus the digitizing logic speed at the comparator outputs. The successive-approximation method is slower because it takes *b device* clock-cycles for a *b*-bit measurement.

The digital voltmeter attains its high accuracy, typically at the expense of a 2×2^b clock cycle measuring time for a *b*-bit measurement. The most commonly used method, called *dual-slope*, integrates the input while a counter establishes a fixed interval, then integrates the reference in the return direction while the counter output is incremented until the integral returns to its starting point. Since the greater the average input, the greater the integral that must be counted down, the ratio of the return count to the initial count is proportional to the ratio of average signal to reference.

Digital Speech and Audio I/O Requirements
Digital *speech* analysis, synthesis, and transmission impose real signal limits and require psychophysical compromise. The pitch of speech goes no higher than about 1.5 kHz, so a 5-kHz bandwidth (10-kHz sampling) passes enough harmonics for speech to be recognizable as that of a given individual. A dynamic range of 72 dB, or 12 bits, is optimal, while 50 dB is recognizable though granular, and less resolution causes listener annoyance. Digital *audio and music* are much more demanding than speech (see Table 5.8). The resolution and speed requirements are set by both the dynamic range and frequency range of human hearing. The dynamic range between "silence" and pain exceeds 100 dB. The range between full orchestra and background noise in a concert hall is closer to 80 dB. A DSP goal of 80 dB is thus adequate for "concert-hall realism" in digital home-audio systems, while digital studio signal-processors seek an 84- to 94-dB (14-to-16-bit) range. The upper end of hearing at about 20 kHz makes 40-50-kHz sampling rates typical in digital hi-fi audio.

Effects of I/O Quantization Error
The quantization process at the converter affects the signal-to-noise level of any further digital processing: a 12-bit quantization does not give a 16-bit output unless time-consuming averaging is used (see SIGNAL AVERAGERS, Section 8.2).

To maximize dynamic range,
>   Maximize input signal level, so information is not lost in the least-significant bit
>   Filter noise above the signal range which might alias down
>   Minimize harmonic distortion which also might alias down
>   Use resolution adequate to keep *granulation* low.

The last item above, *granulation noise*, is discernible when too few bits are available from the a/d converter. ADC quantization can lead to aliasing which is *not* removable by pre-ADC filtering. For example, a sine wave of amplitude just larger than the smallest ADC step is quantized as a square wave (Figure 5.26a). The harmonics of the square wave descend slowly: in 1/3, 1/5, 1/7 ... proportion for 3rd, 5th, 7th harmonics. Harmonics of the quantized signal can wrap around and alias down to lower frequencies (Figure 5.26a). Because this noise is created after the antialias filter, it persists in the digital computation.

Granulation noise can be masked by adding *dither*, noise intentionally added to the signal to ensure that adjacent cycles are not quantized with the same value. Dithering reduces spurious spectral peaks by smearing the quantization noise over the frequency domain. A familiar example of granulation noise is seen in low-resolution graphics displays (Figure 5.26b). The line drawn close to the horizontal displays a jagged form while the 45-

degree line looks smooth. While both lines may be "straight" as stored in memory, an apparent periodicity appears in the low-resolution display, which is annoying because the period changes rapidly with angle of the line. *Dithering* blurs the line slightly by transferring intensity gradually into adjacent pixels. The steps are made less objectionable, at the cost of edge definition. Dithering and related masking filter operations are extensively used in image processing (Chapter 8) and can find a place in instrumentation and audio as well.

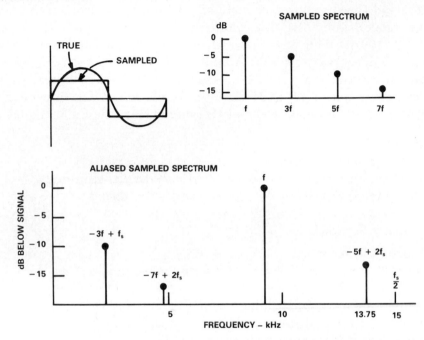

a. Small noise components are quantized into square waves whose harmonics can alias down in frequency.

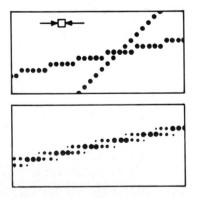

b. Dithering in this graphic example makes quantization noise less objectionable; pixel size is shown by the square.

Figure 5.26. Quantization noise.

5.5.3 EXPANDED I/O DYNAMIC RANGE WITH MINIMAL SPEED LOSS

The speed-resolution tradeoffs may be seen by comparing performance of a full-line manufacturer's series of standard a/d converters (see Sheingold, ed., 1986 for more details on converter design and performance alternatives). In general, speed goes inversely as resolution at a given cost. If more resolution is needed than can be provided by an ADC otherwise well-matched to the system throughput, two real-time scaling enhancements described below (and in further detail in the above reference) provide enhanced resolution without sacrificing speed.

(1) Subranging or Residue Converter. Figure 5.27a illustrates a 12-bit, 10-MHz subranging a/d converter constructed with two flash encoders having resolutions that add up to 13 bits (6 and 7, in this case). The analog signal from the track & hold is applied to a 6-bit flash encoder and a buffered video delay line at the same time. The 6-bit encoder converts the analog signal to binary, producing the 6-most-significant bits. These bits are stored in a register and also applied to a 6-bit d/a converter *having an accuracy of at least 12-bits*. The output of the d/a converter is inverted and subtracted from the delayed track-&-hold output in a summation network, and the resulting "residue" is amplified to the 7-bit converter's full-scale span.

The "residue" is then converted to digital by a 7-bit flash encoder and represents the less-significant portion of information. The outputs of the 6-bit holding register and the 7-bit flash are then combined in an output register, employing digital correction to yield a 12-bit parallel binary output with minimal errors resulting from the 6-7-bit transition.

It would not be at all unusual for converters of this type to exhibit discontinuities around the transition codes, due to mismatch between the front-end and residue encoding circuits. However, these discontinuities are minimized in converters employing digitally corrected subranging (DCS), since the information to accurately characterize the bit-6 transition is already present in digital form because of the accurate d/a conversion and summing, and the 7th bit in the second conversion.

Here's a simplified explanation of how it works: The result of the first conversion—in the holding register—is equal (digitally) to (Input − Residue + 6-bit Error: $I - R + E6$). It is converted to analog and subtracted from Input. Assuming a perfectly accurate DAC and subtraction, the result is

$$I - (I - R + E6) = R - E6$$

This difference is scaled up and converted, giving a digital quantity equal to (Residue—6-bit Error + 12-bit Error). Finally, the two digital words are added, giving (Input + 12-bit Error). That is,

$$(I - R + E6) + (R - E6 + E12) = I + E12$$

The extra bit is necessary because $(R - E6)$ can exceed 1 LSB of 6 bits.

Although the explanation is simple, the execution is not. Fast 6-bit DACs with better than 12-bit accuracy are not easy to make, nor are fast subtracting amplifiers with adequate dynamic linearity. Remember that both the DAC and the output amplifier must settle before the second conversion can be completed.

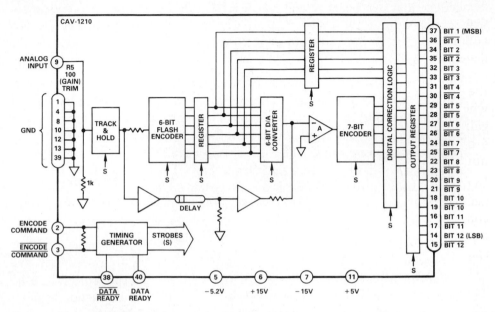

a. Two-stage flash converter with digitally corrected subranging.

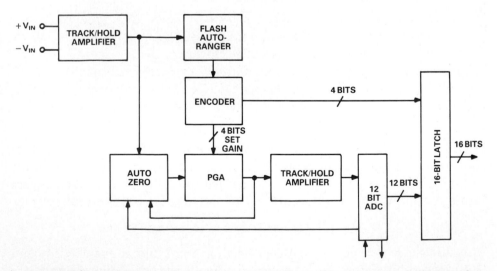

b. Autoranging conversion. Binary gain control keeps the signal dynamically at full-scale of the converter input.

Figure 5.27. Enhanced-range a/d conversion.

However, this technique, incorporating fast emitter-coupled logic, has resulted in 10-bit converters capable of 40-MHz sampling rates and 12-bit converters that can handle 20-MHz sampling, with similar architecture to the one depicted in Figure 5.27a. The technique is promising for even faster and higher-resolution converters implemented with newer high-speed logic families in bipolar silicon or gallium arsenide.

(2) Autoranging Gain Control (Floating-Point Converter). A floating-point converter improves dynamic range by acquiring data in two parts, usually employing a two-step approach. In a typical approach, a binary programmable-gain amplifier (PGA) scales the signal to the proper range, typically between one-half and full scale; the input code to the PGA controls the exponent of 2 (i.e., $G = 2^X$). The scaled signal is converted by an ADC whose output code forms the mantissa, while the digital signal used to set the gain prefaces it with $\log_2 G$, i.e., X. For example, a 12-bit converter with a 4-bit preface and the output code **0100** 1100 0000 0000 would represent a number equal to 0.75×2^4.

In the approach to floating-point-converter design shown in Figure 5.27b, the gain of the PGA is set with a flash autoranger. The 20-bit-dynamic-range conversion circuit shown consists of a pair of track-&-hold amplifiers, a flash octave digitizer (reference levels are at octave ($\times 2$) intervals instead of equally spaced as in Figure 5.25b), a nine-range programmable-gain amplifier, and a fast 12-bit a/d converter.

The first track-&-hold amplifier holds the signal for the flash autoranger, which determines which binary quantum the input falls in, relative to full scale (i.e., $\frac{1}{2}$ to $\frac{1}{4}$, $\frac{1}{64}$ to $\frac{1}{128}$, etc.), and encodes the information in 4 bits. Responding to it, the PGA adjusts its gain to the appropriate level, and the track-&-hold amplifier holds the amplified signal while the two-step a/d converter translates it into a binary number. A 16-bit latch holds the 4-bit output from the encoder and the 12-bit output from the converter.

5.5.4 NONLINEAR I/O DIGITIZATION: COMPANDING AND THE LOGDAC™

Why Nonlinear?
If neither one of these linear methods spans enough dynamic range to represent a given set of signals adequately in digital form, with a limited number of bits, consider incorporating one of the two intentionally nonlinear methods below into the signal path. The first, companding, compresses high resolution into a small word-width while retaining full-speed on the ADC. The second, the LOGDAC, is a very wide range digital attenuator for analog signals. For details, see Sheingold, ed. (1986).

Companding:dynamic-range expander
Companding is a method for expanding the dynamic range of an ADC or DAC. Companding extends the dynamic range by nonlinear quantization: extra bits

are assigned to the small-signal end, and fewer bits at the large-signal end. Quantization intervals more-closely spaced at small signal levels and more widely spaced at high signal levels permit recording with comparable fidelity both gross changes and subtle nuances in the signal. The companding converter has a quasi-logarithmic transfer function, with a quasi-antilog at the other end to restore the original signal. A companding ADC-DAC pair, available as a chip called the CODEC, packs wide dynamic range into a small number of bits. For an input, $|x|$, the compander output is

$$F(x) = \text{sgn}(x) \frac{\ln(1 + \mu |x|)}{\ln(1 + |\mu|)}$$

where $\text{sgn}(x)$ is $+1$ or -1, depending on the sign of input x. The constant μ is the compression parameter; μ governs the dynamic range. With $\mu = 255$, standard in North America and Japan, 72 dB (12 bits of information) are compressed into only 7 bits. On the other end, after decompanding (antilog) the output has a 72-dB dynamic range but only 7 bits (42 dB) of true accuracy for any data point. Companding is successful for speech signals; these specifications provide high-quality recognition of individual speakers while reducing storage and transmission demands. Higher-resolution companders are less satisfactory than other solutions, since enormous importance is placed on the accuracy of the few bits which make the largest output changes.

The LOGDAC^{TM}: Binary Volume Control

A closely related conversion device of very wide dynamic range is the LOGDAC, a digital attenuator chip, for automatic gain control. The transfer function of a typical LOGDAC is an exponential:

$$y = V_{\text{in}} (10)^{-0.375 \, x/20}$$

where V_{in} is the signal input and x is the integer value of the binary gain-control input. The AD7111 LOGDAC, for example, spans 88 dB, nearly the full audio dynamic range, with only 8-bits of control input. The leftmost four bits of x cover a coarse range in 6-dB intervals, while the other four increment between any two 6-dB points in 0.375-dB intervals. The LOGDAC is an excellent high-resolution digital gain control element for analog signals, since the system can use it to automatically control the gain level to remain within a predetermined and digitally controlled range.

It should be noted, though, that while its analog linearity is excellent and resolution is adequate for audio gain control, it cannot be used as a precision element of high accuracy auto-ranging converters, because a 0.375-dB minimum gain step (although below the threshold of audio perception) implies a 4% gain change, with an uncertainty of $\pm 2\%$.

EXERCISES FOR CHAPTER 5

Exercise 5.1: If your calculator automatically switches over to floating point when the result of a computation exceeds the size of the registers, you can determine the size of the internal registers. Square a number entered in fixed point. Keep increasing the size of the number until the answer is returned in floating point. What is the size (in digits or in bits) of the internal accumulator of your calculator?

Exercise 5.2: Write two rows of about six positive or negative two-digit decimal fractions, such as 5.7, or -2.8. Multiply each term in row A by the corresponding term in row B, and add up the results of the multiplications. Do this twice: once truncating to the right of the decimal point of each multiply before summing, and a second time not truncating until the end. Then try this:

Row A: put in numbers which integrate to zero (sum of positives equals sum of negatives).
Row B: put in numbers that are reasonably close in value.

Exercise 5.3: Suppose a 32-point running average is to be performed on a stream of data with 10-bit incoming accuracy. How wide must the accumulator be?

Answer: $10 + 5 = 15$, so 16 bits is safe.

Method: $2^5 = 32$, or see table "Number of Bits Added to Output Accumulator" in Section 5.1.3.

Exercise 5.4: A computer calculation carried out in 24-bit arithmetic is found to span a 10^{38} dynamic range of exponent. (a) How many bits are assigned to exponent and how many to mantissa? (b) How many digits of accuracy are retained?

Answer: (a) 7-bit exponent, 17-bit mantissa. (b) 16 bits $= 1$ part in 64 K $=$ more than 4 but less than 5 decimal digits.

Exercise 5.5: Architectural issues can cause trouble for the programmer. For example, in a FORTRAN program, a nested DO loop can be speeded considerably by reducing the dimensionality of the array. Consider a program with the structure:

```
DO 100 I  = 1,IMAX
  DO 100 J  = 1,JMAX
  .....
  F(I,J) = some mathematical operation on array elements
  .....
100 CONTINUE
```

It will work much faster if done as a one-dimensional vector rather than a two-dimensional array. That is, replace the **F(I,J)** step by a new variable **IJ** and a vector **F(IJ)**:

IJ = JMAX * (I − 1) + J
F(IJ) = some mathematical operation on array elements

Try it for a simple example.

Explanation: The source of the problem is in address generation, in the time to compute the array pointer location. In FORTRAN, the rows are scanned first, then the columns, while PASCAL scans the columns first and then each row. The original form above placed the rows of **F(I,J)** in the outer DO loop and the columns in the inner DO loop, a very inefficient match to the FORTRAN compiler's order. The fix shown replaces the 2-D array by a vector and explicitly computes the needed address in the new 1-D row, to make speed less dependent on compiler structure.

Exercise 5.6: Apply the Mead/Conway rule to the set of memory subsystems of Table 5.4.

Answer:

	Reg Ref	**Mem Ref**	**Paged**	**Disk**
Size	32	64 K	1 M	10 M
$\sqrt{\text{Size}}$	5.7	256	10^3	3×10^3
$\dfrac{\sqrt{\text{Size}}}{\text{Access Time}}$	5.7	85	160	20

The ratio, $\sqrt{\text{size}}$/access time, is *not* a constant. The Mead/Conway rule does not work well when there is a mix of different storage methods (e.g., disk vs. RAM). Within a given technology, such as RAM, access time is determined by number of cycles, (3 for memory reference,) which must first decode an address and then fetch the data. On-chip register reference instructions tend to be faster. That is why small on-chip *cache memory* buffers are an important DSP acceleration factor.

Exercise 5.7: Why does the standard deviation for the rectangular error probability distribution in finite length quantization have the values given in Figure 5.19 and Eq. 5.9?

Solution: In general, the square of the standard deviation, σ^2, is the expected value of the square of the deviation from the mean:

$$\sigma_\epsilon^2 = E(\epsilon - m_\epsilon)^2$$

$$-\int P(\epsilon)(\epsilon - m_\epsilon)^2 d\epsilon$$

For the case of rounding (twos complement) as in Figure 5.19,

$$\sigma_\epsilon^2 = \int_{-\Delta/2}^{\Delta/2} \frac{1}{\Delta} \epsilon^2 d\epsilon$$

$$= \frac{\Delta^2}{12}$$

where $\Delta = 2^{-b}$ for b-bit quantization.

Chapter Six

Real DSP Hardware

6.1 INTRODUCTION – CHAPTER OVERVIEW

Key architecture issues for accuracy and throughput have been dealt with in the previous chapter; we can now consider real systems. This chapter will introduce the building blocks of typical real-world DSP systems, discuss system alternatives such as single-chip and multi-chip processors; and compare alternatives in terms of performance and the cost of hardware and software development. The discussion will include examples of popular present-day solutions and include fixed-point and floating-point alternatives, where appropriate. The instruction sets of both programmable and microprogrammable alternatives will be introduced; but application examples of DSP systems implemented with real-world development tools will be saved for the following chapter.

We will seek to walk the thin line between the over-specific and the overly general. The features of specific chip sets will not be covered in exhaustive detail, since such detailed information is generally found in manufacturers' data sheets and manuals; furthermore, such information has a tendency to become obsolete. On the other hand, generalities are not enough to provide the reader with sufficient information to really understand how systems go together. Therefore we will mention some universal DSP elements, discuss industry-standard parts as prototypes—and describe newer chips as future possibilities. In particular, for the building-block approach we will discuss industry-standard Am29xx Bit-Slice chips and describe the Word-Slice™ approach. For the programmable processor approach we will feature both the pioneering TMS320 family and a recently introduced single-chip micro-processor, the ADSP-2100.

Common Requirements for DSP ICs – Compared to General-Purpose μPs

Addressing: DSP data-address sequences tend to be long, structurally complicated, but regular. Consider the FFT as an example:

  Length: The program cycles through $N/2$ data-addresses $\log_2 N$ times (assuming no parallelism or pipelining).

  Complicated: The pairs of data points that are combined have relative addresses spanning the range from adjacent to halfway apart in the N-point range.

  Regular: The number of passes is predetermined, and the addresses of data accessed are part of a regular structure.

Data Manipulation: Computations are highly repetitive, involve relatively simple combinations of data points, and usually include a multiply operation. The most common DSP "primitive" (using the computer symbol for multiplication, $*$) is

$$Y = M*C + B$$

This simple structure serves for filters, convolution and (in complex-number form) the FFT.

How DSP ICs Meet Those Requirements

DSP Hardware Terminology: To design a DSP system, it is necessary to become familiar with key hardware components, their architecture and key specifications, and instruction-set and program-development issues. Some are familiar from general-purpose microprocessors, but less-familiar terminology also appears. We first define and explain the building blocks functionally in Section 6.2, and then turn (in Sections 6.3 and 6.4) to specific hardware implementations as Bit-Slice or Word-Slice components or as subunits within a DSP microprocessor. The DSP hardware components (review Chapter 5, Section 5.2.2 and Figure 5.7 as needed) are:

  Multiplier or Multiplier-accumulator (MPY/MAC)
  Arithmetic-Logic Unit (ALU)
  Barrel Shifter (BS)
  Address Generator (AG)
  Instruction Sequencer (SEQ)

The functions and desired features of each will be outlined in Section 6.2.

How DSP Components Meet the Goals: Components designed for DSP carry out the goals outlined in Chapter 5:

  Extensive parallelism and pipelining in computational units;
  Computation speed matched from unit to unit so no bottlenecks develop;
  Instruction sequencing and data-address generation carried out with no overhead on data-computation.

In the FFT, for example, a data-address generator computes the decimation address-pair sequences, while an instruction sequencer feeds control information for cycling through each butterfly computation and looping within each pass of the overall sequence shown in Figure 5.13. The barrel shifter contributes power-of-two scaling as needed for accumulator and multiply-accumumulator inputs and outputs, when incipient overflow anywhere in the data array is signalled (block-floating-point mode). The array multiplier or MAC ensures that multiplies take no more time than add/subtracts in the basic $(M*X + B)$ algorithm. Coding is compact, and execution speed is higher by factors of 10 than a general-purpose μP with identical raw cycle speed.

The format for each element of the following discussion will be:
  Function Definition; what are the similarities and differences between a given DSP element and a corresponding general-purpose element?
  Where/why would you use one? Application example?
  Specifications/Functions of importance
  What do those functions let you do for DSP?
  Example of a specific circuit/block diagram

The same computational building blocks appear in both microprogrammed Bit/byte/Word-Slice components and in programmable DSP microprocessors; they will therefore be discussed in common prior to a separate discussion of the two alternatives in Sections 6.3 and 6.4. By the end of the chapter, the reader should have a feel for how to select a system architecture to fit applications of varying complexity and performance. Factors in choice of system architecture include:
  How easily can the chip-set be assembled into a system? (Numbers of added "glue" chips, etc.)
  How easy is the system to program? How easy is it to accomplish standard DSP "primitives" – the most common DSP operations, such as filters and spectrum analyzers?
  How easy is the system to change? What is the total development cycle?
  What is the performance/cost and performance/effort ratio?

6.2 KEY DSP HARDWARE ELEMENTS

6.2.1 MULTIPLIERS

DSP Multiplier Definition and Functions
The DSP multiplier performs rapid parallel ("array") multiplication of two b-bit inputs in a single clock cycle. Format control allows for signed, unsigned, or mixed-mode multiplication operations. Proper rounding (not truncation) is essential.

Desirable Features
Multiplier desiderata are summarized in Table 6.1. The dominant multiplier

specification is speed for a given resolution. DSP multipliers are therefore fully parallel "array" multipliers rather than the clocked "shift-and-add" of software multiplication. The cost in chip area ("real estate") is significant, reduced somewhat by the employment of algorithms (e.g., Booth's algorithm) to eliminate redundant operations when a string of 1's or 0's is encountered. The main benefit is single-cycle multiply speed to match other computational elements and data transfers. It is desirable that the multiplier function include control of input and output data formats: twos-complement or unsigned, and fractional or integer. Mixed-mode capability is a desirable enhancement for double-precision, among other applications.

With stand-alone multiplier chips, the demand for single-cycle multiplication raises issues of bus-capability to handle in parallel the concomitant 4-*b* data bits (X, Y, and double-width Z). While ideally 4 ports would be available (Figure 6.1a), external pinout limitations may force a compromise. In generic designs such as the MPY-16 or ADSP-1016A, the less significant half of output Z, labeled ZL, shares the Y-input bus (Figure 6.1b); there is no bus contention since output Z appears well after input Y is latched, and ZL is often unused (rounded off). Alternatively, Z may be multiplexed out (Figure 6.1c) on a single output data bus. This feature appears for example in the Am29516. A cluster of input *and* output registers also facilitates pipelining for increased throughput—at the cost of increased latency.

Table 6.1 Desirable Features for Multipliers

Speed: comparable to RAM access time; fully parallel operation
Input Formats: Twos complement or unsigned, fractional or integer
Mixed mode: Signed or unsigned—a desirable enhancement
Output Formats: Right-shifted ("Integer") or left-shifted
   ("Fractional")
Extended or double-precision option
Fixed point vs. floating-point tradeoffs
Single-cycle multiply of two inputs X and Y; bussing, pipelining issues
Input and output registers configurable for latching (pipelined
   operation) or transparency (minimum latency)

Bit and Word Formats
The Extra Bit in Twos-Complement Multiplication: The product of two *b*-bit *signed* numbers ($b + 1$ bits) needs only $2b + 1$ bits ($2b$ bits plus sign). However, the product of a pair of *twos-complement* (TC) *b*-bit-plus-sign numbers (Figure 6.2a) produces an extra bit, i.e., $2(b + 1)$ bits in order to be able to handle the (rarely used) square of negative full scale. For all other input combinations, it replicates the sign bit (to verify, see Exercise 6.1). Since the multiplier output is $2(b + 1)$ bits wide for signed numbers, it will accommodate

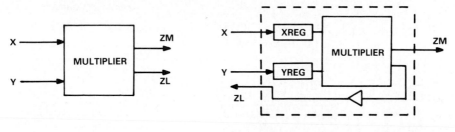

a. 4-port.

b. 3-port (MPY-16 or ADSP-1016A equivalent).

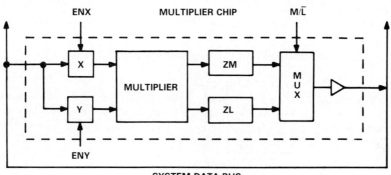

c. Inputs and outputs multiplexed to common data bus.

Figure 6.1. Multiplier chip configurations.

$(b + 1)$-bit *unsigned*-number multiplication ($b + 1$ bits, no sign). Because of these possibilities, generic multipliers contain an internal single-bit shifter to allow the user to select what to do with the redundant bit when performing twos-complement multiplications with negative full-scale excluded (Figure 6.2, a and b).

The user specifies the input format (unsigned-magnitude or twos-complement) by input format control lines (labeled TCX and TCY) in this example. If an *unshifted* result is selected (using the output format adjustment, FA), the least-significant data bit stays at the right in both input and output, as in integer arithmetic, and the sign bit remains extended. Shifting *leftwards* by one bit, and adding a zero on the right, keeps a single sign bit at the left, with the most-significant data bit next to it in both input and output, as in fractional arithmetic. A third alternative (used in some multipliers) replicates the sign bit as the MSB of the less-significant half as well, but this causes loss of resolution.

Shift alternatives are commonly specified by a table of data formats, as in the example of Figure 6.2c. The user can obtain a factor-of-two precision enhancement if one of the two multiplier inputs is an unsigned number, so that there are $2b + 1$ magnitude bits with *no* redundant sign bit in the output.

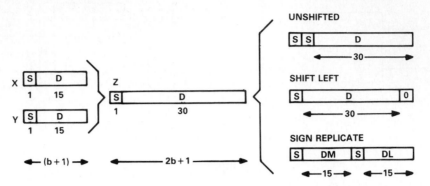

a. Shift options in twos-complement outputs.

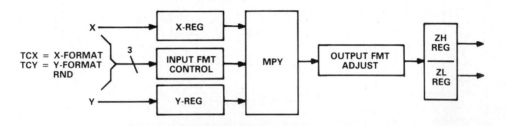

b. Input format specifiers.

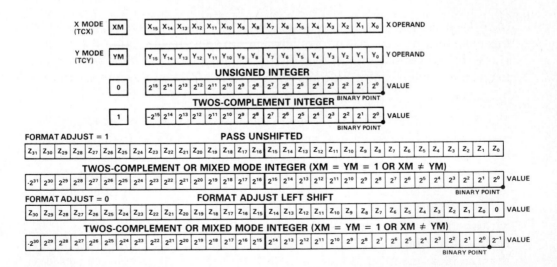

c. Example selected from table of data formats (ADSP-7018).

Figure 6.2. Alternatives for the extra sign-bit in a multiplier output.

Sign Control and Mixed Mode: The availability of two independent input format specifiers (TCX and TCY in Figure 6.2b) offers a desirable option, since it permits input pairs X,Y to be interpreted as any of:

X	Y	Result
signed	signed	signed
unsigned	unsigned	unsigned
signed	unsigned	signed ("mixed-mode")
unsigned	signed	signed ("mixed-mode")

where "signed" means that 2's complement is used. Examples of mixed-mode operations include the multiplication of a signed data value with a set of unsigned filter coefficients or scale transformations. Mixed-mode also facilitates double-precision operations on signed numbers, as shown in Figure 6.3.

Since the most-significant (M) half of each input contains the sign, the uppermost portion $YM*XM$ of the result must be evaluated as a signed multiply, the center portions must be evaluated as mixed mode (one member, M, is signed, the other, L, is unsigned), and the lower portion $YL*XL$ must be evaluated as unsigned. Mixed-mode operations require *independent* format control for both X- and Y-inputs. Input format controls, TCX and TCY, make this easy to do (they are standard on most multipliers), while only a *single* control input to select unsigned or signed multiply format is standard on most multiplier-accumulator chips. The logic for double-precision multiply must also provide for the fact that the sign bits in partial products 2 and 3 fall over a data bit in PP4; they must be sign-extended appropriately to avoid treating the sign bits as data bits, leading to an error.

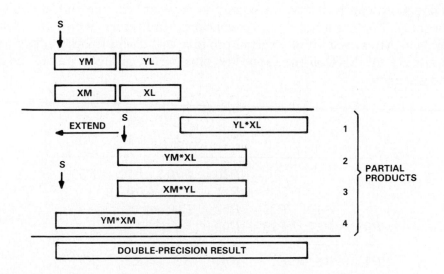

Figure 6.3. Double-precision twos-complement multiply: sign and format control.

Rounding Issues: If only the MSH (more-significant half) of a multiplier output is to be kept, rounding is preferable to truncation, as discussed in Ch. 5. Rounding is implemented in a multiplier by adding 1 to the most-significant bit of the LSH (less-significant half) before truncating. If the LSH is above half of its full-scale range, adding the 1 in its MSB (= ½ full-scale of the LSH) ripples a 1 into the LSB of the MSH. For example,

	MSH	LSH	
	10011011	10101010	Full-width output, LSH > FS/2
	+	10000000	Augment LSH, then truncate →
=	10011100		Rounded output incremented
	10011011	00101010	Full-width output, LSH < FS/2
	+	10000000	Augment LSH, then truncate →
=	10011011		Rounded output unchanged

Unbiased rounding eliminates the small positive net bias that occurs when the LSH happens to fall exactly at 10000000, as discussed in Chapter 5. The unbiased-rounding algorithm is simple: perform the rounding above, and then detect if the LSH is all zeros, signifying that the LSH started out at "half-scale." The unbiased-rounding logic rounds this event *up* half the time and *down* half the time (for example, add 1 to the LSB of the MSH if it is 1, 0 if it is 0).

Speed, accuracy, cost tradeoffs

How do multiplier speed, accuracy, and cost trade off? For a given architecture, logic family, and gate-delay time, the array multiply time scales roughly as the wordlength (not as the square as it would for the shift-and-add algorithm). For example, a full *combinatorial* multiplier (called the Wallace tree) evaluates each bit of a partial product with a ripple-carry adder whose inputs are the AND of two input bits plus partial products just to the right and just above:

				$X3$	$X2$	$X1$	$X0$
				$Y3$	$Y2$	$Y1$	$Y0$
				$P30$	$P20$	$P10$	$P00$
			$P31$	$P21$	$P11$	$P01$	
		$P32$	$P22$	$P12$	$P02$		
	$P33$	$P23$	$P13$	$P03$			
$R7$	$R6$	$R5$	$R4$	$R3$	$R2$	$R1$	

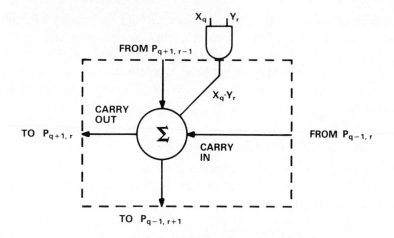

Diagram of Computation for Each Partial Product P_{qr}

The time for a given result, R, to settle is the time for the partial products Pqr to ripple through from the top and from the right; it is proportional to the word width. Multiplier operation is therefore nearly linear in word-width. For example, at 25°C,

Word Length	Chip	Feature size: 5-micron MPY-Time nanoseconds		1.5-micron Replacement Chip	MPY-Time nanoseconds
8 bits	ADSP-1080KD	85	→	ADSP-1080AKD	33
12 bits	ADSP-1012KD	110	→	ADSP-1012AKD	50
16 bits	ADSP-1016KD	145	→	ADSP-1016AKD	70
24 bits	ADSP-1024KD	200	→	ADSP-1024AKG	95

The scaling with word-length is only approximate, since I/O settling adds a fixed overhead, and subtleties in layout may add as much to the scaling with wordlength as does the number of gate delays. Propagation time also scales with lithographic feature size: a shrink in feature size increases the speed, but not quite proportionately. The example shown (ADSP-1016 to ADSP-1016A) underwent a factor of 3 shrink, which resulted in a 2-times reduction in multiply time.

Cost is a complex function of chip maturity: development amortization, production yield, generic-or-proprietary market, etc.; for a given technology, cost can be expected to increase somewhat faster with increased wordlength since chip area goes up nearly as the square of the wordlength (resulting in fewer chips per wafer) and yield decreases as chip area increases.

Pipelined Multipliers

A *pipelined* multiplier (Figure 6.4) can double its throughput by latching a result and then accepting the next input pair while output formatting, shifting, and saturation operations are under way. While such a set of intermediate registers facilitates repetitive computations, it can also add latency at the beginning and end of a loop as the pipeline fills and empties. A pipelining *option* can be the best of both worlds; it would combine a standard architecture with pipelining, which could be made transparent for nonrepetitive computations.

6.2.2 MULTIPLIER-ACCUMULATORS

DSP Multiplier-Accumulator Definition and Functions
An accumulator combined with a multiplier is desirable because it facilitates carring out the ubiquitous

$$\sum b(n)\,x(n - k)$$

operation of filters, Fourier analyzers, and vector operations. The desirable features for multipliers seen in the previous section apply equally to the multiplier-accumulator (MAC). The opportunity also exists in a MAC for built-in feedback paths and intermediate pipelining registers, which enhance throughput because overhead on repetitive calculations is lessened. Internal

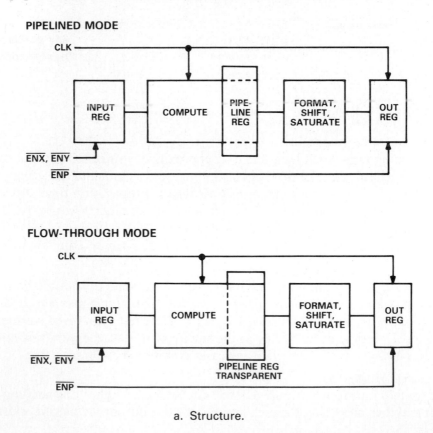

a. Structure.

TRANSPARENT (UNPIPELINED) MODE

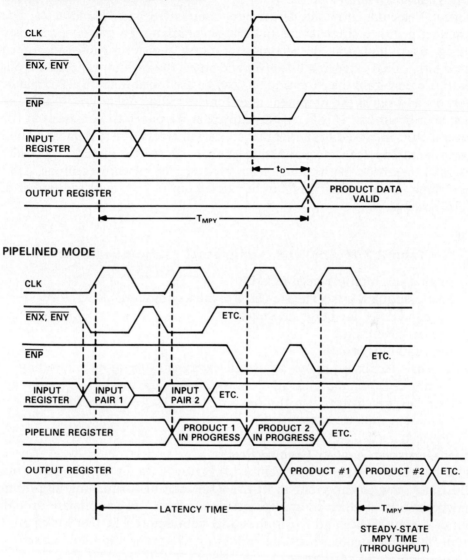

b. Data timing in unpipelined (or transparent) and pipelined modes.

Figure 6.4. Pipelined multiplier with transparent option.

feedback paths can make possible, for example, single-cycle computation of the common DSP operation $M*D + B$. An example will be given later of an enhanced multiply-accumulate chip with enough feedback paths and intermediate registers to permit an on-chip butterfly computation in 6 cycles. The following sections describe the embodiment of these desirable features in generic industry-standard MACs, as well as in DSP microprocessor and Word-Slice chips.

MAC Desirable Features

Multiplier-accumulator desiderata are summarized in Table 6.2. The dominant goal is a single-cycle multiply-accumulate. To preserve accuracy during an accumulation, the accumulator output register includes additional guard bits, called extension bits, over and above the $2b$-bit multiplier output width. Logic within the extension bits can be used to anticipate accumulator overflow and signal the imminent need for a rescaling operation. If only the most-significant half (MSH) of the multiplier output is to be passed to the accumulator, unbiased rounding prevents small errors due to truncation and biased rounding from accumulating. Enhanced MACs include especially designed feedback paths and extra registers to permit common DSP algorithms (DSP "primitives") to be carried out in the smallest number of cycles possible.

Table 6.2 Multiplier-Accumulator Desirable Features

Single-cycle multiply-accumulate
Adequate guard bits (extension bits) in the accumulator (ACC)
Overflow detection before it occurs
Unbiased Rounding
Saturation Arithmetic
Internal feedback paths to lessen overhead on loops, speeding computation common DSP "primitives"
User beware: distinguish throughput (best-case, pipeline full) from latency single-time operation).

Width Issues and Overflow Dynamic Range

The multiply-accumulate structure of Figure 6.5a in its simplest form illustrates how to carry out repetitive DSP number-crunching algorithms independently of other computational elements. A full $2N$ bits are brought out of the multiplier and fed to the adder/subtractor. The feedback path C from output register to other adder input provides the accumulator action:

$$R(n+1) = X(n+1)*Y(n+1) \pm R(n)$$

where the result $R(n)$ of the previous computation is fed back by the pathway labeled C. The ADD/SUB output is widened by M *extension bits*, so that at least 2^M repetitive MPY/ACC operations can be performed without overflow (Figure 6.5a). Industry-standard MACs (e.g., TRW's 1010, AMD's Am29510, Analog Devices ADSP-1010A) contain 16 data bits, 32 multiply output bits, and 35-bit accumulator, so $M = 3$ in this case. However, the trend is towards more extension bits in both DSP microprocessors and Word-Slice components. For example, a 40-bit accumulator with a 32-bit multiplier allows 256 instruction cycles without risk of overflow. This is advantageous,

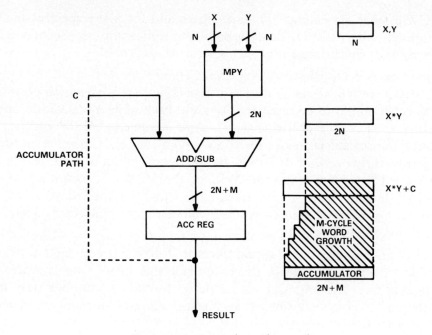

a. Basic structure and word-growth.

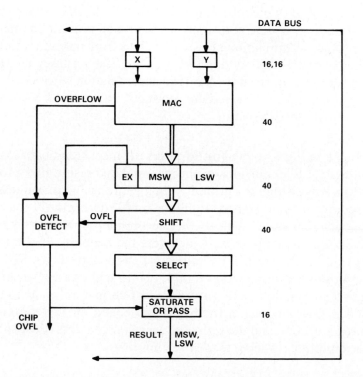

b. Elements of the enhanced MAC structure of the ADSP-1101 Integer Arithmetic Unit.

Figure 6.5. Multiplier-Accumulator structure.

since 256 steps are enough for most filters and for many spectral-analysis operations. Alternatively, floating-point multiplication (see Section 6.2.4) defuses the dynamic-range issue, at higher hardware cost.

Output Structure, Overflow, and Rounding Control

The MAC's output is via a set of registers (Figure 6.5(b)) which permit its extra-width output to be saturated, brought back to data-bus width, and/or transmitted in double-precision as two-word increments, MSW,LSW.* The accumulator register is divided into LSW, MSW, and EX (*extension*) portions. To permit early overflow detection and rescaling action, it is preferable that the *entire* EX-register (plus the MSB of the MSW) be examined by the overflow detector logic. If not all bits are equal (all ones or all zeros), an accumulator overflow can be signalled well before the extension register has actually overflowed.

A shifter at the output fills a dual purpose if both left and right shifts are possible. A *left* shift removes the redundant sign bit of twos-complement multiplication, bringing out a properly formatted number for later calculations. A *right* shift downscales block-floating-point numbers as word-growth is detected.

Rounding is an essential MAC option. If the LSW is to be eliminated, a 1 is added to the MSW with the rounding rules as in the *Multipliers* section above. *Unbiased* rounding, a desirable enhancement not present in many MAC designs, prevents accumulation of a small dc bias from outputs which fall just half-way between adjacent rounded values. Highest accuracy is attained by avoiding rounding until the last step of an accumulation *sequence*, if possible, since information in the intermediate products' and sums' LSWs is thereby allowed to contribute to the result.

Saturation Logic

The final block in Figure 6.5(b) multiplexes the most significant (MSW) and least significant (LSW) portions of the accumulated result out onto the data bus. A saturation-arithmetic option is valuable in this block, since a series of multiply-accumulates on signed numbers could momentarily overflow and then come back "on-scale" with further accumulation without necessitating rescaling. Saturation arithmetic, which sets the contents of the register to maximum value if an overflow or underflow has occurred (see Figure 5.17), avoids the serious error of a wraparound, or—in many cases—even the need to respond to an overflow. The saturation action may respond to the adder or shifter overflow bit, which in turn may depend on the contents of the extension register, EX, and the sign of the register contents (the MSB of EX). Here is an example of response to the overflow flag:

---

*The terms, MSW, LSW (most-significant word, least-significant word), and MSH, LSH (more-significant half, less-significant half) are used somewhat interchangeably. One can make the distinction that MSH, LSH refer to division of a word when a bus is divided, while MSW, LSW might refer to double-width multiplication results as they are separately placed on the bus.

Example: Multiply-accumulate with Saturation Option Selected

OVFL	EX-Reg MSB	R contents after saturation operation	
0	0(POSITIVE)	No change	
0	1(NEGATIVE)	No change	
1	0(POSITIVE)	0111....1111	1111....1111
1	1(NEGATIVE)	1000....0000	0000....0000

Overflow of the accumulator or shifter with saturation logic implemented gives as the MSW output the largest 16-bit positive number if the EX-Reg indicates positive sign, and the largest-magnitude 16-bit negative number if the EX-Reg indicates negative sign. If a shifter is in use, the saturation logic can evaluate whether the *post*-shift value overflows, thus anticipating an overflow by one bit.

Sign-Extend

The MAC must provide properly sign-extended outputs for twos-complement results, and zero-filled results for unsigned data. Format lines (as in Figure 6.2) signify input data formats, and the output format depends upon the operation performed, e.g., twos complement if either input was twos complement, and unsigned if a logical operation was performed.

Extra Registers and Feedback Paths

Expanded computing power results from extra intermediate registers and feedback paths. For example, single-cycle computation of $(Z = M^\star D + B)$ is possible in the MAC shown in Figure 6.6 (at the heart of the ADSP-1101 Integer Arithmetic Unit) via the direct accumulator preload path labeled P from one of the input ports. Since I/O has few gate delays compared to multiplication, it is possible to read *two* inputs in a *single* cycle over one input port. As shown in the Figure, input data D is preloaded through the Y-port into the multiplier YREG on the falling edge of the previous cycle. On the rising edge of the clock, constant B is loaded (via path P) into the accumulator via the same Y-port and coefficient M is loaded into XREG; and multiplication begins. The accumulated result ACC $= M^\star D + B$ is available by the end of a single clock cycle, with both B and D loaded through the Y-port during one cycle.

Feedback from adder output back to multiplier input permits efficient evaluation of simple polynomials for approximation of functions. For example, path R2 in Figure 6.6 facilitates evaluation of

$$Z = 1 + aY + bY^2$$

a polynomial encountered (among others) in quadratic interpolation and in evaluating functions such as the square root. Output quantity, aY, could be fed back to be multiplied by $(b/a)Y$ (where b/a has been pre-computed and stored) to create the 2nd order term, bY^2.

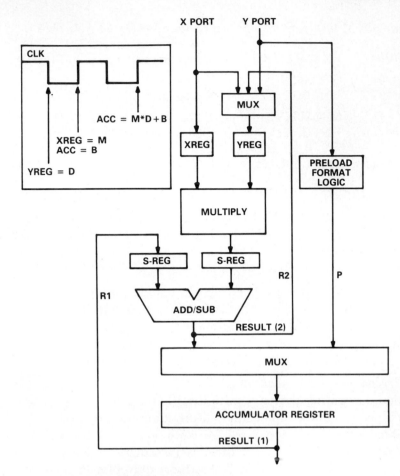

Figure 6.6. Simplified diagram of enhanced multiplier-accumulator integer arithmetic unit (ADSP-1101) with feedback paths and direct preload.

6.2.3 ARITHMETIC LOGIC UNIT (ALU)

ALU Definition and Functions

A DSP ALU is similar to the ALU of a general-purpose μP, such as the one in Figure 5.3. The usual array of arithmetic and logical instructions is performed. What sets a DSP ALU apart are its extra features (similar to those discussed for the MAC, for example) which permit single-cycle instruction execution, and pipelining for repetitive operations but transparency for flow-through operations. In addition, a DSP ALU is often closely associated with scaling elements, such as shifters, which keep data on-scale while preserving maximum dynamic range.

Desirable Features

An ALU oriented towards general-purpose computers has design features optimized for *arithmetic*. An ALU oriented towards controllers will feature *bit manipulation and character handling*. While the former will carry out short sequences of micro-instructions, the latter may carry out long microprograms (e.g., disk controller), that relatively slow controller applications can tolerate. By contrast, an ALU oriented towards workstations, graphics, array processors or DSP emphasizes *speed*: fewer addressing modes and short micro-instruction sequences, so the application can operate more quickly. The DSP ALU must operate in parallel and in synchronism with different computational units (MAC, shifter, etc.), so the basic operations must execute in the fewest possible cycles. Features that are desirable for a DSP ALU are summarized in Table 6.3.

Table 6.3 ALU Desirable Features

Adequate accuracy and chainability for high-precision computation.

ARITHMETIC FUNCTIONS: Add, subtract, add with carry, subtract with borrow, absolute value.

LOGICAL FUNCTIONS: AND, OR, exclusive OR, logical negation.

Output flags: zero, equal, less than, greater than, carry...

Divide and (less important) multiply capability.

Efficient data movement and computation:

Data moves plus arithmetic in *same cycle*.

TWO operands loadable into the ALU on each cycle.

Adequate Register Files:

Dual-set for context switching (for storing data during interrupts).
Or: Register File for fast access and storage of intermediate results.

Conditional operations, e.g., Add + (Shift if Flag Set) are an asset.

Adequate Feedback pathways: Output to input (accumulate); and, if linked to shifter, ALU inputs to shifter or shifter inputs to ALU.

Pipeline or transparent flowthrough option on input registers.

Close association with Shifter for scaling operations common in DSP.

Double-precision capability with minimum lost time.

ALU Functions and Data Formats

Arithmetic and Logic: An ALU performs the usual complement of arithmetic and logical functions:

ARITHMETIC FUNCTIONS: Add, subtract, add with carry, subtract with borrow, absolute value;

LOGICAL FUNCTIONS: AND, OR, exclusive OR, logical negation.

In order for arithmetic-plus-shift operations to be performed, the chip must have the capability of distinguishing twos-complement and unsigned-magnitude data, since a right-shift treats the two differently (sign-extend vs. zero fill—see Section 6.2.4).

Condition Flags: As with a general-purpose ALU, the DSP ALU must signal status information on the computation just performed: Was the result positive, negative, or zero? Was there a carry out from the MSB, or did the ALU overflow? Sign and zero flags may be logically combined to signal less than zero, less than or equal to zero, etc. The special feature of DSP-ALU flags is the way they are used by other units, such as the program sequencer and the shifter, to initiate a block floating-point rescaling, for example.

Rounding: Rounding is relatively unimportant in the ALU, essential only in treating the output of a raw multiply/divide operation, and in floating point operations (see BOX, FLOATING POINT, following Section 6.2.4 below).

Division Capability: Division occurs in many DSP systems, for example in linear predictive coding of speech (e.g., the Durbin recursion: see Chapter 8) and in matrix operations such as matrix inversion. Division also occurs in setting the depth in all *perspective* operations in graphics. Although multiply and divide capability should be among an ALU's "primitives", the algorithm in an ALU is based on shift-and-(add or subtract), so these operations will take many cycles. Multiplication is far better left to the single-cycle array multiplier (above). However, since division is not among basic MPY or MAC operations (although it can be implemented with a MAC circuit), it is a key ALU primitive operation.

Saturation Logic: As with the MAC, saturation logic in the output register may be a desirable enhancement.

Arithmetic Plus Conditional Shift: An ALU which can conditionally shift its result up or down during the same cycle, depending on the state of an input flag, enhances the speed of many DSP operations. For example, Add + (Conditional Shift) is an asset in the FFT butterfly. A butterfly output is nondecreasing, but since amount of growth depends on the signal, an automatic downshift that doesn't depend on actual signal level loses signal-to-noise. A shift conditional on a block-floating-point flag keeps maximum dynamic range, hence optimal signal-to-noise. The lack of extra cycles in a conditional shift is an asset: parallel operations of other units need not halt for data to stay in synchronization.

ALU Architecture Issues

Word Width and Easy Paralleling: ALU width must have sufficient capacity for adequate DSP accuracy: 16- or 32-bit I/O is typical. Since paralleling of ALUs for fast operation in double precision will be common, status flags (carry, sign, etc.) and shifted bits must be readily transmittable between identical ALUs.

Single Cycle Data-Moves-plus-Arithmetic: Data movement and an arithmetic operation can be performed in the same cycle, using *dual-ported registers*, which can provide ALU input or output while simultaneously doing a bus access. Dual-ported registers can be both written-to and read in one cycle. The (previously stored) value is read out of the register at the beginning of a cycle and a new value written in at the end of the cycle. For example (Figure 6.7), on the ALU input side, a register first provides the ALU with an operand; then, on the next half-cycle, it is loaded with a new value from the data bus.

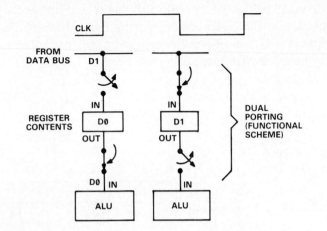

Figure 6.7. Dual-porting is equivalent to timed switching, first disgorging previous data, and then accepting waiting input data.

For maximum speed, an operand from an X-register and an operand from a Y-register should be loadable into the ALU in the same cycle.

Adequate Register Set: A flexible register set is essential. For example, two duplicate register banks permit context switching, which will facilitate interrupt service. *Context switching* is like having a two desks to work on: one task can be put on hold and left unchanged while another task is taken up. Alternatively, addressable *Register Files* (see Figure 6.8) provide fast access to repeatedly used information (e.g., coefficients) and for storage of intermediate results without the need for a memory access.

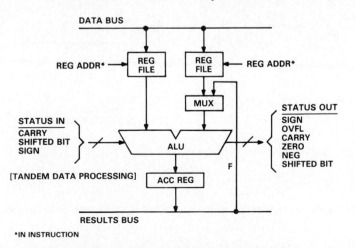

Figure 6.8. Example of DSP ALU structure.

Flexible Internal Data Paths and I/O Pathways: Flexible internal data pathways are an asset. For example, the accumulator output can be an input for the next ALU cycle, with the feedback path *F* shown in Figure 6.8 The ALU may find itself on the same chip as another computational unit. For example, an ALU and Barrel Shifter (described in the next section) are a particularly powerful combination for rapid scaling during computations. It is useful to have data pathways which permit either unit to be an input source or an output destination for the other without accessing the bus.

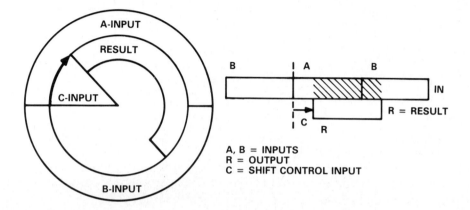

a. A word having width, *R*, is carved out of wider word, A,B, and shifted by amount *C*.

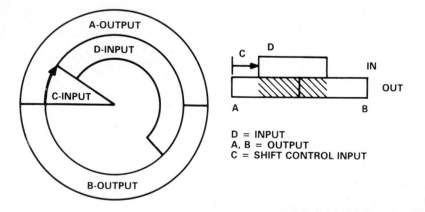

b. Alternative structure: input, D, is placed anywhere (*C*) in the double-wide A,B output field.

	Control Code		Shifter Array Output
HI reference	LO Reference		
+16 to +127	+32 to +127	0000000000000000	0000000000000000
+15	+31	R000000000000000	0000000000000000
+14	+30	PR00000000000000	0000000000000000
+13	+29	NPR0000000000000	0000000000000000
+12	+28	MNPR000000000000	0000000000000000
+11	+27	LMNPR00000000000	0000000000000000
+10	+26	KLMNPR0000000000	0000000000000000
+9	+25	JKLMNPR000000000	0000000000000000
+8	+24	IJKLMNPR00000000	0000000000000000
+7	+23	HIJKLMNPR0000000	0000000000000000
+6	+22	GHIJKLMNPR000000	0000000000000000
+5	+21	FGHIJKLMNPR00000	0000000000000000
+4	+20	EFGHIJKLMNPR0000	0000000000000000
+3	+19	DEFGHIJKLMNPR000	0000000000000000
+2	+18	CDEFGHIJKLMNPR00	0000000000000000
+1	+17	BCDEFGHIJKLMNPR0	0000000000000000
0	+16	ABCDEFGHIJKLMNPR	0000000000000000
−1	+15	XABCDEFGHIJKLMNP	R000000000000000
−2	+14	XXABCDEFGHIJKLMN	PR00000000000000
−3	+13	XXXABCDEFGHIJKLM	NPR0000000000000
−4	+12	XXXXABCDEFGHIJKL	MNPR000000000000
−5	+11	XXXXXABCDEFGHIJK	LMNPR00000000000
−6	+10	XXXXXXABCDEFGHIJ	KLMNPR0000000000
−7	+9	XXXXXXXABCDEFGHI	JKLMNPR000000000
−8	+8	XXXXXXXXABCDEFGH	IJKLMNPR00000000
−9	+7	XXXXXXXXXABCDEFG	HIJKLMNPR0000000
−10	+6	XXXXXXXXXXABCDEF	GHIJKLMNPR000000
−11	+5	XXXXXXXXXXXABCDE	FGHIJKLMNPR00000
−12	+4	XXXXXXXXXXXXABCD	EFGHIJKLMNPR0000
−13	+3	XXXXXXXXXXXXXABC	DEFGHIJKLMNPR000
−14	+2	XXXXXXXXXXXXXXAB	CDEFGHIJKLMNPR00
−15	+1	XXXXXXXXXXXXXXXA	BCDEFGHIJKLMNPR0
−16	0	XXXXXXXXXXXXXXXX	ABCDEFGHIJKLMNPR
−17	−1	XXXXXXXXXXXXXXXX	XABCDEFGHIJKLMNP
−18	−2	XXXXXXXXXXXXXXXX	XXABCDEFGHIJKLMN
−19	−3	XXXXXXXXXXXXXXXX	XXXABCDEFGHIJKLM
−20	−4	XXXXXXXXXXXXXXXX	XXXXABCDEFGHIJKL
−21	−5	XXXXXXXXXXXXXXXX	XXXXXABCDEFGHIJK
−22	−6	XXXXXXXXXXXXXXXX	XXXXXXABCDEFGHIJ
−23	−7	XXXXXXXXXXXXXXXX	XXXXXXXABCDEFGHI
−24	−8	XXXXXXXXXXXXXXXX	XXXXXXXXABCDEFGH
−25	−9	XXXXXXXXXXXXXXXX	XXXXXXXXXABCDEFG
−26	−10	XXXXXXXXXXXXXXXX	XXXXXXXXXXABCDEF
−27	−11	XXXXXXXXXXXXXXXX	XXXXXXXXXXXABCDE
−28	−12	XXXXXXXXXXXXXXXX	XXXXXXXXXXXXABCD
−29	−13	XXXXXXXXXXXXXXXX	XXXXXXXXXXXXXABC
−30	−14	XXXXXXXXXXXXXXXX	XXXXXXXXXXXXXXAB
−31	−15	XXXXXXXXXXXXXXXX	XXXXXXXXXXXXXXXA
−32 to −128	−16 to −128	XXXXXXXXXXXXXXXX	XXXXXXXXXXXXXXXX

c Shifter output vs. *C* for typical shifter. ABCDEFGHIJKLMNPR represents the 16-bit input pattern. X stands for the (sign-) extension bit.

Figure 6.9. Barrel-shifter action.

6.2.4 SHIFTERS: SCALING CONTROL

Shifter Definition and Functions

A shifter scales numbers to prevent underflows and overflows and performs conversions between fixed- and floating point (see BOX at the end of this section). Unlike the bit-per-cycle shift of general-purpose microprocessors, the DSP shifter must be capable of shifting a word by many bits in a single machine-cycle to preserve single-cycle computation speed. It is possible to use a multiplier to perform many shift operations (see Exercises); however, the full complement of shift operations is best performed by a shifter cell or chip designed for the purpose.

The shifter action, shown schematically in Figure 6.9a, maps a portion of a long input word, A,B, onto a bus-width output word, R, with the amount of shift set by the control word, C.* In another use of this topology (b), a bus-width input, D, is shifted onto a location in double-width output, A,B, at a starting point set by shift-control input, C.

The shifter performs arithmetic shifts and logical shifts. Arithmetic shifts scale numbers upwards (left-shift) and downwards (right-shift), as shown in Figure 6.10. Right-shifts *sign-extend* the result, filling new left-bits with the sign-bit, S, to preserve the value for both positive and negative numbers. While single-bit shifts may retain the bit shifted out (for example in the carry flag as shown), multiple right-shifts usually truncate the number. Left-shifts fill with zeros the new bits at the least-significant end; logic is usually included to handle overflow; this differs slightly from shifting in the general-purpose μP. For example, a left-shift in which MSB $\neq S$ sets an overflow flag in the general-purpose μP, while in the DSP shifter the MSB output in register R (Figure 6.9a) is generally saved in a temporary bit-test status register to signal imminent overflow so action can be taken on the next cycle to avoid overflow.

Logical shifts (in masking, for example) assume unsigned data; the left- or right bits of the shifted number are filled with zeros, depending on direction (See Figure 6.10). The barrel shift architecture of Figure 6.9 permits a word *rotation* of from 1 to b bits in only two cycles, shown in Figure 6.10(b), loading the word first into input A, and then into input B.

Basic Shifter Operations

Basic shifter functions include:

    n-bit arithmetic and logical shifts (shift-immediate)

    Word rotations and word merges

    Bit-test/bit-set/bit-clear

---

*Since the left end of A and the right end of B can in principle form a continuum, so that shifts can occur in circular fashion, like a word sliding around the hoop of a barrel, the popular name *barrel shifter* has been applied to this DSP function. However, many "barrel shifters" truncate R instead of rolling it over (i.e., cut the hoop); thus the connotation of the term is a device that produces a rapid (one-cycle) left- or right shift in response to a control word, with or without rollover. Nevertheless, since A,B is at least twice as long as R, rotation can still be produced in terms of R (Figure 6.9c).

Shift immediate returns a value shifted by the control value, C, provided in the instruction. Some shifters limit the choice of C to a fixed subset of shifts (arithmetic only, 1 bit only, 1,2,4 bits only, left only); they lose in generality, and the limitation can be circumvented only at the cost of one or more cycles.

MICROPROCESSOR SHIFT OPERATION

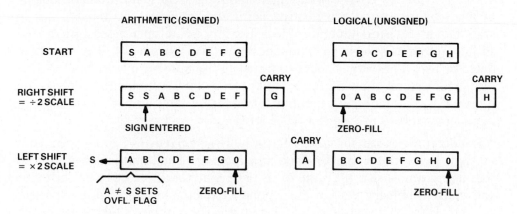

MICROPROCESSOR ROTATES

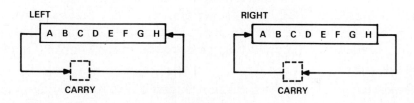

a. Handling of sign extension and zero filling in arithmetic and logical shifts.

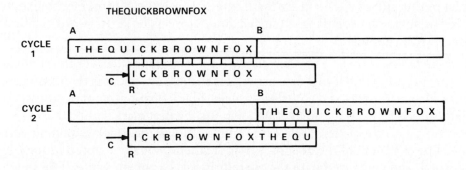

b. Shifters can rotate in two cycles.

Figure 6.10. Rotation in microprocessors and shifters.

Shifter Scaling Functions

Enhancements for Scaling: Desirable shifter enhancements include:
 Normalization
 Denormalization
 Automatic block floating-point

These provide low-overhead exponent handling in normalization and denormalization, for example in floating point and fractional arithmetic, and in scale-factor determination for block-floating-point operations. Enhanced shifter functions involve an exponent detector (Figure 6.11). The exponent detector uses a priority encoder to translate the number of identical leading bits (which signify how far the number can be left-shifted without changing its mantissa) into a binary number representing the number's base-2 exponent—and stores it in a register (EXP REG).

Normalization: Normalization (Figure 6.11b) is a fixed-to-floating point conversion, generating a mantissa and an exponent. The input is left-shifted so that its rightmost sign bit is the MSB. The number of places to be shifted is converted to a binary word by the exponent detector and put into an exponent register (EXP). The hardware which converts the number of bits to be shifted into an exponent word is a *priority encoder*, a standard logic function; for 2^b inputs and b outputs, its output is the binary code corresponding to the address of the highest numbered input that it finds asserted (priority encoders are also used in "flash" a/d converters, see Figure 5.25b). Normalization is a two step operation: the exponent detector first calculates the value of the shift needed, then supplies that value to the shifter control (C) input for the actual shift.

Denormalization: Denormalization (Figure 6.11b), or floating-point to fixed-point conversion, is the inverse of normalization. An exponent loaded from the data bus into the EXP register forms the control code for the shift. The sign is extended by the shifter as shown in Figure 6.11b. Bits below the precision of the shifted mantissa are lost. Extensions of the control mechanism for denormalization permit logical shifts, double-precision shifts, word rotations, and word swapping.

Block Floating-Point: The *block floating-point* operation derives the exponent of the largest member of a data array. This single exponent is then associated with the entire array and controls its scale. Block floating-point exponents are determined by exponent-detection logic. As the first member of the array passes through, its exponent is detected and latched in the Block Exponent register. The exponent of the next data point is then evaluated, stored in the EXP register, and compared with the current Block Exponent value. The smaller (negative exponent) of the two is latched as the new Block Exponent. The process continues until the end of the block, when the Block Exponent register will contain the least-negative exponent value found, associated with the largest member of the array. The BFP exponent-determining operation is an inspection only; no actual shifting takes place, since the block exponent is not known until the entire block has passed through. The example in the box below shows a twos-complement fixed-point signed fractional-binary data array (first and second columns). The exponents of all elements are determined (third column), and the BFP exponent is established as the sequence progresses (-2, fourth column, datum 6). All elements are scaled up and stored (column 5) as a set of (non-normalized) mantissas with a common exponent of -2, to keep the array-filling optimum dynamic range.

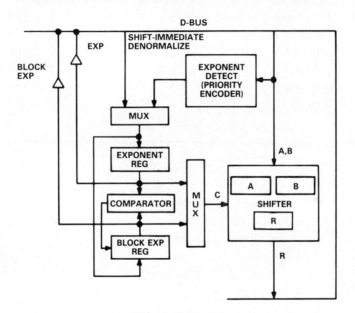

a. Block diagram.

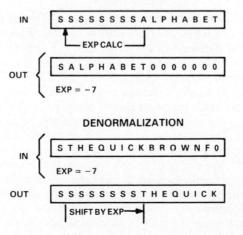

b. Normalizing and denormalizing functions.

Figure 6.11. Shifter with exponent detection and block floating-point capability.

Example of Block-Floating-Point Exponent Determination and BFP Scaling

Array Index	Input Data S	Stored Exponent EXP	BFP	BFP-Adjusted Output Data S
1	00001001	-3	-3	00100100
2	11110001	-3	-3	11000100
3	11111100	-5	-3	11110000
4	11110110	-3	-3	11011000
5	00001010	-3	-3	00101000
6	00010011	-2	-2	01001100
7	00000010	-5	-2	00001000
8	00000101	-4	-2	00010100

$$\text{BFPEXP} = -2$$

Block floating-point preserves maximum dynamic range of the data block without sacrificing accuracy of the mantissa, since no bits are taken away from the data words to represent the exponent, as true floating-point would require. True floating point may be needed when treatment of data as a block is not appropriate, or when higher precision demands 32- to 64-bit calculations. A floating-point convention known as the IEEE Standard 754 (see BOX) has been adopted to promote interchangeable hardware and transportable software.

FLOATING POINT FORMAT: IEEE STANDARD 754

An accepted standard for floating-point computation is essential. A program that runs well when developed on one system can give underflow/overflow errors on another if the dynamic-range limits differ. Even the choice of what constitutes double-precision differs substantially. For example, the DEC default double-precision floating point does not necessarily extend the number of bits assigned to the exponent beyond the 8 assigned in single-precision; thus, computation may still go out of range in the exponent, even though it is very precise in the mantissa. Code generated with a popular "C" compiler for personal computers may compile perfectly on a UNIX standard "C" compiler for minicomputers; however, if the computation defaults to a dynamic range of 10^{38}, even in double precision, programs which utilize the 10^{99} dynamic-range standard adopted for the personal-computer "C" compiler will go out of range when executed on the more powerful minicomputer! Floating-point standards are designed to eliminate such problems, facilitating transportability between systems. Described here is the IEEE-754 floating-point standard, the generally adopted convention for computers, array processors, and floating-point chip sets.

The stated goals of the IEEE floating-point standard are to:
(1) Facilitate movement of programs between computer systems.
(2) Enhance the capability of programs and programmers and safety of numerical results.
(3) Encourage the development and distribution of robust, efficient, and portable numerical programs which produce identical results on all conforming systems;
(4) Support execution-time diagnosis of anomalies, smooth handling of exceptions, and interval arithmetic;
(5) Provide for development of standard elementary functions (exp, cos, etc.), very high precision arithmetic, and coupling of numerical and symbolic algebraic computation.

IEEE-754 FLOATING-POINT BINARY FORMAT

For a binary floating-point number, $SM\ 2^E$, the arrangement of 32 bits in single-precision format is:

1 bit	8 bits	23 bits
SIGN	EXPONENT	FRACTION
S	EEE..EE	FFF.......FF
	MSB LSB	MSB LSB

where S is the sign, E...E is the 8-bit exponent, and F....F is the 23-bit fractional part, which together with S makes up the mantissa. For example, if $n = -15 = -1111_B$, then, in normalized form, the mantissa is 1.111 and the exponent is 3, i.e., -1.111×2^3, which is expressed in the standard format as 1 10000011 11100....0, taking into account the hidden bit and bias— see below. *It is important to realize that the IEEE floating point standard always refers to numbers in signed-magnitude format, not to twos-complement numbers.* Since it is known in advance that all mantissa bits are magnitude bits, not sign bits, an extra bit can be gained by "hiding" a 1 to the left of the implied binary point. From the above example,

Number	$1.11100000 \times 2^{E\prime}$
Stored mantissa:	11100000
Mantissa interpreted as:	1.11100000

Normalized floating-point numbers in the IEEE standard are thus always formatted so that their mantissa magnitudes range from 1.00000...00 to 1.111111...111, with the 1. implied or "hidden" and only the fraction stored.

The arrangement of bits in *double-precision* format is:

1 bit	11 bits	52 bits
SIGN	EXPONENT	FRACTION
S	EEE........EEE	FFF.............FFF
	MSB LSB	MSB LSB

with an 11-bit exponent and 52 bits in the fraction, making up (with sign and hidden bit) a 54-bit mantissa.

The exponent in either case is *biased*, with a constant offset chosen so that the exponent can be expressed as a positive number. The exponent bias is $+127$ in single precision and 1023 in double precision. Thus, for an exponent of 1 in SP, the exponent field will have a value of $1 + 127 = 128$, or 10000000. The mantissa is specified as a binary *fraction*, with (implied) binary point just to the left of the MSB. The *value* of a binary number represented in IEEE format is thus:

$$(-1)^S 2^{(E - \text{bias})}(1.b_0 b_1 b_2 \dots b_p - 1)$$

where the sign bit, S, is 0 or 1, the exponent, E, is any integer between 0 and 255 (single precision) or 2,047 (double precision), and the mantissa has p fraction bits, $b_0 \dots b_{p-1}$.

The standard was designed to provide single- and double-precision formats to fit within 32- and 64-bit word lengths, respectively, as well as extended single- and extended double-precision formats for accurate accumulation of intermediate results prior to roundoff. *Extended precision* means that a full 32 or 64 bits of fractional mantissa are available for mathematical processing; the sign and exponent bits occupy an additional 11- or 15-bit-spaces. The range of IEEE-754 format parameters is summarized below.

RANGE OF IEEE 754 FLOATING-POINT PARAMETERS

		Format		
Parameter	Single	Single Extended	Double	Double Extended
precision p(bits)	24	-32	53	64
max exp E_{max}	$+127$	$+1,023$	$+1,023$	$+16,383$
min exp E_{min}	-126	$-1,022$	$-1,022$	$-16,382$
exp width (bits)	8	11	11	15
total format width (bits)	32	43	64	79

Single-extended-precision provides 8 more bits of dynamic range while executing almost as fast as single precision and much faster than double precision, since accumulators can be readily extended beyond 32 bits while true 64-bit operations require extra cycles. Many DSP algorithms can compute values of adequate accuracy in extended single precision, avoiding the slowdown associated with double precision.

Rounding: The standard includes precise guidelines for rounding. The default option is *unbiased*, i.e., rounds up or down to the nearest representable value. Other options are *directed* rounding: upwards towards $+\infty$, downwards towards $-\infty$, and towards 0.

The standard also defines special quantities to signal under-range, overrange, and invalid or undefined operations. These involve subtleties beyond the scope of our discussion, illustrated briefly by the examples below (for further clarification, see, for example, the IEEE document* or manufacturers' data sheets on floating point parts such as the ADSP-3210/3220).

---

*Reference: "A proposed standard for binary floating-point arithmetic," IEEE Task P754, December, 1982, widely abstracted and published. Now available as an ANSI/IEEE standard; from IEEE: 754-1985, "Standard for Binary Floating-Point Arithmetic."

Special Quantities: The standard includes two infinities, $+\infty$ and $-\infty$, represented as: $\pm$ [exp all 1's] [fraction all 0's]. Since the IEEE standard uses signed-magnitude, not twos-complement, notation, there are also two zeros, since the result of an arithmetic operation always contains a sign. *Denormalized numbers*, resulting from gradual underflow are too small to fit in the defined exponent range with the specified mantissa format, and are represented by an exponent set to its minimum value (0000...00). This is translated into a value whose "extra" bit is set to zero, i.e.,

$$(-1)^S 2^{-127} (0. b_0 b_1 b_2 ... b_{p-1})$$

To signal an invalid or undefined operation, the standard includes a symbol called *NaN*, for *not a number*. NaN is a symbolic entity encoded as [exp all 1's] [fraction nonzero], which signals an invalid operation when it appears as an operand, and propagates quietly during an arithmetic operation in which it is not involved.

Allowed operations and results of uncalculable conditions: The standard defines certain arithmetic rules, such as "the destination format shall be at least as wide as the widest operand" and the range of allowed operations (i.e. square root may be evaluated only for positive numbers). The standard specifies the consequences of exceptions such as overflows, underflows, divide-by-zero, and invalid operations such as $0^\star\infty$, $0/0$, etc. The NaN may be made use of for diagnosis ("quiet NaN") or for trapping ("signalling NaN"), a condition which may be picked up by special hardware to set a flag (hardware response) or jump to an exception handler (software response).

DECIMAL CONVERSION RANGES
The range of decimal numbers which fit within the IEEE standard are, for decimal of the form $(\pm) M \cdot 10^{(\pm)N}$:

DECIMAL CONVERSION RANGE

Precision Format	Decimal to Binary		Binary to Decimal	
	M_{max}	N_{max}	M_{max}	N_{max}
Single	$10^9 - 1$	99	$10^9 - 1$	53
Double	$10^{17} - 1$	999	$10^{17} - 1$	340

COMPARISON WITH DEC FLOATING-POINT FORMAT
Because of the wide prevalence of Digital Equipment Corporation (DEC) VAX computers, floating-point hardware which may be designed into array processors or microVAX-based workstation graphics accelerators also supports the DEC floating-point format, which evolved well before the IEEE standard was developed. Selection between IEEE and DEC formats usually involves asserting a logic level on a single pin. The difference between the two formats involves the hidden bit (DEC value ½, not 1) and the exponent bias values, as well as the handling of exceptions. The differences are not relevant to our context.

6.2.5 DATA-ADDRESS GENERATOR

Address Generator Definition and Functions

An *address generator* has one or more output registers containing numbers that specify the locations in memory for data and coefficients being read or written. The data-address generator (DAG) provides the fast, flexible addressing essential in DSP algorithms which typically demand several data transfers to and from the number-crunching components in each instruction cycle. DAG functions cannot be performed efficiently by the same μP that does number-crunching.

Why address generators? Examples of data-array scanning and generation of addresses for FFT data, among the most common uses in DSP computation, illustrate the speed advantage achieved by removing address-generation tasks from the CPU.

Data-array scanning is perhaps the most common DSP address manipulation. Not all scans are sequential; therefore an address generator with its own ALU makes possible scanning every second, fourth, etc., address, as in decimation operations. Most DSP table lookups are "circular"; the table is of finite length and the program is to loop back to the beginning upon reaching the end of the table. Automatic endpoint testing within the DAG relieves the CPU of this task (and many unnecessary computation and test steps). The powerful digital signal-processing DAG combines these functions within a single compound statement of the form:

$$\mathbf{Y} = \mathbf{Y} + \mathbf{R}; \text{IF } \mathbf{R} \geqq \mathbf{END} \text{ THEN } \mathbf{Y} = \mathbf{BEGIN}$$

where **Y** is the address value, **R** is an offset value, and **BEGIN** and **END** specify table boundaries.

Many programs need to modify the value of the offset, **R**, as they progress. The DAG's ALU can also take over this task, written symbolically:

$$\mathbf{Y} = \mathbf{Y} + \mathbf{R}; \text{modify } \mathbf{R}$$

The modification might be a fixed increment, $\mathbf{R} = \mathbf{R} + \mathbf{B}$, each time, for example. In the FFT case, data from widely differing regions of address space are fetched and combined in the butterfly calculation. Here the DAG's ALU might provide an upshift or downshift of the offset, **B**, to implement a radix-2 butterfly sequence.

Removal of address-generation overhead from the main μP is a recognized solution in Bit-Slice designs; one or more bit-slice ALU's, such as the Am2901 (Figure 6.12), may be dedicated solely to address-generation tasks. Although this has been a popular solution, 4-bit slices must be cascaded to get an adequate address range; also, a general-purpose μP, such as the 2901, lacks such desirable DAG functions as circular-buffer addressing and bit reversal.

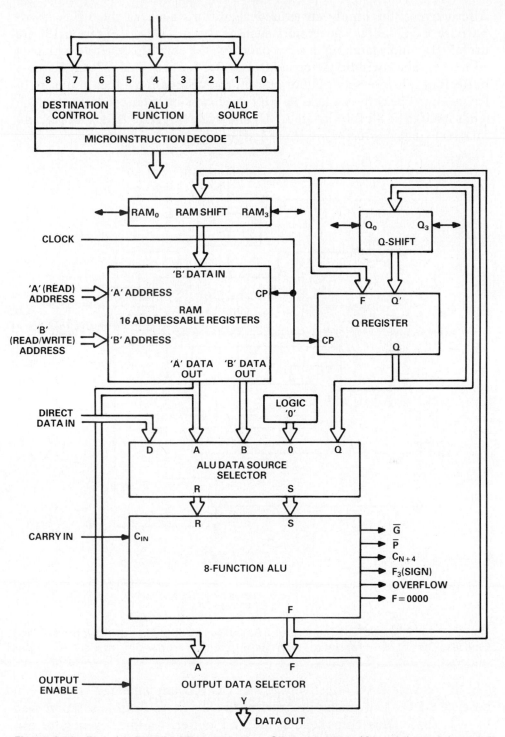

Figure 6.12. Four-bit Bipolar Microprocessor Slice, Am2901. (Copyright © Advanced Micro Devices, Inc., 1985. Reprinted with permission of copyright owner. All rights reserved)

Alternatively, for highly structured algorithms such as the FFT, fixed-purpose FFT address generators such as the Am29540 (Figure 6.13) are useful. By simply latching in a few data bits, the chip can generate both data addresses and twiddle-factor addresses for either Radix-2 or Radix-4 butterflies, bit-reversed input or output, and decimation either in time or in frequency. The chip is a flexible FFT spectrum analyzer component which lends itself to an 8-cycle (single ALU) or 4-cycle (separate ALUs for real and imaginary data) butterfly (see Hoste and Grinde, 1985).

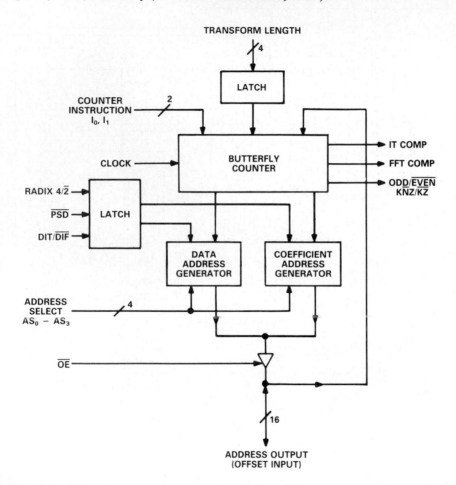

Figure 6.13. FFT address generator, Am29540. (Copyright © Advanced Micro Devices, Inc., 1985. Reprinted with permission of copyright owner. All rights reserved)

The *Word-Slice* DAG effectively replaces DAGs implemented with several bit-slice ALUs, and, being programmable, is a more flexible solution than fixed-purpose DAG chips such as the FFT butterfly address generator. An address generator is itself a microprocessor, since address sequences in DSP algorithms are often not in a simple 1,2,3 sequence. For example:

(1) In an FFT, data pairs combined in a butterfly have addresses separated from one another by increasing powers of two as the computation progresses through the sequence of passes (recall Figure 3.4).

(2) In an FIR filter calculation, address generators set the pointers which index coefficient and data locations during the convolution sum (recall Figure 4.5). Circular buffer and window indexing are also address generator functions.

(3) In a sine-table lookup, interpolation between a small number of stored values increases accuracy with modest memory and computation requirements. An address generator computes the offset address between desired x-value and the nearest stored x-value and outputs the difference for the interpolation computation, as shown below.

Address Generator Desirable Features

A simple but effective address generator structure is shown in Figure 6.14. Each of the registers, R (Result), M (Modification), and L (Length) may be replicated more than once, for multi-purpose applications such as data-address and coefficient-address generation in the same chip).

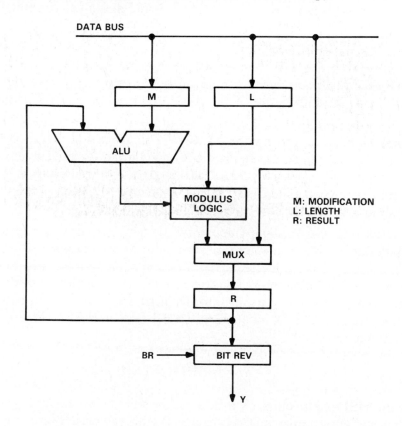

Figure 6.14. General scheme of data-address generator.

Desirable data-address-generator functions include:
Send a precomputed address to data memory
Modify the address by a computed offset value
Perform logical (AND, OR, XOR) operations and shifts
Compare a pointer to a preset value (circular buffer)
Reset pointer to buffer origin when pointer = preset
Reset to origin when pointer $\geqq$ or $\leqq$ preset
Reverse the order of the bits

The rationale for these functions rests in the address computations commonly encountered in DSP algorithms. A typical DAG operation is:

Y = R; modify R contents by a computed offset value

where **Y** specifies the DAG output, **R** specifies the contents of an internal DAG output register, and ";" specifies concurrency, i.e., both operations performed in a single cycle. The computation following the ; occurs after **Y** is updated (**Y = R**). Most data accesses either do a table lookup or scan a data array. Both will benefit from logic capability and modulus arithmetic, as described below.

Address Generator Enhancements
Shift Capability: DAG *shift* capability provides simple FFT twiddle-table or butterfly data-pair addressing, since addresses offsets go up or down by a factor of two (i.e., shift by one binary position) in sequential FFT passes.

Bit-Reversal Capability: In decimation computations such as the FFT, either the input signal is supplied with addresses in bit-reversed order or the output spectrum is returned addressed in bit-reversed order. Ordinarily, a separate computation sequence is required to sort into the required order. *Bit-reversal logic* implemented in a DAG can do this automatically, when enabled, as part of the address generation.

Logic Function and Masking Capability: In table lookups, for example in interpolation or Newton-Raphson recursion, DAG capability to perform logical operations permits the partial masking of an address field to address a stored function value (more significant half); then the less significant half is used as an offset to compute the improvement. For example, with high-resolution argument value **X = [(MSH)(LSH)]** first stored in DAG registers, DAG operation

Y = MSH; R = LSH

feeds the **MSH** of **X** to output **Y** to address the stored table values (1st approximation) and then prepares DAG register **R** with value **LSH** for computation of the 2nd approximation in the next cycle (see Chapter 8, Interpolation, for more details).

Flexible Modulus Arithmetic: In data-array scans, the DAG computes pointer increments and resets to a base address when the end of the buffer is reached. With address offsets or loop ranges preloaded into DAG registers, continuous loops without program overhead are possible, since the address generator keeps track of location in the loop. Implementation of the circular loop so common in DSP programs is called *modulo addressing,* in the sense that the next address larger than the length of the circular buffer is the first address in the table, just as $1111 + 0001 = 0000$ in 4-bit arithmetic. Modulo addressing with an arbitrary modulus (not just a power of 2) is a desirable enhancement to implement the pointer wraparound of a circular buffer. For example:

$$R = B + (R - B + M)MOD(L)$$

R is the DAG address-register content, **B** is a base address, **M** is a computed offset increment, and **L** is the length of the buffer (modulus).

Word-Slice Address Generator Example: ADSP-1410

An example of an emerging generation of full-function address generators, the ADSP-1410, is shown in Figure 6.15. The information that follows is intended only as a cursory overview of functions of a specific DAG chip, significant as the first general-purpose microprogrammable address generator. Manufacturers' data sheets and instruction/application notes should be consulted for essential details.

The ADSP-1410, a CMOS device, includes the general DAG functions listed above, and in addition:

    Pre- or post-update
    Thirty 16-bit registers
    Alternate-instruction execution
    16-bit address generation
    Double-precision (30-bit) address capability
    35 ns clock-to-output time, and 70 ns worst case cycle time
    150-mW power dissipation

In a single cycle, the DAG outputs a previously computed address and computes a new address for the next cycle. The detection of an address boundary and the conditional looping around a circular buffer do not add any cycles.

The registers are organized into four files: 16 *address* (R) registers; six *offset* (B) registers, four *comparison* (C) registers, and four initialization (I) registers. A 10-bit microcode instruction (INSTR input) controls looping, address modification, and register read/writes and internal transfers. An *alternate instruction register* (AIR) permits conditional substitution of a prestored instruction, conserving external microcode transfers when repetitive operations are performed. There are two data ports; addresses go out over the *Y*-port, and register inputs and outputs to set moduli or offset values are made via the bidi-

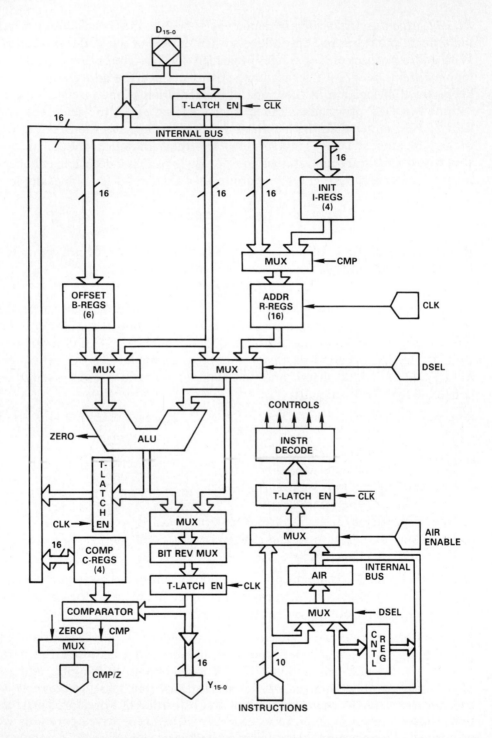

Figure 6.15. The ADSP-1410 Data-Address Generator.

rectional D-port. A pathway from the D-port to the ALU allows for direct data input to an ALU computation, and a pathway from the D-port directly to the Y-output provides a DMA mechanism. Bit-reversal capability (BIT REV MUX) permits bit-reversed FFT outputs to be retrieved in normal order, or FFT twiddle factors to be fetched in bit-reversed order from memory. Double-precision (full 10^9 address-range) addressing is possible in a single cycle with two DAGs. Alternatively, a single DAG chip can provide double-precision addresses in two cycles.

Full modulo addressing, as in the example above (see *Flexible Modulus Arithmetic*)

$$R = B + (R - B + M)MOD(L)$$

is provided in a single cycle by use of the **AIR**. For example,

$$Y = R; \text{IF R} > \text{L THEN AIR}$$
$$\text{ELSE R} = R + M$$

where the AIR contains the alternate instruction:

$$Y = R; R = R - L$$

where **B** is a base address, **M** is the address modification, and **L** is the length of the buffer.

The order of outputting an address and computing a new address is described in terms of pre- and post-updating. *Pre-updating* (the mode discussed above) is the faster mode, since the update address is waiting for the next cycle in a register. However, if external data must be brought on-chip, two cycles are required for a data-read, modification, and address output. In *post-update* mode, external data can be brought on-chip, modified, and outputted in a single cycle. However, the cycle time must be extended appropriately for the update computation. Most DSP algorithms have a regular enough structure to utilize the pre-update mode with address increments fixed by the algorithm.

6.2.6 PROGRAM SEQUENCER

Sequencer Definition and Functions
A *program sequencer* generates the instruction addresses which determine program flow (Figure 6.16a). A program sequencer removes overhead by keeping track of program-counter incrementing (including conditional logic and looping), subroutine handling, and outside-world interrupts. For example, a program may include counters which are tested to determine when a loop is done. These functions are kept track of in a program sequencer by register stacks (*count* stack or *loop* stack). Return addresses are retained in a program counter

(PC) stack for access after a subroutine or an interrupt service request is finished. A program sequencer also accesses status information (signs, overflow, stack empty...) generated by the various computational units for comparison with conditional instructions (JUMP IF...).

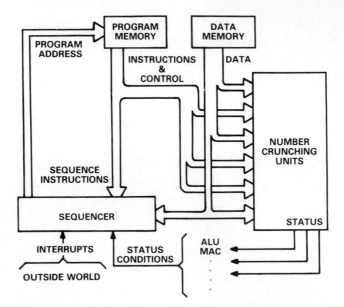

a. Instruction sequencer and its relationship to other DSP system components.

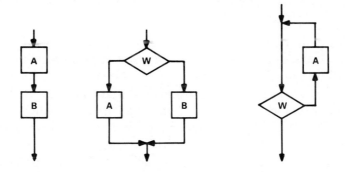

b. Simple sequencer control structures: A then B; If W then A else B; While W do A.

Figure 6.16. Sequencer functions.

The sequencer thus has several main functions:

Handle normal program flow, incrementing the program counter by 1 each cycle

Keep track of subroutine addressing, and manage the return-address stack

Manage loops in an overhead-free manner, controlled by on-chip loop counters

Jump to appropriate exception-handler routines when a data overflow is detected or data rescaling is required

Service interrupts from external I/O devices, jumping to appropriate interrupt service routines and returning to the previous task when done.

Sequencer Desirable Features
A modern program sequencer includes the following components:
  Program Counter
  Conditional Logic Tester
  Stack for:
    Subroutine and Interrupt return addresses
    Counter values
    Jump addresses
  Loop counters

The trend in sequencer design is to include powerful program flow-control functions discussed below, including adequate stack size, zero-overhead program branching, either in software looping or by hardware prioritized interrupts. Desirable enhancements include forcing a jump address directly from the data bus, flexible stack-boundary setting, and zero-overhead conditional branching.

Basic Sequencer Requirements
Adequate Stack Size: The stack must be large enough to handle nested subroutines, and several levels of interrupts.

Branch Conditions and Interrupt Prioritization: Both software testing (flags and counter values) and hardware testing (interrupts) can cause a branch. High-priority interrupts should be able to take precedence over lower-priority tasks such as flag testing. Interrupt priority ranking and interrupt masking are important features.

Sequencer Pipelining Is Not Recommended: While computational elements can be pipelined for enhanced execution speed, the instruction sequencer is best not pipelined, since pipeline delays make difficulties for the software designer and the system as a whole. Chaos would result, for example, if an external interrupt or a detected overflow were not serviced until after several cycles.

Forcing a Direct Instruction Address: While destination addresses are stored on the sequencer in normal execution, it is desirable that a direct address or an address offset be readable through an external port. External direct addressing, for example, is used to force a branch when debugging.

Desirable Sequencer Enhancements
Stack Boundary Settability: Subroutine, jump, and interrupt return addresses may end up stored in the same stack. Since stack size is limited, it would desirable to be able to set the boundaries of each of these regions flexibly, assigning more or less space as the application demands. Stack overflow, where one stack eats into another, is to be avoided at all costs, ideally by a highest-priority non-maskable interrupt.

Zero-Overhead Conditional Branch Capability: Two- and three-way jumps within a single instruction can speed execution and ease programming by eliminating the need to code explicit loop testing. For example, a two-way "test and modify":

IF LE BRANCH; UPDATE C

The sequencer tests for the *end of a loop count* (**LE** asserted as in Figure 6.17b); if not completed, it continues with a normal program-counter (PC) update.

Sequencer Functional Organization Example

The functions of a powerful sequencer are detailed in Figure 6.17. The sequencer carries out next-address selection by combining instruction op-codes (JMP...), condition logic, interrupts, and loop information. The *next-address multiplexer* (Figure 6.17a) can select one of four sources:

Vector address (interrupt service)
PC incremented value (normal next instruction)
PC stack (return from subroutine or interrupt service)
Instruction (e.g., jump address in the code)

A fifth alternative is an indexed jump: the next address is put onto the program address bus externally and then latched into the PC.

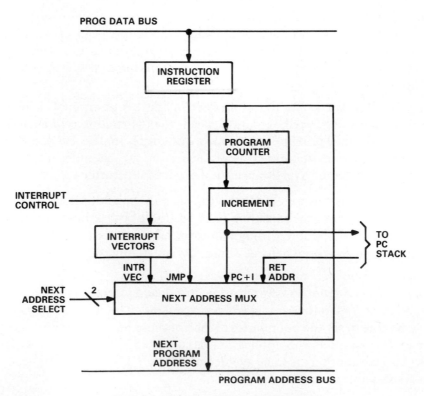

a. Next (program)-address multiplexer.

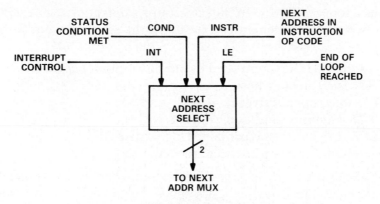

b. Next-address selection.

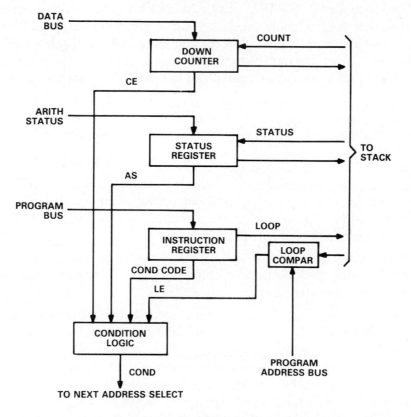

c. Condition logic.

Figure 6.17. Functional Organization of a powerful program sequencer (from ADSP-2100 DSP Microprocessor).

Normal stepping utilizes the *Program Counter* (**PC**) and *PC Stack*. The PC contains the currently executing instruction's address; its output is incremented by 1. The incremented value can address the next instruction or, in the event of a subroutine jump or interrupt vector, it can be pushed onto the

PC stack as a return address. In indexed address operations, a new PC value generated by an address generator appears on the sequencer bus for loading into the PC.

The next-address MUX is controlled by one of 4 inputs (Figure 6.17b) via the *Next-Address Select* block, whose inputs are:

(1) **INT** = interrupt occurrence (if any)
(2) **COND** = a status condition has been met
(3) **INSTR** = the address is contained in the instruction op code
(4) **LE** = the end of a loop has been reached.

In case (3), the op code may specify either an immediate address or an indirect address reference. The *Condition Logic* block (Figure 6.17c) compares status information from computational blocks (ALU, MAC) with conditional portions of instructions (JUMP IF...) and outputs a bit to signal that a selected condition is met. Inputs from the counter (**CE** = count expired) and from the loop stack and loop comparator allow the condition code to be taken from the top of the loop stack if the loop comparator detects that this is the last instruction of a DO loop. This feature permits an overhead-free jump at the end of a DO loop.

The sequencer accesses processor status information stored in the *Status Register*. In the design described here, the Status Register contains the following information:

Arithmetic status (ALU and MAC)
Mode status (operating modes)
Stack status (PC, Count, Status, and Loop stacks)
Interrupt control (edge or level, nesting mode)
Interrupt mask (IRQ enabled or masked)

Key status-register information (arithmetic, mode, and interrupt mask), saved on the status stack, is pushed on and popped off automatically when an interrupt service routine is entered.

The *Loop Stack and Comparator* section provides a zero-overhead looping mechanism. A DO UNTIL instruction contains the address for the end of the loop, as well as the termination condition. When a DO UNTIL instruction is executed, this information is loaded into the loop stack, and the PC value pushed onto the PC stack. The loop comparator compares the end-of-loop address with the next address, and signals the end of the loop when the two are equal. When this happens, the condition logic looks to the loop stack for its condition code rather than from the current instruction register. Depending on the condition code, the next-address selector chooses between the PC stack (jump to beginning of loop) and the PC incrementer (fall out of the loop). If all this happens automatically, looping does not slow computation.

An *Interrupt Controller* determines if an interrupt request has occurred and initiates a jump to an interrupt service routine if that interrupt line is unmasked and has higher priority than an interrupt already being serviced.

Interrupts may be either edge- or level triggered. The edge is latched until it can be serviced and cleared. Edge-triggering is simple to implement, and relatively insensitive to precise interrupt timing since it is latched in as soon as it occurs. However, in a noisy environment, level-sensing may be preferable since a noise spike might be mistaken for an interrupt. When a level-triggered interrupt is selected, the level must remain asserted until the interrupt is serviced.

An interrupt request which passes the interrupt mask is then tested for priority. The highest-priority unmasked interrupt forces a jump to the address set aside for its interrupt service routine. During an interrupt vector, status information is pushed onto the status stack. Interrupts can be made interruptable or not, enabled by a mask status register loaded with a value which can either mask out interrupts of priority equal to or lower than that of the interrupt being serviced, or, alternatively, mask out *all* interrupts until the current one has been serviced.

A Word-slice Program Sequencer Example

An example of a Word-Slice sequencer with many interesting features, the ADSP-1401, appears in Figure 6.18. In addition to the basic features listed above, this chip contains:

    10 Interrupt vectors on-chip

    4 Loop counters

    64-word stack for Subroutine and Jump address as well as Counter storage

    Absolute or relative branching capability

    16-bit address-width for programs of up to 64 K in length

    90-ns instruction-cycle time and 35 ns clock-to-address time

The basic function performed is:

$$Y = \text{New Address}; \; PC = Y + 1$$

where **Y** is multiplexed from one of four sources, depending upon the instructions, conditions, and interrupts:

    Program counter (current address + 1);

    A jump address or return address from the stack;

    An absolute or relative (through the adder) address from an external source;

    A prioritized and maskable interrupt.

A single 64-word stack provides for both extensive subroutine nesting and counter storage. The same stack stores jump addresses for immediate and indirect addressing. In addition, a direct addressing option is available, with the address value fed in via the *D*-port. This latter feature is particularly useful in debugging. The boundary between subroutine stack and jump-register stacks is set flexibly by a stack-limit pointer.

There are 8 external interrupt lines, "wired" to internally stored interrupt vectors. In addition, two internally generated interrupts are available to service counter underflow or stack-limit violation.

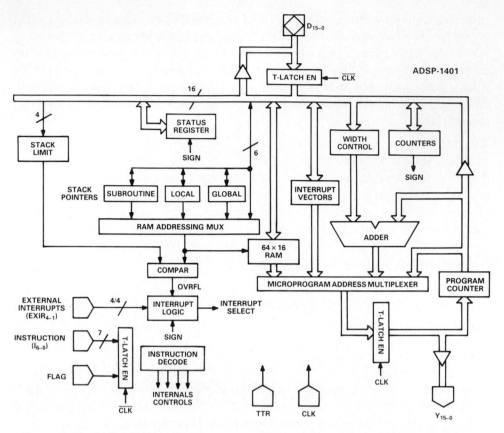

Figure 6.18. A Word-slice microprogrammable program sequencer (ADSP-1401).

This architecture makes possible the equivalent of three-way jumps in a single cycle:

IF Cond1 THEN Branch1; ELSE IF Cond2 THEN Branch2; PC = PC + 1

where **Cond1** and **Cond2** may be end-of-count flags, for example; normal **PC** incrementing occurs if neither branch is executed.

6.2.7 MEMORY ISSUES

Memory Overview

It should be apparent that the choice and configuring of memory for high-throughput processing involves a number of special considerations. They are summarized here in the context of selecting memory chips for data and program storage.

Desirable Features for DSP Memory

Separate memory chips to be addressed by DSP components must obviously match their speed. Yet fast bulk memory used to be relatively uncommon, with production of RAM and ROM (read-write and read-only memory) driven by the volume needs of general-purpose μP applications (e.g., personal computers). Fast static low-power *CMOS* memory was even less common. However, the substantial demand for fast static RAM, driven by DSP and by CAD (computer-aided-design) graphics applications, has led to a response by a number of new manufacturers. In fact, CMOS static RAM has become one of the fastest growing areas. Although priced well above the "jellybean" range of the personal computer market, RAM is now available with specifications (at this writing) such as:

8K $\times$ 8 static R/W CMOS RAM (Cypress CY7C185)

35-ns access time
TTL compatible
Standby power: 220 mW
Power when active: 825 mW
Automatically powers down when deselected

A number of issues that came up earlier in the context of on-chip registers also apply to off-chip bulk memory.

Dual-ported memory has two separate ports, each with separate controls, address, and I/O lines, to permit independent accessing of each element in memory. Dual porting permits a previous value to be read and a new value to be written in the same cycle. For example, an image can be displayed while the contents of the next frame are being updated.

Register files are another way of increasing memory productively without decreasing speed or adding complexity. Devices like the ADSP-3128, with its 5 16-bit ports and 128 × 16 or 64 × 32-bit RAM, act as multi-port scratchpads with very flexibly configured internal pathways capable of inputting a 32- or two 16-bit words, outputting a 32- or two 16-bit words, and either inputting or outputting an additional 16-bit word. In the single-precision mode, up to 6 16-bit transfers per cycle are attainable; in double precision, 5 32-bit transfers can be performed per cycle.

First-in, first-out registers (FIFOs) act as pipelines to match slower memory to faster processors.

An extensive set of *support registers on the DSP chip* lessens the demand for a separate memory access.

Cache memory also lessens memory demands. The most recently used instructions remain resident.

Address generators lessen the demands on memory: The time required to bring a data point into a computational element is the sum of the time to generate the data address plus the time to fetch the data.

$$t(\text{data}) = t(\text{addr. gen.}) + t(\text{data fetch})$$

Since the throughput is dominated by the total time t(data), speeding the address generation phase by using a dedicated address generator puts less demands on the data fetch phases, allowing the use of slower, less-expensive memory for a given t(data). A similar argument applies to instruction fetching from memory.

While not a memory issue as such, insufficient available memory address range in single-chip processors will come to haunt the user if the system architecture is efficient only for the on-chip data memory. When the on-chip memory is used up, throughput slows.

A mixture of fast and slower memory can reduce system costs. Program-memory must be fast, since the system clock speed is limited by the *sum* of sequencer delay and memory-access time. However, data memory can be slower if one is willing to pipeline the data.

6.3 SYSTEM SELECTION

6.3.1 DSP SYSTEM ALTERNATIVES

System Selection is Task Oriented

The type of system to be used depends very much on the task to be performed. The choice will be quite different for an FFT spectrum analyzer than for a vector-graphics workstation. This section outlines alternate ways to combine the hardware elements introduced in Section 6.2: from the building-block microprogrammed chip-set to the fixed-architecture programmable microprocessor. Then examples are shown of flexibly configured microprogrammed systems with paralleled operation for highest throughput.

The discussion of microprogrammed systems here will be brief, because microprogrammed system architectural details depend very much on which family is chosen; also, considerable information for configuring such systems is available from manufacturers; finally, the target for this book is DSP operations on signals in instrumentation systems, for which the programmable microprocessor is usually adequate. Floating-point alternatives are mentioned, since their higher dynamic range lessens the concern for overflow and underflow errors. However, the high cost at present of floating-point chip sets with speed adequate for DSP restricts them to high-end systems such as array processors. Since the DSP microprocessor illustrates most of the issues encountered in processing signals, yet features modest cost and relatively easy programming, it will receive the bulk of the discussion. This section ends with consideration of architecture and instruction sets of recent DSP microprocessors.

Throughput is the dominant factor. Nearly any DSP operation can be performed by a general-purpose μP such as the 8086 or 68000, but the results would not be fast enough for most real-time operations. DSP microcomputers, microprocessors, and configurable chip-sets came into being to met this need.

Microprogrammable vs. Programmable Solutions
The components described below may be organized into distinct frameworks:

The *programmable* DSP microcomputer (DSP μC) and microprocessor (DSP μP) are generalizations of the general-purpose microprocessor, optimized for extensive number crunching. An architecture capable of carrying out $Y = MX + B$ in one instruction cycle is a common goal, because this operation appears in almost all DSP array operations. Although the software is flexible, the architecture is defined.

The DSP *microcomputer* puts computational elements and working data- and program memory on-chip. It is suitable for use in many remarkably sophisticated products. On the other hand, its on-chip memory limits the DSP microcomputer to the audio range of frequencies for most common DSP operations and limits performance even there. A better choice for substantial applications is the DSP *microprocessor*, which puts memory off-chip so programs and data can easily expand. But even here there are limitations: the architecture and instruction set are fixed; bus-width sets a limit on data array length and program size; and instruction word-width and architecture configuration may limit the data manipulations which can be performed at full speed.

The *microprogrammable* chip set is an erector set, the most expensive and most powerful solution for the most demanding fixed-purpose applications. These *Byte-Slice*™ (8-bit slice) or *Word-Slice*® (16-bit slice) chips* are easily paralleled. Since the chips in a given family may be interconnected as desired, the user has the freedom to design specific data paths not available otherwise. A specific task may require higher throughput than a DSP μP can handle, or a specific algorithm may be awkward to implement except by a custom configuration. For example, there is no way that a DSP microprocessor can do a 10-tap FIR filter in one cycle; if that is an absolute requirement, a paralleled array of microprogrammed elements with 10 simultaneous multiplications and rapid 10-term addition must be used. Similarly, the FFT butterfly takes from 12 to more than 30 cycles in popular DSP microprocessors; if only a 4-cycle butterfly will provide needed signal throughput, a microprogrammed building-block solution with several paralleled number-crunchers becomes a viable option.

Microprogram system instructions are created by stringing together in one parallel word the instructions of all chips. Since the system instruction set is not fixed, program development follows a different and more difficult course than for the general-purpose μP or the DSP microprocessor. To remove some of the strain of bit-level microprogramming, special tools allow the user to define instruction mnemonics at a higher level. These tools will be discussed briefly in Chapter 7. There is no easy way to benchmark a comparison be-

*Byte-Slice is a trademark of Advanced Micro Devices; Word-Slice is a registered trademark of Analog Devices, Inc.]

tween microprocessor and microprogrammed solutions, since the architecture of the latter is custom-tailored for the task.

Speed and Power: the Common Denominator
While the chips described below differ in many key specifications, two issues are common to all: speed and power. Speed and power dissipation of several IC technologies are compared in Figure 6.19. Highest speed in silicon-based circuits is achieved with bipolar or emitter-coupled logic (ECL) families. They achieve what advertisers might call "blazing speed," in every sense of the term; they are very fast, but one can fry an egg on a board stuffed with ECL chips dissipating 10 watts apiece. Lowest-power performance is achieved with CMOS, whose speed-power product is superior to bipolar and ECL. CMOS speeds are increasing rapidly, with sub-100-nanosecond cycle times becoming common, and 100 picosecond gate-delays on the sub-micron

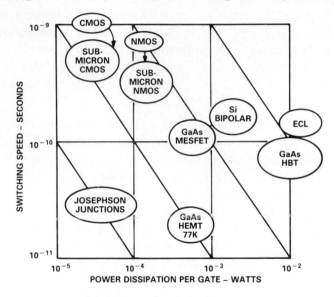

WHAT THE ABBREVIATIONS MEAN

CMOS	Complementary metal-oxide semiconductor
NMOS	n-channel MOS (faster than p-channel)
ECL	Emitter-coupled logic
GaAs	Gallium arsenide, makes faster transistors than silicon (Si)
HBT	Heterojunction bipolar transistor
HEMT	High electron mobility transistor, a GaAs hetero-structure design
Het . . .	Heterostructure: layers of two different materials grown in perfect atomic alignment
Josephson	Superconducting logic based on the Josephson effect
MESFET	Metal-semiconductor FET, or Schottky-gated transistor
MOS	Metal-oxide-semiconductor transistors
SUB-MICRON	Feature sizes below 10^{-6}m

Figure 6.19. Comparison of speed and power-per-gate of common digital IC families. The comparison is a snapshot which evolves as scales shrink; these data are illustrative of 1-to-2 micrometer feature sizes, except where noted.

horizon. Even faster performance at medium power dissipation is achievable with gallium-arsenide (GaAs)-based ICs. Though presently at an early stage of integration—well below VLSI—GaAs ICs have growing importance for moderate word-width (8-12 bit) video or satellite communications, or for IC test equipment which must be faster than the chips it exercises.

Once the logic family is chosen, the speed-power relationship is common to all chips in a family. It will therefore not be dwelt on below; but the reader is cautioned to consider power dissipation as well as speed in selecting any system. The emphasis here will be on CMOS chips, with comparisons to bipolar industry-standard families as appropriate.

Loading: a cautionary note

CMOS chip-speed performance may deteriorate from a specification measured when standing alone if its inputs and outputs are loaded by interconnects or driving other chips in the user's system. Unless care is taken to supply bus-drivers as needed, rise times of a 100-ns-cycle-time chip may slip significantly, and system performance may become unreliable unless clock speed is slowed. Since this may bring throughput below signal requirements, care must be taken that the RAM, I/O, and "glue" chips and surrounding bus interconnects have adequate driving capability. Capacitive loading is the dominant factor, as shown in the following example.

Loading example: Delay rises linearly with output capacitance. A chip with guaranteed capability to drive an 80-pF load with a 10-ns rise-time will slow to almost half the speed with twice the capacitive loading.

Explanation: CMOS circuits draw significant current only during a switching transition; they therefore must have considerable current-driving capability to maintain high switching speed. Since there are few resistive loads, the dominant factor is capacitive loading. A chip specified to drive an 80-pF load (plus internal capacitance) through a 5-V logic transition with a 10-ns rise-time must deliver a peak current of:

$$I(\max) = C\frac{\Delta V}{\Delta t} = (80\,\mathrm{pF} + C_{INT})\frac{5\mathrm{V}}{10\mathrm{ns}} = 40\,\mathrm{mA} + 0.5 \times 10^9\,C_{INT}$$

This maximum current also determines the incremental slowdown with increased capacitive loading:

$$\frac{\Delta t}{\Delta C} = \frac{\Delta V}{I(\max)} = \frac{5\mathrm{V}}{40\mathrm{mA}} = 0.125\,\mathrm{ns/pF}$$

so an added 100-pF load adds 12.5 ns to the rise time. Steady-state *current*-loading demands are secondary unless the CMOS chips are interfaced to TTL. Drawing 4 mA continuously will not seriously deteriorate the performance of a chip designed to deliver 40 mA (but dissipation, $W = VI$, should be checked). Low-power Schottky-TTL with its sub-mA input-current demands is therefore the family of choice for TTL-to-CMOS interfacing.

6.3.2 MICROCODED SYSTEMS

MICROPROGRAMMING EXAMPLE: MUSIC BOX

As a simple instance of microprogramming, a hardwired logic example demonstrates the generation of a microcode instruction sequence and the desirability of data-address generators. A singing music box may be constructed from a pitch-generator that can generate any note plus a dictionary of syllables as pitch and syllable inputs to a voice generator (see Figure 6.20). A time-stretcher on the syllable data output sets the rate of delivery so syllables given

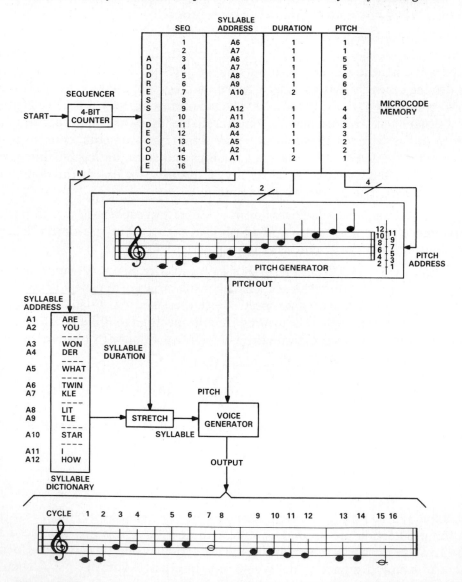

Figure 6.20. Microcoded music box, showing microcode memory and syllable dictionary for "Twinkle, twinkle."

more than one cycle (the half-notes shown) are appropriately slowed. Information to play a given note sequence and a given syllable sequence is stored in microcode memory addressed by a counter which cycles through the tune in 16 counts (in this case). The pitch generator and syllable dictionary, fed the correct pitch address and syllable dictionary addresses in parallel, sound the notes and syllables of the selected song. The microcoded address sequences are shown for "Twinkle, twinkle little star."

These components illustrate the basics of a microcoded system: instruction sequencer, microcoded memory, data memory and data-address generator, and computational elements. Here, the instruction sequence is particularly simple: count from 1 to 15 and then repeat *ad nauseam*. The *sequencer* itself is just a counter, since there are no conditional branches and no interrupts, only a single loop as the counter cycles through its range. The *microcode memory* issues in parallel control information for all components as the sequencer strobes through its addresses. Data memory comprises the notes and the syllables. The "*computational element*" is the voice generator, which modulates the appropriate syllables by the appropriate pitch.

The pitch and duration sequences of music tend to follow rules which can often be inferred. For "Twinkle, twinkle" the rules are simple; they are written in Figure 6.21. It is straightforward to program a simple com-

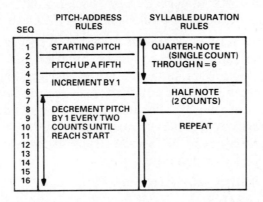

a. Rules for pitch and duration generation for "Twinkle, twinkle."

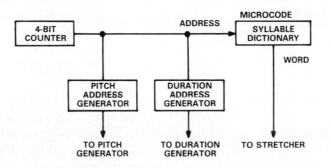

b. Address generators reduce microinstruction complexity.

Figure 6.21. Microcoded music box details.

putational element to output these simple sequences of pitch and of duration. The *address generators* which execute the address sequences feed the data memories and thereby reduce the load on the microcode, since it becomes unnecessary to store by rote any sequence for which an addressing rule can be computed. The microcode memory need only store the syllable sequence to the song.

Microprogramming

The music-box example illustrates microprogramming in simple form, with a binary counter sequencing through the fixed cycle of notes of a single instrument. A real microprogrammed system is more akin to an orchestra. The typical microprogram sequencer of Figure 6.22 (driven by a clock) plays the role of orchestra conductor, directing multiple players who each carry out portions of a preprogrammed score (the microcode). The conductor follows the score (a portion of the microcode also controls the sequencer) and also sets the dynamics of the performance, responding to both the intended sequence of operations in the score, including repeats (loops), as well as to the realities of the performers (processor status conditions) and audience feedback (interrupts from I/O devices). For background reading on micro-programming, see the bit-slice guidebooks (e.g., Mick and Brick, 1980).

The system configuration resembles that shown earlier (e.g., Figure 6.16) except that in a microprogrammed system, program memory has no standard bus-width or instruction set, but contains the microinstructions which supply the parallel combination of 1's and 0's to exercise the functions of each chip in concert (see Figure 6.22b). The instructions for all chips are paralleled to form a system microinstruction or microcode word. The microprogram for the process is a sequence of microinstructions which exercise the joint functions of the DSP components.

Example: Speedy Microcoded FIR Filter Parallels MAC with AG and SEQ

Custom-tailored address generation and program sequencing speed the flow of data through computational elements. For example, consider an N-tap FIR filter which must operate in steady state. Since data is continually streaming through, throughput must match signal bandwidth. The filter carries out the computation ("$*$" means *multiplication*):

$$Y(n) = \sum_{i=0}^{N-1} a(i)*X(n-i)$$

During each computation cycle, the N most recent X-values are weighted by filter coefficients, $a(i)$; their sum becomes the new Y-value, while the oldest X-value is overwritten by a new input. Both X and Y values are to be stored in circular buffers of length $M > N$. The X-buffer is required by the filter computation; the value of M specifies how long an input/output history is kept for display or other action. The design steps of a microcoded system are (a) selecting the hardware organization required to meet speed requirements;

(b) writing the operations of each component in a high-level symbolic way; (c) organizing the bookkeeping of each component's function at each cycle to carry out the hardware operations in (a), and (d) translating the symbolic instructions of (c) into bit-level microcode.

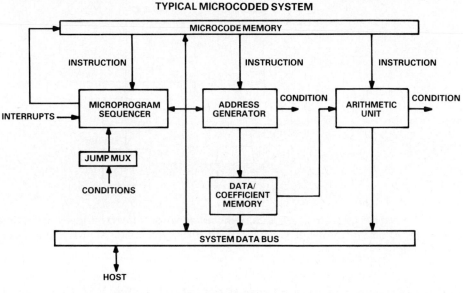

a. Overview.

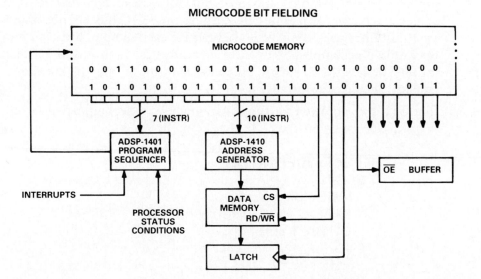

b. Microcode word broken up into sections, each an instruction to an individual component.

Figure 6.22. Microcoded system.

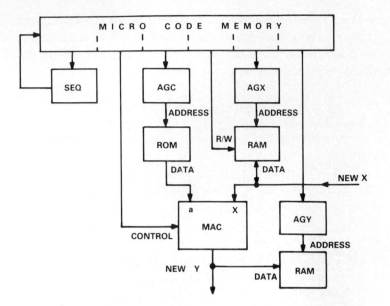

Figure 6.23. Fast FIR filter structure employing microprogrammed parallel computing architecture. [From: P. Toldalagi, CMOS Architecture Speeds Signal Processing (IEEE WESCON 1984)]

(a) Coefficients and data are stored in buffers, as shown in Figure 6.23. Three separate address generators (AGC, AGX, AGY) are employed for coefficient, x-input, and y-output addresses. This parallelism achieves optimal throughput by keeping the MAC busy at all times: an N-tap filter needs only $N + 1$ cycles for each sample. Steady-state operation is accomplished by circular buffers for both X and Y, i.e., buffers of length, M, with pointers, Rn, specifying the locations currently being read or written.

(b) The FIR filter process for the equation above is accomplished as the sequence of operations:

```
REPEAT INDEFINITELY
  BEGIN
    MUL a(0) by X(n)
    REPEAT MAC of a(i) by X(n − i), N − 1 times
    OUTPUT Y(n)
  END
```

(c) The symbolic instructions in a system incorporating a three-port MAC, a sequencer, SEQ, and separate address generators, AGC, AGX, and AGY for coefficient, X, and Y arrays, would have the sequence:

SEQ	AGC	AGX	AGY	MAC
LDCOUNT	RDCOEFF	RDX	WRY	OUT
LOOP	RDCOEFF	RDX	NOP	MUL
BACK	NOP	WRX	NOP	MAC

Here, the address-generator data-read (**RD**) and -write (**WR**) instructions generate addresses within their respective buffers, reset the pointers, **Rn**, to origin, **In**, when the address reaches a buffer boundary, **Cn**, and otherwise increment the pointer by one:

$$\textbf{RDn: Addr} \leftarrow \textbf{Rn; IF (Rn} \geqq \textbf{Cn) THEN Rn} = \textbf{In; ELSE Rn} = \textbf{Rn} + \textbf{1}$$

A read instruction to memory places the data point at the specified address onto the data bus; the coefficient and x-value reads (**RDCOEFF, RDX**) in the previous cycle are latched into the multiplier and operated on in the next cycle (**MUL** on entering the loop, **MAC** later on). The oldest x-value is overwritten (**WRX**) by a new data-point on exiting the loop (**BACK**), and the filtered y-value is output (**WRY**) to memory upon re-entering the loop (**LDCOUNT**). The sequencer keeps track of the pointer within the coefficient buffer.

LDCOUNT:	**Co = N;**	**counter reset**
LOOP:	**Co = Co − 1; IF POS JUMP *;**	**Decrement and loop**
BACK:	**JUMP − 2**	

where * designates a jump to "self"—i.e., repeat the instruction

(d) Each microinstruction corresponds to a binary bit pattern in an individual component. The bit pattern is generated by translating the chip instructions, much like assembling a program for a general-purpose μP.

This example also illustrates the procedure for custom-tailoring the micro-coded instruction set.

The sequence of high-level operations is translated into a bookkeeping of what operation each component must carry out in each cycle. The operations are defined symbolically, e.g., **RDn, LDCOUNT**, etc.

Next, the symbolic operations are translated into specific instructions to the components. Rather than do this at the bit-level, mnemonics which resemble assembly language constructs are defined by the *user*. The line, **RDn**, above is an example.

Finally, the translation of mnemonics into bit-level microcode is carried out by a "meta-assembler," which is basically a cross-reference table of mnemonics and the corresponding bit-level instructions. Further simulation and debug steps are carried out by other development tools to be described in the following chapter.

If the result does not meet throughput specifications, the hardware organization is redefined, for example introducing more parallelism or higher-performance components (next section). Modern flow-control components, such as the data-address generator and instruction sequencer used above, greatly simplify the generation of microcode. Previously, the exact bookkeeping of all component instructions would have to be traced through each cycle of the entire microprogram sequence. The three lines given above replace many lines of bookkeeping.

Example: Microcoded FFT Processor Alternatives
Complex FFT Signal Processor: Block diagrams for typical microprogrammed and programmable DSP systems were seen in Figures 5.6 and 5.7. While the same basic elements are present, their number and arrangement can differ greatly. For example, consider the microprogrammed complex signal-processor shown in Figure 6.24. Here, ALUs are paralleled so complex components can be processed in the same cycle.

While the DSP microprocessor can also be programmed to carry out a complex FFT, the throughput will be lower than for the microprogrammed

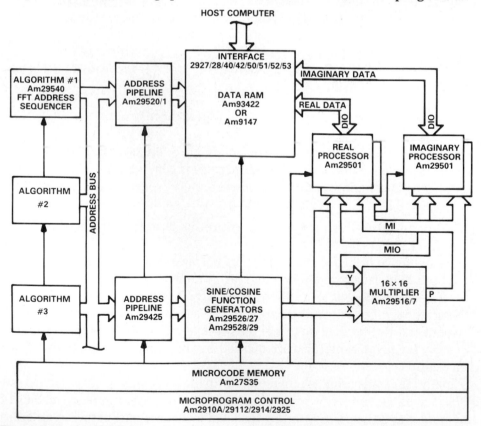

Figure 6.24. Am29500 Array Processor, a high-performance microprogrammed complex signal processor. (Copyright © Advanced Micro Devices, Inc., 1985. Reprinted with permission of copyright owner. All rights reserved)

version, since real and imaginary components will take turns on a common bus. Butterfly computation takes from 12 (second generation) to more than 30 cycles (first generation) DSP μPs (see Section 6.4), while microprogrammed assemblies, such as the one shown here, take as little as 4 cycles.

A microprogrammed system need not carry out a single function; other algorithms which utilize the *same* architecture can be included in the microprogram memory. For example, pieces of code or test routines can be downloaded into a *writable control store* (a fancy way to say program RAM for debugging a microcoded system). The alternate address generators shown in Figure 6.24 (Algorithms #2, #3) could also be configured to perform a sequence of different data-massaging operations: windowing, FFT, phase and magnitude, display (including zoom and pan), Fourier filtering (i.e., bandpass in frequency), and inverse FFT.

Throughput Improvement by Custom Paralleling

Consider the requirements of the FFT butterfly calculation employing the decimation-in-time algorithm. Real and imaginary parts of outputs, A' and B', are computed from inputs A and B:

$$AR' = AR + BR*WR - BI*WI$$
$$AI' = AI + BI*WR + BR*WI$$
$$BR' = AR - BR*WR + BI*WI$$
$$BI' = AI - BI*WR - BR*WI$$

The computation requires:

(1) Input of A and B, two complex or 4 real number inputs;
(2) Four multiplications: $BR*WR, BI*WI, BR*WI,$ and $BI*WR$;
(3) Four summations of the form $(a \pm b \pm c)$, or six simple summations;
(4) Scaling down by two to prevent overflow (better if conditional, but ignore that improvement for now);
(5) Output of A' and B', two complex (4 real) outputs.

Suppose that a system is to be put together with ALUs and multipliers. When the configuration consists of one ALU, one multiplier, and one data-bus to data-memory, a straightforward exercise shows that 8 cycles are required to carry out the above operations. For increased throughput, Byte/Word-slice systems can be configured to process real and imaginary data components separately. Extra chips can be connected to remove bottlenecks where one element is used only a fraction of the time because another is saturated. The tradeoffs in operations needed as a function of number of paths and processors are (Table 6.4).

With one data-memory bus, one ALU, and one multiplier, the butterfly will take 8 cycles; the multiplier is used only half the time. Doubling the number of ALUs and memory bus access (Real/Imaginary ALU and data-memory), speed is increased by a factor of two; all components are 100% utilized. A

Table 6.4 Effect on throughput and resource utilization of paralleled components in an FFT spectrum analyzer.

Resource →	Data-memory Bus		ALU		Multiplier	
Operations needed per butterfly	8		8 (6?)		4	
	Buses	**Cycles in use**	**ALUs**	**Cycles in use**	**MPYs**	**Cycles in use**
Options:	1	8	1	8	1	4
	2	4	2	4	1	4
	4	2	4	2	2	2

further doubling of all resources (separate sine/cosine memory and two multipliers) would bring a further factor-of-two enhancement. However, the first twofold improvement required only a modest component increase and achieved optimal resource utilization, while the second improvement doubles the cost over the first improvement.

The two-ALU processor of Figure 6.24, with separate real and imaginary data-memories, carries out 4-cycle butterfly computation. This example utilizes the special-purpose FFT address generator of Figure 6.13. Of the four cycles available, the DAG uses three to generate addresses (two data-addresses and one sine/cosine pair) and one to increment the butterfly counter. A single general-purpose programmable DAG could also have been used. The data stream in the table below illustrates the operations that must be executed by the microcode to carry out the 4-cycle butterfly.

Cycle	Data Bus = ALU-IN	Multiplier Inputs	Multiplier Output	Real ALU Operation	Imag ALU Operation
1	Read B				
2		BR, WR			
3	Read A	BR, WI	BR∗WR		
4		$\overline{\text{BI, WR}}$	$\overline{\text{BR∗WI}}$	AR − BR∗WR	
5		BI, WI	BI∗WR		AI − BR∗WI
6			BI∗WI	AR + BR∗WR	BI′
7	− − − −	− − − −	− − − −	BR′	AI + BI∗WR
8	Write B′				AI′
9				AR′	
10	Write A′				
11					
.					
.					

Follow operations down a diagonal to observe how each multiply output prepares the next ALU input, etc. The "holes" are cycles where no action is taken by that component; they are just right to be filled in by a pipelined overlap. For example, in cycle 7 the unused spaces in the first three columns (shown dashed) are precisely the places to be filled in at the next butterfly in the pipeline (operations underlined in cycle 3). Interleaving reduces the 8-cycle butterfly to 4 cycles, but requires painstaking microprogramming of each component's operation in each cycle. This example's microprogram is worked out in detail in an excellent application note [Hoste and Grinde, 1985].

Performance is calculated from the cycle time and the number of cycles to complete the transform. For longer arrays, the startup and finish portions are short compared with the time spent doing all the butterflies, so

$$T = t \cdot 4 \cdot (N/2) \log_2 N = t \cdot 2N \log_2 (N)$$

where t is the cycle time. For $t = 100$ ns, this system does a 1,024-point FFT in about 2 ms.

An alternative configuration is possible with only one ALU and dual data memory to produce a compromise 6-cycle butterfly (see Exercises).

Microcoded Pipelining – Pluses and Minuses: The above example makes use of one level of pipelining to interleave the butterfly computation optimally for the computational resources selected. More levels of pipelining provide greater throughput enhancement (especially in such large arrays, where latency is unimportant) and permit interfacing to slower components (e.g., inexpensive mass memory). In addition to the problem of latency in smaller arrays, deep pipelining makes microprogramming even more difficult, since a micro-instruction is selected several cycles ahead of its execution. The interleaving of operations of paralleled computing components must be painstakingly worked out lest small timing differences cause one element to get out of sync with its coprocessors.

6.3.3 SINGLE-CHIP DSP MICROPROCESSOR SURVEY

DSP Microcomputer vs. DSP Microprocessor
The DSP *microcomputer*, typical of first-generation chips such as the TMS32010, emphasizes memory on-chip. Though fast when on-chip data is accessed, performance usually suffers when off-chip data access must be made, especially when the design adds cycles which slow throughput when an external memory reference is made. Yet on-chip memory size is severely limited, given the many other functions that must share the available real-estate. The DSP *microprocessor* is designed to access external memory efficiently, so only data register files and an instruction cache are included on-chip. The area gained by not putting memory on-chip is occupied by functional units which speed address generation, program sequencing, and numerical processing.

What Distinguishes a Powerful Number-cruncher?
A DSP microprocessor should contain the functional equivalents of all key components of Section 6.2:

    3 arithmetic elements: ALU, MAC, SHIFTER,
    plus data fetch and flow control: ADR GEN, SEQ

with an internal organization that allows a flexible combination of arithmetic and data movement in the same cycle.

If any of the three computational units is absent, attaining adequate speed for DSP will involve compromises and kludges. For example, an inflexible shifter makes block floating-point signal scaling difficult. Limitations in the data-fetch and flow-control sections are equally serious: if zero-overhead looping is absent, loop/branch-testing is 2 to 3 times slower; the necessary in-line code generation eats memory at a voracious rate (Chapter 5, Figure 5.15). Section 6.4 will examine in depth the architecture and instruction set of a third-generation prototype, the ADSP-2100 single-chip microprocessor.

Example: TMS-320 series
The Pioneering 32010: The TMS-320 series began with the 32010, whose internal organization and performance were described in Chapter 5. This important pioneer introduced DSP microcomputing to a wide audience, including a great many consumer products. It has found a place in significant instrumentation products, computer products (voice recognition interface), and in automotive applications ("Please buckle your seat belts"). The on-chip memory limitations and the inefficient straight-line coding (because of slow looping) were ameliorated in the later TMS-32020.

The 32020 Enhanced Follow-on: The 2nd generation of the 320 series was introduced in 1985 (see, for example, S. S. Magar et al, *Electronic Design*, Feb. 21, 1985). The 32020 represents a significant improvement over its predecessor, the 32010. Here is a comparison of key specifications:

	TMS32020	TMS32010
Instructions	115	60
Data memory	64 K off-chip	144 words on-chip
		(64-K addressable off-chip)
Program memory	none on-chip	1.5 K on-chip
	64 K off-chip	4 K off-chip
RAM on-chip	544 words	144 words
ROM on-chip	none	1.5 K words
Aux. registers	5	2
Multiplier	16×16 plus shift	16×16
Barrel shifter	16 bits left	16 bits left
Host interface	on-chip	external
Cycle time	200 ns	200 ns
MAC time	200 ns	200 ns
RAM access	1-cycle (repeat instr.)	
	2-cycle (normal)	2-cycle (normal)
	3-cycle (table-read)	3-cycle (table-read)
Address computation	DAG on-chip	incr/decr only
Chip size	45,000 sq-mils	115,000 sq-mils
Power dissipation	1.2 W	0.9W
Package	68 pin-grid	40-pin DIP

The principal changes can be seen by comparing 32020 chip structure shown in Figure 6.25 with that of the 32010 (Figure 5.10). On-chip data memory has

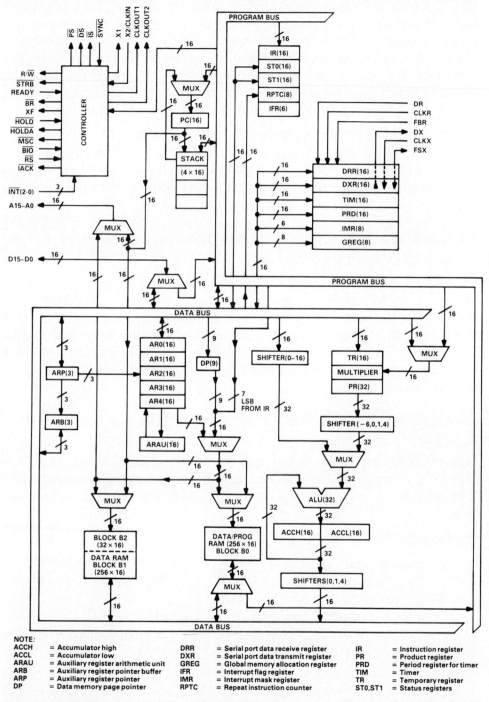

Figure 6.25. Block diagram of the TMS32020 DSP microprocessor. (Courtesy Texas Instruments Semiconductor Group)

NOTE:

ACCH	= Accumulator high	DRR	= Serial port data receive register	IR	= Instruction register
ACCL	= Accumulator low	DXR	= Serial port data transmit register	PR	= Product register
ARAU	= Auxiliary register arithmetic unit	GREG	= Global memory allocation register	PRD	= Period register for timer
ARB	= Auxiliary register pointer buffer	IFR	= Interrupt flag register	TIM	= Timer
ARP	= Auxiliary register pointer	IMR	= Interrupt mask register	TR	= Temporary register
DP	= Data memory page pointer	RPTC	= Repeat instruction counter	ST0,ST1	= Status registers

been expanded to 544 words, enough to hold moderate data buffers. The penalty for going off-chip has been lessened for repeated instructions (IN, IN, IN...), which are single-cycle, though other memory accesses still pay a factor-of-two slow-down penalty. A computing address generator has been added. Program memory has been moved totally off chip. Sequencing of instructions is not appreciably changed, so loops and branch testing require several cycles. A multiple-bit shifter is present, but block floating-point exponent detection is not automatic. The chip is a significant improvement, incorporating many of the desirable functional elements listed earlier.

Not to be ignored is the extensive user support network developed for a successful and maturing first-generation product. Seminars, development systems, I/O boards, second-source hardware and software, and a network of experienced users all combine to help a first-time DSP chip user to get started.

6.4 DSP MICROPROCESSOR: ADSP-2100

6.4.1 INTRODUCTION

The discussion which follows is intended to illustrate the principal features and capability of a third-generation DSP microprocessor, the ADSP-2100. The ADSP-2100 packs nearly all of the desirable functions mentioned earlier into a single chip. This discussion is an overview of the hardware and instruction set, to illustrate what is now possible. Naturally, the prospective user will need to consult appropriate hardware and programming manuals for more-detailed information.

6.4.2 ADSP-2100 MICROPROCESSOR SYSTEM CONFIGURATION

Overview

Figure 6.26 illustrates a typical system that might be built around the ADSP-2100 DSP microprocessor. There are separate buses for program instructions (PM) and data being processed (DM), each with its own address and data lines.

The following labels:

Program-memory addresses:	PMA
Program-memory data:	PMD
Data-memory addresses:	DMA
Data-memory data:	DMD

are used to identify the four external buses. "Data-memory data" refers to the *contents* of the data memory.

Since all buses can carry signals simultaneously, this architecture makes it possible for an arithmetic operation, *plus* a memory reference or I/O operation, to be carried out in a single machine cycle; this implies single-

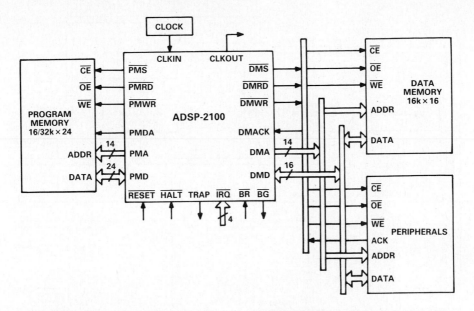

Figure 6.26.   Basic system configuration of ADSP-2100 DSP microprocessor.

cycle multiply-accumulate capability. Yet pipelining is only one level deep, so nonrepetitive operations pay little penalty in latency. CMOS memory is now available with sub-100-ns access times; since it can keep up with the processor, no speed penalty is paid for keeping memory off-chip.

The system operates with an 8-phase clock, running at four times the instruction-cycle rate (i.e., $4 \times 12.5/\mu s$ max in mid-1988). It produces a clock output that defines the processor cycle and hence all instruction cycles. There is no internal microprogramming: the logic blocks shown in Figure 6.27 all exist as shown, and all instructions execute in one cycle (unlike the multi-cycle instruction execution of general-purpose μP's and some DSP microprocessors).

With its multiple-bus structure, the ADSP-2100 supports a high degree of operational parallelism. *In a single cycle*, it can: fetch an instruction, compute the next instruction address, perform 1 or 2 data transfers, update 1 or 2 data-address pointers, and perform a computation.

Program Memory
External program memory will hold 16K words of 24-bit length, sufficient for DSP programs, which are kept short for execution in real time. The architecture is so efficient that almost any signal-processing function can be coded in far less than the memory space available. For example, we will describe later in the Chapter an FIR N-tap filter module that takes only 6 lines of code, and requires $N + 6$ machine cycles. Also shown is code for a complete conditional block-floating-point radix-2 DIT FFT routine that takes only 52 lines of code; a somewhat longer version with more loops to increase speed executes in 12.77 ms for 1,024 points.

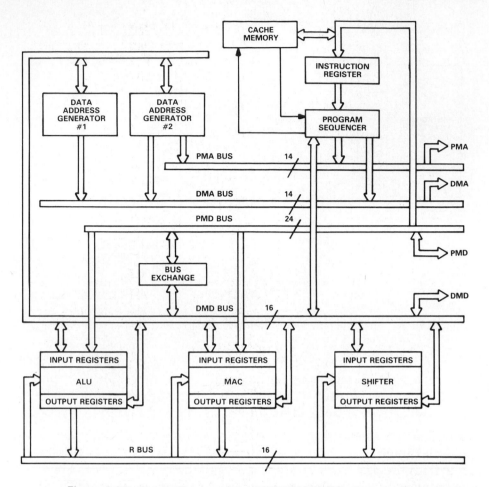

Figure 6.27. Block diagram of the ADSP-2100 DSP microprocessor.

With programs stored off-chip, program memory stored in RAM can be changed dynamically by the host processor to implement demand-paged virtual memory. The available program memory space is actually twice as large, with the PMDA line providing a choice between *two* 16K banks of program-memory. One bank may be permanent instructions in ROM, while the other bank may be RAM for storage of coefficients which may be changed during operation of the processor. The PMDA line is asserted in any non-instruction-fetch cycle, putting a valid address out onto this alternate bank.

The 24-bit instruction word is long enough to allow coding of *multi-function instructions*, which access data and do computations in the same cycle, for high throughput. The processor can enable a program-memory write, set by the Read/Write (PMWR) line; for example, it can be used to update coefficients or even program flow, based on past results.

Data Memory
External data memory addressing is sufficient for a 16-K array of 16-bit

words. Data is bussed separately from the instructions (Harvard architecture), and both program and data memory banks may be accessed in a single cycle.

Input/Output Peripheral Devices

I/O devices are *memory mapped* into the data memory address space. External data therefore need not pass through the host, but are directly accessed by the DSP processor, which is generally faster than the host. *Handshaking* control allows the processor to be interfaced to a slower I/O device; the processor will not leave a data-memory transfer cycle until the DMACK line acknowledges that the external device is done. There are four *interrupt* lines (IRQ), which may be tied to external peripherals. The interrupts are internally prioritized and individually maskable. The ADSP-2100 will halt and relinquish control of both the program and memory interfaces, in response to assertion of BR (bus request), giving direct memory access to an external device; compliance is signalled with BG (bus grant).

Host Interface

The interface connections to the host processor can either upload or download program instructions and parameters, dump a data array from the host, access the results of a computation stored in data memory, or address peripherals. In a typical system, the host exerts master-slave control over the DSP processor, via three signals: RESET, HALT, TRAP.

The *RESET* line forces a clock-phase initialization, resets all stack pointers, clears the cache memory, resets the program counter, masks all interrupts, and clears the memory status register. When the RESET line is released, the processor begins execution of the instruction pointed to by the program counter.

The *HALT* line enables temporary suspension of program execution. The internal state will be frozen, and all bus signals remain static so that they can be probed during a debugging operation. A momentary release of the HALT line executes a single-step operation, also for debugging. If the current instruction is operating on program memory, the HALT will not take effect until the instruction fetch cycle, permitting the host to disable program memory and feed its own instruction into the DSP processor.

The *TRAP* line is asserted when the DSP processor encounters a TRAP instruction inserted in a program by the user during debugging, to signal "I have arrived." The internal state is frozen as in the HALT condition above, so that the host can, for example, read current bus addresses and contents.

6.4.3 ADSP-2100 MICROPROCESSOR ARCHITECTURE

Overview

The block diagram of DSP processor functions (Figure 6.27) illustrates a high degree of parallelism. Each subunit acts as an independent processor for specific operations. Each connects to all four external busses (program memory address and contents, data memory address and contents), as well as an internal bus (the Results or *R*-bus), which transfers results between sub-units. The PMD-DMD Bus-Exchange register allows program- and data-memory contents to be exchanged; this permits the result of a computation to determine what the program does next, or program parameters to be passed from data-memory storage or external I/O devices.

The major computational units are the Arithmetic-Logic Unit (ALU), the Multiplier-Accumulator (MAC), and the Shifter. While data is transferred in 16-bit widths, internal results registers have 32- and 40-bit widths. Data loaded from/to the input registers is pipelined one level deep. This permits each of the three units to be crunching on past/future operands of another unit. Direct feedback of an output register contents passes via the *R*-bus to any other unit without passing through an input register, so feedback occurs without a pipeline lag.

Two Data Address Generators compute the address sequence for data being fetched. DAG1 supplies addresses to data memory, and DAG2 can supply addresses either data or program memory, so that both program- and data-memory can both be accessed in the same cycle.

The Program Sequencer computes the instruction-address sequence, and thereby controls program flow. It takes the responsibility for loops and branching, leaving the computational units (ALU, MAC, Shifter) free to compute on data, not instruction-addresses. The program sequencer takes information from the instruction register (conditions codes, jump address, etc.) The Instruction Register stores instructions prefetched in the previous processor cycle, so there is no waiting for an instruction-decode.

The Instruction Cache Memory automatically retains the previous 16 instructions, saving cycles when repetitive loops are performed.

Each of the processor's computational units will be discussed below in relation to the key functional specifications defined in Section 6.2.

Arithmetic-Logic Unit
The Arithmetic-Logic Unit (ALU) shown in Figure 6.28 performs the usual set of arithmetic and logic functions, enhanced by a *divide* instruction. The ALU has an unusually flexible set of registers surrounding it. A similar set surrounds the MAC and Shifter. The register configuration makes possible single-cycle compound operations.

ALU Registers
Each input register, X0,X1 and Y0,Y1, can be filled with two related or independent 16-bit values. The *Y*-input can access either the data bus or the program bus contents (lower 16-bits), so computations can be performed on information stored in program memory. Either input register can be read back to the data bus, and the two outputs are independent (dual ported) so

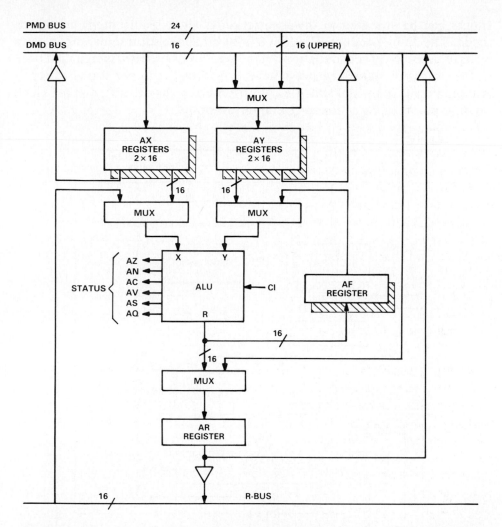

Figure 6.28. Arithmetic-logic-unit block diagram for the ALU block of Figure 6.27.

one register can be feeding the ALU while the other's contents are being returned to the DMD-bus. Direct feedback of an ALU result to become the next Y-input is provided by the AF register without being pipelined through the ALU's input register. ALU outputs pass through the AR register either to the internal R-bus or back to the DMD-bus. The AR register can be enabled to carry out saturation arithmetic.

A complete duplicate ALU register set is present (shown as shadows on the diagram) to permit context switching (save contents upon interrupt).

Multiplier-Accumulator

The Multiplier-Accumulator (MAC), shown in Figure 6.29, carries out high-speed, 40-bit-precision single-cycle (multiply + [add/subtract]) operations.

The MAC is a full array multiplier, with 16-bit inputs and a 32-bit output.

Inputs can be specified in the instruction as signed, unsigned, or mixed-mode. The 32-bit product, P, is fed to a 40-bit adder/subtractor, for an extra 8 bits of precision in accumulation before overflow. Overflow is signalled back to the processor status register whenever any of the upper 9 bits of result is a significant (non-sign) bit (i.e., it differs from the others), so that block floating-point scaling action can be taken promptly.

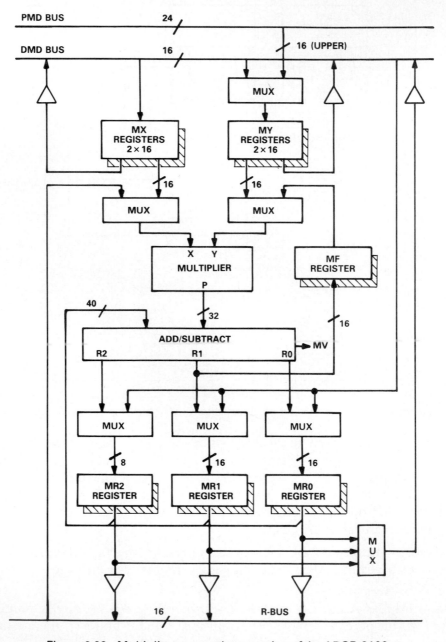

Figure 6.29. Multiplier-accumulator section of the ADSP-2100.

Rounding is available. The rounded 24-bit result appears in registers MR2 + MR1, and, with overflows ignored, as a 16-bit result in register MF as well. Rounding is unbiased.

MAC Registers
The multiplier-accumulator is surrounded by a register set similar in structure to that of the ALU; here's how they differ: The result register, MR, is divided into three sections: MR0 (least-significant) contains bits 0-15, MR1 contains bits 16-31 (more-significant half or rounded result), and MR2 contains bits 32-39 (overflow accumulation). The MR register can also be preloaded, from the DMD-bus for example, to preset the initial conditions in a filter computation or preset the constant term in a power-series calculation. An output from the 8-bit MR2 register to the 16-bit DMD-bus is automatically sign-extended or zero filled, as appropriate. The MAC has a saturation arithmetic option which will set the MR register to plus or minus full scale if overflow has occurred.

Like the ALU, the MAC has a duplicate register set (shown as shadows), which can be activated during an interrupt service routine for fast context switching; a new task can be executed without transferring current states to storage.

Shifter Block and Scaling Functions
The Shifter (Figure 6.30) features unusually flexible and overhead-free scaling to preserve the maximum possible precision with 16-bit (single-) or 32-bit (double-) precision data. While going from fixed to floating point usually sacrifices precision to gain dynamic range, the shifter's facility at block floating-point array operations retains the full bus-width precision of fixed point for optimum signal-to-noise computations. The shifter can execute in one cycle a shift of any number of places in either direction for both double-precision and block floating-point operations: normalization, denormalization, shift immediate, and block exponent adjust.

Shifter Array: The shifter array places a 16-bit input in SI onto a 32-bit field in the results (SR) register. The placement is determined by an 8-bit control code (C) and a high-low reference signal (R). The 8-bit (signed) control code sets the number of places and the direction, right or left. The control code is multiplexed in from one of three sources, depending upon the operation:

Operation	Source of shift control code
Normalization	Negated value of the SE register
Denormalization	SE register
Shift-immediate	Contained in instruction

The HI/LO control sets the point of reference for the shifts, and facilitates double-precision operations since it permits the same control code to set the shift for both halves of the number.

HI/LO	Reference point for shift
high	Upper half of output field
low	Lower half of output field

The 32-bit output of a 16-bit input is zero-filled at the least-significant end and sign-extended on the most-significant end.

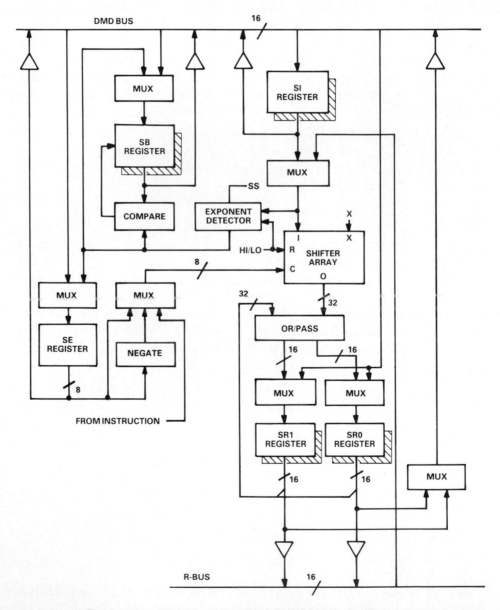

Figure 6.30.   Block diagram of the ADSP-2100's shifter.

OR/Pass Logic: The block labeled OR-PASS below the shifter facilitates multiple-precision shifts. When in the PASS mode, shifter output passes unchanged into the output register SR1/SR0. When set in the OR mode, shifter output is bitwise OR'ed with the previous contents of SR1/SR0. For example, a double-precision number can be shifted in two cycles with a result in the SR output register as if 32-bit data had been shifted in one step, e.g.,

Shifted MSH	1010110110101101	0000000000000000
LSH		1010001110100011
OR'ed result	1010110110101101	1010001110100011

Exponent Detector: To normalize a number, the Exponent Detector examines the shifter input (SI) and stores in register SE a base-2 exponent, whose value depends upon the HI/LO mode setting: HI (for the upper half in double precision, or for single precision), HIX (an extended-precision version of HI), or LO (lower half of a double-precision number).

(HI) Here, the input is interpreted as a single-precision number or the upper half of a double-precision number. The exponent detector determines the number of redundant leading sign bits and produces a code which indicates how many places the number must be left-shifted to eliminate all but one sign bit.

(HIX – HI-extend) This mode looks for ALU overflow due to addition or subtraction (arithmetic overflow status, $AV = 1$), and stores a $+1$ exponent to indicate an extra bit (the ALU carry bit) and prepare for a right-shift to recover from the overflow. If $AV = 0$, the HIX mode and HI modes operate identically.

(LO) In the LO state, the input is interpreted as the lower half of a double-precision number. The exponent detector interprets the shifter sign (SS) bit in the arithmetic status register as the sign bit of the number. The exponent is offset by 16 for operations on the lower half of double-precision numbers.

For example, with S = sign bit, N = first non-sign bit, and D = don't-care bit, the output of a typical normalization operation is:

Mode	Shifter Input	Exponent Output	AV	Previous Sign
HI	SSSSSNDDDDDDDDDD	-4		
HIX	SSSSSNDDDDDDDDDD	-4	0	
HIX	DDDDDDDDDDDDDDDD	$+1$	1	
LO	SSSSSNDDDDDDDDDD	-20		S

Shifter Registers: Shifter input (SI) and result (SR) registers resemble those of the ALU and the MAC functions. Shifter Exponent (SE) and Shifter Block

(SB) registers perform scaling operations. The 8-bit SE register holds the exponent during normalize and denormalize operations (below). The 5-bit SB register holds the block-exponent value from the most recent block-floating-point adjust operation. SE and SB can also be read or loaded from the DMD-bus, and can be compared with one another. Like the ALU and MAC, the shifter has a duplicate set of registers (shadows) for holding data during interrupt service routines.

Shifter Functions: The shifter performs the full complement of normalization, denormalization, shift immediate, and block exponent adjust, as described in Section 6.2.

Data-Address Generators

Addressing modes of the ADSP-2100 feature indexed addressing with the ability to modify addresses for zero-overhead loops and branches, and manage pointers in circular data buffers. Two separate address generators are provided for addressing data stored externally. Both program-memory and data-memory may be accessed in the same cycle. While address generators are principally for data (signal) address generation, one of the two can access program memory—for example, to address stored coefficients. DAG1 also has bit-reverse capability for decimation operations such as the FFT. The internal logic of an address generator is shown in Figure 6.31.

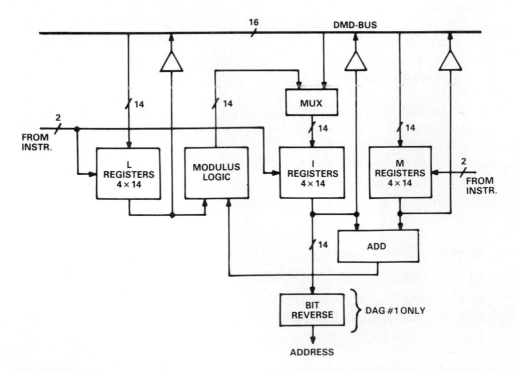

Figure 6.31. Data-Address-Generator Block Diagram. Only DAG1 contains bit-reverse capability. Refer to Figure 6.27 for bus connections of DAG1 and DAG2.

Address-Generator Registers and Modulus Logic: A data-address generator performs an address computation based upon information stored in its registers. The output is fed to an address bus.

$$\textbf{Next Address} = (\textbf{I} + \textbf{M} - \textbf{B})\,\textbf{Modulo(L)} + \textbf{B}$$

Here, **B** is the data-buffer base address, **I** is an index, **M** is a modification, and **L** is the length, for instances where circular-buffer address-wraparound is required, as when cycling through an N-tap FIR filter.

Each DAG register actually contains four independent 14-bit registers, for flexible context-switching of DAG functions (for example, addressing both FFT data pairs and twiddle-factors). The index register, I, provides a starting address, a pointer to be modified by a value, M, the offset. A length register L controls wraparound in a circular buffer. When not operating on a circular buffer, setting L to 0 disables modulo addressing.

All DAG registers are writable and readable from the DMD-bus. Since address outputs are 14-bits wide and the DMD-bus is 16-bits wide, the upper 2-bits of the I-register and L-register are padded with zeros when being read onto the DMD-bus. Since M-register modifications may go either forwards or backwards, the upper two bits are sign-extended when reading it onto the DMD-bus.

Program Sequencer

The ADSP-2100 *program sequencer*, shown in Figure 6.32, generates the instruction addresses which determine program flow. Counters which point to current locations in a data buffer or keep track of location within loops are kept track of automatically by register stacks, such as a count stack or a loop stack. Return addresses are retained in a program-counter (PC) stack for access after a subroutine or an interrupt service request is finished. The program sequencer also includes the status registers which contain status information (signs, overflow, stack empty...) generated by the various computational units.

Status Register: The status register stores the following information
      Arithmetic status (ALU and MAC)
      Mode status (operating modes)
      Stack status (PC, Count, Status, and Loop stacks)
      Interrupt control (edge- or level-triggered, nesting mode)
      Interrupt mask (IRQ enabled or masked)

Key status-register information (arithmetic, mode, and interrupt mask) is saved on the status stack, pushed on and popped off automatically when an interrupt service routine is entered.

Condition Logic: The condition logic block, shown at the center of Figure

6.32, compares status information from calculation blocks (ALU, MAC) with conditional portions of instructions (JUMP IF...) and outputs a bit to signal that a selected condition is met. Inputs from the counter (CE = count expired) and from the loop stack and loop comparator allow the condition code to be taken, not from the current instruction, but from the top of the loop stack if the loop comparator detects that this is the last instruction of a DO loop. This feature allows an overhead-free jump at the end of a DO loop.

Interrupt Controller: There are four external interrupt lines. The interrupt controller determines if an interrupt request has occurred and initiates a jump to an interrupt service routine if that interrupt line is not masked and has higher priority than an interrupt already being serviced. Interrupts may be either edge- or level- triggered. Edge/level interrupt sensitivity is software selectable individually for each interrupt line.

An interrupt request which passes the mask is then tested for priority. The highest-priority unmasked interrupt forces a jump to its interrupt service routine. As status information is pushed onto the status stack, the mask status register is automatically loaded with a new value which can either mask out

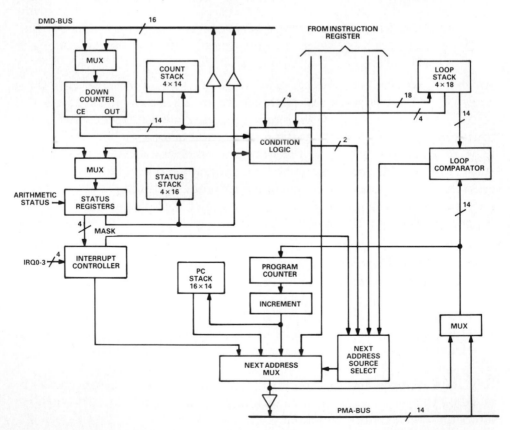

Figure 6.32. Block diagram of the ADSP-2100's program sequencer. Its relationship to the rest of the overall device can be seen in Figure 6.27.

lower-priority interrupts or, alternatively, mask out all interrupts until the current one has been serviced. Thus, interrupts can be made interruptible (nested interrupts) or not, as desired.

An interrupt-nesting inhibit instruction inhibits context switching more than one level deep. If the primary register set has been exchanged for the shadow registers upon an interrupt, and its contents will be needed to resume that task later, the interrupt nest-inhibit prevents any further swaps from destroying register contents.

Program Counter and Stack: Program counter (PC) operation is relatively conventional. The incremented PC value can address the next instruction, or be pushed onto the PC stack as a return address in the event of a subroutine or interrupt. In indirect-address operations, a new PC value is generated by DAG2 and appears on the PMA-bus for loading into the PC.

Next-Address Selection: Next-address selection is carried out by combining instruction op-codes, condition logic, interrupts, and loop information. The next-address multiplexer selects one of four sources:

Vector address (interrupt service)
PC incremented value (normal next instruction)
PC stack (return from subroutine or interrupt service)
Instruction (jump address in the code)

A fifth alternative is carried out when an indexed jump is executed and the next address is generated by DAG2 and put onto the PMA-bus.

Loop Stack and Comparator: This section provides the zero-overhead looping mechanism. A DO UNTIL instruction contains the address for the end of the loop and the termination condition. When a DO UNTIL instruction is executed, this information is loaded into the loop stack, and the PC value pushed onto the PC stack. The loop comparator compares the end-of-loop address with the next address, and signals the end of the loop when the two are equal. The condition logic then looks to the loop stack for its condition code rather than the current instruction register. Depending on the condition code, the next-address selector chooses between the PC stack (jump to beginning of loop) and the PC incrementer (fall out of the loop).

Instruction Cache Memory
Instruction cache memory stores a short history of up to 16 recently executed instructions. As Figure 6.27 shows, it is closely associated with the program sequencer and the instruction register; it can bypass the PMD bus to provide instructions. Once the instruction cache has been loaded with a short program, say a DO loop, the PMD bus is not needed for instructions and can instead be used for data stored in program memory. When data can be read from program memory without interfering with the flow of instructions, the ADSP-2100 becomes, in effect, a Harvard-architecture processor with two simultaneously addressable parallel data buses. For the multiply/accumulate

operations typical of DSP algorithms, this gives significant speed advantages, because both operands for a product can be read from two separate memories simultaneously.

Cache Memory Array and Memory Monitor: Each instruction fetched normally from the instruction register is also written into the cache; this is the "write" mode. It is stored at the cache memory address specified by the four LSBs of the program-memory address. When the PMD bus is busy with a data transfer, the instruction register is loaded from cache; this is the "read" mode. If the loaded instruction is valid, as determined by the *cache memory monitor*, it will be executed on the next cycle.

6.4.4 ADSP-2100 MICROPROCESSOR INSTRUCTION SET

DSP Microprocessor Instruction-Set Overview
Described below is the instruction set through which the user gains access to these unusual features of this DSP microprocessor chip:

Zero-overhead looping;
Single-cycle context-save: swaps the full register set;
Multi-function instructions: arithmetic and data move in one cycle;
Load two operands from off-chip in one cycle;
Parallel access to both program and data operands;
Five addressing modes;
50 general-purpose registers on-chip

The instruction set facilitates computation-intensive programs by allowing data movement between the microprocessor's computational units with minimum overhead. Address generation and instruction sequencing are automatic, seen by the user as simple constructs:

Construct	Usage
IF condition	**Conditional branch;** **tests counters and ALU/MAC status lines**
DO UNTIL	**Loop until termination condition met**

Instructions translate into single 24-bit words. Unlike typical general-purpose and many DSP-microprocessors, every instruction executes in one cycle.*

The flexible register set surrounding the computational units gives great freedom in coding. The registers are dual purpose: they are available for general purpose on-chip storage when not otherwise occupied in computation, saving many memory access cycles.

---

*The only exception is when an instruction accesses *data* (e.g., a coefficient) from program-memory and the next instruction has not yet been loaded into program cache memory. The first pass through such a sequence will automatically add an extra cycle for that instruction fetch. Later passes through the cache execute in only one cycle.

A multi-function instruction combines arithmetic and a data move in the same cycle:

instr1, instr2 ;

where the registers read in the first may be written-to by the second half or do independent tasks, as desired. To write a program for the ADSP-2100, one writes down all the arithmetic steps, separately writes down all the I/O to be carried out, and then combines the two in multifunction instructions.

Subroutines and interrupt service are made compact and speedy by the single-cycle context save: while the alternate register set is enabled, the contents of the primary register set are held unchanged. Subroutines and interrupt-service routines are shorter and faster as a result, being freed from the cluster of PUSHes and POPs common in general-purpose μP coding.

The instruction set is summarized below. The notation is as follows:

Symbol	Meaning
$+, -$	Add, subtract
$\star$	Multiply
$a = b$	Transfer into a the contents of b
,	Concurrency: separates multifunction instructions
DM(addr)	The contents of data-memory at location "addr"
PM(addr)	The contents of program-memory at location "addr"
[Option]	Optional operation, e.g., [IF condition]
$\left\|\begin{array}{l}\text{option a}\\\text{option b}\end{array}\right\|$	Available options list
;	End of instruction

In the instruction descriptions that follow, capital letters and punctuation marks are fixed labels to be included as they appear, while lower-case letters designate user-specified labels. The reader will find that the following sections are heavily condensed. Their purpose is to illustrate the "look and feel" of coding procedures to control the very powerful hardware described above and to squeeze out the full performance capability—while retaining an almost-high-level-language simplicity. Software manuals should be consulted for further details.

Arithmetic Instructions

Instruction Types: Standard, special, and conditional arithmetic instructions operate with the ALU, the MAC, or the shifter. *Standard* instructions, which include the bulk of the arithmetic operations, can be executed conditionally (IF condition...), testing the arithmetic status register, and may be combined with a data transfer in single-cycle multifunction instructions. *Special* instructions are a small subset; for example, saturation arithmetic must stand alone (no combination with data transfer). Examples of standard and special functions for ALU, MAC, and shifter are described below.

Conditional instructions, when used: test whether the previous ALU result is equal to (EQ) zero, not equal to (NE) zero, greater than (GT) zero, greater than or equal to (GE) zero, less than (LT) zero, less than or equal to (LE) zero; examine ALU carry status (AC, not-AC), *x*-input sign (POS, NEG) and ALU overflow (AV, not-AV) or MAC overflow (MV, not-MV); and determine whether counters have reached zero (not count expired—NOT CE).

ALU Instructions

ALU Standard Functions: Standard ALU functions include add, subtract, logic (and, or, exclusive-or), pass, negate, increment, decrement, clear, and absolute value. There are two negate functions: "–" does twos-complement subtraction, while NOT obtains a ones complement. The PASS function passes the listed operand but tests and stores status information for later sign/zero testing. A typical choice of ALU addition instructions would be in the form:

$$[\text{IF condition}] \quad \begin{vmatrix} \text{AR} \\ \text{AF} \end{vmatrix} \quad = \text{xop} \quad \begin{vmatrix} + \text{ yop} \\ + \text{ C} \\ + \text{ yop} + \text{ C} \end{vmatrix} \quad ;$$

for add/add-with-carry. Instructions are in similar form for subtraction and logical operations: AND, OR, XOR. If the options AR and "+ yop" are chosen, and if xop and yop are the contents of MR1 and AF, an unconditional instruction would read:

$$\text{AR} = \text{MR1} + \text{AF} ;$$

ALU inputs, xop and yop, can be selected from ALU input registers AX0, AX1, and AY0, AY1. In addition, yop can be zero or come from the ALU feedback register (AF). Input xop can come from the AR register or, via the *R*-bus, from the multiplier results registers (MR0, MR1, MR2) or the shifter registers (SR1, SR2). This shortens code and speeds execution by eliminating many separate register-move instructions. ALU outputs may go into either the results register, AR, or the feedback register, AF.

When an *IF condition* function is included, the selected status condition is examined and the instruction executed if the condition is true. If false, the cycle is completed with a no-op. If the condition is omitted, the operation is executed unconditionally. The conditions include the complete set of status bits from the previous computation, involving ALU, MAC, or Sequencer loop-counter.

ALU Special Functions: Division is the sole ALU special function; it is executed in two steps: DIVS computes the sign, then DIVQ computes the quotient. A full divide of signed 16-bit divisor into signed 32-bit quotient is accomplished as a DIVS followed by 15 DIVQ's, with the dividend preloaded into AF or AY0 and the divisor from any of the registers indicated above as available to xop.

MAC Instructions

MAC Standard Functions: Standard MAC functions include multiply, multiply-accumulate, multiply-subtract, transfer MR, saturate MR conditionally, and clear. Conditional operations are based upon status register contents. A typical MAC instruction has the form:

$$\text{[IF condition]} \quad \left| \begin{matrix} \textbf{MR} \\ \textbf{MF} \end{matrix} \right| \quad \textbf{= MR + xop * yop} \left| \begin{matrix} \textbf{(SS)} \\ \textbf{(SU)} \\ \textbf{(US)} \\ \textbf{(UU)} \\ \textbf{(RND)} \end{matrix} \right| \quad ;$$

MAC input, xop, may come from the MX0, MX1, or MR0, MR1, or MR2 registers, or via the results-bus (*R*-bus) from the accumulator (AR) or the shifter (SR1,SR0) outputs. MAC input, yop, may come from the MY0 or MY1 register or from multiplier feedback (MF), or it may be zero. MAC outputs can go either to the results (MR) or the feedback (MF) register. The (optional) accumulation input comes from the previous MR contents. A rounding option (RND) and a sign specification (signed-SS, unsigned-UU, or mixed-US or SU) complete the instruction; if none is specified, the default is: *x*-input and *y*-input signed and the result unrounded.

MAC Special Functions: Accumulator saturation is the sole MAC special function.

IF MV SAT MR;

The instruction tests the MAC overflow bit (MV) in the arithmetic status register and saturates the MR register (for one cycle only) if that bit is set.

Shifter Instructions

Shifter Standard Functions: Shifter functions include arithmetic shift and logical shift, as well as FP and BFP scaling operations: derive exponent, normalize, denormalize, and block-exponent adjust. Conditional operations are based upon status register contents. The form is:

[IF condition] SR = [SR OR]	ASHIFT	xop	(HI)	; Arithmetic shift
	LSHIFT		(LO)	Logical shift
	NORM			Normalize
[IF condition] SE = EXP xop	(HI)	;		Derive exponent
	(HIX)			
	(LO)			
[IF condition] SB = EXPADJ xop ;				Block-exponent adjust

Shifter input, xop, may come from the SI register or be fed back from output (SR0,SR1), or via the R-bus from the accumulator output (AR) or any of the multiplier outputs (MR). Arithmetic (ASHIFT) and logical (LSHIFT) shifts do *denormalization*, obtaining the number of bits to shift in the SE register. In an ASHIFT, the extension bit is the sign bit of the input operand. The option, SR OR, causes the shift output to be logically OR'ed with the contents of SR.

The point of reference is set by:

HI: The input is interpreted as a single-precision signed number or the upper half of a double-precision signed number (SR1). Contents of SE will be the negative of the number of redundant bits in the input.

HIX: As with HI, but + 1 is stored in SE if ALU overflows (AV set)

LO: The input is interpreted as the lower half of a double-precision number (SR0). If MSH consisted of all extended sign bits, EXP totals the number of leading sign bits for the DP word and stores it in SE.

Normalization is carried out in two cycles. First, the instruction

$$SE = EXP\,xop$$

brings out the value of the exponent and stores it in the SE register; then the instruction

$$SR = NORM\,xop$$

uses that value to shift the operand that many places.

Block floating-point scaling is carried out by the instruction

$$SB = EXP\,ADJ\,xop$$

which conditionally transfers the exponent from register EXP to register SB if the exponent of xop is larger than the value presently in SB.

Shifter Special Function: Shift-immediate is the sole shifter special function. The number of places (exponent) to shift is specified in the instruction word.

SR = [SR OR]	ASHIFT	xop BY ⟨data⟩	(HI)	;
	LSHIFT		(LO)	

for $-128 < ⟨data⟩ < 127$

Data-Transfer (Move) Instructions

This class includes register-to-register and memory read/write instructions for both program- and data-memory. A large cluster of nearly 50 registers is accessible directly via the DMD-bus. This includes registers surrounding the ALU, MAC, and Shifter, which hold data inputs and outputs, as well as Address-generator and Sequencer registers. There are 10 classes of data transfer instruction. Their format and operands are described below. Except for immediate instructions, data addresses are computed by the Data-address generators via the contents of their index (I) and modify (M) registers. The data movement choices include

Register (to register) move
Load register immediate
Data Memory read (or write), direct (or indirect) address
Program memory read (or write), indirect address

Terminology:
reg = permissible register accessible to the DMD bus
dreg = input and output registers of ALU, MAC, and shifter; subset of "reg"
immediate value = 16-bit number contained in the instruction field.
immediate address = 14-bit address contained in the instruction field.

Load Register Immediate
reg = ⟨data⟩ ;   **⟨data⟩ = immediate value.**

Register-to-Register Move
reg = reg ;

If the registers are of different lengths, a larger register is right-justified and the higher-order bits are dropped when moving into a smaller register; a smaller register is sign-extended if signed (or zero-filled to the left if unsigned) when moved into a larger register.

Immediate-Address Data-Memory Move: The address is provided by a 14-bit immediate address in the instruction-word. Since no registers are tied up in generating addresses, any accessible register can be the operand of an immediate-address move.

Data-Memory Read
reg = DM(⟨addr⟩) ;   **⟨addr⟩ = immediate address**

Data-Memory Write
DM(⟨addr⟩) = reg ;

Indirect-Address Memory Move: The indirect-address transfer instructions cause data to be moved between data registers (*dreg*) and either data-memory or program memory. The memory address for the current operation is provided by one of the four *I*-registers (Im) of a data-address generator. The value of *I* is then stored back again, after being modified by the contents of one of the four *M* registers (Mn) and of the buffer length stored in one of the four *L* registers (Lm). The operation is:

  Read/write at address specified by Pointer:
    Memory address = Im
  Then modify pointer
    Im = (Im + Mn)mod Lm

Indirect addressing may access either data-memory or program memory. The Index and Modify registers, Im and Mn, are I4 through I7 and M4 through M7 (DAG2) for program memory, and I0 through I3 and M0 through M3 (DAG1) and I4 through I7 and M4 through M7 (DAG2) for data memory. The modulus registers, L0 through L7, are paired with the Index registers, i.e., L5 is the modulus register for the memory location indexed at I5.

  Data- or Program- Memory Read into Data Register
$$dreg = \text{DM(Im,Mn)} ;$$
$$dreg = \text{PM(Im,Mn)} ;$$

  Data-Memory Write: Immediate or from Data Register
$$\text{DM(Im,Mn)} = \begin{vmatrix} dreg \\ \langle data \rangle \end{vmatrix}$$

  Program-Memory Write from Data Register
$$\text{PM(Im,Mn)} = dreg ;$$

Multifunction Data- and Program-Memory Read: A powerful combined Move instruction reads data into a pair of ALU or MAC input registers from both data-memory and program-memory in the same cycle:

$$\begin{vmatrix} AX0 \\ AX1 \\ MX0 \\ MX1 \end{vmatrix} = \text{DM}(\begin{vmatrix} I0 \\ I1 \\ I2 \\ I3 \end{vmatrix}, \begin{vmatrix} M0 \\ M1 \\ M2 \\ M3 \end{vmatrix}), \begin{vmatrix} AY0 \\ AY1 \\ MY0 \\ MY1 \end{vmatrix} = \text{PM}(\begin{vmatrix} I4 \\ I5 \\ I6 \\ I7 \end{vmatrix}, \begin{vmatrix} M4 \\ M5 \\ M6 \\ M7 \end{vmatrix}) ;$$

Multifunction Arithmetic Instructions
Multifunction instructions execute more than one operation in a given cycle, combining arithmetic with an independent data transfer. Only certain combinations of arithmetic and data-move instructions are possible within

the constraints set by the 24-bit instruction-word limit:

Arithmetic portion: Only the standard functions are allowed (ALU/MAC/SHIFT); conditional execution is disallowed; shift immediate is not available; and division is not available.

Data Transfer portion: Memory access must use indirect addressing; only data registers (*dregs*) can be operands.

Arithmetic multifunction classes are:
Arithmetic with Register-to-Register Move
Arithmetic with Data- or Program-Memory Read
Arithmetic with Data- or Program-Memory Write
Arithmetic with Data- and Program-Memory Read

Combined instructions are summarized below. ALU, MAC, and SHIFT stand for any member of the standard arithmetic instruction set, without conditional "IF"s. The format is two instructions, separated by a comma:

Do first,   Do second

where the arithmetic operation and the data move may be in either order:

"Destination" = "Source",   Arithmetic operation on operands ;
Arithmetic operation on operands ,   "Destination" = "Source" ;

For each class of operation, one of these orders is preferred. Combined instructions are interpreted as executing sequentially from left to right, but the order may be exchanged in some conditions, depending on the relationship between the "destination," "source," and operands.

Arithmetic with Register-to-Register Move: In the preferred version, the data in the "Destination" register may be first used as an arithmetic operand before the "Destination" register is overwritten by the contents of the "Source" register. This version is written as:

$$\left| \begin{array}{l} \langle \mathbf{ALU} \rangle \\ \langle \mathbf{MAC} \rangle \\ \langle \mathbf{SHIFT} \rangle \end{array} \right| , \ \textit{dreg} = \textit{dreg} ;$$

In the other version, data in the "Source" register is first copied into the "Destination" register, and then the operands, which may include the "Source" register, are operated on arithmetically. This operation is written as:

$$\textit{dreg} = \textit{dreg}, \left| \begin{array}{l} \langle \mathbf{ALU} \rangle \\ \langle \mathbf{MAC} \rangle \\ \langle \mathbf{SHIFT} \rangle \end{array} \right| ;$$

If the functions do not have a register in common, either order will give the same result. But multifunction instructions quite often involve the same register twice, and in that event, these restrictions must be observed:

1. The "Destination" *dreg* in the first function cannot be used as an operand in the second; a "Source" *dreg* in the first *can* be written over by the result of the second.
2. Both operations cannot have the same *dreg* as a destination.

Arithmetic with Data- or Program-Memory Read

$$
\left| \begin{array}{c} \langle \mathsf{ALU} \rangle \\ \langle \mathsf{MAC} \rangle \\ \langle \mathsf{SHIFT} \rangle \end{array} \right| , dreg = \left| \begin{array}{c} \mathsf{DM\,(Im\ ,\ Mn)} \\ \mathsf{PM\,(Im\ ,\ Mn)} \end{array} \right| ;
$$

Here, the arithmetic operation is performed; then a transfer from memory to a *dreg* takes place. The Index and Modify registers, Im and Mn, are I4 through I7 and M4 through M7 (DAG2) for program memory, and I0 through I3 and M0 through M3 (DAG1) and I4 through I7 and M4 through M7 (DAG2) for data memory.

This operation may also be performed in the reverse order. If the same register label appears both in the arithmetic and data-transfer sections, the above rules apply. The previous data in *dreg* may be an operand for the arithmetic before it is overwritten by data from DM. In the reversed order, data in *dreg* may not be used as an operand for the arithmetic operation.

Arithmetic with Data- or Program-Memory Write

$$
\left| \begin{array}{c} \mathsf{DM\,(Im\ ,\ Mn)} \\ \mathsf{PM\,(Im\ ,\ Mn)} \end{array} \right| = dreg , \left| \begin{array}{c} \langle \mathsf{ALU} \rangle \\ \langle \mathsf{MAC} \rangle \\ \langle \mathsf{SHIFT} \rangle \end{array} \right| ;
$$

The transfer from the indicated *dreg* to memory is made, and the arithmetic operation takes place. The Index and Modify register options are as above. The operations may be written in reverse order, but the result of the arithmetic operation cannot be moved to memory in the same cycle.

Single-Cycle Arithmetic plus Data- and Program-Memory Read: This instruction combines an ALU or MAC operation with access to both program- and data-memory in the same cycle.

$$
\left| \begin{array}{c} \langle \mathsf{ALU} \rangle \\ \langle \mathsf{MAC} \rangle \end{array} \right| , \left| \begin{array}{c} \mathsf{AX0} \\ \mathsf{AX1} \\ \mathsf{MX0} \\ \mathsf{MX1} \end{array} \right| = \mathsf{DM} \ \left(\left| \begin{array}{c} \mathsf{I0} \\ \mathsf{I1} \\ \mathsf{I2} \\ \mathsf{I3} \end{array} \right| , \left| \begin{array}{c} \mathsf{M0} \\ \mathsf{M1} \\ \mathsf{M2} \\ \mathsf{M3} \end{array} \right| \right) , \left| \begin{array}{c} \mathsf{AY0} \\ \mathsf{AY1} \\ \mathsf{MY0} \\ \mathsf{MY1} \end{array} \right| = \mathsf{PM} \ \left(\left| \begin{array}{c} \mathsf{I4} \\ \mathsf{I5} \\ \mathsf{I6} \\ \mathsf{I7} \end{array} \right| , \left| \begin{array}{c} \mathsf{M4} \\ \mathsf{M5} \\ \mathsf{M6} \\ \mathsf{M7} \end{array} \right| \right) ;
$$

Here, in the second-half cycle, the input registers of either the ALU or the MAC are loaded with data, which may then be operated on in the first half of the next cycle. There is a restriction in this combined instruction: feedback registers AF or MF cannot receive the results of the arithmetic, since they are employed to carry out the single-cycle juggling of data.

Program Flow-Control Instructions

This class of instructions directs the program sequencer. In the absence of a program flow-control instruction, the sequencer automatically fetches the next contiguous instruction for execution. Program flow control provides:

Jumps to interrupt service routines and calls to subroutines,
Returns from interrupts and subroutines
DO loops.
TRAP instruction

An (optional) "IF" condition can first test any of the status conditions defined above.

JUMP and CALL: The instruction format is:

$$[\text{IF condition}] \quad \begin{vmatrix} \text{JUMP} \\ \text{CALL} \end{vmatrix} \begin{vmatrix} \text{I4} \\ \text{I5} \\ \text{I6} \\ \text{I7} \\ \langle \text{addr} \rangle \end{vmatrix} \quad ;$$

A JUMP transfers execution to the instruction found at the new address, as in interrupt service routines. A CALL instruction brings in subroutines, and pushes the present PC onto the PC stack as the return address. The address for a JUMP or a CALL has one of two sources: a label ($\langle$ addr$\rangle$), written explicitly in the program, or an index-register accessible to DAG2. An I-register referenced by a CALL or JUMP is *not* modified by the M and L registers of the address generator. If the condition is omitted, the jump or call is performed unconditionally.

RETURN The format is

$$[\text{If condition}] \quad \begin{vmatrix} \text{RTS} \\ \text{RTI} \end{vmatrix} \quad ;$$

There are two possible returns: RTS from a subroutine, and RTI from an interrupt service routine. In either case, a RETURN pops the return address from the PC stack. A return from an interrupt also pops the status stack back, returning arithmetic and mode status and interrupt mask registers to the values they had prior to the interrupt. No separate status POP instruction is required as with many general-purpose and DSP microprocessors. If the condition is omitted, the return is performed unconditionally.

Loop Instructions: The zero-overhead DO UNTIL instruction has the form:

DO ⟨addr⟩ [UNTIL condition] ;

The label ⟨addr⟩ designates the last instruction in the loop. The "condition" determines the termination of the loop: a counter going to zero, or a flag being set. When the DO is entered, the address, ⟨addr⟩, and the termination condition are pushed onto the Sequencer loop stack, and the current PC + 1 is pushed onto the PC stack to become the next address after loop termination. The looping hardware automatically checks the condition code whenever execution passes through the address-value, ⟨addr⟩.

Trap Instruction: A trap instruction, inserted while a program is debugged, conditionally halts the processor and asserts the hardware TRAP. The format is:

[IF condition] TRAP   ;

Omitting the condition causes the TRAP to be performed unconditionally.

Miscellaneous Instructions
This final class modifies index register values, pushes and pops stacks, and switches between the two alternate register banks.

Modify Address Register

MODIFY (Im,Mn)   ;

The value of the address in Index register, I, is modified by M (modulo L) as in indexed addressing, but without any data transfer to or from memory. For either data address generator, the value of Mn is added to Im, then processed through the modulus logic with buffer length as determined by Lm, the L register corresponding to Im; the result of the address-pointer calculation is stored in Im. The choices of Index and Modify registers, Im and Mn, are among I0 through I3 and M0 through M3 (DAG1), and I4 through I7 and M4 through M7 (DAG2).

Stack Control: This is a manual PUSH and POP of the stacks. The format is:

$$\begin{bmatrix} \begin{matrix} | \text{PUSH} | \\ | \text{POP} \quad | \end{matrix} \text{STS} \end{bmatrix} [\, , \text{POP CNTR}][\, , \text{POP PC}][\, , \text{POP LOOP}];$$

A PUSH or POP on the status stack replicates the normal general-purpose μP instruction but need not be used often, since subroutine and interrupt service take care of the status stack automatically. The counter and loop PUSH/POP are special in the DSP microprocessor; they find their use, for example, in a conditional DO UNTIL loop where a counter value or loop index needs saving when the condition is satisfied. A POP CNTR puts the counter stack into the down counter. A POP LOOP destroys the bottom loop-stack entry. This combination of stack control facilitates nested loops with multiple subroutine CALLs, and also JUMPs where the return address is a different location (after processing an overflow response, for example).

Mode Control: Mode control switches between the two identical sets of registers (main and "shadow"); it also enables or disables the bit-reversal register, the ALU overflow status latch mode, and the AR register saturation mode. Any number of these modes can be redefined in a single cycle, and will remain unchanged until again redefined.

$$\left[{\begin{matrix} \text{ENA} \\ \text{DIS} \end{matrix}} \; \text{BIT\_REV} \right]\left[, {\begin{matrix} \text{ENA} \\ \text{DIS} \end{matrix}} \; \text{AV\_LATCH} \right]\left[, {\begin{matrix} \text{ENA} \\ \text{DIS} \end{matrix}} \; \text{AR\_SAT} \right]\left[, {\begin{matrix} \text{ENA} \\ \text{DIS} \end{matrix}} \; \text{SEC\_REG} \right] ;$$

Bit-Reverse when enabled, causes addresses generated by DAG1 to be output in reverse order (e.g., $00 \ldots 11 \rightarrow 11 \ldots 00$).

ALU Overflow Status Latch, when enabled, causes the AV bit in the arithmetic status register to stay set once an ALU overflow occurs; DIS will clear it.

ALU AR-Register Saturation Mode, when enabled, causes the AR register to saturate if an ALU operation causes an overflow.

Alternate-Register-Data Bank: The data-register bank-select bit determines which set of registers is currently active (0 = primary, 1 = secondary). The interrupt-nesting-inhibit function inhibits context switching more than one level deep. If the secondary register set has replaced the primary register set upon an interrupt, and the primary register set's contents will be needed to resume that task later, the interrupt nest inhibit prevents any further swaps from destroying register contents.

Example of Actual Instruction Coding:
The ADSP-2100's instruction set has been described in terms of mnemonics that show as clearly as possible the functions performed. This serves as source code to a compiler whose output object code is a 24-bit binary word; the disposition of bits in that word is the microcode that actually produces the parallel action of the DSP μP. Figure 6.33, for example, shows the coding for an ALU or MAC one-cycle operation with data- and program-memory read.

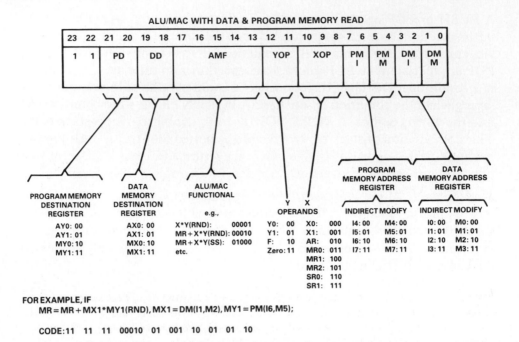

ALU/MAC WITH DATA & PROGRAM MEMORY READ

FOR EXAMPLE, IF
MR = MR + MX1*MY1(RND), MX1 = DM(I1,M2), MY1 = PM(I6,M5);

CODE: 11 11 11 00010 01 001 10 01 01 10

Figure 6.33. Coding for ALU or MAC operation with data- and program-memory read.

Subroutine Examples

The examples of program listings that follow illustrate the features described above: compact coding, multifunction instructions, rapid throughput, and a relatively readable instruction set. Table 6.5 is an example of an FIR filter module, and Table 6.6 is an example of a decimation-in-time FFT subroutine. These examples are intended to illustrate the compact structure (hence fast execution) of the instruction set. The stages and methods of actual program development for modules such as these will be described in the following chapter.

Table 6.5 Single-precision FIR transversal-filter program module. It requires a total of N + 6 cycles for a filter of length _N_; at an 8-kHz sampling rate and an instruction-cycle time of 125 nanoseconds, this permits a filter of 900 taps with 94 instruction cycles of other operations.

```
.MODULE     fir_sub;

{
            FIR Transversal Filter Subroutine

            Calling Parameters
                I0 --> Oldest input data value in delay line
                L0 = Filter length (N)
                I4 --> Beginning of filter coefficient table
                L4 = Filter length (N)
                M1,M5 = 1
                CNTR = Filter length - 1 (N-1)
```

```
                         Return Values
                            MR1 = Sum of products (rounded and saturated)
                            I0 --> Oldest input data value in delay line
                            I4 --> Beginning of filter coefficient table

                         Altered Registers
                            MX0,MY0,MR

                         Computation Time
                            N - 1 + 5 + 2 cycles

                         All coefficients and data values are assumed to be in 1.15 format.
        }

        .ENTRY       fir;

        fir:         MR=0, MX0=DM(I0,M1), MY0=PM(I4,M5);
                     DO sop UNTIL CE;
        sop:             MR=MR+MX0*MY0(SS), MX0=DM(I0,M1), MY0=PM(I4,M5);
                     MR=MR+MX0*MY0(RND);
                     IF MV SAT MR;
                     RTS;
        .ENDMOD;
```

Table 6.6 Complete conditional block-floating-point radix-2 decimation-in-time FFT subroutine. The constants *N* and log2N are the number of points and the number of stages in the FFT, respectively. To change the number of points, these constants are modified. Instructions that write butterfly results to memory are in boldface.

```
.MODULE      fft;

{            Performs Radix-2 DIT FFT

             Calling Parameters
                inplacereal = Real input data in scrambled order
                inplaceimag = All zeroes (real input assumed)
                twid_real = Twiddle factor cosine values
                twid_imag = Twiddle factor sine values
                groups = N/2
                bflys_per_group = 1
                node_space = 1

             Return Values
                inplacereal = Real FFT results in sequential order
                inplaceimag = Imaginary FFT results in sequential order

             Altered Registers
                I0,I1,I2,I3,I4,I5,L0,L1,L2,L3,L4,L5
                M0,M1,M2,M3,M4,M5
                AX0,AX1,AY0,AY1,AR,AF
                MX0,MX1,MY0,MY1,MR,SB,SE,SR,SI

             Altered Memory
                inplacereal, inplaceimag, groups, node_space,
                bflys_per_group, blk_exponent
}
.CONST       log₂N=10, N=1024;                    {Set constants for N-point FFT}
.EXTERNAL    twid_real, twid_imag;
.EXTERNAL    inplacereal, inplaceimag;
.EXTERNAL    groups, bflys_per_group, node_space;
.EXTERNAL    bfp_adj;
.ENTRY       fft_strt;
```

```
fft_strt:    CNTR=log₂N;                             {Initialize stage counter}
             M0=0;
             M1=1;
             L1=0;
             L2=0;
             L3=0;
             L4=%twid_real;
             L5=%twid_imag;
             DO stage_loop UNTIL CE;                 {Compute all stages in FFT}
               I0=^inplacereal;                      {I0 -->x0 in 1st grp of stage}
               I2=^inplaceimag;                      {I2 -->y0 in 1st grp of stage}
               SB=-2                                 {SB to detect data > 14 bits}
               SI=DM(groups);
               CNTR=SI;                              {CNTR = group counter}
               M4=SI;                                {M4=twiddle factor modifier}
               M2=DM(node_space);                    {M2=node space modifier}
               I1=I0;
               MODIFY(I1,M2);                        {I1 -->x1 of 1st grp in stage}
               I3=I2;
               MODIFY(I3,M2);                        {I3 -->y1 of 1st grp in stage}
               DO group_loop UNTIL CE;
                 I4=^twid_real;                      {I4 --> C of W⁰}
                 I5=^twid_imag;                      {I5 --> (-S) of W⁰}
                 CNTR=DM(bflys_per_group);           {CNTR = butterfly counter}
                 MY0=PM(I4,M4),MX0=DM(I1,M0);        {MY0=C,MX0=x1 }
                 MY1=PM(I5,M4),MX1=DM(I3,M0);        {MY1=-S,MX1=y1}
                 DO bfly_loop UNTIL CE;
                   MR=MX0*MY1(SS),AX0=DM(I0,M0);     {MR=x1(-S),AX0=x0}
                   MR=MR+MX1*MY0(RND),AX1=DM(I2,M0); {MR=(y1(C)+x1(-S)),AX1=y0}
                   AY1=MR1,MR=MX0*MY0(SS);           {AY1=y1(C)+x1(-S),MR=x1(C)}
                   MR=MR-MX1*MY1(RND);               {MR=x1(C)-y1(-S)}
                   AY0=MR1,AR=AX1-AY1;               {AY0=x1(C)-y1(-S),}
                                                     {AR=y0-[y1(C)+x1(-S)]}
                   SB=EXPADJ AR,DM(I3,M1)=AR;        {Check for bit growth,}
                                                     {y1´=y0-[y1(C)+x1(-S)]}
                   AR=AX0-AY0,MX1=DM(I3,M0),MY1=PM(I5,M4);
                                                     {AR=x0-[x1(C)-y1(-S)],}
                                                     {MX1=next y1,MY1=next (-S)}
                   SB=EXPADJ AR,DM(I1,M1)=AR;        {Check for bit growth,}
                                                     {x1´=x0-[x1(C)-y1(-S)]}
                   AR=AX0+AY0,MX0=DM(I1,M0),MY0=PM(I4,M4);
                                                     {AR=x0+[x1(C)-y1(-S)],}
                                                     {MX0=next x1.MY0=next C}
                   SB=EXPADJ AR,DM(I0,M1)=AR;        {Check for bit growth,}
                                                     {x0´=x0+[x1(C)-y1(-S)]}
                   AR=AX1+AY1;                       {AR=y0+[y1(C)+x1(-S)]}
bfly_loop:         SB=EXPADJ AR,DM(I2,M1)=AR;        {Check for bit growth,}
                                                     {y0´=y0+[y1(C)+x1(-S)]}
                 MODIFY(I0,M2);                      {I0 -->1st x0 in next group}
                 MODIFY(I1,M2);                      {I1 -->1st x1 in next group}
                 MODIFY(I2,M2);                      {I2 -->1st y0 in next group}
group_loop:      MODIFY(I3,M2);                      {I3 -->1st y1 in next group}
               CALL bfp_adj;                         {Compensate for bit growth}
               SI=DM(bflys_per_group);
               SR=ASHIFT SI BY 1(LO);
               DM(node_space)=SR0;                   {node_space=node_space × 2}
               DM(bflys_per_group)=SR0;              {bflys_per_group= }
                                                     {bflys_per_group × 2}
               SI=DM(groups);
               SR=ASHIFT SI BY -1(LO);
stage_loop:    DM(groups)=SR0;                       {groups=groups × 2}
             RTS;
.ENDMOD;
```

EXERCISES FOR CHAPTER 6

Exercise 6.1. *Sign replication in twos-complement multiplication*. A b-bit twos-complement number, $-R$, has the value, $-2^{b-1} + (2^{b-1} - R)$; the format is:

$$\boxed{-2^{b-1} \mid +2^{b-2} \mid +2^{b-3} \mid \ldots \mid +2^2 \mid +2^1 \mid +2^0}$$

i.e.,for an eight-bit number (7 bits plus sign), the 0's and 1's are associated with binary weights:

$$\boxed{-128 \mid +64 \mid +32 \mid +16 \mid +8 \mid +4 \mid +2 \mid +1}$$

For positive numbers, the first bit is zero. For negative numbers, it is asserted; hence it serves as a sign indication. For example, the number $+6$ would be 00000110. Its negative, -6, would be $(-128) + (+128 - 6)$, or $(-128 + 122) = 11111010$.

The value of the product of R_1 and R_2 in twos complement is

$$\underbrace{-2^{2b-1}}_{\text{sign bit}} + \underbrace{(2^{2b-1} + R_1 R_2)}_{\text{2's complement}}$$

If the product is negative (R_1 and R_2 of differing sign), the expression in parentheses is equal to the twos complement of $R_1 R_2$; if the product is positive, the result is simply $R_1 R_2$. In the case where $R_1 = R_2 = -2^{b-1}$ (i.e., negative full scale), the result will be $+2^{2b-2}$, putting a 1 bit next to the sign bit, i.e., 0100... This is a valid computational result, because only the first bit is the sign bit. When unity, it indicates an actual negative magnitude (-2^{2b-1}) in twos complement computations. For all other values of R_1 and R_2, the second bit is 0 or 1 (extending the sign bit), for the simple reason that nothing else is as big as the square of negative full scale; since this is an unlikely value (there is no positive counterpart); the extra bit is largely redundant.

Perform the multiplications of these b-bit signed numbers:

$$
\begin{array}{ll}
01111 * 01111 & \{2^{b-1} - 1\}^2 \\
01111 * 10001 & \{2^{b-1} - 1\}\{-(2^{b-1} - 1)\} \\
10001 * 01111 & \{-(2^{b-1} - 1)\}\{2^{b-1} - 1\} \\
10001 * 10001 & \{-(2^{b-1} - 1)\}^2
\end{array}
$$

They represent the largest normally used products, and all will be found to have identical first and second bits (00..., 11..., 11..., and 00...), irrespective of the rest of the number. For example, the result in the fourth case is 0011100001, i.e., $(-15)^2 = 225$. However, for the $(-\text{full scale})^2$ case

$$10000 * 10000 = 0100000000 \qquad (-16)^2 = 256$$

Any other product will be smaller than this, and the first two digits will be identical.

Exercise 6.2. *Double-precision bit manipulation.* Verify that the partial products must be shifted as shown in the double precision mixed-mode example of Figure 6.3 to correctly line up the bits in the final summation. Try an (unsigned) multiply of 4×4 numbers such as:

$$
\begin{array}{r}
1111 = 15 \\
\times \quad 1111 = 15 \\
\hline
11100001 = 225
\end{array}
$$

Break the inputs up into (MSH,LSH) = (11,11), evaluate the 2×2 partial products, and shift as shown in the text before adding.

Exercise 6.3. *Unbiased vs. normal rounding.*

   Normal rounding rule: To round a 32-bit number down to 16 bits, add one-half the radix (i.e., 1) into bit 15 (the most-significant bit of the lower half) of the adder chain, let the carries ripple through the upper portion of the result, and truncate all bits below bit 16. Verify that this corresponds to the rule for rounding decimals, where one-half the radix is 5; i.e., 1.506 and 1.505 round to 1.51 while 1.504 rounds to 1.50.

   Unbiased rounding rule: If the lower 16 bits of a 32-bit number are at the halfway mark (1000 0000 0000 0000), add bit-16 (the LSB of the upper half) into bit-15 and truncate. Otherwise, perform normal rounding. How does the unbiased rounding rule work?

   Answer: The LSB of the MSH has a 50-50 chance of being a 1 and forcing an upward round.

   Examples:

1XXXXX00 10000000   rounds to 1XXXXX01 (normal round)
                             1XXXXX00 (unbiased round)

1XXXXX01 10000000   rounds to 1XXXXX10 (normal round)
                             1XXXXX10 (unbiased round)

Exercise 6.4. *Sign extension.* When a 16-bit signed twos-complement number is sign-extended to 32-bit width, a zero in bit 15 signals that the 16 bits to the left should all be zero, while a 1 brings the leading 16 bits to all 1's. Show that this guarantees the proper placement of 16-bit signed numbers in the 32-bit number field.

   Method: Imagine the number field written on a strip of paper. Now make the strip twice as wide. What must be done to create the new leading bits so the original set occupies the region on the strip closest to zero?

Example:

8-bit negative number: 11111011

Sign-extend to 16-bits: 1111111111111011

Exercise 6.5. *Using a multiplier to do shifts.* A multiplier chip can perform a wide range of shift operations if a separate shifter chip is unavailable. Demonstrate this action on input and output data, using a diagram in the format of Figure 6.11b, for the following situation:
Multiplier input B: input data to be shifted
Multiplier input A: amount of the shift (2^n, i.e., a 1-bit in the 2^n position; for example, if $n = 3, 2^3 = 1000_2$). Show that the multiplier output is equivalent to a barrel-shifter output, left-shifted by n bits. Show that the shift is arithmetic if B is signed and logical if B is unsigned.

Exercise 6.6. *Microcoded 6-cycle FFT butterfly with only one ALU* The butterfly computation is reduced to 6 ALU operations by a trick. Once the new A values have been computed, the new B values can each be computed in a single cycle (rather than two) using these new A values:

$$AR' = AR + BR*WR - BI*WI$$
$$AI' = AI + BI*WR + BR*WI$$
$$BR' = AR - BR*WR + BI*WI = 2AR - AR'$$
$$BI' = AI - BI*WR - BR*WI = 2AI - AI'$$

Show how to configure one ALU, one MAC, and data and coefficient memory access as needed to achieve a 6-cycle butterfly. Include the interleaved instruction sequence in your solution. This solution is possible with generic architecture parts. (For another solution with one MAC and *no* ALU, see Dintersmith and Nuttall, 1984). This solution makes use of special features in the ADSP-1101 *integrated arithmetic unit* (extended-MAC) chip. The ADSP-1101 features internal feedback of outputs such as AR' to input, shift-and-add capabilities, and a *dual* input-register set and a *dual* output accumulator, so the computation can ping-pong back and forth between real and imaginary parts without doubling up on hardware.

Exercise 6.7. *DSP Microprocessor* In a multifunction instruction, a register being moved can also be an arithmetic operand. If both sides of the register-move are operands in the arithmetic, the combined operation is not written in a way which clearly indicates the outcome. The hardware outcome is definite and predictable; the problem is in understanding the causality. For example, consider the feedback loop (Figure 6.34) implied by the instruction:

AR = PASS AX0, AX0 = AR ;

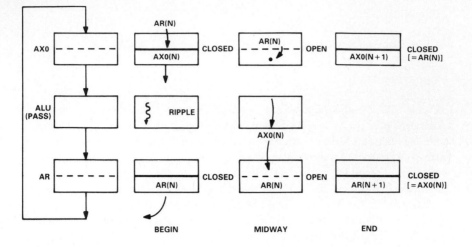

Figure 6.34. Example of a dual-function instruction whose outcome is clear only when the master-slave nature of the registers is understood. The instruction is: AR = PASS AX0, AX0 = AR.

The ALU takes as its input a register, AX0, which is in turn fed by the ALU output, AR. The outcome is quite definite, even though the instruction looks indeterminate. Each register can be both read and written in the same cycle. When this happens, the value read from the register is the old one. The new value will be loaded into the register at the end of the cycle; it cannot be read out until the next cycle. The operation is somewhat like that of two registers in series, with the connection between them disabled except midway through an instruction cycle, as in a "break-before-make" electrical switch. This feature allows an input register to provide an operand into the ALU while simultaneously being loaded with the next operand from memory. It also allows the result register to be stored in memory while simultaneously being loaded with the results of a new computation. If the cycles could be labeled so as to indicate causality, the instruction could be read without confusion:

$$AR(N+1) = PASS\ AX0(N),\ AX0(N+1) = AR(N)\ ;$$

Chapter Seven

Software Development
for the DSP System

7.1 INTRODUCTION

Software development is the main hidden cost of digital signal processing. As with the general-purpose microprocessor (GP μP), the hardware capabilities of DSP systems are only fully realized with the investment of considerable time in developing optimal programs. Software development tools make the programmer more productive by shortening the time to write, translate, and debug code. The examples below illustrate the trend in DSP software support to look much like traditional microprocessor development systems: the user is insulated by several layers from coding 1's and 0's.

This chapter covers development tools for translating and debugging source code of DSP systems. The emphasis is on the programmable DSP microprocessor (DSP μP); as a concrete example, the tools available for the ADSP-2100 are introduced. While the instructions, program examples, and development-system commands are necessarily specific to this particular DSP microprocessor, the functions and development steps are typical of what is desirable for efficient software development in third-generation DSP μPs.

Issues encountered in code development for the other main class of DSP systems, i.e., those built around microprogrammed chips, will be introduced briefly and contrasted with those of the GP μP.

The flow of section 7.2 follows the development cycle. First the *target system* is specified: I/O interfaces to the outside world, and the anticipated program- and data-memory resources. Translating higher-level mnemonics into binary object code (Figure 6.33) is the task of an *assembler*, just as in the GP μP. Next, a *linking loader* combines assembled subroutines into an executable binary program. To debug both program and system hardware, a variety of simulation and emulation tools are important. *Instruction-level simulation* maps target-system registers over into development-system memory locations, which permits tracing through the operation of a simulated system. *Circuit emulation* tools operate at the hardware level, replacing the DSP μP with hardware modified to permit the development system to access all internal registers and buses.

To keep the treatment application-driven, we will consider two specific examples. First, a signal processor consisting of an FIR filter module with analog input and output will be followed through all phases of software development within the ADSP-2100 DSP μP environment. Then a signal processor oriented towards speech recognition will be discussed as an example which fully utilizes the resources of this third generation DSP μP.

The chapter closes with a brief overview of development tools for microprogrammed systems. Microprogrammed system architecture varies greatly from one application to the next, so development tools are more primitive. The chapter introduces some of the nomenclature and describes tools such as the *meta-assembler* (a user-defined assembler), which at least takes microprogramming beyond the bit level.

In Chapter 7, standard software tools are applied to DSP, with appropriate modifications. For example, the parallel processing of several operations in one cycle in the modern DSP microprocessor alters the look of the code from standard assembly language syntax. The custom user-defined architecture of the Bit-, Byte-, and Word-Slice microprogrammed systems forces introduction of new software tools such as the meta-assembler, with its user-defined software terminology to fit a custom architecture yet avoid bit-level programming. The reader unfamiliar with conventional microprocessor program development will find it useful to review that subject in order to gain a better understanding of Chapter 7.

7.2 SOFTWARE DEVELOPMENT TOOLS FOR THE DSP MICRO-PROCESSOR

7.2.1 DEVELOPMENT SYSTEM PHILOSOPHY AND FUNCTIONS

The Development Cycle
A development system permits the user to carry out the following functions:
  a. Specify or *build* the final target hardware configuration: list memory resources and I/O devices.
  b. *Assemble* source program segments or *modules*: translate program statements (mnemonics) into binary code.
  c. Combine or *link* modules from step 2 with other subroutines and with library routines.
  d. On the host system, *simulate* operation of the target system at the instruction level (lacking true I/O) for program debugging.

    e. *Emulate* operation on the target system, including real I/O with inter-
       rupts and real memory resources.

    f. Divide (*split*) object code into modules to fit ROM chips, and program
       (*burn*) the ROMs.

The steps gone through in a typical DSP system development cycle are shown
in Figure 7.1.

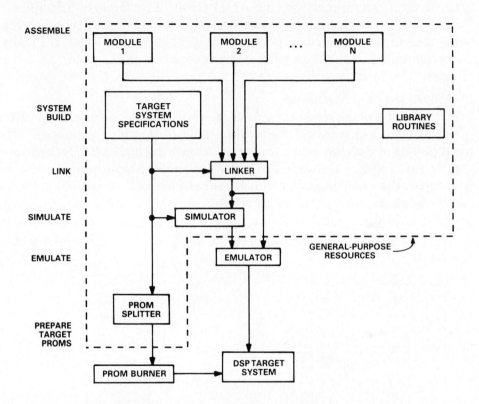

Figure 7.1. Typical application development cycle, showing the modules of a DSP
development system.

As system development moves through this cycle, the software is progres-
sively freed from grammar and syntax errors, then logic errors; then it is fitted
into the real system configuration and exposed to the I/O environment with
which the target system will have to cope. The reader is probably familiar with
most of the above concepts from conventional microprocessor applications;
however, some aspects of DSP system development differ from conventional
microprocessor development:

● High-throughput DSP chips will have parallel processing, so debugging
must be able to access status and contents of concurrent processes. The
architecture and capability of the target DSP μP must be accurately
represented, especially during the simulation phase.

• With interrupt-driven I/O, which is essential to real-time data processing, an interrupt can occur at any point during program execution. The DSP development system should realistically simulate interrupts occurring at widely differing rates.

While the discussion in this chapter is intended to be as general as possible, the concepts are illustrated by examples from the development system for a specific DSP microprocessor, the ADSP-2100. The familiar FIR filter is followed through the development cycle listed above and incorporated, along with input-output, as a module of a signal-processing system. Finally, a more ambitious example, a portion of a speech recognizer, is introduced as an example of interrupt-driven I/O.

Development System Desiderata

The interaction of development system host with target system and emulator is summarized in Figure 7.2. General-purpose computing facilities are used in all phases of development except emulation, which necessarily involves a real target system. Unless many software development projects can be scheduled, the expense and potential obsolescence of a standalone development system should be avoided, if feasible, by integrating system-building, assembly, linking, and simulation into the user's existing facilities. In order to efficiently code and debug a DSP system, the development system should be:

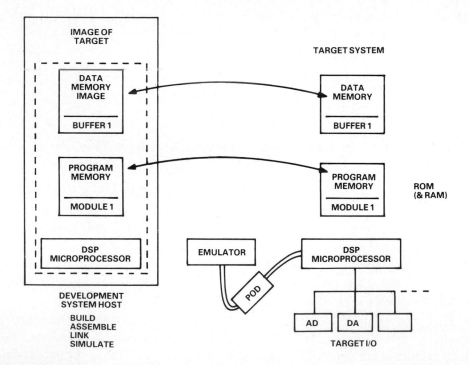

Figure 7.2. Images of the target system's memory map and the DSP μP are replicated in the host during software development.

Portable – The results of program development will reside in a *target system* consisting of a DSP microprocessor and a mix of program ROM, data RAM, and I/O interfaces. The target will usually look more like a signal-processing instrument than a general-purpose computer. With the exception of the emulation phase, software development is therefore best done on a host system: a mainframe computer—or engineering workstation—with more-flexible system software and more-powerful mass storage than the target will have. Development modes will therefore involve *cross-software*, i.e., software written in a well-accepted high-level language, which runs on a wide variety of hosts bearing no relationship to the target hardware or language.

Linkable – The final code will reference specific absolute addresses of the target's memory banks and I/O addresses. To increase transportability between applications, memory and I/O references in the source code should be symbolic. This permits modular structured subroutine design and the incorporation of "canned" subroutines, all of which remain relocatable until linked.

Configurable – Some aspects of the DSP-μP architecture and of the target hardware configuration necessarily require specification. In the dual-bus Harvard architecture, there will be both program memory and (separate) data memory. While program memory is typically ROM and data memory normally RAM, coefficients and data may also be stored in program memory to gain speed advantage from a dual-bus structure. The user will have freedom to configure the mix of program and data-memory, and specify the RAM and ROM portions. The I/O devices are given symbolic names and mapped to specific memory locations. Both memory and I/O choices are specified during the *system-building* phase of the development system.

Debuggable – Development of a target system follows a sequence in which the target gradually becomes more independent of the development system. As with child-rearing, the command: "RUN FRED" is normally preceded by first teaching Fred how to crawl and then walk with decreasing parental guidance.

1. In the *simulation* phase, cross-software running on the host will debug target programs, using the host's memory and simulated I/O. Breakpoints and trace functions will determine program flow and register/bus contents occurring in the neighborhood. But simulation—especially of high-speed target systems—does not execute at real-time speed, and it only approximates inputs and interrupt demands—which is not adequate performance in most applications. Hardware emulation is therefore a virtually essential option.

2. In *emulation*, the actual target system is exercised (and bugs exorcised) with an umbilical cord replacing the target microprocessor. The umbilical ends inside an emulator pod which contains a microprocessor identical to the target, but in an environment where the emulator can take

control of the bus and test target functions one by one.

3. As the target's microprocessor is graduated to working independently, program modules residing in development system RAM can be shifted to target memory. The dividing of programs into specific chips (ROM splitting) is done by the development system prior to permanent transfer of memory contents (ROM burning) on a conventional PROM programmer.

7.2.2 THE TARGET SYSTEM SPECIFICATION PHASE

The *system build* phase specifies the configuration of target system's memory and I/O to the development system. The available absolute memory-address ranges are listed block by block for P-memory and D-memory along with the intended attributes. While data-memory is normally RAM, the program-memory address-range may be filled by ROM for permanent programs or constants, as well as RAM blocks for coefficients or programs or program-segments which may change. The memory-mapped addresses of I/O devices are also specified at this stage.

For example, an ADSP-2100 system builder input file for a signal processor to do FIR filtering might look like:

```
.SYSTEM                    signal_processor {system source file for signal_processor}

.SEG/ROM/ABS = 0/PM/CODE   program_mem[8192];      {declare coeff table}
.SEG/RAM/ABS = 8192/PM/DATA coefficients[2048];    {declare coeff table}
·SEG/RAM/ABS = 0/DM/DATA   data[16384];            {declare data memory}

.PORT/ABS = 16382          a_d_converter;          {declare input port}
.PORT/ABS = 16383          d_a_converter;          {declare output port}

.ENDSYS;
```

System-builder commands (prefaced by a period) appear in the left column with user-defined logical names appearing in the right column. In this example, the system labelled PROGRAM MEMORY will allow for 8,192 words of program ROM beginning at PM address 0. Also in the program-memory map are 2,048 words of ROM beginning at address 8192 for coefficients. Data-memory allocation allows 16 K of RAM space for data buffers. Signal inputs from an a/d converter and display outputs to a d/a converter are mapped to the upper end of the data-memory map. Commands .*SEG* label memory SEGments, while I/O port specifications are prefaced by .*PORT*. Options prefaced by / specify: program or data memory (/pm, /dm), whether the contents will be a program module or data (/code, /data), whether the physical memory will be ROM or RAM, and the absolute starting address (ABS = #xxxx).

The output of the system-build phase is a file passed to the linker to specify where program modules and data buffers can reside—as well as the translation of I/O logical names to physical port addresses.

7.2.3 THE ASSEMBLER

Assembler Constructs

A DSP microprocessor cross-assembler translates assembly-language mnemonics into binary object code, with special constructs to adapt it to the processor's architectural features and the user's specific target configuration.

The inputs to an Assembler are the user-written source code statements for a given module. Frequently referenced code blocks may be specified as *macro*s to shorten the source listing. Large code blocks referenced by several modules may be read from a separate file by an INCLUDE statement. The Assembler's outputs are the binary object files and a listing with diagnostic information for the user. A collection of assembled subroutines or modules is then passed to the Linker.

Assembler Construct Requirements

Available assembler statements, or "constructs," must be adequate to
    specify the structure of programs (who CALLs whom)
    declare variables (any variable must be defined before it is used)
    identify where referenced variables and I/O ports will reside in the target
    fully represent any microprocessor instruction clearly and intelligibly.

Assembler constructs, passed on to the linker, must be defined with care, since any class of possible target systems must be describable. The ancient Egyptians whose hieroglyph for "North" was the same as the one for "downstream", based on experience with the Nile, had difficulty writing home news of the discovery of the Euphrates river, which flowed the other way. The written language had not been set up to anticipate all possible events (or to relate the pole-star to river directions). Correspondingly, the structure of the development system assembler must have the flexibility to make full use of the dual-memory and multiprocessing architecture of the ADSP-2100 and the geography of the specific target system while not trying to do the impossible (e.g., write data into an array specified to be ROM). Assembly directives accordingly specify the nature and placement of both data buffers and of modules:

Data Buffer Requirements – Assembler statements must be able to:
    declare that labeled data-buffer (temporary data-storage) areas reside in
        either Program-memory or Data-memory, whether a buffer is
        to be RAM or ROM, and (if necessary) absolute address boundaries
    declare that a buffer is circular, and set its boundary
    set initial values of buffers and set scaling of binary fractions.

Program Module Requirements – The assembler/linker must be able to declare that a program module is to reside in ROM or RAM and (optionally) set a specific module starting address.

Processor Instruction Constructs – The instruction set described in the previous chapter and the examples below illustrate how processor instruction constructs are high-level in form, unlike the relatively cryptic mnemonics of older microprocessor assemblers. Operands, operations, and instructions read like a high-level language, which promotes rapid learning and transportability between users. For example,

```
DO loop UNTIL CE;                          {do until countdown}
IF AV JUMP label + 1;                      {conditional jump on alu overflow}
MR = MR + MX0*MY0, PM(I4,M7) = SI;         {multiply plus data move}
ENA AUX REGBANK;                           {switch context}
IF MV CALL blk__flt__pnt;                  {overload response}
AR = AX0 + AY0, AX0 = DM(I0,M0), AY = PM(I0,M0);
                                           {multiple data move}
```

In this syntax, a complete instruction ends with a semicolon, while portions of multiple-operation instructions are separated by commas. Comments may appear anywhere within brackets [{}]. The assembler does not distinguish between upper and lower case; however, for ease of interpretation, we will adopt the convention that all *fixed* instructions and labels are upper case; user-defined names (such as labels, variables, and data buffers) are lower case.

A program module consists of assembly language source-code segments, surrounded by assembly directives. Assembly directives define the local and the global variables, establish which variables and parameters can be passed between modules, set program labels and symbolic references to loops and pointers, and set initial values. A module has the following structure:

> Module beginning directive
> Declaration Sections
>> Include and External Module declarations
>> Variable and I/O declarations
>>> External and Global variable specifications
> Code Blocks
>> Macro block(s)
>> Conditional assembly block(s)
> Module termination directive

This structure will be illustrated by the simple example of a continuous throughput FIR filter, and then by the more sophisticated example of a real-time speech recognizer.

Example: Transforming FIR Filter Subroutine into Assembler Module

While it would be beyond our scope to give a complete list of assembler directives and their function (such as those contained in the *ADSP-2100 Cross-Software Manual*), an example of an assembler module illustrates the whys and hows. A familiar FIR filter, somewhat like that introduced in the previous chapter as an example of the instruction set, will now be surrounded by the assembly language directives needed to inform the development system about program structure and target-system configuration.

The FIR filter is selected as the first example because the simple algorithm permits a quick introduction to the Assembler style and syntax. This section examines the FIR code, then shows how to incorporate it as a subroutine within a real-time data acquisition main program. The FIR delay-line block diagram and moving average memory-map of Figure 7.3 correspond to the subroutine example shown in Table 7.1.

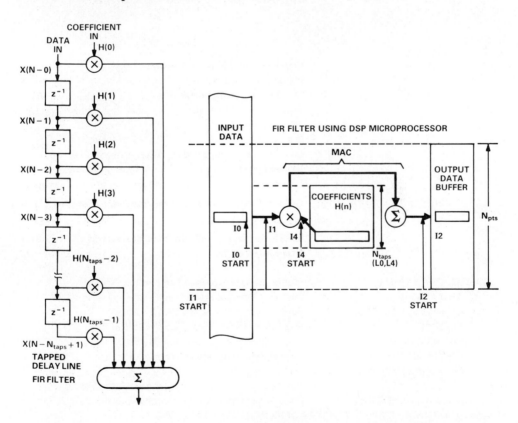

a. Tapped delay-line.

b. Moving average of variable weight. Labels refer to the corresponding program listing of Table 7.1.

Figure 7.3. FIR filter.

Data Buffer Requirements – Assembler statements must be able to:
  declare that labeled data-buffer (temporary data-storage) areas reside in
    either Program-memory or Data-memory, whether a buffer is to be
    RAM or ROM, and (if necessary) absolute address boundaries
  declare that a buffer is circular, and set its boundary
  set initial values of buffers and set scaling of binary fractions.

Program Module Requirements – The assembler/linker must be able to declare
that a program module is to reside in ROM or RAM and (optionally) set a
specific module starting address.

**Table 7.1 FIR Filter Subroutine for the ADSP-2100. The comment header
summarizes the algorithm, input, output, and performance.**

Algorithm: $$y(N) = \sum_{K=0}^{L-1} H(K) * X(N-K)$$

Input:	I0 points to starting place in delay line
	I1 points to input buffer
	I2 points to output buffer
	I4 points to coefficients
	M0 contains 1
	M4 contains 1
	L0 = Ntaps contains order of filter, L (number of taps)
	L4 = Ntaps also contains order of filter (2nd pointer)
	Cntr = Npts contains number of output points desired
Output:	Output buffer contains desired number of samples.
Cycles:	(Ntaps + 6)∗Npts + 7
	Npts is the desired number of output samples
	Ntaps is the order of the filter

```
STARTFILTER: AY0 = L0;          {put number of taps into Temp = Acc Y input}
       AR = AY0 − 1;            {Ntaps−1 into Acc results reg}
       DO outloop UNTIL CE;     {end of loop initialized by CNTR = Npts}
       SI = DM(I1,M0);          {load 1st data point into Temp = Shifter Reg}
       DM(I0,M0) = SI;          {copy 1st point into start of delay line}
       CNTR = AR;               {preset inner loop count = Ntaps − 1}
       MR = 0, MX0 = DM(I0,M0), MY0 = PM(I4,M4);
                                {1st data point into MX and 1st coeff into MY}
       DO taploop UNTIL CE;     {Inner loop: Ntaps (data) × (coeff) accumulate}
taploop: MR = MR + MX0 + MY0, MX0 = DM(I0,M0), MY0 = PM(I4,M4);
                                {Mpy/Acc (data) × (coeff) in delay line}
       MR = MR + MX0*MY0(RND);  {add last tap and round}
outloop: DM(I2,M0) = MR1;       {FIR output point into output buffer}
                                {continue until Ntaps output samples computed}
       RTS;                     {return from this subroutine}
```

Comments on the FIR Filter Subroutine: There are two DO loops: an outer loop (DO outloop) *Npts* in length, and an inner loop (DO taploop) *Ntaps* long. For each output data point, *Ntaps* most-recent input points are multiplied by filter coefficients and accumulated in the inner loop (taploop). A pointer is incremented to the next output point and repeats. The process continues until the required *Npts* filtered output points have been put into the output buffer; then the subroutine exits from the outer loop (outloop).

Data-address generator registers are preloaded with index pointers (I), address-modification (M) increments, and length (L) of circular buffers. Index pointers mark the starting points of the delay-line (I0), the input data buffer (I1), and the output data buffer (I2). Address increments within the loops (M0, M4) are, in this example, equal to 1. Program sequencing is initialized with a loop counter, *cntr*, preset with a value equal to the number of filtered output points desired.

This example illustrates the general-purpose nature of the ADSP-2100 register set. The labeling A, S, or M on a register-memory transfer does not necessarily imply that the register is involved in arithmetic, multiplication, or shifting; it may simply be used for temporary storage. For example, the shift register is not used for computation at all in this module, so register SI is available as a temporary register to carry out in two steps the equivalent of

$$DM(I0) = DM(I1)$$

a move which is not possible in one step directly.

The nested DO loop illustrates how the ADSP-2100 count stack and over-head-free looping work. The down counter is preset with value *Npts*, which sets how many filtered output data points to compute. Just before the inner (delay line) loop is initiated, *cntr* is reset to contain *Ntaps* – 1, the highest-numbered filter coefficient, which determines the termination of the inner loop. The value of *Npts* for the outer loop counter is not lost, since entering the first DO UNTIL statement pushes the counter value onto the count stack. The value *Npts* pops down automatically [Review: Chapter 6] each time the inner loop is terminated, to serve again as the end-of-loop condition for the outer loop. No reinitialization is required, saving an extra step in execution time.

FIR Filter Subroutine Performance: Given that each line of code in Table 7.1 for the ADSP-2100 takes 1 cycle, the number of cycles between entry and return in the FIR subroutine is

$$N(cycles) = (Ntaps + 6)*Npts + 7$$

The basic product-sum (taploop) takes only one cycle per multiply-accumulate; six other steps within the outer loop (outloop) prepare data and constants for the next output-data computation. For each data point processed, only one cycle is added for each additional tap, so little penalty is paid for high-performance multi-tap filters or for multiplexing several filter sections with different sets of coefficients, or for performing other computations in the unused portion of the DSP μP's duty cycle. The ADSP-2100's dual memory, with coefficient storage in P-memory and signal-data storage in D-memory, makes this possible; note the combined single-cycle reference to both DM and PM for data in the key computational step, taploop. Throughput is given by

$$\text{Throughput} = \frac{\text{clock frequency}}{\dfrac{N(\text{cycles})}{Npts}} = \frac{f(\text{clock})}{Ntaps + 6 + \dfrac{7}{Npts}}$$

If *Ntaps* and *Npts* are both much larger than the 6- and 7-cycle overhead terms, the throughput in steady state operation is

$$\text{Throughput} \simeq \frac{f(\text{clock})}{Ntaps}$$

No higher throughput is attainable with any single processor, since nearly one-cycle-per-tap operation has been attained.

Adding I/O and incorporating the FIR Filter into a DSP System: The next step is to incorporate the FIR filter subroutine into a general-purpose signal processing system. An a/d input converter and a d/a output converter are included in the system-building phase as above. First, the general-purpose subroutine of Table 7.1 is converted into a ready-to-assemble subroutine in Table 7.2. The real-time "on-the-fly" subroutine gets its input from and dumps its output to the memory-mapped locations corresponding to AD_INPUT and DA_OUTPUT, rather than input and output data buffers in memory.

Comparing Table 7.1 above to Table 7.2,

```
              SI = DM(I1,M0);              {load 1st data point into shifter reg}
becomes
              SI = DM(ad_input);           {input sample to shifter reg}
while
         outloop:DM(I2,M0) = MR1;          {fir output point into output buffer}
becomes simply
              DM(da_out) = MR1;            {fir output to da converter}
```

The outer loop, outloop, is eliminated, since an output buffer is not needed in real time.

The code is prefaced by a series of declarations, prefaced by periods (.). Declarations specify the nature and location of user-defined labels for variables (.var) and I/O operations (.port). Declarations also specify the linkages allowed between modules (.external, .global, .entry). A list of declarations can be found in the *ADSP-2100 Cross-Software Manual*. The example of Table 7.2 illustrates some of the declarations and means of passing information between modules. The module itself is declared as ROM-resident, with entry-point STARTFILTER. I/O port names are declared, and a constant file, CONST.H, is included so that parameters such as the number of taps, NTAPS, can be passed. The arrays: DATA, which holds the convolution sum, and COEFFICIENTS, which holds the filter coefficient values, are declared to be external; they will be declared as variables in the main program. Note the use of unresolved constants such as $^\wedge$DATA. The $^\wedge$-prefix denotes the *starting address* of buffer DATA, and will not be set until the linker assigns an address later on.

Table 7.2 ADSP-2100 FIR___Filter Subroutine with on-the-fly analog input and output.

```
.MODULE/ROM      fir-subroutine
.INCLUDE         <const.h>                        {hex constant file}
.PORT            ad__input
.PORT            da__output
.ENTRY           start__fir
.EXTERNAL        data, coefficients

STARTFILTER:     {subroutine code section}
                 I0 = ^data                       {point to data buffer address}
                 I4 = ^coefficients               {point to coefficient address}
                 CNTR = NTAPS                      {passed via const file}
                 SI = DM(ad__input);               {input sample to shifter reg}
                 DM(I0,M0) = SI;                   {copy point to delay line}
                    MR = 0; MX0 = DM(I0,M0), MY0 = PM(I4,M4);
                                                   {data into MX and coeff into MY}
                 DO taploop UNTIL CE;
taploop:            MR = MR + MX0*MY0; MX0 = DM(I0,M0); MY0 = PM(I4,M4);
                                                   {MAC (data) × (coeff) in delay line}
                 MR = MR + MX0*MY0(RND);           {add last tap and round}
                 DM(da__output) = MR1;             {FIR output to DA converter}
                 RTS;                              {return from this subroutine}
```

This is incorporated, in Table 7.3, into a MAIN program which in turn calls various data I/O and processing modules. A circular data array (DATA) is declared in RAM to hold the convolution sum, and another (COEFFICIENTS) is declared in ROM for the filter coefficients. Both arrays are made global so that other modules, such as fir__subroutine, can reference them. Subroutines fir__subroutine and other__stuff are declared to be externals. Circular buffer lengths (L0, L4) are initialized; the % prefix designates the (as yet unspecified) length of an array. This gives maximum

Table 7.3 ADSP-2100 Signal-Processor main program example.

```
.MODULE/ROM/ABS = 0  main program;
.INCLUDE             <const.h>;                  {include constants}
.VAR/DM/RAM/CIRC     data[ntaps];                {convolution buffer and length}
.VAR/PM/ROM/CIRC     coefficients[ntaps];
.GLOBAL              data, coefficients;
.EXTERNAL            fir_subroutine, other_stuff;

                     {Initializations}
                     L0 = ^DATA;                 {buffers}
                     L4 = ^coefficients;
                     M0 = 1;                      {modify-register increment}
                     M4 = 1;
                     I0 = ^DATA;                  {data pointer}
                     CNTR = ^DATA;                {data buffer length}
                     DO zero UNTIL CE;            {clear data buffer}
zero:                  DM(I0,M0) = 0;

mainloop:            CALL fir_subroutine          {data grab, filter, output}
                     CALL other_stuff             {other processing}
                     JUMP mainloop
```

flexibility. The user need only declare the *existence* and type of variables and constants in source programs, leaving specific values, such as filter length, *Ntaps*, and data-array length, *Npts*, to be selected for a given application at system-build time and passed to the Linker.

This example illustrates the incorporation of modules into a main program and the techniques of passing information and coupling modules at the instruction level.* It would not be a good real-time DSP system because the timing of the signal sampling is set in software by the length of mainloop. An interrupt example will be given below in which the I/O sampling is set by a separate fixed clock which signals an interrupt request at fixed intervals.

Assembler Outputs
Assembler outputs include a list file, for the user to refer to in debugging, and an object file.

The List File: The list file is an assembler output which shows on a line-by-line basis the source-code translation and where it will reside in memory. The list contains the source-program statement or assembler directives, the hexadecimal object code into which a statement was translated, line numbers for convenience in reading the listing, the type of memory (pm for Program, dm for Data) into which code is to be placed, the allocation or grouping of arrays and variables, the location where the code will reside, and the symbolic address (e.g., Start Filter) corresponding to the address shown. Allocations with the same number will be linked as a group, and memory locations prefaced by an apostrophe (') are relative, with actual location left to the linker.

---

*The example is illustrative; not all the details of the code are explained. The reader who would go deeper should consult the manufacturer's manuals.

For example, the first line of the FIR filter subroutine of Table 7.2 might appear in a list output as:

Line	Mem	Alloc/Loc	Code	Statement/Directive
57	pm	6'0019	0D148	START FILTER AY = I0;

Line 57 will reside in program memory, will be linked to other lines with group allocation #6, and will have (relative) location 0019.

Object file: The binary object file, which contains the code shown in the list file, is passed along to the Linker together with object files from other modules.

Error Handling: The Assembler can detect three types of run-time errors: software failures, syntax errors, and module structure errors.

Software Failure – e.g., when a table search fails for an element known to be in the table.

Syntax Errors – as in mistyping or in illegal statement syntax.

Module Structure Errors – when, for example, a variable name is first declared in a VAR statement ("you will find me in *this* module) and then an EXTERNAL statement (look for me *outside* of this module) follows with the same variable name.

Assembler Exercise: The FIR subroutine given in Table 7.1 and portrayed in Figure 7.3 envisions input data pre-stored in a buffer indexed by I1. But if data is output dynamically to a graphics display, input data need not be stored for more than the number of coefficients, *Ntaps*, and output data are put into a circular buffer, the moving window, which is updated continuously. Modify the program to make this possible.

Method: Pointer I1 stays as is, but now a new loop counter needs to be defined. The size of the moving window output is set by the display device: how many data points are to be viewed at one time—several hundred to 1,000 are typical for graphics resolution. You have a choice between two presentation formats:

(a) Update mode: the picture on the screen is frozen except that as time goes on, the oldest data is written over by the latest data.

(b) Moving-window mode: The data moves across the screen, and the window is always the most recent N-point set, where N is the window width.

Show how each of (a) or (b) is accomplished by the looping.

7.2.4 THE LINKER

LINKER FUNCTIONS

What a Linker Does: The Linker is like the hotel room-clerk who matches guest requirements, specified by the space "reservations" requested at Assembly time, to available rooms, as specified in the architectural description of the System Build File. The Linker couples together relocatable object modules prepared by the Assembler, including library subroutines, and produces absolute memory images for run-time execution. The memory images intended for the target system will then be passed on for software Simulation and hardware Emulation, then partitioned into physical memory segments in RAM or to be burned in ROM. Refer back to Figure 7.1 for an outline view of Linker inputs and outputs, and for more detail, consult the appropriate software development manual.

While the Assembler can detect errors in syntax, it is the Linker which, looking up the target architecture configuration, determines how software modules are partitioned into physical memory resources: into what chips at what addresses. Symbolic addresses are assigned to absolute memory locations, and cross-references between modules are verified for correct structure. Linker outputs include the memory images, a symbol table to be passed on to Simulator/Emulator debugging, and, (if selected) a memory map for the user.

Inter-Module Linkage: The syntax of modern assemblers, such as that of the ADSP-2100, permits total freedom in the naming of variables and labels within in a module, without the need to worry if the same name is used in a different locality for a different purpose. This means that symbols defined in a module (with a .VAR declaration) can only be used locally within that module unless included in a .GLOBAL or an .ENTRY statement in that module and also referred to in an .EXTERNAL statement in another module. The etiquette is shown in Figure 7.4. Symbols declared to be global are intended to pass

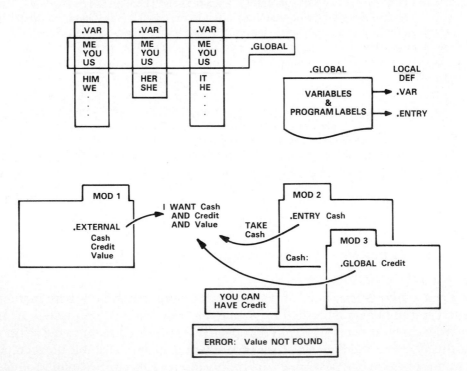

Figure 7.4. Valid Assembler syntax contrasted with valid linkages.

between modules. A symbol must also be defined locally by a .VAR for variables or an .ENTRY for symbolic addresses or subroutine entry points.

Correct syntax does not ensure correct linkage (Figure 7.4). Symbols Cash, Credit, and Value are declared as .EXTERNAL in Module 1, so the linker knows they are supposed to be found outside Module 1. Cash is an .ENTRY program label in module 2, and Credit is a .GLOBAL variable name in Module 3. The Linker sees the .EXTERNAL declaration and looks through all the other modules to find labels Cash, Credit, and Value. The .ENTRY Cash and .GLOBAL Credit allow those labels to be passed to Module 1. No label "Value" is found by the linker, which therefore reports an error message.

Since a declaration is required to pass information to another module, a label may be named locally without concern for label duplication in other modules.

Linker Allocation of Memory Space: The source code of any module may specify memory features: program memory or data memory, RAM or ROM, and buffers to be made circular. The linker parcels these requests among resources specified in the Architectural Description File. The linker will detect, as shown in Figure 7.5(a), if an absolute address referenced by a module does not exist in the architecture description file, or if a module tries to allocate *module code* (i.e., program material) into memory space which has a *data-*only attribute.

The most straightforward linker allocation algorithm (Figure 7.5(b)) operates on a first-come, first served basis, with modules picked up in the order they are encountered and put into available memory with the specified attributes. An error message is given if space runs out in the allocation. Clearly, this mode may not make optimum use of available resources, since no attempt is made to make a best fit of modules into available memory banks. An alternative mode sorts the modules by size and finds the smallest vacancy still left in available memory with the specified attributes. This mode is clearly optimal but also takes longer to carry out.

Linker Outputs and Error Diagnostics
The linker outputs the absolute memory images, a map file, and a debug symbol table.

Linker Memory-Map File: The memory map file tells what goes where and cross-references all symbol usage. The format is self-explanatory (see, for example, *ADSP-2100 Cross-Software Manual*). The memory map includes:

    Memory sector address listings, with attributes (pm/dm, RAM/ROM, data/ code;

    Sequential memory allocations by module, together with associated symbols; and

    Symbol cross-reference table: what symbol is referenced by what module(s)?

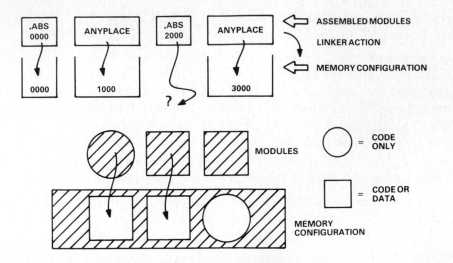

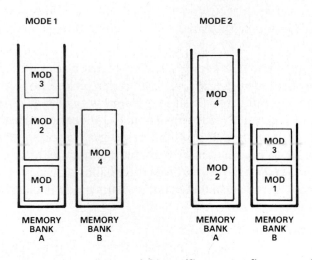

a. The Linker keeps track of finding physical memory resources to match the address requirements and code/data attributes requested in the source code with the resources available.

b. Memory allocation modes: (Mode 1) Direct (first-come first-served), and (Mode 2) Optimized.

Figure 7.5. Linker action.

Sequential memory allocations are made whenever a group of symbols have been declared within a single .VAR statement; the programmer thereby specifies that these elements are to be kept together

Debug Symbol Table: If the "DEBUG" option is selected when the Linker is invoked, the Linker generates a Debug Symbol Table. Each symbol encountered is listed along with the Module(s) that reference it, the type of memory that is referenced, and the absolute address of that symbol.

Linker Error Handling: Three types of run-time errors are caught by a Linker.

Software Failure – when a table search fails, or when a referenced symbol is not found in any module;

Memory Allocation Errors – insufficient memory or, as in the examples of Figure 7.4, reference to an absolute address that isn't there.

Illegal Symbol Cross-Reference Errors – an attempt to reference an undeclared symbol, or to improperly cross-reference a symbol out of its module.

Here is a software failure example. Suppose a module contains the line:

JUMP Sierra;

but the module itself does not contain symbolic address Sierra. The linker searches all .EXTERNALS to see if Sierra was meant to be found outside the module, searches all memory references to find Sierra, and if Sierra is not found will report a linker error.

As an example of an illegal cross-reference error, the following two modules may be legal in Assembly but will not Link:

.Module	A;		.Module	B;
.ENTRY	Tango;			
.EXTERNAL	Uniform;		.EXTERNAL	Tango;
Tango:	JUMP Uniform;		Uniform:	JUMP Tango;
etc.			etc.	

Assembly will work with no errors reported, because symbolic addresses (Uniform in Module A, Tango in Module B) which lie outside a module are declared in that module to be EXTERNAL. Uniform in Module B knows to jump outside Module B to find Tango, thanks to the .EXTERNAL Tango statement, and will find it in Module A thanks to its .ENTRY Tango. However, a linker error occurs. Even though Tango in Module A knows that Uniform lies outside Module A (.EXTERNAL Uniform), Tango will not know where to jump to find Uniform, because Uniform was not declared as .ENTRY or .GLOBAL in Module B, and can not be accessed outside the module. To fix the problem, add .ENTRY Uniform to Module B.

7.2.5 THE SIMULATOR
Simulator Functional Desiderata
What a Simulator Does: The Simulator permits programs to be fine-tuned in software prior to making the significant effort to bring up the target system in hardware. The Simulator acts as much like the target system as the nature of software permits. Programs under development are run on a host machine. For simulating input/output, physical I/O devices are replaced by host

memory locations. Breakpoints can readily be set so that when a program arrives at a specified program or data location, there is a display of all register contents when the breakpoint occurred or a Trace of system activity before the breakpoint. Sections that cause a crash or dead-ended loops can readily be identified. The host's memory plays the role of memory in the target system, acting as ROM when so specified. Even interrupts can be simulated by timers set to trigger a simulated interrupt line. This last feature is relatively unusual in microprocessor simulators, yet crucial in debugging interrupt-driven DSP target systems.

Simulator inputs and outputs are shown in Figure 7.6. The inputs are the Program/Data memory images and debug symbol table created by the Linker, simulated input data buffers downloaded from the host, and simulated inputs. Simulator outputs are processed signals uploaded to the host and

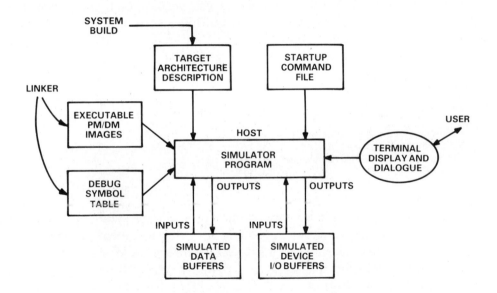

Figure 7.6. Simulator inputs and outputs.

simulated outputs to I/O ports. The primary Simulator diagnostics are displays of program execution (including a disassembly that translates the binary code back into assembly language mnemonics), register contents and chip status, stacks, program- and data-memory contents, etc. The ADSP-2100 Simulator described below is interactive, so that registers, instructions, and memory contents can be examined and also changed by the operator during a debug session.

Simulator Functional Performance: The Simulator carries out the functional performance of a specified DSP microprocessor target system. For example, for the ADSP-2100 DSP-*MP*, the target can have:

Data memory: up to 16K, 16-bit (non wait-stated)
Program memory: up to 16K, 24-bit, RAM/ROM (non wait-stated)
I/O Ports: up to 8, mapped into D-memory,
    with optional wait-states
Interrupt sources, up to 4, user-configurable

Although target memory is assumed fast enough to keep up with the high-performance DSPμP, I/O devices such as an A/D converter may be inherently slow compared to the DSP system. The DMACK control line (Figure 6.26) provides for this handshaking in the ADSP-2100 hardware; note the lack of corresponding line on the P-memory side. In simulation, a read or write to a D-Memory address mapped to an I/O device may therefore provide an optional number of *wait-states* (NOP cycles) to simulate the time required for the device to provide a valid input or register a valid output. *True* handshaking is left for hardware emulation.

Simulated Memory – Target system configuration provided in the Architecture Description File is passed along to the Simulator as a framework for allowable operations, e.g., which memory segments exist and whether a given segment is RAM or ROM. Any unused portions of the PM/DM address range will effectively cease to exist after memory-image description is passed to the Simulator. An attempt to write to ROM or to read/write a nonexistent memory area will be flagged by the Simulator as an error.

Simulator Input-Output – Empty buffers for output data are created as blank upload files. I/O simulated on the host should behave as much as possible like real-world I/O, with real data buffers replaced by inputs from simulated data files and processed outputs uploaded back to output files. Interrupt capability and masking should simulate that of the target system. To enable an I/O device, the following must be specified:
    Port number and memory-mapped name of the device
    Time to wait for a device to complete a Read or Write operation
    Whether the device is input-only, output-only, or input/output
    File names on the host in which simulated input data is to be found or
        where output data to be placed
    Interrupts, their priority and their frequency.

Simulation of an interrupt-driven I/O device at timed intervals only approximates the asynchronous nature of a true interrupt, since its timing may be independent of the DSP system clock. However, most sampled signal inputs will occur at precisely timed intervals, so the approximation is not unrealistic.

SIMULATOR EXAMPLE
Simulator Diagnostic Outputs: A Simulator session is an interactive dialogue. An example of a Simulator screen display is shown in Figure 7.7, but the interactive effect of watching it update in simulated real time is necessarily

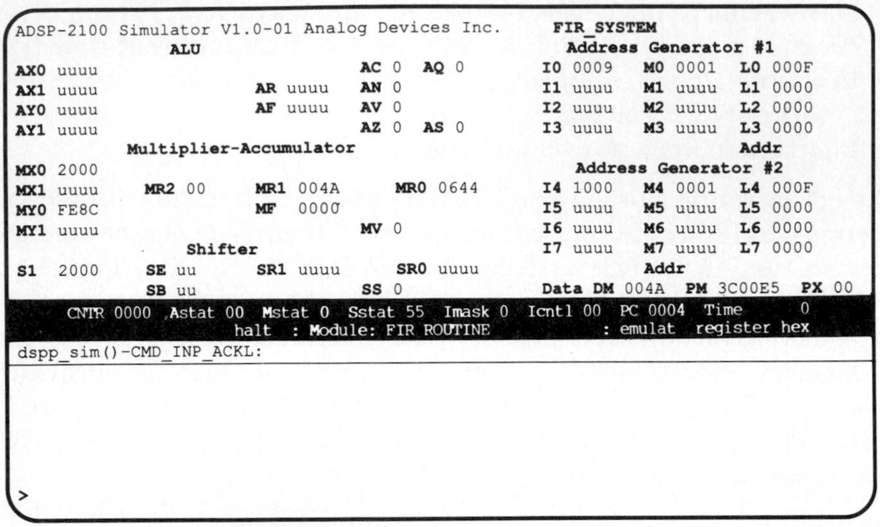

```
ADSP-2100 Simulator V1.0-01 Analog Devices Inc.     FIR_SYSTEM
                ALU                                  Address Generator #1
AX0 uuuu                          AC 0  AQ 0     I0 0009  M0 0001  L0 000F
AX1 uuuu              AR uuuu     AN 0           I1 uuuu  M1 uuuu  L1 0000
AY0 uuuu              AF uuuu     AV 0           I2 uuuu  M2 uuuu  L2 0000
AY1 uuuu                          AZ 0  AS 0     I3 uuuu  M3 uuuu  L3 0000
            Multiplier-Accumulator                                   Addr
MX0 2000                                             Address Generator #2
MX1 uuuu     MR2 00    MR1 004A       MR0 0644   I4 1000  M4 0001  L4 000F
MY0 FE8C               MF  0000                  I5 uuuu  M5 uuuu  L5 0000
MY1 uuuu                          MV 0           I6 uuuu  M6 uuuu  L6 0000
            Shifter                              I7 uuuu  M7 uuuu  L7 0000
S1 2000    SE uu    SR1 uuuu      SR0 uuuu                         Addr
           SB uu                  SS 0           Data DM 004A  PM 3C00E5  PX 00
      CNTR 0000  Astat 00  Mstat 0  Sstat 55  Imask 0  Icntl 00  PC 0004  Time      0
              halt : Module: FIR ROUTINE              : emulat  register hex
dspp_sim()-CMD_INP_ACKL:




>
```

Figure 7.7. ADSP-2100 simulator screen display. Example of register contents and chip status.

missing. The command line at the bottom (> : prompt) is the normal entry point to specify I/O ports, examine and change memory or register contents, and select between simulation display modes.

This particular example centers on (a) register contents and chip status. Other choices display
  (b) Contents of specified sections of program memory and the corresponding disassembled instructions
  (c) Contents of specified sections of data memory and corresponding buffer names
  (d) Contents and stack-pointer locations for count, status, loop, and program-counter stacks
  (e) Program-cache memory contents and pointers
  (f) A Trace buffer dump of recent data- and program-memory activity and disassembled instruction.

The screen of Figure 7.7 is split into three windows.
  Top Window: Information requested by the user; this example shows header information (i.e., Simulator name, version number, and system name assigned by the user) and register contents (hex) for all computational elements, address generator, program counter.
  Middle Window: This window, which remains the same in all display options, shows processor status. The processor status information displayed includes program sequencer registers (e.g., CNTR) which determine program flow, arithmetic-, mode- and stack-status registers, interrupt mask and control register, and program counter.

Lower Window: Simulator messages and command inputs to the Simulator. The lower window may also contain a disassembled instruction sequence. The user selects the information to be displayed and controls the simulation by typing into this window when the Simulator is ready to accept a command, indicated by a prompt >. In execution, the window is filled by the present (top) and immediate future instructions, which the target will carry out.

Information requested can change in "simulated real-time" in the upper window, the status is continuously updated in the middle region, and the executing program statement may scroll upwards in the lower window. The values of registers, stacks, memory contents, and cache "hits" correspond to the consequences of the top line of code displayed in the lower window, so system contents can be checked against what the currently executing line was intended to do. The actual rate is of course slow compared to the 100-ns cycle time of the target system; it depends upon native speed of the host operating the Simulator program and the operating mode selected by the user (below).

Simulator Modes and Operation: From the Command line, the user can set the displays or operating modes, directly preset any register or memory location, set a breakpoint or trace condition. Other commands can show the breakpoints, trace-triggers or contents, dump the trace buffer to the screen, read dummy data or a program into the simulated system, write a data or program file, reset with selected breakpoint and trace conditions, and of course run the program under test.

Modes of simulation can be selected from among three choices:

Emulator mode – the system is free-running with occasional display updates and no disassembly of instructions until a halt or breakpoint is encountered. This is the startup mode and is the fastest (but still only about 100 instructions per second when operating on a VAX-class host). The lower window is available for command lines to be typed in.

Extend mode – the system updates the selected display on a line-by-line execution basis, with disassembly, and scrolls the currently executing code. This mode is much slower than the Emulator mode.

Single-Step mode – executes one instruction and updates the selected displays.

The mode can be switched flexibly back and forth from Emulator to Extend to Single-Step whenever simulation is halted.

Breakpoint and Trace Usage: A *breakpoint* brings the Simulator to a halt at a predetermined point so that all register, memory, stack, and status contents at that point can be examined. A *trace* is a window into system actions, like an oscilloscope, storing what happened in a series of instructions before the selected "trigger" condition. A breakpoint stores what happened only at the selected point, but all system contents can be looked at. A trace shows more history but less contents; only the PM and DM data, addresses, and disassembled instruction are displayed.

A trace can be entered anytime the simulation is halted, by a TRACE request from the command line. The trigger for a breakpoint can be set to occur either when reference is made to a specified address in program-memory or a specific location in data-memory. Although more-sophisticated Simulators allow combinations of address and contents with bit-masking, a Simulator is best for instruction-level debugging. Deeper access to the system busses is better left for the hardware emulation phase.

7.2.6 THE PROM SPLITTER

The PROM splitter performs the function for target physical memory chips that splitting a log into pieces performs for a fireplace. Chunks of code destined for ROM are divided up into sizes that fit into regions of physical memory chips and put into the correct format to be readable by common PROM burner* units.

The PROM splitter program combines PM/DM memory images from linker outputs with available ROM address boundaries passed by the Architecture Description File generated by the System Builder. The output is a series of upload files, one for each PROM chip, expressed in 4-character HEX, with start and stop information following relatively standard PROM burner formats.

7.2.7 TARGET EMULATION

What an Emulator Does

> *Emulate: To strive to equal (or excel), especially through imitation.*

While a Simulator provides an environment for software debugging and produces absolute memory images to be written onto a Target system, at some point the real hardware world must be faced, with its unpredictable noise, flaky interfaces, and unanticipated input/output conditions. An Emulator is a *real-time* debug tool that takes over where the Simulator leaves off. The real-time execution speed can also be an asset if the debug requires execution of 100,000 instruction cycles, which might take all night to compute in simulation on a general-purpose host—but only a few seconds in real-time.

When a program compiles and loads but doesn't work in a real-world environment, it is usually unclear whether problems lie in *software* logic (perhaps an interrupt-handler subtlety) or in *hardware* that is malfunctioning or mis-

---

*The term PROM "burner" goes back to early ROMs whose images were recorded by sending high drive-currents to evaporate selected fusible links. The term has stuck even though PROMs now are electrically programmed by much gentler means (and some are even re-programmable).

matched to the application. The emulator is a brain-transplant to the target DSP microprocessor, so that trouble in a module can be identified with the offending real-world culprit: the handshake never reciprocated, the interrupt that vectored off into a service routine that didn't really accept the I/O, the signal noise spike that caused the program to malfunction, etc. As shown in Figure 7.8, the Emulator connects to the target and also talks to the development system and to the programmer via RS-232 serial ports.

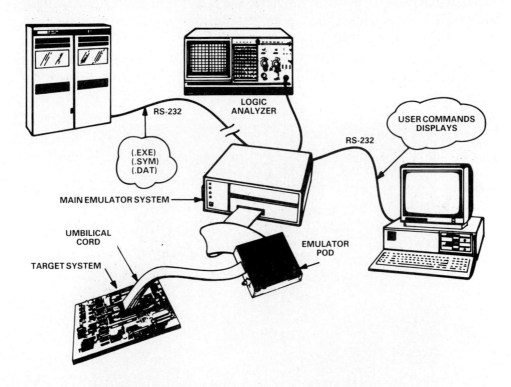

Figure 7.8. On-Target Emulator: the final handholding link between development on the host and independent target-system operation.

The Emulator is the final handholding link between software development on the host and independent target system operation. The umbilical, plugged in to where the target's DSP microprocessor would go, connects to a functionally identical microprocessor. The emulator DSP μP is placed directly adjacent to the target connector, so that a standard part operating under conditions not too different from those of a complete target system may suffice; the alternative is a processor selected for extra-high speed to overcome propagation delays and with physically buffered outputs to drive load capacitance. Memory residing more distantly in the Emulator box must be selected for higher speed to compensate for propagation delays (1ns/ft or 3 ns/m) due to lead length of the umbilical.

DSP Emulator Functions

The Emulator shares the Simulator command structure and uses Simulator-compatible upload/download files. Displays on the Emulator terminal resemble those of the Simulator, except that the Emulator can run standalone while the Simulator runs only on the host. The Emulator has the capability to set software breakpoints, upload and download PM and DM to and from the host, inspect and change all DSP μP registers, halt execution and single-step, and bank-switch memory between Emulator and target. The Emulator's DSP μP makes all bus contents available for examination, signals key events, and allows the emulator to take control of the target's activities.

The layout of the ADSP-2100 Emulator hardware is shown in Figure 7.9. The hardware comprises two portions; one half deals with DSP μP functions, the

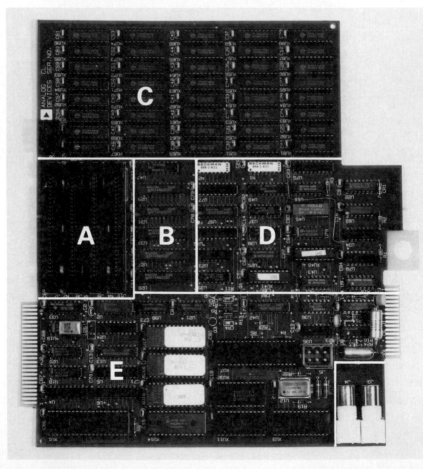

Figure 7.9. ADSP-2100 Emulator layout. A: Umbilical cable to pod containing the Emulator's ADSP-2100. B: High-speed state machine controlling the bus of (A). C: RAM to map full Target program-memory. D: Interface between the DSP μP and the Emulator's own GP μP. E: Emulator's GP μP, in this case an 8088.

other half with GP μP functions. In the DSP μP half there is the umbilical cable to a pod containing the Emulator's ADSP-2100; a high-speed state machine which generates the instructions to take control of the program-memory bus of the ADSP-2100; and high-speed RAM to map full Target program-memory. The two portions communicate via a double-buffered tri-stated register set. The slower GP μP half contains the Emulator's GP μP, in this case an 8088, and its associated Emulator ROM, timers, and communications chips.

Controlling the Split Brain – A key problem in designing an Emulator for a DSP μP is that a general-purpose μP will not have the speed and certainly not the efficient architecture and performance of the DSP μP. While a GP μP is sufficient to manage many Emulator functions, such as uploading, downloading, and memory management, it cannot realistically exercise the DSP μP functions. The DSP μP resident in the Emulator pod (a) is therefore self-emulating. The Emulator's managing GP μP notes when specified functions (such as TRAP, HALT, RESET, DMACK) occur in DSP μP operation and takes control of the DSP μP's bus until requisite information is passed. The internal status and contents of the DSP μP are passed through a buffered-register set (d) to the Emulator by a *full context save*, a program executed by the GP μP; it maps a full image of the DSP μP into the GP μP. This technique makes it possible to access the insides of the DSP μP from only the external pins of the program-memory data-bus, by a sequence of DSP μP program steps (not necessarily simple!) which read the various registers and pop various stacks onto the bus for transfer to the GP μP. The high-speed state machine (b) controls the precise timing of DSP μP bus states to carry out the needed functions with the rapid timing of the DSP μP, which the GP μP alone could not accomplish.

The Learning Curve – Bank-switching from Emulator to Target Memory— The Emulator contains a full complement of program memory (c) which can be mapped into the Target's address space, and provides for bank-by-bank switching of program-memory functions from Emulator to Target, as debugging takes place and Target ROMs are prepared. But *data* memory must be Target-resident, since the Emulator's purpose is to finalize program execution in the Target system. Likewise, Target I/O must be fully implemented so that the Emulator can exercise it.

Simulator Functions Missing From the Emulator – By the nature of self-emulation, some information deep in the DSP μP can not be accessed by the same processor while it is carrying out in real-time Target DSP μP program steps. While programs can be written to dump most internal information onto the bus, a few pieces of information can not be self-accessed without destroying the information, since those very registers are used by the transfer program. By contrast, the Simulator memory maps all internal DSP μP information; a Simulator display is of the mapped images, not the true contents. For example, information about program cache memory is absent from the ADSP-2100 Emulator displays, and the Stack Display Mode cannot get at the loop stack. These are, however, a very small fraction of the DSP μP contents.

Transparent Mode to Access Host Resources – The transparent mode turns the host system into a smart terminal, in effect. The Emulator's terminal is frozen out; the host terminal takes its place and can carry out any of the above functions. In addition, the transparent mode provides for downloading and uploading between Emulator/Target and host. If a bug is found, code can be fixed, reassembled, linked, and downloaded rapidly to the target. Regions of Target/Emulator DM or PM can be uploaded for examination by the host to see if anticipated action occurred. The Emulator is thereby kept relatively modest in scope by sharing a portion of the host's full complement of features.

7.2.8 DSP SYSTEM EXAMPLE: REAL-TIME ISOLATED WORD RECOGNIZER

Introduction: Word-Recognizer System Overview

Speech recognition is a challenging problem in real-time digital signal processing. To illustrate the phases of software development, we will use the example of an isolated spoken-word recognizer implemented with an ADSP-2100. The system and the steps it takes in word recognition are shown in Figure 7.10. It has one input channel, an a/d converter, and will make use of Interrupt I/O. Data is sampled by an independent timer at 10kHz and stored in frames of 144 samples each, or 14.4ms per frame—an interval long enough to contain spectral features, yet not so long that the features change significantly within the window.

A quadrature mirror filter, which detects the energy in a series of frequency bands in like manner to an FFT, performs spectral analysis. A module called the Silence Detector tests each frame in turn, looking for periods of silence which define word endings. The spectral content of all frames in the word forms the fingerprint, which is then compared against words in a feature dictionary, first coarsely (near-matches) and then more precisely on the near-matches.

The recognition algorithm selected is called *dynamic programming*. A distance matrix is computed between features on the amplitude vs. frequency plots for all frames within the encoded utterance (spoken sound) and the template of features of each utterance in the dictionary. The dictionary template which gives the smallest net error in the sum of squares (or "distance") between the features of all frames is the best match for that word. Dynamic programming minimizes the total residual error of the entire utterance rather than that of any individual frame.

To keep the discussion here brief, it is limited to the data-acquisition and spectral-analysis functions, which amply illustrate the stages of system

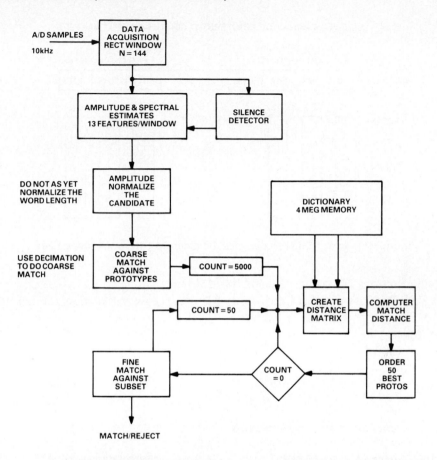

Figure 7.10. Real-time dynamic-programming-based 5K-word discrete utterance recognizer (DUR) using spectral energies and a DSP microprocessor.

configuration and software development. For more details on the recognition process, see Morgan and Leary, 1986.*

The system executive cycles through four phases: (1) data grabbing, including silence detection, (2) spectral analysis, (3) feature extraction, and (4) coarse and fine matching against the dictionary, as shown in Figure 7.10. To avoid software-dependent timing, inputs are interrupt-driven, so this example also illustrates how an interrupt service routine is incorporated.

Data is handled, as shown in Figure 7.11, by a circular array of pointers to an array of input buffers.

The Left Pointer marks the frame currently being filled by A/D inputs. The Right Pointer marks the frame whose spectrum is currently being analyzed. Both step rightwards in time; to keep the recognizer in real time the analysis pointer must stay ahead of the frame-filling pointer.

*D. Morgan and K. Leary, "5-K-Word Discrete Utterance Recognizer in Real Time using ADSP-2100 Digital Signal Processor," *Proceedings of IEEE ICASSP* (1986).

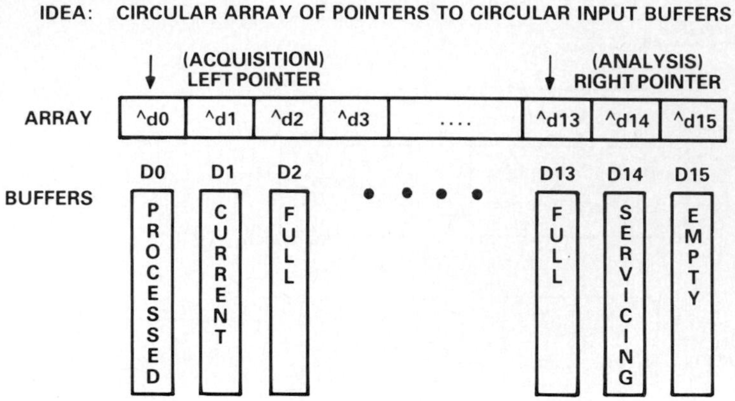

IDEA: CIRCULAR ARRAY OF POINTERS TO CIRCULAR INPUT BUFFERS

Figure 7.11. Input data handling for the discrete-utterance recognizer.

Discrete-Utterance Recognizer System Build Phase
An ADSP-2100 system builder input file for the word recognizer has the
form:*

```
.system DUR_SYSTEM;

.seg/pm/module/rom/abs=0          PMAREA1[4096];
.seg/pm/data/ram/abs=16384        PMAREA2[16384];

.seg/dm/data/ram/abs=0            DMA1[16383];

.port/abs=h#3FFF                  ADCONVERTER;

.endsys;
```

The Discrete Utterance Recognizer (DUR__SYSTEM) will have 4,096 words
of program ROM beginning at PM address 0. Also in the program-memory
map are 8,192 words of RAM beginning at address 16384 for coefficients and
temporary working files. Data-memory allocation allows 16 K of RAM space
for input-signal and output-spectra buffers. Signal inputs from an a/d
converter are mapped to the upper end of the data-memory map.

Defining ADSP-2100 Interrupt I/O
Interrupt I/O is the norm for a DSP μP, because data needs to be sampled
at precisely timed intervals. An external programmable timer is typically

---

*It was noted earlier that upper-case/lower-case don't matter to the assembler program; the use of case is a
convenience to set off user-defined names (labels, variables, data buffers) from keywords. Everywhere else
in this chapter, the convention followed for tabulations relating to ADSP-2100 applications is that used
in the manufacturer's *Cross-Software Manual* and *Applications Handbook*, i.e., keywords are
capitalized, user-defined names are lower case. In this section, however, the use of case is reversed
to agree with that employed by the paper's authors for the photocopied listing in Table 7.4.

employed, which sends out interrupt request signals to the DSP μP at intervals based on a stable quartz crystal. How are these requests sent to the corresponding interrupt service routine?

The ADSP-2100 hardware interrupt structure assigns the lowest four program memory locations permanently to interrupt requests:

Interrupt Level	Vector Address
IRQ0	**0000**
IRQ1	**0001**
IRQ2	**0002**
IRQ3	**0003**

Since the hardware forces a jump to one of these locations when an external device requests an interrupt, the user must place a jump address into these locations pointing to an appropriate interrupt service routine. Unused interrupts should have NOPs (or equivalent) in place of a jump address. The placement of jump addresses is accomplished in user software with the appropriate JUMP instructions as the *first four* instructions (immediately following the declaration and initialization sections) in the *main program*. For example, referring to Table 7.4 in the vicinity labeled *Start of Program Area*, the (single) interrupt location for this example is designated by the INT statements:

```
.init       input: ^d1, ^d2, ^d3, ^d4, ^d5,
              ^d6;
.init       delay_list: quad1, ^d39, quad2, ^d40, quad2, ^d41,
              quad3, ^d42, quad3, ^d43, quad3, ^d44, quad3,
              ^d45;

{* * * * * * * * Start of Program Area * * * * * * * * * * * * *}

INT0:       jump SERVICE_AD;
INT1:       nop;
INT2:       nop;
INT3:       nop;
```

The symbolic labels INT0:, INT1:, etc., are optional but useful as reminders of the function of these jump commands and the need to fill unused interrupt locations with NOPs, not other code. The interrupt service routine, SERVICE__AD, appears later in the source code of Table 7.4.

```
        SERVICE__AD   ena alt__reg;   {INT0 has highest priority}

    etc.
```

When the source program is assembled and linked, the absolute address of the SERVICE__AD routine will replace the symbolic reference in the INT0: JUMP command, and the JUMP commands will be the contents of pm(0000, 0001, 0002, 0003), ensuring that a hardware interrupt occurrence will be serviced correctly:

> **[IRQ0 line asserted by AD timer]** →
> **[hardware JUMP to PM 0000]** → **[contents = JUMP service__ad]** →
> **[SERVICE__AD module entered]**

No problem with a hardware RESET occurs by this assignment, since the RESET is wired to force a restart of the program at location 0004, the first line in the main program after the interrupt jumps.

Data Acquisition and Spectral Energy Computation for the DUR
Program Overview: The program in Table 7.4 is the main calling program of the word recognizer. The portions included are for data acquisition and spectral energy determination. The data acquisition portion illustrates interrupt I/O: Note how the interrupt jumps are defined (see Start of Program Area) and how the AD interrupt is serviced (SERVICE__AD). The mirror filter is a rapid way of determining spectral energy; it resembles the FFT in its decimation operations. A 3-stage version is shown here for brevity; the actual program (Morgan and Leary, 1985) uses a four-stage version.

Detailed programming manuals would of course have to be consulted for a full understanding of Table 7.4, which is included here to illustrate the compact (hence fast) program structure possible with a modern DSP μP. Multifunction instructions are used extensively to increase throughput. For example, in the quad filter section, the instruction

```
EVEN_TAPS   mr=mr+mx0*my0 (ss), mx0=dm(i0,m2), my0=pm(i7,m7);
```

not only performs a single-cycle multiply-accumulate step with data previously put into the multiplier input registers, mx0 and my0; in the same clock cycle it also loads mx0 and my0 registers with values from the next data memory and program memory locations in preparation for the next MAC step.

Because of this compactness, system performance is quite spectacular. The three phases of data grabbing, word filling, and word-match decision making are cycled round, each in about 1/4 ms. Since an average word occupies 40 frames, or about 0.6 seconds, the entire word-gathering and processing must be completed in less than this time for the operation to be real-time.

Table 7.4 Selected modules from the Discrete Utterance Recognition and Spectral Energy Determination program. This program takes data from an ADC and processes into frames of 144 points. Each frame is then represented by a log-spectral and an rms value. The input data is left justified. (courtesy David Morgan)

```
{ *************************** Declarations *************************** }

.include            <durlib.h>;
{ defines:          samples=144; one_ov_samps=h#E4; features=9; }
{                   buff_len=6; quad1=64, quad2=32, quad3=16, }

.external           QUAD_FILTER;
.external           INIT_DELAY_LINE;

.port               A_D_Converter;

.var/pm/ram         storage[1024];
.var/dm/ram         d1[samples], d2[samples], d3[samples];
.var/dm/ram         d4[samples], d5[samples], d6[samples];
.var/dm/ram/circ    d39[quad1];        .var/dm/ram/circ      d40[quad2];
.var/dm/ram/circ    d41[quad2];        .var/dm/ram/circ      d42[quad3];
.var/dm/ram/circ    d43[quad3];        .var/dm/ram/circ      d44[quad3];
.var/dm/ram/circ    d45[quad3];

{ globals }
.var/dm/ram         ri_pt, li_pt, box_ear, raw_data;
.var/dm/ram/circ    input[buff_len];

{ quad }
.var/dm/ram/circ    delay_list[14];

.global             raw_data, box_ear, delay_list;

.init               box_ear : ^storage;
.init               ri_pt : ^input;
.init               li_pt : ^input;
.init               input : ^d1, ^d2, ^d3, ^d4, ^d5, ^d6;
.init               delay_list : quad1, ^d39, quad2, ^d40, quad2, ^d41,
                        quad3, ^d42, quad3, ^d43, quad3, ^d44, quad3, ^d45;

{ ********************** Start of Program Area ********************** }

INT0:               jump SERVICE_A_D;
INT1:               nop;
INT2:               nop;
INT3:               nop;

EXECUTIVE:          imask = 0;        { disable ints }
                    call INIT;
                    icntl = h#01;     { int 0 is edge trigerred }
                    imask = 1;        { enable int 0 }

MAIN:               ax0 = dm(ri_pt);              { ri inc after area full }
DATA_HANDLER:       ay0 = dm(li_pt);
                    ar = ax0 - ay0;
                    if eq jump MAIN;
NEW_FRAME:            i0 = dm(li_pt); l0 = buff_len;
                      ax0 = dm(i0,m1);         { m1 should always be 1 here }
```

```
                              call QUAD_FILTER;        { raw_data and box_ear are ptrs}
                    {         call REST_OF_PROGRAM; }
                              jump MAIN;

INIT:                         m0 = 0;   m1 = 1;   m2 = -1;
                              m4 = 0;   m5 = 1;   m6 = -1;
                              px = 0;
                              ena alt_reg;
                              i3 = ^dl;          i3 = samples;
                              ay0 = samples;              { init backrnd reg to samples }
                              af = pass ay0;
                              dis alt_reg;
                              call INIT_DELAY_LINE;
                              rts;

{ ***************** Interrupt Serivce Routines ********************** }

SERVICE_A_D:                  ena alt_reg;              { INT0 has highest priority }
                              ay1 = imask;
                              imask = 0;                { disable all ints }
                              ax0 = dm(A_D_Converter);
                              dm(i3,m1) = ax0;          { using alternate register set }
                              af = af - 1;              { alt af is never disturbed }
                              if ne jump RESTORE;       { i3 is circ, it wraps around }
                               af = pass ay0;           { ay0 holds samples }
                               axl = m1;                { save m1 }
                               m1 = 1;  i3 = buff_len;
                               i3 = dm(ri_pt); ar = dm(i3,m1); dm(ri_pt) = i3;
                               ar = dm(i3,m1);          { we want next buffer area }
                               i3 = ar; i3 = samples;
                               m1 = axl;                { restore m3 }
RESTORE:                      imask = ay1;              { restore imask }
                              dis alt_reg;              { affects mstat reg }
                              rti;                      { which pops sts }
.endmod;
```


```
.module/rom            MIRROR_FILT;
{*********************************************************************}
{       This program finds the amplitude and spectral energies for 12  }
{ spectral regions in the candidate feature vector using using a       }
{ quadrature mirror filter.  This filter is much like a binary tree.   }
{ Here the orthoganality of the tree structure and the decimation is   }
{ taken full advantage of.  By obtaining the 8 energy estimates this   }
{ way, rather in the naive manner, a factor of eight saving is         }
{ obtained.  The problem that remains is to design symmetric filters   }
{ for the regions of interest.                                         }
{*********************************************************************}

.include               <durlib.h>;
.include               <rms.dsp>;      { which calls sqrt }
.include               <sqrt.dsp>;
```

```
                             m0 = 0;   m4 = 0;   m2 = -1;

SPEC_ENERGY:                 my1 = h#7FFF;               { ~ 1 in s.15 }
                             si = 17;   il = dm(raw_data);   cntr = 8;
                             mx1 = dm(il,m1);          { get 1st data point }
                             call SUMMATION;
                             dm(box_ear) = 15;         { store pos in feature vector }
                             rts;


{ *****************************  Subroutine Summation  ******************** }

SUMMATION:                   do ENERGY until ce;
                               mr = 0;
                               cntr = si;
                               do SUM_SPEC until ce;
SUM_SPEC:                        mr = mr + mx1 * my1 (ss), mx1 = dm(il,m1);
                               mr = mr + mx1 * my1 (ss);
                               call LOGMR;   { returns value in 4.28 format }
ENERGY:                        pm(15,m5) = srl;       { store energy est }
                             rts;


{ *****************************  Subroutine Quad  ************************** }

QUAD:                        i4 = i2;                  { i2 is the high output }
                             modify(i4,m4);            { i4 is the low output }
                             ay0 = dm(i6,m5);          { get place in delay line }
                             i0 = ay0;                 { input delay line --> i0 }
                             modify(i0,m2);            { modify delay line by 2 }
                             do FILTERING until ce;    { 2 points --> delay line }

                               mr = 0,  mx0=dm(i0,m2), my0=pm(i7,m7);
                               cntr = si;                      { s.15 input * s.15 }
                               do EVEN_TAPS until ce;
EVEN_TAPS:                       mr=mr+mx0*my0 (ss), mx0=dm(i0,m2), my0=pm(i7,m7);
                                 mr=mr+mx0*my0 (ss),  mx1=dm(i1,m1), my0=pm(i7,m7);
                                 mr=mr+mx1*my0 (RND),  dm(i0,m3) = mx1;
                                 if mv sat mr;         { if product > 32 bits sat }
                                 af = pass mrl, my0 = pm(i7,m5);     { inc i7 }

                               mr = 0,  mx0=dm(i0,m2), my0=pm(i7,m7);
                               cntr = si;
                               do ODD_TAPS until ce;
ODD_TAPS:                        mr=mr+mx0*my0 (ss), mx0=dm(i0,m2), my0=pm(i7,m7);
                                 mr=mr+mx0*my0 (ss),  mx1=dm(i1,m1), my0=pm(i7,m7);
                                 mr=mr+mx1*my0 (RND),  dm(i0,m3) = mx1;
                                 if mv sat mr;         { if product > 32 bits sat }
                                 ar = mrl - af, my0 = pm(i7,m6);     { dec i7 }

                               dm(i2,m1) = ar, ar = mrl + af;        {store high (-)}
FILTERING:                     dm(i4,m5) = ar;        { decimation is done w/in loop}
                             m2 = -2;
                             modify(i0,m2);            { i0 is circ, have to modify }
                             m2 = 2;                   { reset m2 }
                             ar = i0;
                             modify(i6,m6);            { store location left off in }
                             dm(i6,m7) = ar;           { delay line in array list }
                             rts;

.endmod;
```

```
    .external                LOGMR;              { log base 10 of mr reg }

    .external                delay_list, raw_data, box_ear;

    .var/dm/ram/circ         data_ping[samples];
    .var/pm/ram/circ         h_coef[quad1];  { only declare 1 circ def per line }
    .var/pm/ram/circ         g_coef[quad2];
    .var/pm/ram/circ         f_coef[quad3];

    .init   h_coef : <coef1.dat>;           .init   g_coef : <coef2.dat>;
    .init   f_coef : <coef3.dat>;


INIT_DELAY_LINE:            i6 = ^delay_list;  l6 = 14;      { 402 cycles }
                            l0 = 0;
                            ax1 = dm(i6,m5);        { get delay length }
                            cntr = 7;
                            do SET_UP until ce;
                              ay0 = dm(i6,m5);      { ^d_line --> i0 }
                              i0 = ay0;
                              cntr = ax1;
                              do CLEAR_LINE until ce;
CLEAR_LINE:                   dm(i0,m1) = 0;
SET_UP:                       ax1 = dm(i6,m5);      { get delay length }
                            rts;


QUAD_FILTER:                i1 = dm(raw_data);    l1 = samples;
                                  m2 = 2;
                                  m3 = 3;
                            i5 = dm(box_ear);     l5 = 0;
                            i6 = ^delay_list;     l6 = 14;
                                  m7 = 2;
                            modify(i6,m5);              { skip over length to ^d_line }


AMPLITUDE:                  RMS_VALUE ( samples, i1, one_ov_samps );
                            mr1 = sr1;  mr0 = sr0;  { output is a s.15 in sr1,sr0 }
                            call LOGMR;             { find log base 10 of amp }
                            pm(i5,m5) = sr1;        { store log amplitude }


BEGIN_FILTER:               l2 = samples;  l4 = samples;


STAGE_1:                    i1 = dm(raw_data);      { i1 : input area }
                            i2 = ^data_ping;        { i2 : output area }
                            i7 = ^h_coef;   l7 = 64;        l0 = 64;    si = 30;
                            m0=72;  m4=72;   cntr = 72;      call QUAD;


STAGE_2:                    i1 = ^data_ping;
                            i2 = dm(raw_data);
                            i7 = ^g_coef;   l7 = 32;        l0 = 32;    si = 14;
                            m0=36;  m4=36;   cntr = 36;      call QUAD;
                            modify(i2,m0);   cntr = 36;      call QUAD;


STAGE_3:                    i1 = dm(raw_data);
                            i2 = ^data_ping;
                            i7 = ^f_coef;   l7 = 16;        l0 = 16;    si = 6;
                            m0=18;  m4=18;   cntr = 18;      call QUAD;
                            modify(i2,m0);   cntr = 18;      call QUAD;
```

FULL WORD PROCESSING TIME (milliseconds)

INT sampling overhead	8.6
INT New word	16.4
Silence detector	4.0
Quad mirror filter	67.0
Amplitude normalization	0.2
SUBTOTAL Acquire and spectrum	96.2
RECOGNITION SEARCH	
Coarse match	297.0
Feature download	424.0
Fine match	72.0
Feature download	3.0
TOTAL	510.2

The recognition is complete within the time it takes to say another word. This performance is quite spectacular considering the large (5,000-word) dictionary size.

7.3 MICROCODED SYSTEM SOFTWARE DEVELOPMENT

The first half of this chapter dealt with DSP μP software development—designed for DSP, but with a fixed architecture. Now we continue with a quite different challenge—software development for microcoded systems whose architecture is configured by the user.

7.3.1 DIFFERENCES BETWEEN GP AND WORD-SLICE SOFTWARE DEVELOPMENT

The steps in Bit-, Byte-, and Word-Slice microcoded software development follow the principal phases of GP μP or DSP μP software development, with significant differences due to the variety and flexible choice of hardware that might be used.

The assembly phase is replaced by a *meta-assembler*; its preliminary definition phase allows the user to define mnemonics which, in assembly, translate to microcode bit-strings for the specific hardware complement. Meta-assemblers are becoming readily available commercially and will be preferred by most users.

The link-load phase is similar to that of the DSP μP.

Software *simulation* on a host system—and hardware *emulation* boards—are less likely to be encountered in microcoded systems, since the architecture configuration is target-specific, while for a GP μP or DSP μP the architecture is fixed and simulation/emulation tools are therefore cost-effective to develop. The labor required to write a Simulator program which specifies the architecture, interconnects, operations, and timing sufficient to realistically model a target system operation is prohibitive if it has to be repeated for each new target hardware configuration. Likewise, a hardware debug which takes control of the system buses is unlikely since the hardware configuration changes with each microcoded system; development and debugging of a new debug board for each case would be prohibitively expensive in most cases.

A memory-stepping debug is the norm: the Target microcode PROM is unplugged and replaced by a plug connected to the development system, which exercises the microprogram from a RAM image in the host. While stepping through a microprogram can certainly speed software development, access is limited to information which appears on the microcode instruction bus in normal target program execution. A full map of target registers, for example, is not obtained.

7.3.2 META ASSEMBLERS: FROM USER-DEFINED MNEMONICS TO MICROCODE BITS

What is a Meta-Assembler?

The Meta Assembler allow the user to define mnemonics for each microcode word. Skordalakis (1983) gives an excellent introduction, plus a comparison table illustrating key figures-of-merit to judge in selecting a commercially available Meta Assembler. The Meta Assembler translates the mnemonics into the corresponding string of 1's and 0's. Virtually all microcode system designers will employ a Meta Assembler if possible, since mnemonics can be remembered much more easily than binary strings.

The typical Meta Assembler operates in two stages, Definition and Assembly. In the Definition phase, the user defines instruction mnemonics to be associated with the binary microinstructions. The microinstructions are broken up into fields (associated with different functions within a chip, or to independent chips, in the case of a fully microcoded system). The fields are filled with variables, constants, or "don't care" bits. The variables, which are also defined in this first phase, are the possible operations corresponding to the appropriate bit field, and are filled at execution time by the selected operation in the source code. The Definition phase produces a file which the Assembler refers to when assembling the user's program.

The Assembly phase resembles a conventional fixed-architecture Assembler except that most of the mnemonics are not fixed, but are looked for in the definition file. Meta Assemblers may include directives which add power, such as MACROs and Conditional Assembly. MACRO capability permits frequently used instruction sequences to be inserted at assembly time wherever the given MACRO label is encountered in the code. The MACRO definition includes variable names for parameters which may be passed during execution, just as in a high-level language subroutine call. Conditional assembly permits special-purpose programs to be put together from a generic package of subroutines, merely by specifying a few constants at the start of the code. The constants are referenced in the conditional assembly statements, ASMIF and ENDASM, which surround modules that may or may not be needed in a given application program:

```
{assembly constant declarations – this application}
.CONST          CASE1 = 0;
                CASE2 = 1;
                CASE3 = 1;
                CASE4 = 0;
{conditional section}
.ASMIF (CASE1);
    {follow with code section of macro}
.ENDASM;

.ASMIF (CASE2) AND (CASE3);
    {follow with code section or macro}
.ENDASM;
```

Simple Circuit Examples: From Truth Table to Microinstruction
To understand the Definition phase of a Meta Assembler, we begin with the example of converting the truth tables of two simple digital operations, a 2-1 multiplexer and a 1-bit latch, into higher-level instruction mnemonics. The circuits are shown in Figure 7.12.

Multiplexer

Truth Table:	SEL	OUT
	0	IN1
	1	IN2

Meta Assembler Instruction Definition:

Instruction		Opcode
MUX: OUT = IN1		DEF B#0
OUT = IN2		DEF B#1

Here, the output (OUT) equals either input 1 (IN1) or input 2 (IN2), depending upon whether the select (SEL) input is 0 or 1. The corresponding instructions, OUT1 = IN1 and OUT = IN2, in the group labeled MUX: are defined by the Meta Assembler using the DEF statement common to any assembler: whenever the mnemonic OUT = IN1 is found in a program, that line will be replaced by a binary 0 on the single microcode control line attached to the multiplexer enable input.

Similarly, a latch followed by a tri-state output would have the truth table and instruction set:

Latch plus tri-state output

Truth Table:	EN	TRI	Q	OUT
	X	1		Q
	1	X	D	

Here, enabling the tri-state passes latch output Q, while enabling the latch stores input D at flip-flop output Q. The X symbol is "don't care." This translates to microcode instructions:

Meta Assembler Instruction Definition:

	Instruction	Op Code
LATCH:	Q = D	DEF B#X1
	OUT = Q	DEF B#1X

Here, in the assembler output microcode, the rightmost bit is connected to the tri-state enable line and the leftmost bit is connected to the flip-flop enable line. The presence of the X ("don't care") indicates that the hardware has more than one function (both a latch and a tri-state gate). Microcode bit patterns will be OR'd together to completely specify hardware operation and eliminate the "don't care" bits by the time final code is generated.

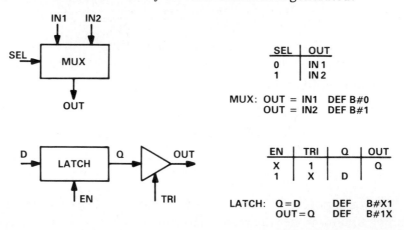

Figure 7.12. Micro-instructions of simple circuits.

Meta-Assembler Code Generation for Word-Slice DSP Chips
This pattern of mnemonics defining microcode is readily extended to more-complex chips and then to chips operating together as a microcoded system. For example, the set of microcode instructions for a single-port multiplier-

accumulator, the ADSP-1110A, is shown in Table 7.5. The microcode instruction field has 6 bits. Upon examination of the table, one can readily see that bit 5 signals a multiply or MAC operation when it equals 1, and a non-multiply operation, such as input/output, when 0. Within the Output instruction group, bit 0 can be seen to enable a left shift (sl) on the output when 0, no shift when 1. Bit 1 signals saturation arithmetic when 1 (sat), etc. Within the Multiply operation group, bits 3 and 4, when 1, specify that X and Y, respectively, are to be interpreted as twos complement (X_{TC}, Y_{TC}), etc.

The identifications in Table 7.5 form the basis for the Meta-Assembler instruction definitions shown in Table 7.6. Consider the Output Instructions (lines 36-39):

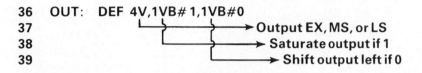

```
36   OUT:   DEF 4V,1VB# 1,1VB#0
37                              ──────► Output EX, MS, or LS
38                              ──────► Saturate output if 1
39                              ──────► Shift output left if 0
```

Bit groups with different functions are separated into fields by commas. The letter V specifies a variable of width specified by the number to the left: 4V is a 4-bit variable. The choice of appropriate variables will be specified later in the user's source code. The variables are also defined in this phase. The leftmost 4 bits in the OUT instruction signify which portion of the 40-bit output is put into the 16-bit bus. The choices, MS, LS, and EX, are defined in the table (lines 25-27). For example:

$$\text{MS:} \quad \text{EQU} \quad \text{B\#0110}$$

The most-significant (MS) half of the multiplier output is among the possible variables for the OUT operation, and the value 0110 clearly corresponds to the MS entry in the microcode, Table 7.5.

Fixed-value bits are specified as B#0 and B#1 for fixed 0 and 1. But a V to the left of a B# specifies that this is also a variable, and the B#x specifies whether a 1 or 0 is "TRUE." For example, 1VB#0 is a 1-bit variable which is TRUE when its value is 0. Note the correspondence with this variable in the rightmost field and the microcode specification of Table 7.5: a left-shift is specified by a 0 in the rightmost bit.

Similar explanations apply to the multiply group of Meta Assembler instructions. For example,

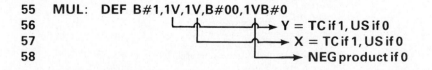

```
55   MUL:   DEF B#1,1V,1V,B#00,1VB#0
56                          ────► Y = TC if 1, US if 0
57                          ────► X = TC if 1, US if 0
58                          ────► NEG product if 0
```

Table 7.5 ADSP-1110A Instruction Set.

Instruction Group	Instruction			Microcode Instruction 5 4 3 2 1 0	Comments
Miscellaneous	NOP			0 0 0 0 x x	No Operation
	CKMR			0 0 0 1 x x	Clock MR
Input	X = BUS			0 0 1 0 x x	
Preload	LS = BUS			0 1 0 0 0 0	
	MS = BUS			0 1 0 1 x 0	
	EX = BUS			0 1 0 0 1 0	
Transfer	LS = MS			0 1 0 0 0 1	Sets SLE register
	MS = EX			0 1 0 1 0 1	
Sign Extend	EX = SIGN EXT MS			0 1 0 0 1 1	
	MS = SIGN EXT LS			0 1 0 1 1 1	
Output	BUS = EX			0 0 1 1 0 1	All output instructions are asynchronous
	BUS = EX (sl)			0 0 1 1 0 0	I5 – I2:
	BUS = MS			0 1 1 0 0 1	
	BUS = MS (sl)			0 1 1 0 0 0	0011 = EX
	BUS = MS (sat)			0 1 1 0 1 1	0110 = MS
	BUS = MS (sl,sat)			0 1 1 0 1 0	0111 = LS
	BUS = LS			0 1 1 1 0 1	I1 – I0:
	BUS = LS (sl)			0 1 1 1 0 0	01 = to bus
	BUS = LS (sat)			0 1 1 1 1 1	00 = to bus shifted
	BUS = LS (sl,sat)			0 1 1 1 1 0	10 = to bus shifted w/saturation
					11 = to bus w/saturation
Multi-Operation	Y = BUS; CKMR;	$X_{US}*Y_{US}$		1 0 0 x 0 0	Require two cycles to complete.
	Y = BUS; CKMR;	$-X_{US}*Y_{US}$		1 0 0 x 0 1	Other instructions can be executed
	Y = BUS; CKMR;	$X_{US}*Y_{US}$	+MR	1 0 0 0 1 0	on the second cycle.
	Y = BUS; CKMR;	$-X_{US}*Y_{US}$	+MR	1 0 0 0 1 1	
	Y = BUS; CKMR;	$X_{US}*Y_{US}$	−MR	1 0 0 1 1 0	I5 = Multiplication/MAC operation
	Y = BUS; CKMR;	$-X_{US}*Y_{US}$	−MR	1 0 0 1 1 1	I4 = Y twos complement
	Y = BUS; CKMR;	$X_{TC}*Y_{US}$		1 0 1 x 0 0	I3 = X twos complement
	Y = BUS; CKMR;	$-X_{TC}*Y_{US}$		1 0 1 x 0 1	I2 = Subtract previous result
	Y = BUS; CKMR;	$X_{TC}*Y_{US}$	+MR	1 0 1 0 1 0	I1 = Add/subtract previous result
	Y = BUS; CKMR;	$-X_{TC}*Y_{US}$	+MR	1 0 1 0 1 1	from product
	Y = BUS; CKMR;	$X_{TC}*Y_{US}$	−MR	1 0 1 1 1 0	I0 = Negate product
	Y = BUS; CKMR;	$-X_{TC}*Y_{US}$	−MR	1 0 1 1 1 1	
	Y = BUS; CKMR;	$X_{US}*Y_{TC}$		1 1 0 x 0 0	
	Y = BUS; CKMR;	$-X_{US}*Y_{TC}$		1 1 0 x 0 1	
	Y = BUS; CKMR;	$X_{US}*Y_{TC}$	+MR	1 1 0 0 1 0	
	Y = BUS; CKMR;	$-X_{US}*Y_{TC}$	+MR	1 1 0 0 1 1	
	Y = BUS; CKMR;	$X_{US}*Y_{TC}$	−MR	1 1 0 1 1 0	
	Y = BUS; CKMR;	$-X_{US}*Y_{TC}$	−MR	1 1 0 1 1 1	
	Y = BUS; CKMR;	$X_{TC}*Y_{TC}$		1 1 1 x 0 0	
	Y = BUS; CKMR;	$-X_{TC}*Y_{TC}$		1 1 1 x 0 1	
	Y = BUS; CKMR;	$X_{TC}*Y_{TC}$	+MR	1 1 1 0 1 0	
	Y = BUS; CKMR;	$-X_{TC}*Y_{TC}$	+MR	1 1 1 0 1 1	
	Y = BUS; CKMR;	$X_{TC}*Y_{TC}$	−MR	1 1 1 1 1 0	
	Y = BUS; CKMR;	$-X_{TC}*Y_{TC}$	−MR	1 1 1 1 1 1	

Mnemonic Definitions

=	Assign right side to left.
BUS	16-bit external data bus used for all I/O operations.
X	Input register for multiplier.
Y	Input register for multiplier.
EX	8-bit extension register for accumulator.
MS	16-bit most significant product register.
LS	16-bit least significant product register.
MR	40-bit accumulator comprising EX, MS and LS.
sl	Shift left.
sat	Conditional on overflow, saturate the outputted value.
TC	Two's complement number.
US	Unsigned magnitude number.
SIGN	Sign bit (MSB) of specified register.
CKMR	Clock product into EX, MS and LS.
*	Multiply
x	Microcode instruction bit can be either a 0 or 1.

Table 7.6 Instruction-definition source code for the ADSP-1110A Single-Port MAC.

```
LINE

   1    TITLE  ADSP-1110 INSTR DEF SOURCE CODE: 2/23/85,B
   2    LIST   X                       ; list cross ref table
   3    ;
   4    WORD   6                        ; set ucode word size
   5    ;
   6    ;
   7    ; Instructions Without Arguments
   8    ;
   9    NOP:    DEF     B#000000        ; no operation
  10    CKMR:   DEF     B#000100        ; clock the memory register
  11    LDX:    DEF     B#001000        ; load x register
  12    PEX:    DEF     B#010010        ; preload EX register
  13    PMS:    DEF     B#010100        ; preload MS register
  14    PLS:    DEF     B#010000        ; preload LS register
  15    TEX:    DEF     B#010101        ; transfer EX to MS
  16    TMS:    DEF     B#010001        ; transfer MS to LS (sets SLE register)
  17    SMS:    DEF     B#010011        ; sign extend MS into EX
  18    SLS:    DEF     B#010111        ; sign extend LS into MS
  19    ;
  20    ;
  21    ; Output Instruction
  22    ;
  23    ; accumulator registers...
  24    ;
  25    EX:     EQU     B#0011          ; extension
  26    MS:     EQU     B#0110          ; most significant
  27    LS:     EQU     B#0111          ; least significant
  28    ;
  29    ;
  30    ; left-shift and saturate switches
  31    ;
  32    SHL:    EQU     B#0             ; uses negative logic
  33    SAT:    EQU     B#1             ; uses positive logic
  34    ;
  35    ;
  36    OUT:    DEF     4V,1VB#1,1VB#0
  37                                    ; 1st field = EX, MS, or LS
  38                                    ; 2nd field = SAT (default is no SAT)
  39                                    ; 3rd field = SHL (default is no SHL)
  40    ;
  41    INVALID EX,1X,SAT               ; no SAT allowed on EX
  42    ;
  43    ;
  44    ; Multi-Operation Instructions
  45    ;
  46    ; arguments...
  47    ;
  48    TC:     EQU     B#1             ; twos complement notation
  49    US:     EQU     B#0             ; unsigned notation
  50    SUB:    EQU     B#1             ; subtract previous result
  51    NEG:    EQU     B#1             ; negate product
  52    ;
  53    ; multiply without accumulate
  54    ;
  55    MUL:    DEF     B#1,1V,1V,B#00,1VB#0
  56                                    ; 1st var = TC or US for Y input
  57                                    ; 2nd var = ditto for X
  58                                    ; 3rd var = NEG prod (default is NO NEG)
  59    ;
  60    ; multiply/accumulate
  61    ;
  62    MAC:    DEF     B#1,1V,1V,1VB#0,B#1,1VB#0
  63                                    ; 1st & 2nd var = same as for MUL
  64                                    ; 3rd var = SUB accum from prod...
  65                                    ; ...(default is ADD TO)
  66                                    ; 4th var = NEG prod (default is NO NEG)
  67    ;
  68    END
```

The correspondence to the multiply group in the microcode can be seen: leftmost bit = 1 (B#1 above) identifies a multiply operation, the next two bits specify Y and X formats, etc.

The actual assembly phase follows once this Definition file exists. The Assembler syntax supports the appropriate set of constructs, for example

OUT   MS

would be translated to the appropriate source code by looking up the definitions of operation OUT (line 36) and variable MS (line 26).

Code Generation for a Multi-Chip Microcoded System
Meta-Assembler mnemonic generation is thus quite clear for any given existing chip; it is not significantly different from the process of extending an Assembler to incorporate new functions. For example, when the 8080 was upgraded to the 8085, the few new 8085 instructions could be incorporated into an existing 8080 assembler with DEF statements like the above. With fixed chips, it is not unreasonable for the vendors of microcoded chips to supply the microcode definitions as part of the package, since the definitions are, if anything, much easier to read than the bit-fielded microcode truth table.

The Meta Assembler really shows its value in putting together a set of Bit/Byte/Word-Slice chips. The process is an extension of concepts described above, with a microcode bit-field wide enough to accomodate all of the instruction fields of all chips in the system, including glue chips. For example, the ultra-high-speed microcoded FIR filter of Chapter 6 (Figure 6.23) incorporates a Word-Slice Sequencer, three Address Generators, and a MAC chip. Each has its set of mnemonics and corresponding microcode. The filter sequence:

SEQUENCER	ADDR GEN:COEFFS	ADDR GEN: X	ADDR GEN: Y	MAC
LDCOUNT	RDCOEFF	RDX	WRY	OUT
LOOP	RDCOEFF	RDX	NOP	MUL
BACK	NOP	WRX	NOP	MAC

is translated by stringing together the appropriate mnemonics into a "super mnemonic," such as:

field:	SEQ	AGC	AGX	AGY	MAC
contents:	LDCOUNT	RDCOEFF	RDX	WRY	OUT

With 10 or so instruction bits per chip, it is clear why wide address fields occur in microcoded systems. The Meta Assembler must have the capability to handle an adequate microcode field width.

Issues in Choosing a Meta Assembler

This brief discussion of software development for microcoded systems concludes with a few figures of merit in choosing a Meta Assembler. The user must carefully choose a Meta Assembler adequate to deal with the scope of intended applications. Issues in choosing a Meta Assembler include:

How wide a microinstruction bit-field is supported? The wider the better, for multi-chip systems; 100 bits is not unlikely.

Are Macros supported?

Is a PROM formatter available to split code into modules of the correct size to fit in PROM chips?

Is the Meta Assembler portable? Is it written in a standard high level language? Which mainframes will it run on?

How are variable length microinstructions dealt with? How are micro-instruction overlays performed, to eliminate the "don't care" bits from multi-function chips?

How are symbols declared? Globals, locals, etc? How are parameters passed?

How large is the Meta Assembler program? How many passes?

Which Assembler addressing modes are supported: symbolic, relative, paged, etc.

Is Assembler output absolute or relative? Is a Linker/Loader available for making modular subroutines?

Is a debug facility provided? High-quality diagnostics?

Chapter Eight

DSP Applications

8.1 INTRODUCTION

Chapter Overview

This chapter is an applications survey, intended to encourage applications in the reader's own field. The coverage is limited to a few selected examples where DSP has had an impressive impact, since thorough coverage of the depth, breadth, and variety of DSP applications can fill a bookcase. Frequently experienced DSP algorithms or processes are illustrated by examples in selected application areas in the sections that follow, as summarized in Table 8.1. Program-module examples can be found in detail in manufacturers' literature—for example, the *ADSP-2100 Applications Handbook* series from Analog Devices.

The grouping of a given process with a given example is to some extent arbitrary, and there are many overlaps; for instance, digital filtering and spectral analysis appear in many applications. One-dimensional signal filtering and detection involve the same set of computations as two-dimensional image enhancement. Indeed, much is to be gained by cross-fertilization between areas. Advances in 2-D image processing are useful in 1-D audio and speech processing—and vice versa. For example, the concepts of feature-extraction and modelling used in machine-reading characters in reading machines for the blind have been adapted to modelling musical instruments with high information compression and very high fidelity.*

---

*Reading machine: Kurzweil Computer Products, Cambridge, MA 02139. Music synthesizer keyboard: Kurzweil Music Systems, Waltham MA 02154.

Table 8.1. Key DSP processes and typical application examples.

DSP Process	Application Example
Digital filtering	Ubiquitous: HP, LP, BP, comb
Signal averaging	Signal averager, digital oscilloscope
Coherent detection and correlation	Lock-in, Noise testing
Spectral analysis and estimation	Sunspots, pulsars, radar, audio... Extra-terrestrial-intelligence search
FFT, Decimation	Spectrum-analyzer front end, frequency zoom
Interpolation	Signal generation with ultra-high resolution
Data compression and modelling	Speech synthesis, telecommunications
Equation solving	Function generation
Signal generation	Stimulus generation, Music synthesis
2-D Image enhancement	Medical ultrasound, radiography; geological
2-3-D Array operations	Reconstructive tomography

Most of the real-world-data examples would not have been practical in real time prior to the availability of digital signal processing in VLSI. Available space and constraints on our mission limit coverage here of areas which lack the real-time aspect or have little involvement with real-world I/O. While image enhancement is covered, graphics generation is not. While array operations are included, array processors and their mainframe applications are not, since relatively few applications have a real-time component. Nevertheless, key DSP computation primitives and many DSP components find their way into these applications.

These DSP primitives are analogous to the columns that support a structure. The actual computations carried out for the common DSP processes listed will be reviewed in Section 8.2. Table 8.2 lists other applications; some applications that do not fit neatly in a single category are listed as *integrative*: a system whose net capability synthesizes parts from several DSP domains. Acting independently, functions such as graphics generation and sound generation or recognition are combined in a real-world feedback loop, which often includes a human operator:

- The blind reader puts the textbook under an image-recognizing scanner and listens to the verbalized text.
- The pilot, screened from background noise by echo-cancelling earphones, is assisted by verbal and visual clues processed from multiple data inputs and can view on the windshield a computer-graphic map of the scene.

- The translating telephone allows a business transaction in Japanese in Tokyo to be reported in English in New York.

None of these examples is far-fetched. The text-to-speech reader is commercially available, the "pilot's assistant is a DOD-ARPA defined VLSI goal, and the translating telephone is under development in Japan.

The Chapter covers three main operations on signals: detection, generation (or synthesis), and signal analysis. The review of DSP *primitives* in Section 8.2 discusses the major elements common to many—if not most—DSP systems, using the digital transceiver as a prototype. Common issues in signal detection and key detection methods are critically compared here: correlation, coherent detection, spectral analysis, and spectral estimation, along with the use of decimation in frequency-band isolation and frequency zoom.

Section 8.3 illustrates real-time measurement instrumentation, with (a) examples based on correlation principles, (b) matched-filter examples relying on some prior knowledge of the underlying signal, (c) lock-in or coherent detection techniques, (d) a spectrum analyzer, and (e) spectrum estimation. In-depth application examples given are the search for gravity waves with a matched filter and the search for extra-terrestrial intelligence with an ultra-high-resolution spectrum analyzer. Section 8.4 contrasts methods for the analysis, storage, transmission, and regeneration of signals by (a) modelling the source and then manipulating only the source parameters, and (b) *source coding* the actual signals and then compressing the data. Examples are given from speech synthesis and telecommunications.

Closely related Section 8.5 covers real-time function generation, showing how DSP facilitates fast function computation, with examples from digital electronic music. Section 8.6 focuses on the processing of two-dimensional array information to generate new information, with examples from image enhancement, computer vision, and image reconstruction, e.g., the CAT scanner. In each section, the examples (which would have been difficult or impossible prior to DSP) are close to one or more DSP fundamentals, are of wide appeal, and will lead to an appreciation of design issues: the requirements, the implementation alternatives, and how to design a selected solution with DSP hardware and software.

8.2 MAJOR ELEMENTS OF A DSP SYSTEM

8.2.1 THE DIGITAL TRANSCEIVER

System Elements

The elements of a digital transceiver system are shown in Figure 8.1. The digital transceiver combines elements, covered in earlier Chapters, which are common to nearly every DSP system application. *Transceiver* is a *communications* term, but we are using it in a much more general sense; in fact, it is potentially interactive (see comments about *feedback* below).

Table 8.2 Applications of Digital Signal Processing

Real-Time Measurement Instrumentation
  Digital filters; fixed and adaptive
  Spectrum analyzers and estimators
  Phase-locked loops; coherent detectors; correlators
  Transient digitizers and analyzers
  Signal averagers, exponential smoothers
  Time domain reflectometry and pulse analysis
  Signal generation
  Chemical instrumentation: mass spectrometers, chromatographs, etc.
  Observational astronomy: multiple-mirror telescope
  Search for extra-terrestrial intelligence (SETI)
  Thermography: IR imaging of temperature maps

High-Speed Control
  Servo links; position and rate control; guidance: missiles, etc.
  Robotics; remote sensing and feedback
  "Skid-eliminator" adaptive and context-sensitive control
  Disk-drive head positioners
  Engine control—integrated with smart sensors

Speech Processing
  Speech analysis
  Speech synthesis; vocoders...; speech compression
  Speech recognition/speaker authentication
  Voice store and forward
  Speech enhancement, noise cancellation

Digital Audio and Music Processing
  Analysis of musical instrument sounds
  Music synthesizers
  Digital recording studio and digital home reproduction
  Hi-fi compression; error correction; dynamic noise reduction
  Reverb and ambiance enhancement; pitch transposers

Communications
  High-speed modems
  Adaptive equalizers; digital repeaters of analog signals
  PCM companding: μ/A law conversion
  Modulation/demodulation: amplitude, frequency, phase
  Data encryption and scrambling
  Linear-phase filtering; echo cancellation
  Spread-spectrum communication
  Pulsed-echo acoustic hologram (bat sonar)
  Radar and sonar processing; electronic countermeasures

Number-Crunching
  Array processors for mainframes
  Floating-point accelerators for microcomputers
  Vector and matrix processors for supercomputers
  Sparse matrix algorithm processors
  Transcendental functions, iterative-solution architectures
  Artificial intelligence: Lisp machines, parallel processors
  Seismic earth and sonar sea-floor mapping
  Weather prediction and atmospheric modelling

Image Processing
  Image enhancement; pattern recognition; computer vision
  Reconstruction: CAT, PET, MRI tomography
  Radar and sonar image processing
  Digital TV—video DSP processors
  Satellite images; earth resource survey

Graphic Image Generation
  Image management: shapes, sizes, contours, shading, highlights
  Vector manipulation; graphics "engines"
  CAD/CAM workstations; 3-D image generators
  Flight (and other transportation) simulators; arts and film sequences

Real-World Interface Applications
  Transducer linearization; dynamically linearized loudspeakers
  "Smart sensors:" intelligent transducers
  Solid-state camera with DSP enhancement or preprocess
  Flat panel display with local pixel-region processing

Medical and Biotechnology
  Ultrasonic Imaging; digital x-ray: enhanced image allows lower dose
  Patient monitoring (intensive care, EKG...)
  Prosthetic IC Implants (artificial ear....)
  CAT, PET, MRI tomography (see under image processing)
  Fourier-transform IR spectrometer

Integrative Applications
  Printed-text-to-voice converter for the blind
  Multilingual translating telephone
  The intelligent pilot's assistant (DARPA thrust area)
  Unmanned mobile watchdog or armored weapon ("Think-Tank")

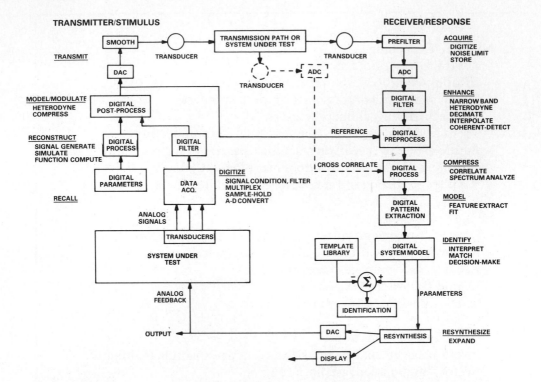

Figure 8.1. Elements of a digital transmitter-receiver or stimulus-response measurement system. Data moving away from the transmission path or system under test progresses towards higher levels of abstraction and lower data rates.

Specific implementation examples in instrumentation (Section 8.3), telecommunications and speech (Section 8.4), fast function generation (Section 8.5), and 2-D image processing (Section 8.6) share many of these DSP steps.

The Stimulus: Many digital processing systems include a stimulus side. In communications, the transmitter portion massages data on the way to the transmission path. In instrumentation, the stimulus excites a response in the system under test from which its key characteristics can be mapped. In *control*, the stimulus and response are compared and the system adjusts variables and parameters to produce the desired form of response. A stimulus which can be correlated at the response end makes possible true coherent detection of an evoked response (below) for extra detection sensitivity.

The stimulus could be a pulse train, white noise, a ramp sweep of a variable, a modulation signal (typically a sine wave), or an "optimum" stimulus which attempts to mirror system response characteristics for signal-to-noise enhancement. The choice of stimulus depends upon what is possible: difficult to simulate a weather system, easy to modulate a linear system in the lab. The choice also depends on what system characteristics are to be mapped: e.g., noise testing may be chosen in order to determine a transfer function. Examples of these options will be discussed.

The Response: A typical receiver might digitize the signal (after antialias prefiltering), digitally filter using whatever knowledge of the signal is available (frequency range, arrival-time estimate, etc.) to eliminate noise bands which might otherwise saturate later processing, and then subject the signal to one or more digital processing steps:

> Signal averaging
> Correlation (auto- or cross-)
> Coherent detection
> Spectral analysis or estimation

The choice among digital processing techniques depends upon the characteristics of the signal and the degree to which the system stimulus is controllable, as discussed below.

The Feedback: Closing the loop between response and stimulus (see Figure 8.12 bottom) gives a feedback loop like that often found in analog systems. Indeed digital control is a lively and important field, ranging from guidance and control of airplanes to controllers for hard-disk computer peripherals and laser-disk players, etc. Because many digital closed-loop control applications do not require the computation-intensive throughput of the DSP applications described in this book, they will not receive extensive attention here. Closed-loop applications will also be found in communications, with differential coding to send only changes in a signal, and in the digital lock-in for laboratory instrumentation—which coherently detects a system's response to digitally generated stimuli.

The DSP operations used here have in common a recurring pattern of computations involving the basic multiply-accumulate step. The corresponding mathematical operations (filtering, spectrum analysis, correlation, decimation, and interpolation) are summarized for convenience in Table 8.3.

Basic Steps in the Digital Transceiver

The major steps in a digital transceiver or in a digital measurement system are largely direct transcriptions (though at enhanced performance) of blocks familiar in analog signal processing.

Transmission or Stimulus Side

Recall or digitization: Digital parameters stored in memory are selected; if an analog signal is to be transmitted digitally, it is digitized by an a/d converter.

Digital reconstruction: Digital parameters pass through a digital signal-generation step, usually employing a lookup table or a simulation. A digitized

Table 8.3 DSP primitive mathematical operations.

OPERATION	COMPUTATION
1. Filtering FIR	$y(n) = \displaystyle\sum_{k=0}^{M-1} a_k\, x(n-k)$
IIR All Pole	$y(n) = \displaystyle\sum_{k=0}^{N-1} b_k y(n-k) + x(n)$
General	$y(n) = \displaystyle\sum_{k=1}^{N-1} b_k y(n-k) + \sum_{k=0}^{M-1} a_k x(n-k)$

OPERATION	COMPUTATION
2. Spectrum Analysis DFT	$X(k) = \displaystyle\sum_{n=0}^{N-1} x(n)\, e^{-j2\pi nk/N}$
FFT Butterfly $X_0+jY_0 \rightarrow$ ☐ $\rightarrow X_0{}'+jY_0{}'$ $X_1+jY_1 \rightarrow$ ☐ $\rightarrow X_1{}'+jY_1{}'$	$X_0{}'=X_0+R, \quad Y_0{}'=Y_0+I$ $X_1{}'=X_0-R, \quad Y_0{}'=Y_0-I$ $R=X_1C-Y_1S, \quad I=X_1S+Y_1C$

OPERATION	COMPUTATION
3. Correlation Auto	$C_{xx}(m) = \dfrac{1}{N} \displaystyle\sum_{m=0}^{N-1-m} x(n)\, x(m+n)$
Cross	$C_{xy}(m) = \dfrac{1}{N} \displaystyle\sum_{n=0}^{N-1} x(n)\, y(n+m)$
Power Spectrum	$P_{xx} = \mathrm{DFT}\,(C_{xx})$
Cross-Spectrum	$P_{xy} = \mathrm{DFT}\,(C_{xy})$

OPERATION	COMPUTATION
4. Decimation = downsampling	$\begin{aligned} x_D(n) &= x(n/M),\ n/M = \text{integer}\\ &= 0 \qquad \text{otherwise}\\ X_D(f) &= X(Mf) \end{aligned}$
Interpolation = upsampling	$\begin{aligned} x_I(nM) &= x(n), \quad nM = M, 2M, 3M, \ldots\\ &= \text{Interpol}\,[x(n), x(n+1)],\\ &\quad \text{between } nM \text{ and } (n+1)M\\ X_I(f) &= X(f/M) \end{aligned}$

analog signal is passed through a filter (perhaps also digital) to smooth and limit what is to be transmitted.

Digital modeling and modulation: Source-coding or modelling is used to reduce transmitted bandwidth; digital heterodyning shifts the signal to a carrier frequency chosen to fit the communications pathway.

Digital-to-analog conversion: The signal is returned to the analog world through d/a conversion with smoothing, to fit signal path or system response.

Receiver or Detector Side

Acquire: The received signal or response undergoes a/d conversion, with prefiltering, and is stored in a buffer.

Enhancement
 (1) Heterodyning stages with local filtering narrow the bandwidth for later analysis by subtracting out the carrier frequency.
 (2) Decimation/interpolation in function or filter lookup tables further restricts the bandwidth or fills in missing points.
 (3) Detection algorithms zero-in on the signal and suppress noise via correlations, coherent/incoherent detection, and spectrum analysis.

Preprocessing/compression: Further preprocessing is performed to enhance S/N ratio and minimize error rate during reception:
  Enhancement filtering
  FFT and Fourier filtering

Modeling/pattern extraction: The spectrum is fit to a sequence of peaks, or the data is fit by a set of functions, reducing the demands on later storage or computation.

Identification: The much-reduced parameter set is compared against a template library of possible patterns; the lowest-error fit provides the identification.

Data expansion/resynthesis: To provide for control, interpretation, or further computation, the identified parameters may be re-expanded. The synthesis is done from the coefficients of the best-fit model.

Digital-to-analog conversion: For control or perhaps display, D/A conversion with smoothing returns the received/detected signal to the real world.

Higher Levels of Abstraction at Lower Data Rates
The discussion that follows will relate the distinct case-studies discussed in depth in Sections 8.3 (Signal Detection in Instrumentation), 8.4 (Signal Modelling), 8.5 (Signal Generation), and 8.6 (2-D Signal Processing). The operations performed on the Transmission (Stimulus, Analysis) side and on the Receiver (Detector, Synthesis) side are remarkably similar even though the terminology may differ from field to field.

System blocks in Figure 8.1 progress towards higher levels of abstraction and lower data rates (though not necessarily easier computations) away from the transmission path or system under test. Consider data flow on the receiver side (Figure 8.2). The digitization process stores not just the desired signal but noise and competing signals as well. A selective filter cuts down the bandwidth and hence the spectral range needed in later processing. If the signal is centered about some characteristic frequency, processing can focus

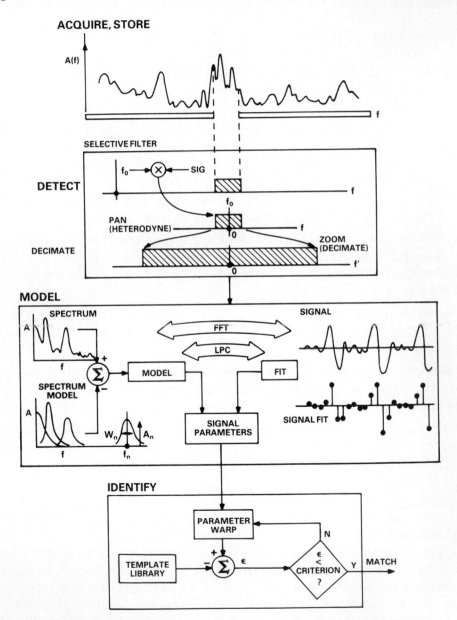

Figure 8.2. Data flow through a typical digital receiver-analyzer.

in on that range by heterodyning and decimation, just as a photographer pans in to the center of the scene and zooms up the magnification. Further analysis using an FFT may produce a spectrum and then model it by a series of peaks, whose characteristic frequencies, amplitudes, and widths represent the data with a much smaller number of parameters in the same way as, for example, the frequency, amplitude, and phase (3 numbers) compactly characterize the infinite progression of a sine wave in time.

Alternatively, one may choose to fit the data to a set of trial functions in the data domain; the coefficients of the fit are a sufficient but much-compressed representation of the data. A data set with 50,000 digitized points per second may be reduced to 10,000/s by a 10:1 decimation zoom, Fourier-analyzed, and then fit to the dominant 5 peaks in the spectrum (15 parameters/second). The behavior of the data in time, or identification of system behavior, may then be much more realistically computed than with the original data set. This progression is familiar in speech synthesis and recognition (Section 8.4) as the modelling of the vocal tract, and in image processing (Section 8.6) as the progression from the enormous number of points in a pixel image to the global shape-recognition and object interpretation. Or the parameters may be resynthesized into signals (with greatly reduced noise) for display, decision-making, or control. Computer-enhanced images from space exploration or altered speech/music synthesis following "sampling" of real inputs are common examples of resynthesis after massaging.

Pattern Matching, Templates, and System Identification
System recognition is an important topical problem. For example, a trainable voice-recognition system follows the progression of Figure 8.2, but a noisy environment or a change in pitch or a sore-throat can raise the error rate. Speaker-*independent* voice recognition is far from solved, involving high-level semantic issues that transcend the concrete manipulations of DSP. Radar and sonar interpretation tries to rapidly identify objects as friend or foe from the signature of the parameters in a model as in Figure 8.2, but there are nontrivial problems. For example (perhaps apocryphal), during a naval engagement in the South Atlantic, a British warship was sunk by an Argentinian Exorcet missile because the radar's computer was said to have correctly identified the signature as that of a French-made rocket (and thus "friendly").

In system identification, the signal parameters modeled by an FFT peak-fit or a data/function-fit are passed to a pattern recognition block. The parameters are compared against "templates" stored in memory of known images, and analysis algorithms steer the fit through a multi-dimensional parameter space, perhaps warping the parameters to allow for unknown distortions and pitch shifts, until a minimum-error identification is made. Since a matrix inversion is normally involved in a minimum-error fit, the reduction of the number of parameters by using a model makes identification computationally feasible. To invert a 15×15 matrix of a speech-parameter-template fit is a realistic real-time performance goal, but there would be little hope of working with the original 10–50,000 points per second.

The template concept also underlies many more-primitive DSP operations. Any filter is a template, "carving" the spectrum to shape. The sine-wave reference in coherent detection or at any point in a spectrum analysis is a template of arbitrarily precise frequency.

In audio testing, using a white-noise stimulus, the values of equalizer parameters (usually in octave steps) necessary to flatten out the received data spectrum form the template which characterizes the audio attenuation spectrum of the system under test. The deconvolution of broadened peaks sharpens information by dividing the peaks by the template of a measurable instrumental broadening function. The following sections include examples of extracting system key parameters by manipulation with such templates.

8.2.2 DIGITAL DETECTION
Optimizing the Response: Signal-Detection Issues
What Do You Need to Know?
The degree of optimization available depends not only on the system, the stimulus, and the noise, but also on the search strategy. For example, suppose a transmitted signal pulse, such as the one shown in Figure 8.3, is received at a detector some time later in a deteriorated state. The choice of DSP detection technique depends on what questions are asked. Increasing refinements are required as the information content required progresses from simple yes/no binary information to waveshape analysis:

> Was any signal received at all in the detection interval?
> When did it arrive? → Range or velocity information
> What was its amplitude? → Attenuation information
> Received waveshape or bandwidth? → Physical properties of propagation
> path

The objectives of maximizing the probability of detecting a signal, yet also minimizing the probability of a false alarm, are in conflict (Figure 8.3), a concept familiar to anyone with a home burglar-alarm system. In practice, an acceptable false-alarm rate is decided upon, the system sensitivity is increased as much as possible within this constraint, and detection algorithms are employed to sort false alarms from real events. In Section 8.3, a few examples from the richly developed literature [See Whalen, 1971, Schwartz and Shaw, 1975] will serve as illustrations.

Yes/No vs. Continuous Measurement
There are really two classes of detection problems: (1) Some measurements result in a finite set of possibilities, e.g., 1/0 or yes/no as in binary

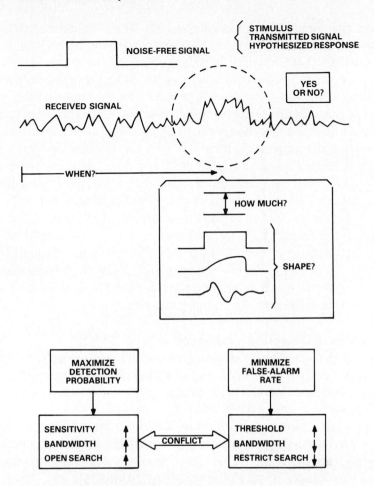

Figure 8.3. In detection of a signal in noise, the information required may be not just a Yes or No, but also When? How much? What shape?

communications or radar detectors, or discrete counting of the number of objects encountered. (2) Other detection problems require estimation of parameters that can vary continuously, like amplitude or frequency. This class corresponds to linear measurements: Doppler radars (which measure speed), radio receivers, and most laboratory measuring-instruments. Often, a measurement whose desired answer is binary requires continuous-signal processing before a black-and-white answer comes out. A modem, for example, includes a quantized a/d converter and digital filtering to extract the 1s and 0s from the analog received signal. Machine vision simplifies a complex digitized image at successively higher levels of abstraction to arrive at a yes/no answer to questions such as "is the cup on the table?."

Coherence vs. Incoherence
Some measurements are naturally incoherent. The receiver of Figure 8.2 tunes its local oscillator close to the transmitter carrier frequency, narrow-

band filters the signal to eliminate noise outside the signal bandwidth, and demodulates down to eliminate the carrier frequency. If the local oscillator which is used for heterodyning is independent of the transmitter's carrier oscillator, it is called *incoherent* or *asynchronous*. When transmitter carrier and local oscillator detector are made *synchronous* (the line marked Reference in Figure 8.1), the measurement is called *coherent*. Coherent generation and detection permit enormous enhancement of signal-to-noise, in the same sense that lasers are more intense than light bulbs: not because of higher average power, but because the coherence allows the power to be concentrated in a narrow range. Coherence permits a resonant buildup of information or energy: an infantry troop marching across a bridge in step is said to be able to cause a coherent excitation of a bridge vibrational frequency and possible destruction of the bridge; walking at a random pace is quite unlikely to do so. Coherent detection strategies enhance S/N for the same reasons, but only when signal and detector can be synchronized. Even in the absence of true coherence, modern receivers may employ a phase-locked loop to actively detect and lock onto the carrier frequency and phase.

Choosing a Signal-Detection Technique
Signal Averaging, Correlation, Coherent Detection, Spectral Analysis/Estimation
How does one decide which technique of signal detection is best for a given application? Section 8.3 will provide guidance for the choice and pointers for their application.

Signal averaging sums repetitions of a signal, using an accurate time reference, called the *trigger* or *stimulus*, for synchronization; attached to each frame, it indicates when the signal starts or ends. The signal-to-noise ratio improves as $\sqrt{N}$, where N is the number of repetitions.

Correlation looks for relationships in time between two signals (cross correlation) or between points on the same signal (autocorrelation).

Coherent detection recovers a modulated signal from noise by cross-correlation. The received signal is mixed with a coherent reference signal (the modulation stimulus of Figure 8.1) and summed over time. Coherent detection offers very-narrow-bandwidth detection as well as the chance to move the signal away from noisy regions of the spectrum.

Spectral analysis looks for periodicities as peaks in the signal's power-vs.-frequency spectrum. *Spectral estimation* tries to find the best fit of the data to a finite set of trial functions in the data domain.

When to pick which recovery technique? As sketched in Figure 8.4, *filtering* enhances signal-to-noise only when the signal band is well separated from noise bands ("good candidate"). If noise and signal bands overlap ("poor candidate"), filtering will diminish the signal's amplitude and, worse yet, distort its form, since different frequencies contained in the signal will be attenuated and phase-shifted differently by the filter.

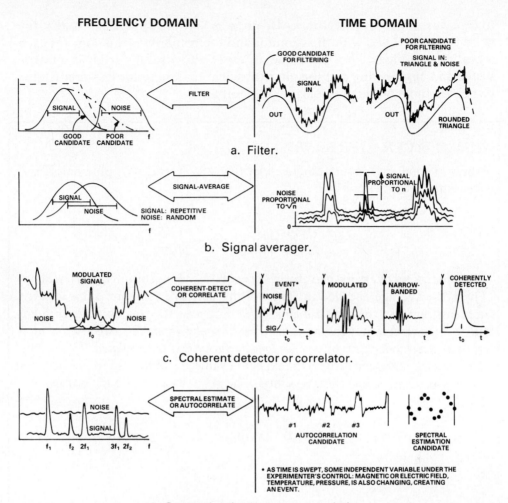

FREQUENCY DOMAIN

TIME DOMAIN

a. Filter.

b. Signal averager.

c. Coherent detector or correlator.

d. Spectral estimate or autocorrelation.

Figure 8.4. Selection among signal recovery techniques depends upon the relationship of signal spectrum to noise spectrum. (Left) Frequency-domain view; (right) time-domain view of the same signals and noise, selected as candidates for the recovery techniques named.

Signal averaging can help in case (b), where the energy of a repetitive signal remains localized in the window while that of noise shifts randomly from one measurement to the next. Even though the signal and noise in any pair of measurement windows may be comparable in magnitude, the sum of coherent signals builds up linearly with the number of measurements, N, while noise power builds up only as $\sqrt{N}$ (see BOX). Thus the average of N repeated windows gives the signal, $s(t)$, enhanced in SNR by $\sqrt{N}$. Signal averaging requires a signal that can be repeated in a precisely timed way, so the averaging is in phase; however, it is not necessary that the repetition be periodic.

In signal averaging, the S/N ratio will grow as $\sqrt{N}$ only to the extent that the noise is uncorrelated with the stimulus and the response is the same for each repeated stimulus. The technique is especially useful for deterministic systems in engineering and the physical sciences, less so for less-predictable systems in biology and physiology, for example.

SIGNAL-AVERAGER BASICS

A noisy signal, $r(t)$, containing a repetitive waveform, $s(t)$, plus noise, $n(t)$, is measured N times and averaged:

$$s_{avg}(t) \;=\; \frac{1}{N}\sum_{i=1}^{N} r(t_i) \;=\; s(t) + \frac{1}{N}\sum_{i=1}^{N} n(t_i)$$

The average recovers the waveform, $s(t)$, but how does the noise, $n(t)$, add up? The noise average is called in statistics the *standard deviation of the average*, in measurements the *standard error* (*SE*). Suppose the rms noise level in any single measurement window is σ. *The standard error, SE, of a series of N repeated measurements is $\sigma/\sqrt{N}$, as long as the noise is uncorrelated.* To see this, note that the standard error is the expected value, E, of the difference between the averaged signal and the true signal. The net noise power $(SE)^2$ is thus:

$$(SE)^2 \;=\; E\,[s_{avg}(t) - s(t)]^2 \;=\; E\left[\frac{1}{N}\sum_{i=1}^{N} n(t_i)\right]^2$$

$$=\; \frac{1}{N^2}\sum_{i=1}^{N}\sum_{j=1}^{N} E\,[n(t_i)\,n(t_j)]$$

The only nonvanishing terms in the above cross-correlation of uncorrelated noise are those with $i = j$, so

$$(SE)^2 \;=\; \frac{1}{N^2}\sum_{i=1}^{N} E\,[n^2(t_i)] \;=\; N^{-2}N\sigma^2 \;=\; \frac{\sigma^2}{N}$$

$$SE \;=\; \frac{\sigma}{\sqrt{N}}$$

Thus, after N repeated windows, the signal averager output gives a signal-to-noise ratio enhanced by $\sqrt{N}$.

What Coherent Detection Does for Signal-to-Noise

Why does a lock-in bring high Q and large noise-reduction? Modulated signal information is centered around the carrier frequency, f_{sig}. Multiplication of a signal at frequency, f_{sig}, by a signal at the reference frequency,* f_0, (Figure 8.5a) shifts the spectrum to the sum and difference points: $(f_{sig} + f_0)$, which is usually in the vicinity of $2f_0$, and $(f_{sig} - f_0)$, which is in the vicinity of 0 (i.e.,

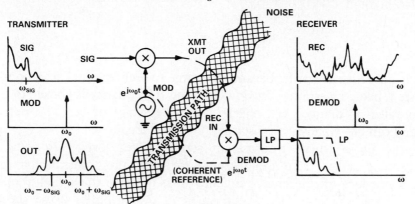

a. Modulation of signal to be transmitted and detection or demodulation of the received signal.

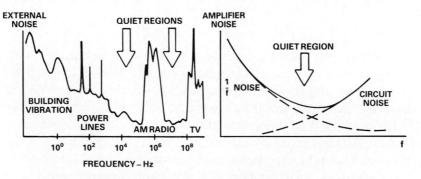

b. Selecting the carrier frequency for minimum noise.

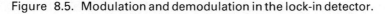

Figure 8.5. Modulation and demodulation in the lock-in detector.

dc). The output low-pass removes the sum-frequency term, leaving the signal downshifted to the dc range, where it can easily be stored or analyzed further. Since convolution in the time domain is equivalent to multiplication in the frequency domain, the convolution theorem can be used to explain the effects of heterodyning and low-pass filtering. The operation is:

$$y(t) = [x(t)\, e^{j2\pi f_0 t}] \star h_{LP}(t) \tag{8.1}$$

---

*In the following sections and their associated figures, the discussions in terms of frequency will use f and angular frequency, $\omega(= 2\pi f)$, essentially interchangeably, but in their correct relationship to time.

For real $x(t)$ this is equivalent to

$$Y(f) \; = \; [X(f-f_0) + X(f+f_0)]\,H_{LP}(f)$$

If the magnitude of $H_{LP}(f)$ equals 1 in the signal region and 0 above a cutoff frequency, f_c, selected to retain only the key information in the signal prior to modulation, then

$$Y(f) \; = \; X(f-f_0) \qquad\qquad\qquad (8.2)$$

There is no frequency drift, because the same reference signal also modulates the system being measured. The low-pass cutoff frequency is chosen to follow system rates of change, yet reject higher-frequency noise. The detection passband can be established by setting the cutoff frequency, f_c, of H_{LP} at just above the point where information starts to get lost. Additional S/N enhancement comes from the freedom to select the transmission carrier frequency or lock-in reference frequency so as to move the signal to optimal regions of the ambient noise spectrum or of the amplifier noise spectrum (Figure 8.5b). The coherent receiver multiplies the received signal by the carrier signal to shift the information back to the original frequency band.

Choosing Among Detection Alternatives for DSP

As Figure 8.4 shows, *coherent detection* gives better S/N enhancement than signal averaging when the signal's frequency can be phase-locked to that of the detector (c). For example, in the instrument known as the *lock-in detector*, the *same* sine-wave that modulates the system response also forms the reference signal for computing a product-sum output. The lock-in is a special case of cross-correlation (see Table 8.3): the product-sum is computed over one period of the sine-wave, and the correlation time-lag variable is replaced by the phase value, ϕ, of the reference sine wave. S/N is enhanced more rapidly than with signal averaging because (1) only the frequency region close to the carrier frequency is passed, so broadband noise is rejected as in a radio receiver, and (2) non-coherent signals, even if very close in frequency, are eventually averaged to zero, as will be discussed below.

When the signal's spectrum is unknown in advance or the system cannot be repetitively stimulated, then neither signal averaging nor coherent detection will work. In that event, *autocorrelation* (Figure 8.4d) can be used to find the power spectrum, if a long-enough record is available to adequately define the spectral features. If it is not, *spectral estimation* will adaptively find the best-fit frequencies, amplitudes, and bandwidths without window weighting problems—which become severe with a short data record. *Autocorrelation* can be done even on systems which cannot be repetitively stimulated for signal averaging or modulated for coherent detection. "Optimal" signal to noise improvement can be approached by matched filtering, i.e., the correlation of the received signal with a second signal selected to bring out the features of most interest, such as arrival time, amplitude, waveshape, frequency, phase, etc. Several examples of matched filters are given in Section 8.3

While these signal-detection concepts antedate DSP, digital methods have brought them to full flower. *Signal averaging*, which uses simple computations, was the first to profit from digital storage and wide-dynamic-range averaging. The modern digital oscilloscope typically includes signal averaging for S/N enhancement. Since signal averaging *per se* is not a very computation-intensive DSP application, it needs no further discussion here. *Correlation* was a direct beneficiary of digital techniques. Early analog correlators using capacitors for storage of analog voltage samples had limited dynamic range, and the relatively high cost of capacitor cells limited the number of samples that could be stored. Digital correlation, using mainframe computers, showed its power in sea-floor mapping, oil prospecting, and space exploration; but it was typically performed off-line. With the advent of DSP chips, and their advantages of high speed and compactness, correlation is finding its way into real-time portable field instruments as well. *Coherent detection*, employed in analog lock-in amplifiers, brought several Nobel prizes because of its enhanced resolution and improved signal-to-noise compared to conventional amplifiers. But *coherent real-time spectral analysis* (coherent detection with a sine-wave reference, to be discussed below), had to await the multiplication speed of DSP. Sophisticated *spectral estimation* and the detection of signals at unknown frequencies has finally come to full fruition with DSP, which combines digital heterodyning with adaptive or predictive filtering. Examples of these techniques will be discussed in the sections that follow.

8.2.3 DIGITAL HETERODYNING, DECIMATION, AND INTERPOLATION

Modulation and Demodulation

Digital Heterodyning
While the analog mixer may consist of anything from a diode to a true analog multiplier, the digital heterodyne step of Figure 8.2 is performed by a DSP multiplier. Since a digital filter operates on relative frequencies, sensing only data-point index, n, heterodyning can make a single filter serve as a multiband front end, for example for a wide-range spectrum analyzer (Section 8.3.4). The computation is just the multiply-accumulate step of a DFT (Table 8.3) at a single frequency, with output into a RAM data buffer. The heterodyning waveform is most commonly stored in a ROM lookup table.

Quadrature Heterodyning
In many telecommunications applications, bandwidth limitations of the communications channel require the carrier to occupy the same bandwidth region as the signal. If any portion of the signal spectrum, $A(\omega_{sig})$, reaches

above the local oscillator frequency, ω_0, then that portion wraps around $\omega = 0$ (dc axis) in the difference-frequency modulated output ($\omega_0 - \omega_{sig}$), as shown in Figure 8.6, since cosine modulation has a peak at both ω_0 and $-\omega_0$. Quadrature heterodyning provides two-phase heterodyned signals needed to cancel the wraparound image, equivalent to heterodyning by a true exponential, $e^{j\omega_0 t}$, which has no negative-frequency component. The mathematical relationships are straightforward and can be found in communications reference texts. Quadrature heterodyning finds a place also in lock-in amplifiers and spectrum analyzers (Section 8.3).

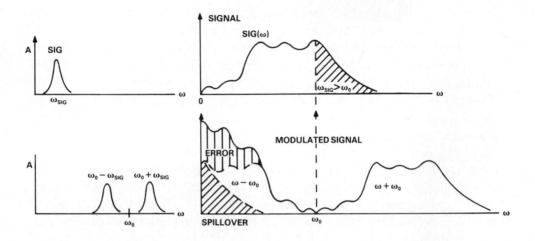

Figure 8.6. False wraparound images occur when the signal spectrum extends past the modulation frequency. Quadrature heterodyning is a common DSP operation to eliminate the problem.

How Does Coherent Detection Differ from Heterodyning?

Coherent detection implies heterodyning with a true phase-lock between modulated signal and the demodulator or mixer. An analog lock-in whose mixer is a simple chopper (Figure 8.7a) illustrates the principle. The advantange of coherent detection is that noise, however close in frequency to the reference frequency, will eventually drift out of phase and average to zero in the low-pass filtered mixer output (b) while truly coherent response stays synchronous and gives a low-pass filtered output (c) proportional to the system's response, R, and relative phase between signal and reference:

$$y(avg) = R \cos \phi$$

Decimation and Interpolation in DSP Systems

Decimation (also called *downsampling*) and *interpolation* (also called *upsampling*) manipulate the frequency of data as it is being input or scanned in from lookup tables, as introduced in section 3.6.1; examples of the technique include *spectral zoom* of input data or *frequency synthesis* of stored waveforms. The decimation-by-N operation keeps one of each N samples of the input signal, and effectively ups the frequency of the output signal, x_D, by the decimation index, N.

There are Nyquist constraints: data frequencies above f_s/N must first be removed, or else the decimated spectrum will alias back across the edges of the rescaled frequency interval. Interpolation or smoothing fills in the gaps in data or a lookup table. The DSP computation is a fit (usually linear) to initial value and slope to generate data values between the available addresses. Interpolation applications will be discussed under Fast Function Generation, Section 8.5. Decimation applications to the front end and "zoom" display of a spectrum analyzer will be described in Section 8.3. These techniques are key elements of digital "VCOs" and PLLs, with decimation and interpolation setting the frequency and an initial value setting the phase. The computations of decimated addresses are conveniently performed by a programmable address generator, such as the ADSP-1410.

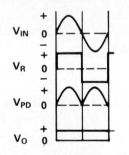

a. Chopper-type lock-in: Input V_{IN}, chopper reference, V_R, phase-detected signal, V_{PD}, and LP-filtered $V_{PD} = V_O$.

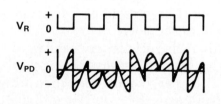

b. The coherent detector averages to zero noise, even if very close in frequency to the reference signal.

c. The dc output (shaded area) varies with phase difference between signal and reference.

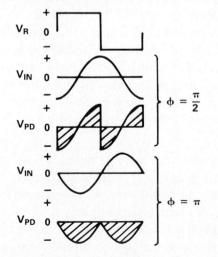

Figure 8.7. Coherent detection. *Source: Higgins, 1983.*

8.3 REAL-TIME DETECTION

8.3.1 EXAMPLES BASED ON CORRELATION PRINCIPLES

Correlation Overview

Correlation is the search for similarities between two sets of data—either a pair of signals (*cross*-correlation) or self-similarities in the same signal at different times (*auto*correlation). Correlation of signals requires the computation of a sum of time-shifted products (see Table 8.4 at the end of this section). Because of speed limitations, it was not a viable application for real-time digital signal-processing prior to the advent of DSP multipliers and MACs.

Correlation finds uses in applications such as:
Sonar, radar images
Identifying systems without disturbing them
Vibration analysis and nondestructive fault detection
Traffic flow analysis and bottleneck detection
Evoked response in neurophysiology
Heartbeat signature analysis in medicine
Seismic exploration for oil
Acoustic absorption and acoustic enhancement
Noise transmission-path detection and elimination
Radio astronomy
Modelling of vocal tracts and word-recognition in speech
Correlator-based instrumentation for physical and chemical sciences
Error detection and correction in digital audio
Noise and error reduction in digital communications

The following sections summarize the basic principles and apply them to a few of the above applications. Correlation analysis presents in the time domain the same information that spectral analysis presents in the frequency domain. Thanks to the FFT, correlation and spectral analysis are comparable in computational demands. The choice depends upon which domain is the more natural for display of information in a given application. Correlation is selected in acoustics when, for example, the time lag between paths needs to be identified in order to eliminate an undesired echo or noise transmission path. Some fundamentals which underlie correlation strategies are summarized for handy reference in Table 8.4.

Example from Acoustics using Cross-Correlation

Cross-correlation helps detect a particular waveform in the presence of noise. A stimulus, $x(t)$, with a distinctive waveform—a seismic explosion, radar pulse, or sonar tone burst—is emitted at repeated intervals. The response, $y(t)$, is received, sampled, and *correlated*, i.e., multiplied by a sampled, delayed $x(t)$ and summed, as in in the block diagram of Figure 8.8b. The form of $y(t)$ contains much information: the delay determines the range, and a Doppler shift from the carrier frequency of the stimulus measures the velocity of the target. A multiple echo signals that $x(t)$ has bounced off several objects or layers. The aim is to unfold target location, velocity, and physical properties such as density, shape, and size.

For instance, correlation analysis in oil prospecting determines which layers beneath the earth's surface have the right density signatures to contain oil; in sonar, correlation helps establish the location and identity of submarines, undersea wrecks, etc. Figure 8.8a shows an example from acoustics. The delay between stimulus and response shows up as a peak at delay time, τ, in C_{xy}. Here, path 1 through the wall is the most direct (shortest time), and is attenuated. Path 2 over the wall takes longer but arrives with more amplitude, while unintended indirect path 3 travels through the floor and vibrates the microphone stand. In noise abatement or acoustics improvement, the peaks are identified from their positions and shapes via distances and physical properties; desired reflections are optimized and undesired ones eliminated.

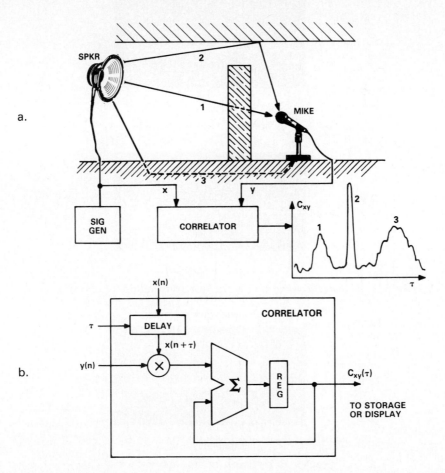

Figure 8.8. Correlation example from acoustics: noise transmission paths determined via portable DSP correlator. a. System and waveforms; b. Digital cross-correlator circuit.

Finding Hidden Periodicities Using Autocorrelation

The *auto*correlation function, computed by correlating a signal with itself, is a test for hidden periodicities in systems which cannot be stimulated for *cross*-correlation. Autocorrelation originated in the nineteenth century as a technique developed in response to curiosity about whether sun spots came and went periodically; it has continued to be a key "discovery" technique in scientific instrumentation: pulsars, neutron stars, brain waves, etc.

Autocorrelation is also one of the principal methods of extracting the parameters that model speech [e.g., see Oppenheim, 1978]. An example is shown in Figure 8.9. The peaks in the autocorrelation function are at periods less evident in the signal itself; in the process of shifting, multiplying and accumulating, signals that are not self-similar are averaged to zero. The

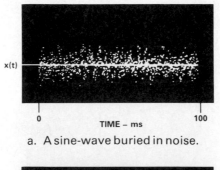

a. A sine-wave buried in noise.

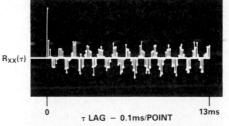

b. The recovered sine wave (autocorrelation function).

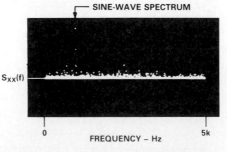

SOURCE: GR EXPERIMENTER, JULY/SEPT 1970 p. 5

c. Spectrum of the recovered sine wave (*auto-spectral* function).

Figure 8.9. Autocorrelation brings out hidden periodicities (Courtesy GenRad Inc.)

autocorrelation of random noise is zero except near $\tau = 0$; only there is noise self-similar. This is why the technique brings out hidden periodicities: the noise is pushed to the $\tau = 0$ end.

But over a *finite* window, W, the total energy in the noise sample can become comparable to that of the hidden periodicity; a small sample will show false periodicities. The window must be wide enough to cover the correlation periods expected in the signal. However, the wider the window, the greater the amount of computation, increasing as the square of the number of samples; prior to the discovery of the FFT, the computational demands of the multiply-accumulate step limited the use of real-time digital correlation.

The results of spectral analysis and autocorrelation are equivalent, since *the autocorrelation function is the Fourier transform of the power spectral density* (Compare Figure 8.9b and c and see Table 8.4). How can spectral analysis, which distributes white noise everywhere in the spectrum, be as effective as correlation, which makes noise "go away"? The integrated power of a signal equals the integrated energy under its spectrum (a consequence of the Wiener-Kinchine theorem of Chapter 2; see Table 8.4). Thus, if $S/N = 1$ for a sine-wave plus white noise, the sine-wave concentrates the same area in its peak that the noise background spreads out over the spectrum. The periodic signal's peak will stick up out of the white noise spectrum, since both have the same area when integrated. The equivalence of autocorrelation and spectral analysis make it possible to do long correlations rapidly using the FFT, as shown below.

The FFT Speeds the Computation of Large Correlations
Correlation and spectral analysis give equivalent results, and either is speeded by the time savings of the FFT. An approach employing FFTs to perform cross-correlation is summarized in Figure 8.10. The correlation of two functions can be obtained as a simple product of DFT's:

$$C_{xy}(n) = \text{DFT}^{-1}[X(k)\,Y^*(k)] \tag{8.3}$$

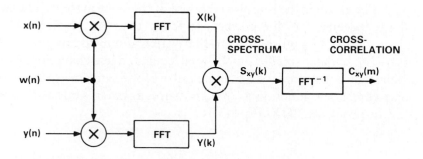

Figure 8.10. Using FFTs to perform correlation and spectral measurements.

using the convolution theorem.* For long-enough sequences, the $N \log_2 N$ multiplications of the FFT are considerably fewer than the N^2 multiplications of the direct convolution-sum, even though the FFT has to be performed thrice. Windowing, $w(n)$, is introduced to avoid end-effects. Correlation via the FFT method demands an understanding of windowing and circular convolution effects due to the DFT (see Chapter 3; a lucid discussion is also given in Rabiner and Gold (1975), Chapter 6, or Oppenheim and Shafer (1975), Chapter 11).

Identifying System Parameters using Noise Correlation
Correlation facilitates identifying an unknown system without disturbing it. A pulse-echo measurement gives the system's impulse response, but the response may be buried in noise unless the applied pulse has a large amplitude. However, a pulse of large amplitude may momentarily overload the system, distort the response of a loudspeaker, or alert the enemy submarine. The cross-correlation of the response to a suitably chosen *continuous* transmitted signal can characterize the system at much smaller signal levels. For acoustic and vibration analysis, the test signal is often a pseudorandom noise source, as shown in Figure 8.11. The signal is applied at low enough levels to avoid exciting any nonlinearities. The theory (see Table 8.4) shows that for a noise bandwidth much greater than the passband of the system under test, *the correlation of the test signal with system response is precisely equivalent to the system's impulse response*. The cross-correlation thus builds up over a period of time a picture of the impulse response $h(t)$, from which the transfer function, $H(f)$, can be obtained as the Fourier transform of $h(t)$.

Example: Speaker/Room Optimization. The response of a listening environment (speaker + room) can be readily obtained (Figure 8.8) by noise testing. Send a fast sequence of short pulses, at random intervals, through a speaker, and pick up the response with a microphone. There are two alternatives: (1) Fourier analyze. The result is the (speaker + room) frequency response, from which room absorption and speaker resonances can be identified and corrected. (2) Correlate the response with the noise sequence. The result is the impulse response of the same system. The two are equivalent, because of the power spectrum theorem. Noise testing is equivalent to passing a pulse through the system and measuring the impulse response—but with far smaller input signal levels and less chance of distorting the system than is suffered with a single pulse of the same integrated energy. The noise-correlation function, $C_{xy}(t)$, measures a system's impulse response, $h(t)$ [for the proof, see BOX, Eq. 8.5]:

$$h(t) = \frac{C_{xy}(t)}{X(0)} \tag{8.4}$$

*The correlation integral is like the convolution integral, except for the time reversal, $(t' + t)$ rather than $(t' - t)$, which corresponds to the complex conjugate of the Fourier transform.

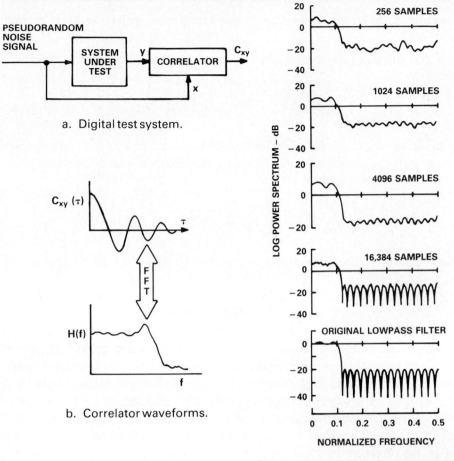

a. Digital test system.

b. Correlator waveforms.

c. Power spectral density computational results for a system consisting of a 39-tap low-pass filter.

Figure 8.11. Correlation of the response of a system under test with the identical noise-test signal builds up a picture of system's transfer function. (Data from Rabiner and Gold, *Theory and Applications of Digital Signal Processing,* ©1975 AT&T Bell Laboratories and reprinted with permission)

where $X(0)$ is the test noise spectral density in volts^2/Hz at the low-frequency limit. The system's transfer function, $H(\omega)$, follows from an FFT of $h(t)$. Figure 8.11c shows a demanding example, the transfer function of a 39-tap filter system measured by noise correlation. While not as precise as a swept-frequency determination, correlation is much faster than measuring the response to sine-wave inputs at a sequence of fixed frequencies. Noise testing also avoids phase distortion errors, which occur if a sine-wave test signal's frequency is changed before the system's impulse response dies out. By contrast, it is not necessary to delay adjacent noise pulses long enough for the impulse response to decay.

Measuring a system's impulse response from noise testing using correlation

The output of the cross-correlation system shown in Figure 8.8 for a given value of τ is:

$$C_{xy}(\tau) = \frac{1}{Mt_s} \sum_{m=0}^{M-1} x(mt_s - \tau) \, y(mt_s) \qquad (8.5)$$

for an arbitrary input, $x(t)$. But $y(t)$ is the convolution of the system's impulse response with the excitation, $x(t)$. Writing this explicitly and interchanging the order of summation gives:

$$C_{xy}(\tau) = \left[\sum_{n=0}^{N-1} h(nt_s) \right]\left[\frac{1}{M} \sum_{m=0}^{M-1} x(mt_s - \tau) \; x(mt_s - nt_s) \right]$$

The second sum is the autocorrelation of the stimulus, $x(t)$, with argument $(\tau - nt_s)$, so

$$C_{xy}(\tau) = \sum_{n=0}^{N-1} h(nt_s) \, C_{xx}(\tau - nt_s)$$

If the bandwidth of the noise stimulus, $x(t)$, is chosen much greater than the system passband, C_{xx} is concentrated in the vicinity of $t = 0$ (noise is self-similar only at near-zero lag). Normally, $h(nt_s)$ is slowly changing over this region and can be brought outside the sum. The sum which remains is, by the autocorrelation-spectral density relationship, the stimulus's spectral density, $X(f)$, evaluated at $f = 0$. As a result, to measure a system's impulse response, just measure its noise-correlation function:

$$h(t) = \frac{C_{xy}(t)}{X(0)}$$

where $X(0)$ is the amplitude of the test noise spectral density in volts^2/Hz at the low-frequency limit.

Table 8.4 Summary of correlation and matched-filter fundamentals

Cross-Correlation Function $C_{xy}(k)$

$$C_{xy}(k) = \frac{1}{\sigma_x \sigma_y N} \sum_{n=0}^{N-1} x^*(n) \; y(n+k)$$

C_{xy} is normalized or scaled to the variance, σ_x, σ_y, of signals $x(t), y(t)$.

Cross-correlation's salient feature: $C_{xy}(k)$ *is an optimum test for similarity of* $x(t)$ *and* $y(t)$.

Lock-in or Coherent Detector (special case of cross-correlation)

$$L(\phi) = \sum_{n=0}^{N-1} x(n) \sin\left(2\pi \frac{n}{N} + \phi\right)$$

The phase of $x(n)$ can be determined by adjusting ϕ until L is at a maximum.

Autocorrelation Function, $C_{xx}(\tau)$, and Spectral Density

A reminder from Chapter 2 (Section 2.3.4): *The autocorrelation function is the Fourier transform of the spectral density.*

$$C_{xx}(\tau) = \sum_{-N/2}^{+N/2} S_{xx}(f) \, e^{j2\pi\tau f_s/N}$$

where $S_{xx}(f)$ is the magnitude-squared or power in the Fourier spectrum of $x(t)$. The relationship goes both ways, i.e., the spectral density can be obtained by a Fourier transform of the autocorrelation function. This relationship is known as the *Wiener-Kinchine Theorem.*

Corollary: *The integrated power of a signal over the measuring interval in time equals the integrated energy under its spectrum* (Parseval's theorem, Chapter 2).

Transfer Functions from Noise Power Spectrum

The correlation of a random-noise test signal with the system's response gives the system's impulse response, $h(t)$.

$$h(t) = \frac{C_{xy}(t)}{X(0)}$$

where $X(0)$ is the amplitude of the test noise spectral density in volts^2/Hz at the low-frequency limit.

Corollary: *Noise testing measures the transfer function $H(\omega) = $ FT $[h(t)]$, as long as the noise bandwidth is greater than the system passband.*

Convolution Sum

$$y(\tau) = \sum_{n=0}^{M-1} x(t) \, h(\tau - nt_s)$$

for an M-tap filter.

Matched Filter

A generalized matched or optimum filter for amplitude detection of a signal of known form, $s(t)$, has the transfer function:

$$H_{opt}(\omega) = \frac{S^*(\omega)}{P(\omega)}$$

where $S^*(\omega)$ is the complex conjugate of the signal's frequency spectrum and $P(\omega)$ is the noise power spectrum measured at the same point in the detection system.

8.3.2 EXAMPLES BASED ON A-PRIORI KNOWLEDGE OF THE UNDERLYING SIGNAL

Matched-Filter Overview

A *matched filter*, $H_{opt}(\omega)$, seeks to maximize the chances of detection of a signal in noise, subject to a specified false-alarm rate. The "optimum" criterion varies with signal and noise characteristics as well as the parameters to be determined. A rich matched-filter lore has evolved (unfortunately with a largely non-overlapping vocabulary) in diverse fields such as chemical kinetics, defects in solids, gravity waves in astronomy, radar, sonar, and telecommunications. The use of DSP matched filters is illustrated here by a recent example from the physical sciences, the search for gravity waves. But first, to see the idea, let one parameter at a time be the unknown: amplitude, phase, or time of arrival, as in the detected signal of Figure 8.3. (See Whalen (1971), Chapters 6 and 10, for the rigorous math.)

Estimate the Amplitude of a Given Signal in White Noise

Suppose the arrival time and transmitted waveshape $s(t)$ are known and the amplitude, A, of the received signal, $r(t)$, is to be determined. The measurement situation and sample waveforms are sketched in Figure 8.12 for a signal $s(t)$ of exponential form. The theory shows that the optimum estimate of the signal's amplitude, A_{opt}, is obtained by correlating (in discrete time) the received signal, $r(n)$, with the transmitted waveshape, $s(n)$. The correlation must cover a measuring interval, t_N (N samples), of duration greater than the (pulselike) signal.

$$A_{opt} = \sum_{n=1}^{N} s(n)\,r(n); \qquad r(n) = s(n) + \text{white noise} \qquad (8.6)$$

Suppose that, instead of performing a *correlation*, we wish to determine the *matched filter*, $h_{opt}(t)$, over the same N-point data interval.

$$y(M) = \sum_{n=1}^{N} h_{opt}(n)\,r(M-n) \qquad (8.7)$$

As a guess, try

$$h_{opt}(t) = s(-t) \qquad (8.8)$$

This gives the same final output amplitude as the matched correlator:

$$y(M) = \sum_{n=1}^{N} s(M-n)\,r(M-n)$$

$$= \sum_{n=1}^{N} s(n)\,r(n) \;=\; A_{opt} \qquad (8.9)$$

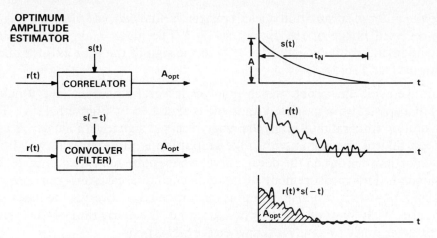

a. Amplitude, for an exponential signal.

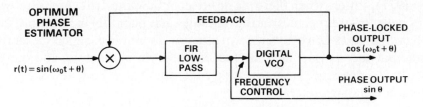

b. Phase, for a sine-wave signal (= phase-locked loop).

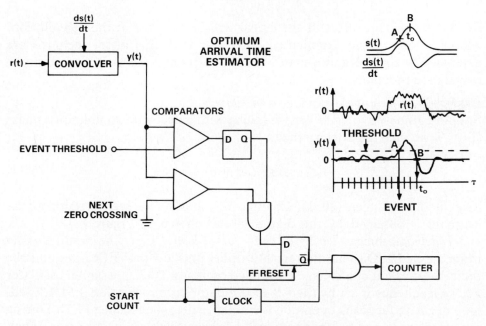

c. Arrival time, for a pulse signal.

Figure 8.12. Matched DSP estimators.

since the order of summation doesn't matter. Thus, *matched filter* $h_{opt}(t)$ is the time-reversed image of the signal, $s(-t)$.★ The guess makes sense, since correlation and convolution differ only in the sign of the time axis for one of the inputs (see Table 8.3).

While we avoid the proof, which involves integral equations (see Whalen, 1981, Chapter 10), a graphic argument is sketched in Figure 8.12(a). The convolution integral multiplies the noisy received signal with a clean image. The portion of the signal with large amplitude is preferentially weighted, while portions far out on the decay curve, where noise tends to dominate, are attenuated. This measurement situation is encountered often—for example in kinetics—when a pulse stimulus is applied to observe the decaying response. With the time constant of h_{opt} varied, the decay time of the system can be determined precisely, even when the S/N is poor.

This matched filter, commonly implemented in convolvers (e.g., Miller et al, 1975, 1977), suffers when there is a dc offset or baseline drift. Any filter which passes dc or low frequencies, as this one does, has poor S/N in its output unless the baseline drift is removed. An attempt to remove signal baseline by subtraction adds noise, since the baseline is down in the noise and therefore difficult to determine. Instead, use a filter function which averages to zero over the correlation interval:

$$\sum_{n=0}^{N-1} h(n) = 0 \qquad (8.10)$$

so that a constant offset in the signal will be rejected in the convolution. Modifications of the exponential correlator with this offset included are discussed by Crowell and Alipanhi (1981), and compared with other common correlation functions.

Estimate the Phase of a Given Signal in White Noise
Suppose a sine-wave signal arrives (along with noise) with an unknown phase shift, θ, which is to be determined.

$$r(t) = A\sin(\omega t + \theta) + \text{noise} \qquad (8.11)$$

The theory (Whalen (1971), Chapter 10) shows that the optimum phase estimate is obtained by the digital circuit shown in Figure 8.12(b). *The matched phase estimator is a digital phase-locked loop (PLL).* The circuit is easily implemented in DSP hardware: a multiplier plus low-pass FIR filter plus the digital equivalent of a voltage-controlled oscillator (VCO), a digital oscillator whose frequency is set by the LP accumulator output (Section 8.5). Closed-loop negative feedback operation is similar to that of an analog PLL. During the transient period before phase-lock has been established, the error signal,

---

★This is true numerically only after the interval, M, of the correlation or convolution sum has passed. Also, the signal is assumed of finite duration, so the matched filter has a finite impulse response.

$E(t)$, fed to the LP input is:

$$E(t) = A \sin(\omega t + \theta) \cdot \cos(\omega' t + \theta') \tag{8.12}$$

The feedback loop rapidly homes in on the value which minimizes the LP dc value, reaches zero when $\omega = \omega'$ and $\theta = \theta'$. This circuit may be used to phase lock a digital oscillator onto a noisy signal, as with the analog PLL. The phase-locked low-pass output is a direct measure of the phase shift of the received signal:

$$E_{dc} = \sin \theta \tag{8.13}$$

Estimate the Arrival Time of a Given Signal in White Noise
Here, the waveform is known or its form is suspected, and the question is when (if at all) such a signal is received. Arrival time is not an uncommon objective; its matched filter is *not* the same as the matched amplitude estimator. *The optimum arrival-time estimator is the derivative of the transmitted signal* (Whalen (1971), Chapter 10).

$$h_{opt}(t) = \frac{ds(t)}{dt} \tag{8.14}$$

The matched-filter DSP circuit and typical waveforms are sketched in Figure 8.12(c). The algorithm resembles a pulse arrival-time circuit of digital electronics; such circuits detect when the *derivative* of the pulse increases past an arbitrary threshold of zero slope (point A), then seeks the next zero-crossing (point B)—which occurs when the pulse is at its peak. This strategy, computationally faster than a convolver, is readily implemented in a DSP system, and may suffice for many applications. However, with low S/N, the zero-crossing time will be jittery, and the true matched filter of Eq. 8.14 will give superior performance. This matched filter drives the output to zero except when $r(t)$ contains significant signal energy; it makes better use of signal energy than simple gating functions to define the arrival time in the presence of noise. The logic shown in Figure 18.12c, which can be implemented in software in the DSP processor, starts counting when the START input is asserted. The first time the convolver output crosses an event threshold (set to avoid triggering with noise), the first comparator "arms" the logic to turn off the counter the next time the convolver output crosses zero, thereby measuring t_0.

Example: The Search For Gravity Waves
Suppose the noise is not white? The *generalized* matched amplitude filter has a transfer function:

$$H_{opt}(\omega) = \frac{S^*(\omega)}{P(\omega)} \qquad \textit{matched filter} \tag{8.15}$$

where $S^*(\omega)$ is the complex conjugate of the signal's frequency spectrum and $P(\omega)$ is the noise power spectrum measured at the same point in the detection system. While the *signal*'s spectrum is not always known exactly, it can be often estimated from properties of the system, detector, and transmission path. The *noise* spectrum characteristics are separately measurable in the absence of the signal. (The complex conjugate, S^*, reverses the time sense of the inverse FT, as in the matched filter, $h(t) = s(-t)$, above.)

Only when noise is "white" does $P(\omega)$ become a constant and drop out, as in previous examples. But noise is rarely "white." Many amplifier noise sources and random noise bursts in transmission have a $1/f$ or "pink noise" spectrum. Random fluctuations due to defects in transistors give step-like noise bursts ("popcorn noise") with the noise power spectrum, $P(f)$, of a band-limited system:

$$P(f) = \frac{4U^2 f_T}{(f_T)^2 + (2\pi f)^2} \tag{8.16}$$

where U^2 is the mean-square noise level and f_T is the average fluctuation rate.

The properties of the *detector* also shape the matched filter by altering the received signal's spectrum, $S(\omega)$. For instance, a tuned detector will have bandpass filter response; as an example of its application, a resonant detector with matched filtering was applied with considerable success [Tyson and Giffard, 1978] to *disprove* the earlier observation of gravity waves, at least within the sensitivity of available detectors. Gravity waves, which are thought to be emitted from rotating double heavy stars (just as a rotating charge dipole emits electromagnetic waves) have been the subject of an extensive search. Weber, a pioneer in the search [see Tyson for references], reported observation of dozens of apparent gravity-wave "events" per month. The detectors are massive high-Q mechanical resonators, brought into excitation by waves of comparable frequency. Background noise from passing trucks, earthquakes, etc., causes false alarms unless rejected; to accomplish this, filter optimization is used, along with coincidence counting with several identical but widely separated detectors. The matched filter's application here is summarized in Figure 8.13. Since the detector is resonant, with natural frequency, ω_0, and decay time, τ, the response after demodulation at ω_0, is:

$$|S(\omega)| = \frac{2\tau}{(\omega\tau)^2 + 1} \tag{8.17}$$

The corresponding matched time response for *amplitude* detection is the Fourier transform:

$$s(t) = e^{-\frac{|t|}{\tau}} \tag{8.18}$$

The exponential above is *not* the matched filter, since the goal is detection of arrival time. The convolver function is thus the derivative of Eq. 8.18 (shown in Figure 8.13). The derivative puts a factor of ω in the numerator of the matched filter equation, which forces a zero in the filter's dc response.

$$|S_{opt}(\omega)| = \frac{2\tau\omega}{(\omega\tau)^2 + 1} \tag{8.19}$$

The matched filter in time is, by the derivative theorem of Fourier transforms:

$$s_{opt}(t) = SGN(t)\, e^{-\frac{|t|}{\tau}} \tag{8.20}$$

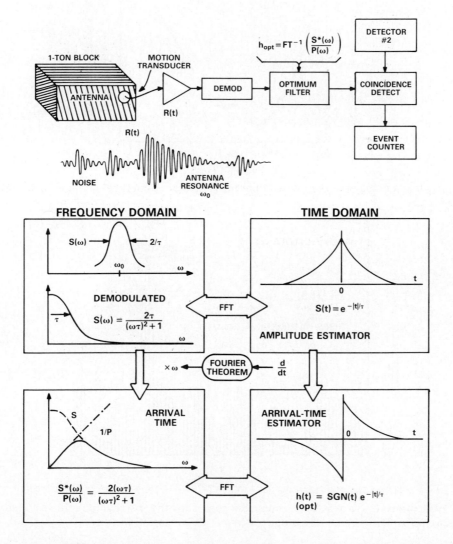

Figure 8.13. Matched filter for detecting arrival time in a band-limited system—the search for gravity waves.

The time constant, τ, in the DSP filter can be adjusted to minimize detector noise level, as demonstrated in Figure 8.14(a). When this matched filter was applied, no significant "events" were observed (Figure 8.14b) in a two-detector coincidence search. A true gravity-wave event would appear as a peak with zero lag; however, as the histogram shows, there appear to be no peaks above the random fluctuation level, except for the calibration pulse applied to both antennas with a 2-second offset. The earlier report by Weber was doubtful, probably because the filter and false-alarm rejection system were less than optimum.

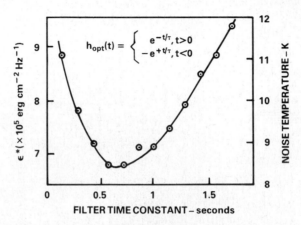

a. Minimum noise observed when filter time constant equals resonant detector time constant.

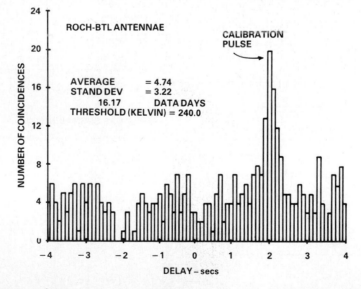

b. Histogram of two-detector coincidences, showing no events above random fluctuations. The peak at $+2$ s is due to an electronic calibration pulse.

Figure 8.14. Results of a search for gravity waves using a matched digital filter. (Coutesy of J. Anthony Tyson, AT&T Bell Laboratories)

8.3.3 COHERENT DETECTION (LOCK-IN) TECHNIQUES

Coherent-Detector Overview

The following section on digital lock-in design for instrumentation illustrates that digital multipliers eliminate the harmonic and intermodulation distortion of analog mixers, that a digital lowpass FIR filter offers superior stability and sharper cutoff for improved signal-to-noise ratios, and that real-time harmonic analysis can be achieved by a multichannel digital lock-in. The lock-in amplifier, or coherent detector (Figure 8.15), is an instrument that can benefit greatly by improvements in DSP. The system under test is stimulated by a variable, $V(t)$, which sweeps through a range of values (wavelength of light, magnetic field value, bias voltage, etc.) A small ac modulation signal, dV, is *summed* with the sweep variable, not multiplied. If the system response, x, varies with V, the extra stimulus creates an extra response:

$$x(V + dV) = x(V) + \frac{dx}{dV}\,dV + \frac{1}{2}\,\frac{d^2x}{dV^2}(dV^2) + \ldots \qquad (8.21)$$

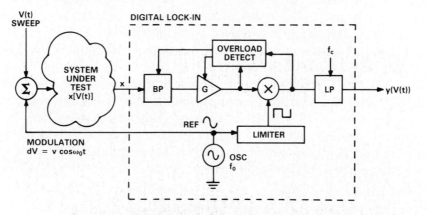

a. System components; dashed line shows functions to be done digitally.

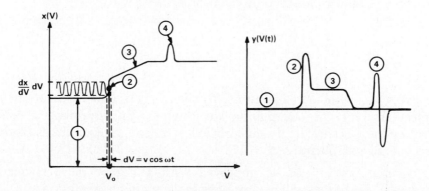

b. The lock-in response at ω_0 is the derivative of the system's response to the stimulus.

Figure 8.15. DSP lock-in amplifier or coherent detector for instrumentation.

When the modulation is a small oscillation, $dV = v \cos(\omega_0 t)$, with $v/V \ll 1$, x will contain ac components:

$$x(t) = x(V) + \frac{dx}{dV}\, v\cos(\omega_0 t) + \frac{1}{2}\,\frac{d^2 x}{dV^2}(v^2 \cos^2(\omega_0 t))\ldots \qquad (8.22)$$

spectral terms: ω_0 $2\omega_0$

where each higher power of the cosine contains a higher harmonic term. The *coherent* detector, or digital lock-in, multiplies this response by a reference cosine wave of precisely identical frequency, followed by a low-pass filter to pass the difference frequency and reject the sum frequency of the product. With the reference frequency set at ω_0, the measured response $y(\omega_0, V)$ is the *derivative* of the system's response, $x(V)$. This derivative action suppresses the baseline and brings out features of the response, as seen in Figure 8.15. Baseline offset (1) is totally suppressed, as is gradual baseline drift. A shoulder (2) which might signal an important transition in behavior is brought out clearly as a peak, a rate-of-change (3) can be measured precisely as a constant level, and a peak (4) gives a double-peak signature. With careful design of bandpass prefilter to eliminate noise outside the signal band, a good lock-in can furnish more than 100 dB of relative (coherent-signal-to-incoherent-background) enhancement. The lock-in also is a good analyzer of harmonic distortion: with detector frequency set at twice the modulation frequency, the output is proportional to the second harmonic distortion in the signal, etc.

It follows from consideration of the spectrum in Figure 8.5 (receiver section) that the combination of coherent detector and lowpass filter is equivalent to a highly selective bandpass filter; its Q is equal to the ratio of carrier frequency, f_0, to lowpass cutoff frequency, f_c.

$$Q = \frac{f_0}{f_c} \qquad (8.23)$$

For example, with lowpass cutoff, f_c, set to 1 Hz and modulation, f_0, of 1 megahertz, Q's of one million are possible. Improvement in S/N can therefore be spectacular: a signal buried in noise a million times larger can be recovered with output $S/N = 1$

The analog lock-in can be improved with a higher performance DSP processor. These ideas will be developed in the sections that follow:

A true multiply function at the mixer or demodulation step removes the odd-harmonic error of the conventional square-wave mixer.

Wide dynamic range can be maintained, even for signals whose level changes greatly, by overload-detection algorithms and auto-gain adjustment at each stage. The result is equivalent to block floating point, with the gain parameters serving as the block exponent.

An output lowpass filter or input bandpass prefilter synthesized with DSP has superior performance: higher Q and improved noise rejection. Also, replacing the analog IIR RC lowpass with a digital FIR lowpass will be shown to avoid distortion of signals whose amplitude is changing with time.

Digital Lock-in Design Example

Removing the Odd-Harmonic Error of the Analog Lock-In

If the system under test is nonlinear, sine-wave modulation generates harmonics which may be a significant portion of the response. This distortion can be studied precisely by coherent detection at higher harmonics of f_0: detection at $2f_0$ gives the 2nd harmonic distortion, at $3f_0$ the 3rd harmonic distortion, etc. (recall the power series above). However, these amplitudes are contaminated in an analog lock-in, whose mixer is usually not a true multiplier but a switch driven by a square-wave generated from the modulation signal. Harmonic contamination occurs because each component of the *reference* square wave heterodynes down to dc its matching frequency component in a harmonic-rich *signal*. Since the square wave has odd harmonics of amplitude in the proportion, 1, 1/3, 1/5..., the demodulated signal at f_0 is:

$$Y(f_0) = X_1(f_0) + \frac{1}{3}X_3(f_0) + \frac{1}{5}X_5(f_0) + \ldots \qquad (8.24)$$

where the X_n are the harmonics in the input signal. The true-multiply function in the digital lock-in removes this odd-harmonic error of the analog lock-in, since the only term demodulated down to dc is $X(f_0)$.

Auto-Gain Control: Block Floating-Point Equivalent

Overload can occur at various points in the lock-in process: the preamplifier, the mixer or phase-sensitive detector (PSD), and the output amplification stage. While analog lock-ins include an overload indicator, it does not detect all overload possibilities. It is straightforward to build into the local memory buffers of the digital lock-in a noise-handling sequence with a flow chart similar to that shown in Figure 8.16. Besides avoiding overflow, data is always stored at maximum dynamic range. The product of the gain factors is stored as a block exponent. Once the lock-in data is gathered, the data and the block exponents can be converted to floating point for further analysis over the full dynamic range.

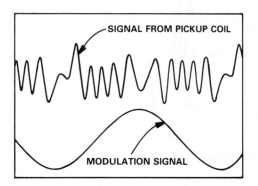

a. Richly nonlinear response of a frequency-modulated system, in this case a crystal exhibiting de Haas-van Alphen oscillations.

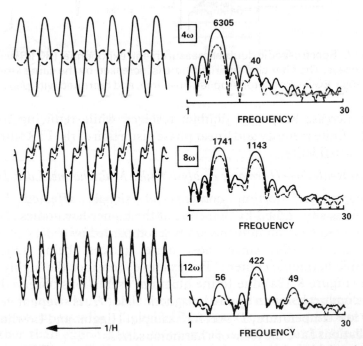

b. Bessel functions on a log scale show how harmonic enhancement goes as the *n*th power of the fm amplitude for the *n*th Bessel function.

c. Resulting enhancement of harmonic content by detecting a higher harmonics. Dashed curves are quadrature components.

Figure 8.18. Higher harmonics in frequency modulation. (From: Alles and Higgins ©1974 American Physical Society and reprinted with permission)

as a function of $1/H$. The harmonic amplitudes, A_r, contain useful electron dynamics information, and the frequency, F, measures conduction electron density. When H includes a small ac modulation frequency, the modulation principle of Eq. 8.22 leads to the familiar equations of frequency modulation:

$$x(H + h \cos \omega t) = A \sum_{n,r} \mathcal{J}_n(\alpha_r) \; \sin(n\omega t) \; \sin(nF/H) \qquad (8.26)$$

where the modulation index, α_r, is a measure of modulation amplitude, h, relative to the period of the rth harmonic of $x(H)$:

$$\alpha_r = 2\pi \frac{hrF}{H^2} \qquad (8.27)$$

Enhancement of weak components, A_r, at large r is provided by detection at large multiples, $n\omega$, of the modulation frequency, since the Bessel function of nth order begins climbing initially as the nth power of α. For example (Figure 8.18b), the relative amplitude of component A_2 is enhanced by 2^{12}, or 72dB, when the 12th harmonic is detected! Note the suppression of A_1 (square points) relative to A_2 (circles) at higher order $\mathcal{J}_n$. Figure 8.18c demonstrates how harmonic content, invisible in the signal and barely discernable in the spectrum at 4ω, is strong at 8ω and dominant at 12ω, for the same modulation amplitude, $h(t)$.

Simultaneous multichannel lock-in operation would speed such harmonic response measurement, but paralleling many analog lock-in instruments can be prohibitively expensive. However, if f_0 is low enough to permit multiplexed operation, a multiple-channel lock-in is possible in real-time. The idea was demonstrated in software with a chopper-equivalent mixer to avoid multiplications (Alles and Higgins (1973)), with consequent odd-harmonic error problems. This was improved with a hardware PLL to strobe input samples at sub-multiples of the modulation period and true sine-wave mixing (Templeton (1976); but the PLL frequency multiplication made frequency adjustment delicate, and processing speeds were limited to nearly one second per data point. A parallel-computing host-slave system (Matheson and Higgins (1978)) upped multiple-harmonic modulation frequencies to several hundred Hz, limited only by multiply time of that generation's minicomputers. This lab instrument idea is ripe for implementation with a single set of high-speed DSP hardware. The operation is equivalent to a real-time spectral analysis but is faster than an FFT, since the only channels computed are the actual harmonics desired. Windowing is also not necessary, since the modulated data is commensurate with the window width; there are no window-edge discontinuities.

DSP Enhancement of Other Lock-In Computations

With DSP, further useful lock-in calculations, that are less feasible with analog techniques—and much less accurate—can be easily performed and displayed in real time: vector-component recording (two-phase lock-in via quad heterodyning, as in Section 8.2), magnitude and phase computation, optionally including log output computation for a dB scale. A real-time display of phase can signal an unanticipated phase shift which would be lost if only one phase component were recorded. An auto-tracking loop can adjust the reference phase dynamically to the signal's phase, and the reference frequency can also be set to auto-track the signal's carrier when the two are independent. A log output can help resolve weak structure, especially if the block-floating-point scheme described above were used during fixed-point data gathering.

8.3.4 SPECTRUM ANALYZER

Overview

Since spectrum-analysis applications have already been discussed in detail (Chapter 3), the discussion below will be somewhat limited in scope. Outlined here are a real product example, a spectrum-analyzer instrument using a TMS-320, and front-end design issues involving decimation filtering; the section closes with a review of key issues in windowing and opportunities, such as averaging, for enhancing spectrum-analyzer performance.

Spectrum Analyzer Squeezes High Performance from TMS-320

A commercial spectrum analyzer, the Hewlett-Packard 3561A, well-illustrates the issues faced and performance obtained when a DSP microcomputer or microprocessor is incorporated into a high-performance signal processing instrument. With a basic frequency range from 1 millihertz to 100kHz, the instrument plots spectral responses with wide (80dB) dynamic range and moderately high speed (7.5kHz real-time data rate). For close scrutiny, a band-selectable zoom is available for frequency analysis with 640μHz resolution (see: *Hewlett-Packard Journal*, December 1984).

Easy to control, the portable instrument (Figure 8.19a) features the popular

a. The instrument.

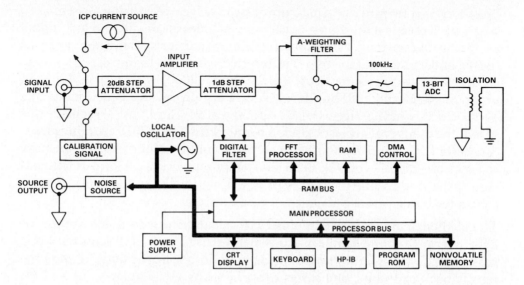

b. Block diagram of host-slave multiprocessor system.

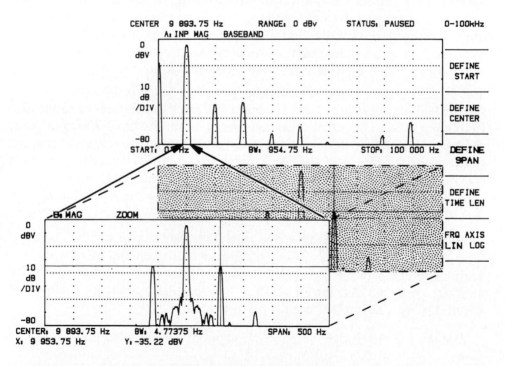

c. Spectrum and zoom feature.

Figure 8.19. Commercial audio spectrum analyzer based on a DSP microcomputer, the TMS-32010. (©Copyright 1984 Hewlett-Packard Company. Reproduced with permission)

"soft-key" front panel of a microprocessor-driven measuring instrument, with on-screen menu-driven selection of measurement mode, digital averaging parameters, data capture-buffer parameters, frequency analysis bands, and window parameters. The internal structure (Figure 8.19b) centers around a main microprocessor, an 8MHz M68000, with nearly 400K of program ROM and 128K of data RAM. FFT processing is delegated to a TMS-32010, which performs a 1,024-point real FFT in 42ms. To achieve this speed, the number of accesses to main memory from the 32010 are minimized. A radix-8 FFT algorithm is used; it requires only 1/4 of the off-chip memory accesses of a radix-2 version. Butterfly computations are performed with on-chip cache RAM while the processor is waiting for I/O to the data memory, controlled by a separate DMA memory-bus arbiter.

The challenge in 16-bit fixed-point FFT calculations is to retain amplitude dynamic range. For example, a 1,024-point real radix-8 FFT requires 4 or 5 passes. Each pass multiplies 16-bit data by 16-bit coefficients, rounds the results back to 16-bits, and stores them in RAM. Accumulation of a 31-bit result 8 times (radix-8 butterfly) could add 3 bits, which could overflow the TMS-320's 32-bit accumulator. Blind downscaling would lose 2 bits (34 − 32) per pass, destroying dynamic range in the 16-bit rounded result. Instead, a block scaling algorithm divides the data down by appropriate factors of two based on the largest data point in the previous pass, avoiding overflow and retaining 80 dB (14-bit) dynamic range in the spectrum.

The FFT is often thought of as having poor frequency-resolution, since the spectral points are evenly spaced over the full sampling frequency range. But frequency zoom is possible, as demonstrated by the factor-of-200 example in Figure 8.19c. The technique, as discussed in Section 8.2, involves modulation (local oscillator shown in the block diagram), followed by a decimation or selection of data points to be transformed. The frequency resolution, Δf, is then limited only by the total time, T_W, available for the measurement ($\Delta f = 1/T_W$)

Decimation Filter Front-End for a Spectrum Analyzer
The capability to pan over to a spectral region of interest and zoom up to full-scale in the display for high-resolution spectral analysis is a key spectrum-analyzer feature. As introduced in Section 8.2, *panning* is done by heterodyning, and the DSP *zoom* is performed as:

ZOOM = [(Heterodyned signal)$\star$(Lowpass)] $\times$ (Downsample)

This combination of heterodyning with decimation was shown earlier in Figure 8.2. We now show how to implement it with DSP hardware. Consider the front end processor, which sets the center frequency and bandwidth of a spectrum analyzer. A typical FFT-based spectrum analyzer heterodynes the input signal with a local oscillator in order to bring the frequency range to be analyzed down to a (fixed) region near dc. Because FFT outputs are uniformly spaced over the computed frequency interval, narrow-band analysis is possible only by decimating the input data to widen its spectrum out to the full analysis bandwidth. The process is called *decimation filtering* because the necessary antialias lowpass is combined with the decimation step. Here is an instrumentation design example:

Input sampling rate: 50 kHz (full audio)
Analysis bandwidths: 50, 100, 200, 400, 800, 1,600, 3,200, 6,400 Hz

The ranges from coarsest to finest resolution may be handled by 8 stages of a divide-by-two decimation filter ($6,400/50 = 128 = 2^7$) as shown in Figure 8.20(a). The decimation steps are performed by the divide-by-two counters, which clock into the next stage every other output of the prior stage. Any one of the outputs, OUT_n, becomes the input to the FFT (not shown), depending on the user-selected analysis bandwidth; all decimated outputs are available at any time. To prevent aliasing, the filter cutoff is set at $6,400/50,000 = 0.128$ (normalized to the sampling frequency). This same normalized cutoff is kept in the antialias low-pass of each decimation stage. Aliasing is automatically avoided, since the cutoff is less than 0.25 of the *decimated* sampling rate, and only one high-performance LP design and one coefficient lookup table are required.

If the computations are interleaved (each successive decimation filter requires only half the computations-per-second of the preceding one), they can be multiplexed so that *all* decimation filter stages are performed by the same hardware. The hardware and computation interleaving are shown in Figure 8.20b and (c). Heterodyning is performed efficiently by the *same* hardware, reading the sine-wave out of the lookup table rather than the filter coefficients at that phase of the computation. The outputs of each stage are stored in an 8-bank RAM; each bank contains enough past history for the next FIR filter stage. Because each succeeding decimation filter operates on only half the data points of the preceding one, the entire set of 8 can interleave in only twice the total computation time of a single filter. Only one of the eight filters is computing during each input data point and heterodyning step, with the appropriate filter strobed by a single bit of a 8-bit counter (see Figure 8.20c).

Hardware address-generators, such as the ADSP-1410 (introduced in Chapter 6), solve the traditionally difficult addressing task of stepping through circular data buffers with arbitrary step size in a decimation filter. The input samples are stored in a circular buffer of length M; $M > (N \times b)$, where N is the order of the filter, and b is the decimation step-size. The samples are retrieved with spacing, b, which reflects the degree by which decimation modifies the effective sampling rate—and weighted by filter coefficients, $a(i)$.

$$Y(n) = \sum_{i=0}^{N-1} a(i) X(n - ib) \qquad (8.28)$$

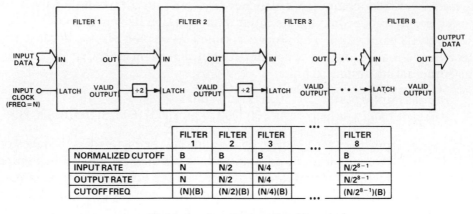

	FILTER 1	FILTER 2	FILTER 3	...	FILTER 8
NORMALIZED CUTOFF	B	B	B		B
INPUT RATE	N	N/2	N/4		$N/2^{8-1}$
OUTPUT RATE	N	N/2	N/4		$N/2^{8-1}$
CUTOFF FREQ	(N)(B)	(N/2)(B)	(N/4)(B)		$(N/2^{8-1})(B)$

a. Divide-by-two decimation filter chain.

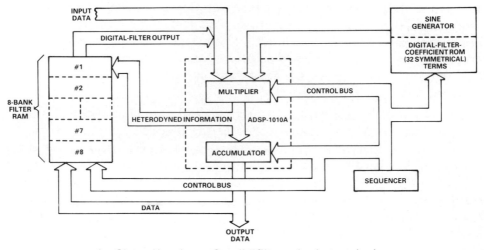

b. Shared hardware for all 8 filters plus heterodyning.

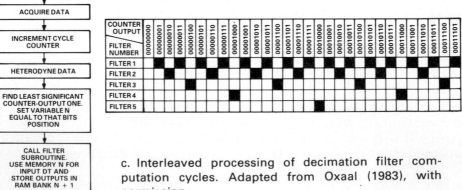

c. Interleaved processing of decimation filter computation cycles. Adapted from Oxaal (1983), with permission.

Figure 8.20. Decimation filter for spectrum analyzer front end.

To implement generation of an address, XAdr, using an ADSP-1410 (Figure 6.15), offset register, B1, is preloaded with the decimation interval, b; register B2 is preloaded with the length, M, of the input data-buffer, x, and comparison register, C1, is preloaded with filter address range, bM. An alternate instruction feature of the ADSP-1410 carries out overhead-free modulus addressing:

$$\text{XAdr: Adr} \leftarrow \text{R1; IF (R1} > = \text{C1) THEN AIR; ELSE R1} = \text{R1} + \text{B1}$$

where R1 is a results register, and the alternate-instruction register, AIR, is preloaded with:

$$\text{AIR:} \quad \text{Adr} \leftarrow \text{R1; R1} = \text{R1} - \text{B2}$$

Since either the normal increment-by-b operation or the alternate pointer-reset instruction takes only one cycle with the ADSP-1410 address-generation hardware, the tightly interleaved multiband filter of Figure 8.20 can be carried out at high speed.

The actual heterodyning step is more complicated than in the lock-in case above, since any portions of the signal spectrum $A(\omega_s)$ which exceed the local oscillator frequency form low-frequency aliases (beyond $\omega = 0$) at negative frequencies in the $(\omega_0 - \omega_s)$ term. *Quadrature* heterodyning (Section 8.2, above) provides two-phase heterodyning signals, which can be combined in an FFT algorithm to cancel the aliased image (see, for example, Oxaal (1983)). The same hardware can multiplex this extra operation, which requires just two table lookups and an extra multiplication per sampled point, plus appropriate extra data buffer storage for the second computed component.

Other Spectrum Analyzer Issues: Windowing, Averaging

Windowing

A good choice of window functions is desirable. Window-weighting functions, usually essential to avoid spectral leakage due to the finite data interval, determine the spectrum shape in the frequency domain and will influence *spectral separation*: how accurately can the spectrum analyzer distinguish nearby peaks and how well can small spectral terms be resolved from the sidebands of larger terms (see Chapter 3). A spectrum-analyzer instrument will have a limited—but carefully selected—set of windows, chosen for their performance in typical real-world situations. For example:

   Uniform★ or *rectangular* (no window-weighting): This case gives the best frequency resolution but suffers from severe sidelobe ringing. However, the uniform window can be used to advantage if an integer number of single-frequency (and harmonics) signal periods are contained in the data

---

★Referred to as "uniform" (rectangular shape) because all points are weighted equally.

window. In that case, the sampled spectral points off the main peaks all fall on the zeroes of $\sin(x)/x$, the sidelobes then vanish and ideally sharp frequency-determination results. The uniform window may be useful for a transient that decays away within the data window (no spillover in the DFT) or for noise signals where windowing has little effect in any case because the spectral structure is broad.

Flat-top: A flat-topped (Tukey) window has unit value within a specified fraction of the window while tapering to zero towards the edges, usually with a cosine function. The flat-top gives good amplitude accuracy, since the data amplitude is unaffected except near the window edges, but at the expense of frequency resolution, which is poor.

Other Spectral Windows: As detailed in Chapter 3, other window functions trade off spectral resolution, amplitude resolution, and computation speed. The Hamming window (raised cosine) gives good spectral resolution but poor sidelobe decay; the Blackman window adds another cosine to reduce the sidelobe problem at the expense of peak broadening. The Kaiser window is in many ways the most flexible, since an adjustable parameter can trade main lobe peak width against sidelobe attenuation.

Exponential Window: This is a special case for "matched-filter" types of impulse-response measurements. The decay time of the exponential is adjusted to best fit the decay rate (for S/N optimization) of the data.

Transient Signals and Averaging

Averaging modes can enhance the signal-to-noise ratio of a noisy spectrum. The choice to be made depends upon the statistical properties of the data. For random signals, the variance of the spectrum can be reduced by repeated measurements, followed by Fourier transforms, with rms averaging of the frequency *spectrum*. If the signal can be repetitively strobed (non-random case), *time* averaging reduces the noise prior to the FFT step.

8.3.5 SPECTRAL ANALYSIS AND ESTIMATION

Spectral Analysis and Spectral Estimation—Two Examples
Spectral analysis with a digital Fourier transform (almost invariably carried out by an FFT algorithm) computes a uniformly spaced set of N spectral amplitudes from an N-point data window. High-resolution determination of amplitudes, frequencies, and phases requires that N be "large enough" for the chosen resolution criteria. The window-weighting choice is necessarily a compromise between frequency resolution and suppression of sidebands. Very high frequency resolution by heterodyning and decimation can be obtained in spite of the resolution limit (1 in N frequency bins) *if the signal continues long enough* (see Decimation, above). The example to be discussed, a special-purpose DSP parallel processor built in the search for extra-terrestrial intelligence, yields 8 *million* frequency channels with a resolution of less than 1 Hz in a carrier frequency of 1.6GHz, or 1 part in 10^{12} resolution!

Spectral estimation is useful when the spectrum is rapidly changing, so that the number of points in a data window is too limited for FFT spectral analysis; for adequate performance, one is forced to *estimate* the spectrum "on the fly". Radar and speech signals, for example, fit this category. While *radar* signals can have basebands in the MHz, the rapidly pulsed or chirped or Doppler-shifted received signals, and the limitations of the best available real-time hardware, limit a given window to relatively few (typically < 256) data points. *Speech* spectra are also not stationary in time. An FFT of speech over an interval short enough to be meaningful contains poor resolution of characteristic vocal tract features. But a spectral *estimation* from the same short signal can successfully reproduce speech by modelling the vocal tract itself. In *spectral estimation*, one finds the set of trial functions and their coefficients, which best fit the data. Even for static spectra, spectrum estimation is useful when, instead of using sinusoids to fit the signal, one uses any arbitrary set of trial functions (usually orthogonal). There is a close connection between spectral estimation and modelling, since the coefficients in the fit are the parameters of the model, as will be shown in Section 8.4. The results of fitting a variety of minimum-error strategies to a single challenging spectrum are shown below.

If enough points are available to determine the spectral features adequately, spectral *analysis* is the usual choice, because it is straightforward; spectral *estimation* requires human experience, judgement, and perhaps even creativity; the choice of estimation technique varies with the characteristics of the signal and the way it was produced, and the information to be extracted. However, a signal with a limited number of points or an especially challenging spectrum (a small peak swamped by a nearby large one) favors spectral estimation. Indeed, the two approaches can be intelligently combined recursively. Typical schemes extract the major features with an FFT and fine-tune the dominant amplitudes, frequencies, and phases with a minimum error fit. Then these terms are subtracted and another FFT is used to expose any fine structure, which is no longer overwhelmed by the bigger peaks (for example, see Alles and Higgins (1973).

Spectral-estimation example

Questions resolved by spectral estimation are familiar: What will be the period of the next business cycle? Are sun-spots periodic, and if so, what is the period? What are the key parameters to model the vocal tract (or a bass fiddle, or a loudspeaker...), given only a response of the system excited by a pulse sequence? Is that a submarine in the (passive) sonar signal? Was that a micro-earthquake or was it a nuclear blast (and where did it occur?) Spectral estimation differs from cross-correlation or coherent detection: there is no controllable stimulus. One seeks to map the spectrum (or range, velocity, or other system dynamics) of a source from its spontaneous emission of signals, since the system is unavailable for stimulation with a sonar burst, noise source, or continuous modulation.

Spectral analysis by the DFT/FFT only defines a frequency to within one spectral channel (bin), with a resolution of 1 part in $(f_s/2)T_W = N(pts)$. By contrast, spectral estimation keeps the basis set small but lets the frequency values and other parameters be finely resolved. When to use spectral estimation rather than a Fourier transform? When the data set or window width is so small that the DFT/FFT spectral peaks do not give amplitude, frequency, or phase information with the needed precision, or because window-weighting broadening of large terms can mask small spectral terms.

Instead, *model the system* that produces the signal and minimize the error between model signal and actual signal. The improvement over FFT processing can be quite spectacular for short data records, because one can make more realistic estimates for the signal outside the window than those made by periodic repetition (as in DFT/FFT); the need for window weighting to smooth window-edge discontinuities is therefore eliminated. Spectral estimation needs more computation than FFT spectral analysis, unless the number of model parameters can be kept small—where spectral estimation works best.

An illuminating comparison of the performance of modern spectral analysis methods has been given by Kay and Marple (1981). Some twelve methods are compared for a test signal of devilish subtlety, comprising these three challenges:

   (i)   One high-energy peak and a peak of comparable energy
        very close (5%) in frequency; plus
   (ii) One peak well-resolved in frequency from (i) but weak
        (20-dB down) in energy; plus
   (iii) A broad band of noise.

The challenge and the results are shown in Figure 8.21. While many of these approaches would have given a "good fit" to the signal in the data domain, the frequency domain displays a clear distinction. The log scale of the

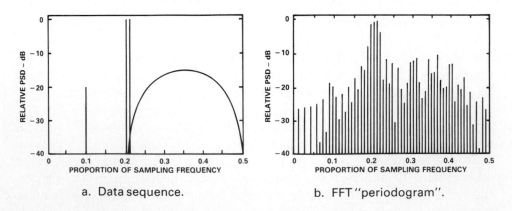

a. Data sequence.   b. FFT "periodogram".

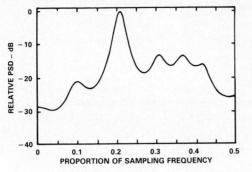

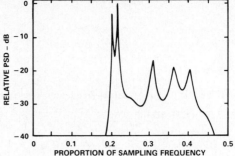

c. Autoregression (AR) or all-pole filter model.

d. Maximum entropy extrapolation of AR method.

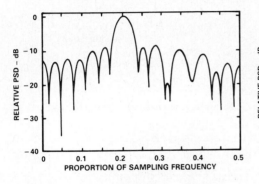

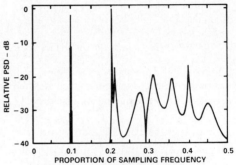

e. Moving average (MA) or all-zero filter model.

f. Harmonic decomposition fit to optimal set of exponentials.

Figure 8.21. Spectral estimates by various methods for the same 64-point sample. (Source: Kay and Marple, ©1981 IEEE—see Bibliography)

spectrum shows how well or badly small signals are recovered, and any broadening of the estimated spectral peaks shows quite clearly.

This is a challenging example. A comparison of the properties of each technique of Figure 8.21 on this short (64-point) signal is given in Table 8.5. It is not an exhaustive performance comparison; other signals might cause one or another feature to be weighed differently.

The magnitude-squared of the signal's FFT, historically known as the *periodogram*, Figure 8.21b, cannot distinguish nearby frequencies which fall within the $\sin(x)/x$ peak of the (unweighted) transform, but the small, well-separated signal is visible among the sidelobes. Window weighting would remove the sidelobe structure, improving the discrimination of challenges (ii) and (iii), but worsening the discrimination of challenge (i).

Table 8.5. Comparison of spectral estimation techniques

Comparison of selected spectral estimation techniques for a 64-point signal (compare Figure 8.21, consult Kay & Marple for details).

Challenge: METHOD	(i) resolves nearby frequency	(ii) detects weak terms	(iii) represents broadband noise	Computational complexity (multiplications)
FFT "periodogram"	No	Yes	Yes	$\dfrac{N}{2}\log_2 N$
Autoregressive (AR) (IIR all-pole)	No	Yes	Moderate (peaks look like system poles)	$NM + M^2 + MS$
Max. Entropy Autoregressive	Yes	No (Yes for improved algorithms)	Same as AR	$NM + M^2 + MS$
Moving Average (FIR)	No	No	No	$>NM + MS$ (nonlinear step)
Harmonic Decomposition (exponential fit)	Yes	Yes	Worse than AR	$M^2 + NM + M^3 + MS$ (polynomial root)

N = number of data samples per window

M = order of model of number of autocorrelation lags

S = number of spectrum samples computed; FFT could be used to shorten the generation of PSD.

The computation sequence for the autoregression (AR) method, shown in Figure 8.22, illustrates the approach in modelling methods. The AR method models the system as an all-pole IIR filter. The signal is assumed to be the response of a system (whose coefficients are to be determined) to a random noise-pulse excitation. First, the autocorrelation function, R_{xx}, is estimated from the signal, and the model coefficients, a_i, are obtained one by one via a method called the Levinson recursion. The mathematical details [see Kay and Marple (1981)] are similar to those developed later in this section for LPC speech analysis (Section 8.3); they need not be repeated here. The recursion loop is iterated until the model represents the signal's spectrum within a set error budget. While the $(N/2)\log_2 N$ advantage of the FFT is absent, modelling will involve fewer computations than the FFT as long as the N-point signal can be fit with an M-coefficient model for which $M \ll N$. The results of this approach are seen in Figure 8.21c.

The autoregression estimate (Figure 8.21c) does not resolve (i) but sees the low-energy peaks reasonably, though the all-pole fit is forced to put in false peaks to fit the broadband noise. There is a characteristic smearing, equal for all peaks, since the autocorrelation estimate is truncated at a number of lags equal to the number of filter parameters, as in LPC speech coding (see Section 8.4).

An extrapolation of the autocorrelation to lags which are not known can improve the spectral estimate. The maximum-entropy method (MEM) finds the time series which has the correct autocorrelation lags over the $(p + 1)$

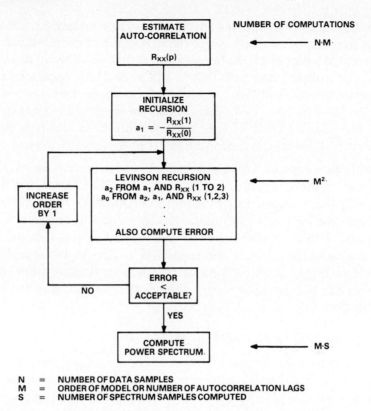

N = NUMBER OF DATA SAMPLES
M = ORDER OF MODEL OR NUMBER OF AUTOCORRELATION LAGS
S = NUMBER OF SPECTRUM SAMPLES COMPUTED

Figure 8.22. Computation sequence for the autoregression (AR) method, via the Levinson recursion. The number of computations in each step are shown to the right for N data samples, a model of order M, and S computed spectral points.

coefficients but is random otherwise. This method "whitens" the net error: it distributes the difference between actual spectrum and the model uniformly over the frequency band. As shown in Figure 8.21d, the AR-MEM has sharp resolution of close peaks (challenge (i)); the small signal (challenge (ii) is barely visible; and the noise power region (challenge (iii) is clearly visible though it is fit by peaks rather than by a smooth spectrum.

Another extrapolation enhancement (not shown) is the *maximum-likelihood method* (MLM), popular in seismic data analysis. The MLM finds a spectral estimate which minimizes the error in the frequency band of interest and has zero response elsewhere; it thus avoids trying to fit poles to regions of pure noise. The MEM method has sharper resolution of actual spectral peaks but can ring due to sidelobes. The MLM method is more stable with respect to noise and has fewer sidelobes, but poorer resolution, (see for example, Baggeroer (1978)).

Moving-average (MA) methods (Figure 8.21e) fit the signal to an FIR filter (all-zeros, no poles). The autocorrelation estimate gives the model coefficients from a nonlinear set of equations, solvable by a procedure called the "method of moments". If only the power-spectrum estimate is desired, the nonlinear-equation step can be avoided, since the MA spectral estimate can be shown (see Kay and Marple (1981)) to be equal to a simple and classic autocorrelation estimate due to Blackman and Tukey. MEM or MLE enhancements of the moving-average method are possible in principle but rare in practice (especially for real-time DSP) because of the nonlinear-equation step and because too many FIR coefficients are required to sharpen the spectral estimate; the AR method is computationally much more efficient.

A quite different approach to modelling is *harmonic decomposition* (also known as the *Prony* method), typified by Figure 8.21(f). The *signal*, not the system, is modeled, and assumed to be a sum of exponentials with differing decay times. Since the frequencies can take on any value, the frequency resolution is excellent. Either narrow-band or wide-band features can be estimated, though the noise spectrum is approximated in a peaky fashion. Even more-accurate determination of true signal frequencies and amplitudes can be made by suppressing the damping terms, though the modelling of the noise band is sacrificed.

Case Study in the Search for Extra-terrestrial Intelligence

"E.T., Call Harvard": Is there intelligent life out among the stars? Although chances of success are hotly controversial, a serious search for extra-terrestrial intelligence (SETI) is being conducted by scientists at several institutions. The discussion below will focus on one project, the "suitcase SETI" of Prof. Paul Horowitz at Harvard and his colleagues at the NASA Ames Laboratory. The project, so-named because of a powerful DSP processor for ultra-high-resolution spectral analysis built into a portable package, has now outgrown its suitcase. The design strategy will be discussed in three steps: (a) where to look, i.e. if "they"—the extra-terrrestrials (E.T.'s)—are broadcasting signals, what is the most likely search region? (b) What search bandwidth and resolution are required? (c) How can a spectrum analyzer be designed to meet these criteria?

If there is life out there, look in the water hole.

The most logical part of the electromagnetic spectrum in which to base a search is the microwave region (Figure 8.23). Noise sources within the galaxy decrease approximately as $1/f$, as do most random phenomena in nature; towards 1GHz, they approach a noise equivalent temperature of about 1K ($-272°C$), well below detector sensitivity. On the high-frequency end, the uncertainty principle places a quantum limit on the noise from likely transmitters such as masers; the absorption by atmospheric water also cuts down sensitivity. Within this region of minimum noise are located two emission lines from H and OH, components of water, hence the name "water hole."

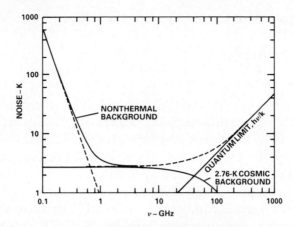

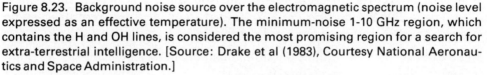

Figure 8.23. Background noise source over the electromagnetic spectrum (noise level expressed as an effective temperature). The minimum-noise 1-10 GHz region, which contains the H and OH lines, is considered the most promising region for a search for extra-terrestrial intelligence. [Source: Drake et al (1983), Courtesy National Aeronautics and Space Administration.]

Two emission frequencies likely to be available to any water-based life form are the hydrogen line at 1.42GHz and the OH line at 1.66GHz. A hydrogen or an OH maser is a likely carrier-frequency choice for an ET to communicate on. But the emissions from a narrow line are Doppler-shifted by the relative motion between observer and the star system being surveyed. Guessing the shift and correcting for it is a key question of search strategy. A "typical" strategy assumes an intelligence smart enough to "chirp" or frequency-shift their transmission from the H or OH frequency so it is at rest with respect to the galactic "rest frame," the remnants of the big bang at the creation of the universe. Other strategies assume less intelligence on the part of the transmitter, but all agree that a search bandwidth of less than about 200kHz could easily miss a transmission. The problem is how to search systematically at each possible frequency in the search range.

Requirements in Bandwidth, Search Region, and Number of Channels
A transmission at any particular frequency is likely to be broadened slightly due to scattering by ionized matter in the galaxy. Estimates place the minimum receiver bandwidth required at less than 0.1Hz. This combination of a many-kHz search bandwidth and a requirement for less-than-1-Hz resolution embodies the SETI DSP challenge.

SETI DSP Spectral Analyzer Requirements

Most likely carrier frequency: 1 to 2GHz

Possible search bandwidth due to Doppler shifts: 10Hz (best case) to 250kHz (most likely limit)

Minimum receiver bandwidth necessary: 0.01Hz to 0.1Hz

Spectral Resolution desired:
$$N(\text{pts}) = (200\text{kHz})/0.05\text{Hz} = 4\,\text{Million channels}$$

The bandwidth range would be only 10Hz if E.T. knows *our* sun's relative velocity (not likely); but it is more than 200kHz if E.T. subtracts the relative motion of its star from the cosmic rest frame (most likely) in determining the carrier frequency to be used. The received signals are brought down to an analysis frequency range by heterodyning with a local oscillator with "rock-solid" frequency stability at the frequency in question. The requirement is then to cover a search bandwidth of 200kHz in channels as narrow as 0.1Hz. This places severe restrictions on the FFT, which must have $N(\text{pts}) = (200\text{kHz})/0.05\text{Hz} = 4$ Million if done directly, requiring some $N\log_2(N) = 10^8$ multiplies per window.

Since each of these 4 million possible channels must be examined for long enough to detect a message (say 10 seconds), a single-channel receiver would take thousands of hours to search the water hole for each selected star. Clearly, a parallel processing strategy is called for (to be discussed below). What should be stored? Because of the enormous volume of spectral data, only the frequency and amplitude of an unusually high-energy spectral peak is saved. What will E.T. have to say? In the absence of a better conjecture, signature analysis might initially be limited to a predictable frequency offset in adjacent FFT windows as our sun moves with respect to the galactic rest frame, and an amplitude envelope corresponding to a plausible transmitter antenna pattern. A preliminary attempt called the Sentinel (Horowitz, 1984) tested some of the above search strategies with a single 64-K FFT, which was chirped to remove the earth's motion with respect to the cosmic rest frame and then scanned over the necessary bandwidth. The Sentinel covered a 2kHz bandwidth at any one time in 0.03Hz channels, chirping over 10Hz or 30 channels per minute. A key result of this preliminary experiment (which reported no extra-terrestrial observations) is that chirping to tune in to signals embodying likely E.T. broadcast strategies effectively zeros out all sources of noise from our solar system. The system described below builds on the Sentinel experience; it combines parallel processing by 128 identical 64K FFT processors, which are fed by a novel 128-channel digital bandpass comb filter.

An 8-Million-channel spectrum analyzer

Overview and Front-end Design: Among the largest FFT spectrum analyzers ever undertaken is the 8-million-channel unit in the META project of Prof. Paul Horowitz of Harvard. Building on prior experience with project Sentinel (above), Horowitz and colleagues at NASA and at Stanford have designed a 128-in-parallel FFT processor which incorporates virtually all basic DSP

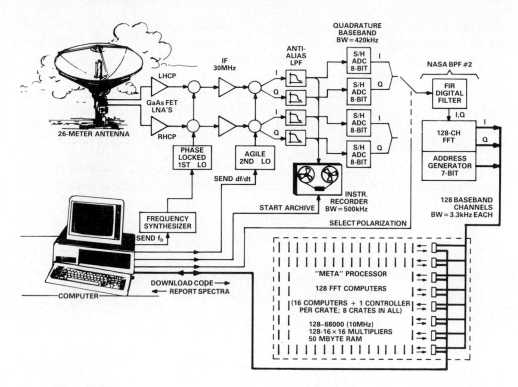

Figure 8.24. Figure 8.24. Signal processing in the META project in the search for extaterrestrial intelligence. (Courtesy Paul Horowitz, Harvard University)

techniques into one system: filtering, demodulation, decimation, and the FFT. The block diagram in Figure 8.24 shows the steps: demodulating the 1.5GHz signal to a 30MHz intermediate frequency, demodulating again to the 400kHz search band, and then dividing the 400kHz search band into 128 parallel sub-bands for fine-scale examination by 128 identical 64K FFT processors (the META processor) operating in parallel.

The received signal (with left- and right- circularly polarized components, LHCP and RHCP) is demodulated in two steps by heterodyning with a carefully fixed carrier (for example 1.66GHz, the hydrogen maser line) and then by a 30-MHz IF frequency (marked "agile" 2nd LO), which can be chirped to remove our sun's motion with respect to the galactic rest frame.

The second demodulation separates the signal into two-phase components, I and Q (in-phase and quadrature), to speed later FFT processing. The 400-kHz search range is then split into 128 channels of 3.2-kHz bandwidth and brought to a dc-origin by decimation and low-pass filtering, using a clever DFT trick described below.

Comb Filter 128-Channel Demodulator using the DFT: When combined with quadrature heterodyning, decimation makes possible a multichannel bandpass comb filter of high selectivity with only a single lowpass filter-design. The NASA comb bandpass is a decimator, followed by a lowpass, in turn followed by a DFT, as shown in Figure 8.25. A DFT over 128 frequencies, nf_0 ($n = 1$ to 128), in effect translates each of its outputs into a new dc origin if each channel of the output, $Y(f)$, is considered as a band of width, f_0. Careful low-pass filtering is included to avoid spillover between channels. This system is equivalent to a 128-channel bandpass comb filter. Each channel has been heterodyned down to center around dc for the final step, a 64-K fine-scale FFT.

Since the method is a useful one, the math will be summarized below.* The filter executes an FIR-lowpass operation on a decimated subset of the data.

$$y(n,k) = \sum_{p=0}^{NM-1} x(n-p) \, h_2(p,k) \qquad (8.29)$$

where h_2 is a lowpass, h_1, followed by a modulation, which together make a bandpass:

$$h_2(p,k) = h_1(p) \, e^{j2\pi pk/N} \qquad (8.30)$$

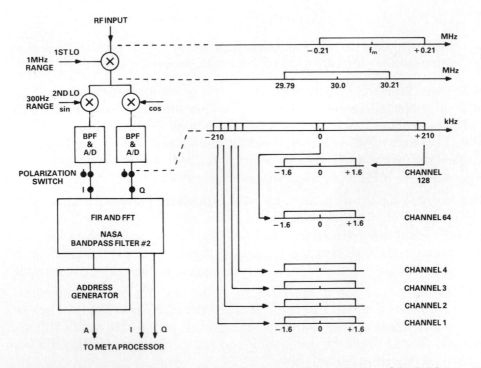

Figure 8.25. Block diagram (left) and frequency-domain operations of the NASA multi-band bandpass filter. (Courtesy Paul Horowitz, Harvard University, original figure drafted by John Forster)

*For more details, see the SETI Working Group Report, NASA Technical Paper 2244 (1984).

This may be rearranged to the form of the actual computations performed:

$$y(n,k) = \sum_{k=0}^{M-1} e^{j2\pi mk/N} \sum_{m=0}^{M-1} x(n-mN-k) \; h_1(mN+k) \quad (8.31)$$

which may be recognized as a decimated lowpass (inner sum), followed by a DFT. Thus, the DFT over a coarse grid brings down each of its channels (k) to a center at dc about which further fine-scale processing (n) can be performed.

The 64-K real-time FFT: The 64-K FFT parallel processor subunit of Figures 8.24 and 8.25, while conventional in design, has high performance and a compact modular structure to allow for 128 replicas to be built and to operate simultaneously. A given Fourier processor does a 64-K complex FFT in integer arithmetic in 14 seconds, using a M68000 CPU and a CMOS multiplier to speed the FFT. With the DSP components described in earlier chapters, this could be speeded at least 10-fold. Another 6 seconds are set aside for power-spectrum computation and peak searching. Vector magnitudes, M, are computed from the FFT components, X, Y, by a trick that avoids squaring operations:

$$M = \max(|X|,|Y|) + 0.5 \min(|X|,|Y|) \quad (8.32)$$

This magnitude estimate, based on the power-series expansion of the Pythagorean triangle rule for small phase angles, is accurate to within 10% for any phase (try it for 45°, the worst case). The search for FFT peaks (the hoped-for E.T. messages) is done locally within a given processor because it is awkward to move 8 million numbers from 128 parallel FFT processors to an archiving processor every time a search window is done. While all the results are not yet in, the exercise elegantly illustrates the application of DSP principles and hardware to a topical problem in scientific instrumentation.

FURTHER READING ON DIGITAL DETECTION STRATEGIES
(refer to Bibliography for complete listings)

Crochiere and Rabiner (1981) on decimation and interpolation basics.

Crowell and Alipanhi (1981), optimum correlators for transients

Delves and Samba (1982), optimum deconvolution algorithm separates seismic echoes from noise.

Langston (1982) for issues in modem design using a DSP microprocessor.

Oxaal (1983) for elementary-level heterodyning and decimation.

Schwartz and Shaw (1975), general background on signal detection.

Tyson and Giffard (1978), optimum filter in the search for gravity waves.

Whalen (1971), theory behind optimum filters, especially chapters 6 and 10.

8.4 MODELLING IN REAL TIME: TELECOMMUNICATIONS AND SPEECH

8.4.1 WHY MODELLING?

Section Overview

Modelling is a way to analyze or synthesize a signal more compactly than by direct operations on the waveform itself. If a model of the signal *source* can be constructed, only the *parameters* of the model need be processed, transmitted, or stored. Modelling in the time domain attempts to fit a finite sample of the signal to a set of coefficients which can *predict* immediate future behavior, as in extrapolation. Modelling in the frequency domain fits the frequency, amplitude, and width of dominant spectral peaks, and then *regenerates* the signal which that simplified spectrum would produce. An example of these two interrelated methods, linear predictive coding of speech, will be treated here in some detail.

Modelling and prediction are related both to *data compression*, which seeks to reduce bit rates in transmission, processing, and storage of digital information, and to *adaptive equalizers*, which actively enhance S/N in telecommunications. These DSP applications will be discussed briefly below. Special-purpose DSP chips now available for these applications replace earlier more-complex and expensive circuits in modems, modulators, and other communications circuits—making higher performance available at lower cost. The single-chip DSP processor is well-suited to modelling, analysis, synthesis, compression, differential transmission and adaptive filter applications, with higher performance than most special-purpose consumer chips.

These ideas are illustrated in terms of telecommunications, but because digital *image* communication demands even higher rates of transmission, processing, and storage, it too benefits from data compression. An image contains much redundant information; the intensity at a given pixel is related to that at nearby pixels; the *differences* are most significant. Digital image communication with reasonable bandwidth is possible only with high

---

Examples of Modelling and Data-Compression Applications

Domain	Application
Telecommunications	Adaptive modulation; Adaptive filtering
Speech and music	Voice and music synthesizers
Digital audio	Compact disk storage; Digital recording studio
Video	Image Bandwidth compression

performance differential data compression, at computation rates beyond the throughput rates that telecommunications chips can handle. The same data-compression ideas introduced below will prove useful in designing custom DSP microcoded architectures that are well-suited to important application areas such as digital TV. Graphics communications over fiberoptic phone links is another application for which the cost of uncompressed transmission is prohibitive.

"Too Much to Say, Too Little Time"
Recognizable digital speech or high-fidelity digital audio are demanding of bandwidth.

	Amplitude Quantization	Sampling Rate	Resulting Bit Rate
Voice	10 bits	7kHz	70 kbits/s
Hi Fi	14 bits	40kHz	560kbits/s

High-quality digital voice telecommunications, if done directly, would take about 20 times the bandwidth required to transmit the same 3.5-kHz analog signal, with proportionate reduction in the number of calls per channel and consequent higher costs. Voice-grade phone lines, which pass *sine waves* of only up to about 5kHz, are ruled out since they distort binary *pulses* above about 1kbit/s. While long-distance lines have much wider bandwidths (100MHz for coaxial cable, several GHz for optical fibers), their high cost requires multiplexing as many calls onto the same line as possible.

Good-quality digital hi-fi, if stored directly, would consume a megabyte of storage for each 14 seconds (1Mbyte)/((560kbits/s)/8) recorded. While digital optical disks have storage density high enough to permit direct permanent storage of reasonably sized records (massaged for error avoidance and error correction), erasable storage in RAM or ROM of more than a few seconds length is effectively ruled out. In addition, for real-time synthesis or analysis, one must consider the throughput required to achieve 0.5-MHz signal bit-rates directly.

Differential and adaptive techniques, introduced below under *Telecommunications*, are a key to reducing these requirements without compromising signal quality or adding error. If *differences* between sampled data points are encoded for transmission, the bit-rate required to handle only the changes is much smaller than that required to encode (for example) a slowly changing measurement repeatedly without reference to the previous one. *Adaptive receivers* improve the reconstructed signal by adjusting quantization step size for minimum error, while *adaptive filter* systems compensate for phase delays, amplitude errors, and echoes in a communications channel by dynamically altering filter coefficients.

8.4.2 TELECOMMUNICATIONS

Modelling in Data Compression

Comparison of Data Compression Methods

How might one transmit and receive (or store and play) high-quality speech or music at bit rates lower than would be required for direct digitization? Table 8.6 summarizes the main methods, many of which have now been implemented in special-purpose DSP chips. Recalling the general digital transceiver of Figure 8.1, these methods find use in the modelling block on the transmitter side, or the compression block on the receiver side.

Methods called *source coding* send the parameters of a model of the signal in time or frequency; the linear-predictive-coding (LPC) method, popular in speech analysis and synthesis, will be discussed under SPEECH. These methods give the lowest bit rates and high intelligibility, but their quality is more machine-like than in waveform encoding.

Methods called *waveform coding*, adequate for more-demanding digital audio and "sampling" synthesizers, utilize differential and adaptive PCM to reduce

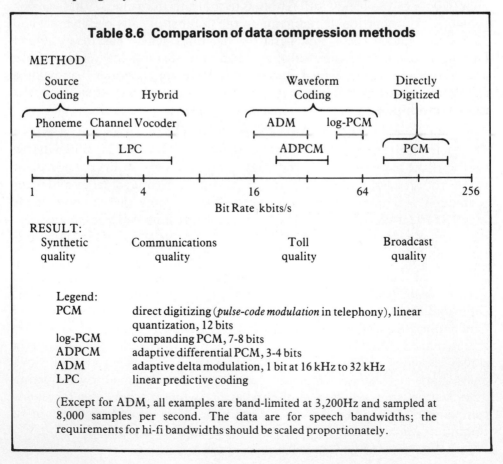

Table 8.6 Comparison of data compression methods

METHOD

Source Coding — Hybrid — Waveform Coding — Directly Digitized

Phoneme   Channel Vocoder   ADM   log-PCM

LPC   ADPCM   PCM

Bit Rate kbits/s

1   4   16   64   256

RESULT:

| Synthetic quality | Communications quality | Toll quality | Broadcast quality |

Legend:

PCM	direct digitizing (*pulse-code modulation* in telephony), linear quantization, 12 bits
log-PCM	companding PCM, 7-8 bits
ADPCM	adaptive differential PCM, 3-4 bits
ADM	adaptive delta modulation, 1 bit at 16 kHz to 32 kHz
LPC	linear predictive coding

(Except for ADM, all examples are band-limited at 3,200Hz and sampled at 8,000 samples per second. The data are for speech bandwidths; the requirements for hi-fi bandwidths should be scaled proportionately.

the bit-rate.* Differential PCM (DPCM) lowers the bandwidth by sending only differences between recent samples, which are integrated back when received. Differential methods reduce the chance for annoying "pops" when a high-order bit error occurs in straight PCM. To minimize accumulated error, a predictor on both source and receiver extrapolates recent data and subtracts it from the signal to make a smaller differential more easily quantized. The predictor is a digital filter. If the filter coefficients are allowed to change, the technique is called adaptive differential PCM or ADPCM.

Differential Data Compression Methods

Since adjacent values of a signal are related to one another, much higher speed and resolution can be attained within a given finite signal-bandwidth if only differences between samples are digitized and transmitted, as in the differential voltmeter of instrumentation or in the tracking ADC. The simplest differential PCM (acronym DPCM) transmits the quantized value, $e(n)$, of the change between adjacent samples of the input signal, $x(n)$. This difference is reconstructed into the received signal, $y(n)$, by an accumulation:

$$e(n) = x(n) - x(n-1) \quad \text{transmitter} \qquad (8.33)$$
$$y(n) = e(n) + y(n-1) \quad \text{receiver}$$

To prevent discrepancies due to quantization noise between input, $x(n)$, and $y(n)$ from accumulating during reconstruction, an equivalent to the received signal, $y(n)$, is reconstructed at the transmitter by replicating the receiver's accumulator. Thus, $y'(n-1)$, generated from $e(n)$ in the same way that y will be generated from $e(n)$ at the receiver, is used instead of $x(n-1)$ to generate the difference signal. That is,

$$e(n) = x(n) - y'(n-1) \qquad \text{transmitter, improved}$$

This form of differential waveform coding in telephony reduces bit rate by a factor of about 8, yet it retains dynamic range adequate for good-quality speech (65dB). A more-realistic prediction (Figure 8.26) extrapolates the received $y(n)$ from the difference signal and a sequence of p previous output samples, in a summation known as a *predictor*:

$$y(n) = e(n) + \sum_{k=0}^{p-1} h(k) \ y(n-k) \qquad \text{extrapolation}$$

where p is the order of the predictor.

---

*PCM stands for *pulse-code modulation*. In telephony, the term means digitizing an analog signal; it stems from the transmission, in serial form, of a pulse-sequence for each digitized data point. The term *modulation* is retained here in deference to the telecommunications acronyms in common use, even though *digitization* more closely fits the present meaning.

The values of $h(k)$, which determine the nature of the extrapolation, are recognizeable as the coefficients of an FIR filter. The same predictor, replacing the accumulator, appears on the transmitter side to generate the differential error estimate, $e(n)$, to be transmitted.

Higher-order predictors correspond to higher-order extrapolations. Performance of the differential modulator is set by its clock frequency, f_s, and the step size; thus it depends upon both frequency and amplitude of the signal, as shown by the reconstructed waveform of Figure 8.26c. Quantization noise is most apparent for slowly changing signals, while rapidly changing signals reach a limit called slope overload. Differential modulation is popular in in telecommunications because:

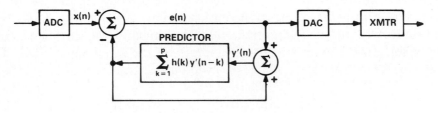

a. Transmitter.

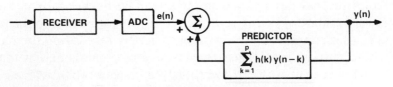

b. Receiver.

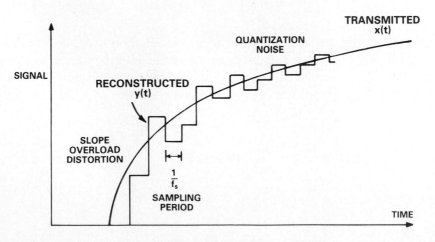

c. Transmitted and reconstructed signals.

Figure 8.26. Differential or DPCM compression with a pth order predictor.

For speech signals, the voice spectrum falls off as $1/f$ at higher frequencies; this just matches the slope-overload criterion of the slope-sensitive modulator, so a human voice is unlikely to push the differential modulator into slope overload.

Differential modulation has a negligible chance of the large error which occurs in PCM in the event of a high-order bit error due to noise in transmission. For example, in transmitting the sequence, 1000 1101 1100 1001, loss of the leading "1" would cause a huge "pop" of noise (1/2 full-scale), while if only *updates in the LSH* (1100 1001) are transmitted, even if the most-significant bit is lost, the error in the reconstructed signal would be only 1/256 of full-scale.

Adaptive Data Compression Methods

A predictor is called *adaptive* if the values of its coefficients (Eq. 8.34) are estimated from the data itself; the nature of the extrapolation changes as the data becomes more or less stationary. With the error minimized by least-squares methods, the $h(n)$ can be shown to be the solution of a set of linear equations, the *same* equations that model speech by linear predictive coding (below).

An ADPCM (*adaptive* DPCM) algorithm is readily implemented using a DSP processor (see, for example, Boddie (1981)) to achieve good quality at 32-kbit rates, with a further reduction to 8 kbits when combined with a μ-law compander. The noise-reduction benefits of digital transmission are achieved with quality indistinguishable from that of analog phone-lines for speech bandwidths. Special-purpose chips (e.g., the NEC μPD7751) incorporate the ADPCM algorithm to give high quality at 14-20K bits, so only about 2K bytes of memory are required per second of information transmission.

8.4.3 CODING OF SPEECH

Vocal-Tract and Speech-Synthesis Models

Speech analysis and synthesis methods

Speech synthesis methods differ in bit-rate and storage requirements as well as in "fidelity" to human sounds. Some popular methods are compared in Table 8.7. We give below a brief summary of an already well-developed field (see Oppenheim (1978) or Rabiner and Shafer (1978)). Speech *analysis*, the inverse process of speech *synthesis*, involves quite similar computations. Speech *recognition*, however, depends on template-searching algorithms which go beyond the scope of a DSP book. We describe in detail the linear-predictive-coding (LPC) method, whose processing computations can to be done in real time, yet is sufficiently powerful for realistic results.

What is the nature of the system that the DSP speech synthesizer is trying to model? The human vocal system (Figure 8.27) combines an excitation source with a time-varying filter, the vocal tract. For *voiced* sounds (the

HUMAN MODEL

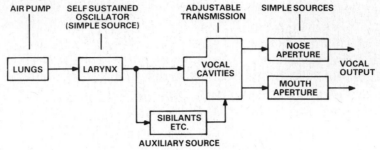

a. Human voice box and excitation.

MACHINE MODEL

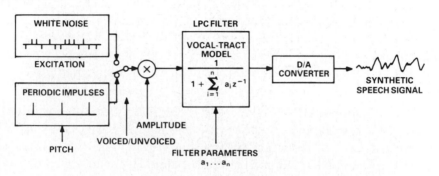

b. Artificial speech generation by analogous components.

Figure 8.27. The vocal system (Adapted from Bucy et al, 1984, Courtesy Texas Instruments).

vowels and tonal parts of consonants), the excitation is a high-frequency pulse string from the vocal cords, vibrating at a time-varying pitch. For *unvoiced* sounds (the onset of "p", "b", "s", "t"...) the excitation is a noise source produced by turbulence at the lips or within the vocal tract. A similar separation of excitation and resonant cavity is an essential part of acoustic musical instruments (violin, flute, piano, etc.). The wide range of expression in the human voice, when compared to a musical instrument, is due to the (literal) flexibility of the resonant cavity. Figure 8.27 also shows a synthesizer equivalent. The output is generated by passing the pitch (voiced) or noise (unvoiced) excitation through the filter, $H(f)$, which convolves each pulse with the filter impulse response, thus simulating the formation of characteristic sounds (called *formants*) by the vocal tract.

The spectra of three vowels are shown in Figure 8.28. The peak pattern corresponds to the vocal tract resonances; it is a signature which gives each its characteristic sound. The high-resolution spectral "zoom" shown in the inset displays peaks at multiples of the vibrational pitch of the vocal cord pulses.

VOWEL CONTROL PARAMETERS

Vowel	Formants (Hz)			Formant Bandwidths (Hz)		
	f_1	f_2	f_3	B_1	B_2	B_3
a	620	1,220	2,550	80	50	140
u	450	1,100	2,350	80	100	80
i	310	2,020	2,960	45	200	400

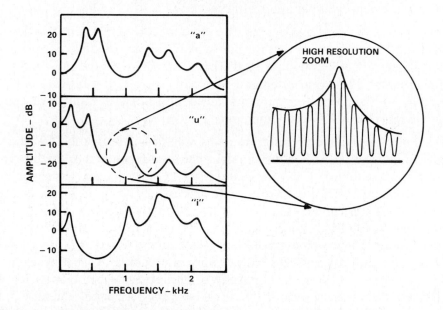

Figure 8.28. Spectrograms of three vowels. The magnified inset shows on a finer frequency scale the multiples of the pitch.

To avoid the formidable storage requirements of direct digital recording and reproduction, speech-synthesizer designers try to approximate the spectral intensities (vocoder), model the system which produces it (LPC), or compress the speech waveform itself (waveform coding).

The *channel vocoder* analyzes speech by passing the signal through a bank of bandpass filters and extracting the amplitude in each frequency band (channel) as a function of time. In vocoder *synthesis*, the excitation components (pitch and noise) excite the bandpass filter bank at the same fixed frequencies, with gains set by the stored amplitudes.

The vocoder's advantage is that it makes it possible to store a fixed yet complete "vocabulary" of phonemes (smallest units of speech that distinguish one utterance from another in a given language); they can be combined to form *any* word, with rules to govern the pitch, inflection, stress, and timing. The vocoder therefore lends itself, for example, to text-to-voice translators for the blind with an infinite vocabulary yet no need for voice *input* coding of specific words. Compact recording of parameter information is readily accomplished with the same bandpass filters by running the system backwards, so vocoders can record as well as reproduce speech.

Waveform coding methods record true speech but compress the record, using DPCM or ADPCM as above. The playback is very realistic, though the vocabulary is limited to the exact words already spoken into it. The storage demands are relatively large: at 64K bits per second, waveform coding requires 0.5 megabytes per minute of speech.

Linear-predictive coding lessens the amount of stored information by modelling the vocal tract in real time so as to follow (predict) the principal spectral peaks. LPC methods are thus very compact in storage and bandwidth requirements. The storage demands are quite low: with a bit-rate of 1,200 baud, as in the Texas Instruments Speak-and-Spell, a minute of speech requires less than 64K bytes of memory.

LPC is not limited to a preprogrammed vocabulary: automatic LPC *analysis* algorithms have been demonstrated in speaker-dependent trainable voice recognition (see Bucy (1984) for example). LPC speech *quality* is superior to that of vocoders, with less hardware or processing; it requires only as many filters as peaks in a given sound rather than as many filters as channels in the spectrum. While LPC speech is inferior to waveform coding, the memory requirements are about 40-times less. LPC analysis and synthesis can be carried out with a single DSP processor, and such LPC synthesizers are finding their way into low-cost products or product enhancements such as the electronic voice-alert system ("Please fasten your seat belts. Thank you."), employing TI voice-synthesizer chips and available in many cars. The background for LPC speech analyzer and synthesizer design will be developed below.

Table 8.7 Speech synthesis methods and performance

Method:	LPC	Vocoder	Waveform Coding
Bit rate, bits/s	2,400	4,800–9,600	32K-64K
Hardware requirement	Low (DSP processor)	Medium (bandpass)	Moderate (ADPCM)
Software algorithm	Voicebox model	Spectral analysis	Compression
Speech quality	Good	Fair	Excellent
Example	Speak and Spell (TI)	Votrex or Votan	Digitalker (National)
Best Domains	(1) Low-cost units (2) Trainable	Text-to-speech	Highest fidelity

Linear Predictive Coding

Basics★

The elements of a linear-predictive-coding system are shown in Figure 8.29, which may be compared with the digital transceiver of Figure 8.1. To represent a sound, take a Fourier transform of a segment short enough to include just one phoneme, yet long enough to give spectral peaks characteristic of the particular vowel or consonant. Rather than dealing with the entire spectrum, fit the dominant peaks with the parameters of bandpass filters which represent what the formants do in the vocal tract. Instead of storing information in the form of a spectrum, store the model parameters: frequency, amplitude, and width of the peaks. This small set is all that is stored in LPC analysis.

To get the signal back again, let those coefficients define the properties of a real-time digital filter bank, driven by an excitation, $x(n)$, with amplitude, G. The excitation may be either a periodic pulse sequence (voiced sounds) or a noise source (unvoiced sounds). If the filter parameters were well-

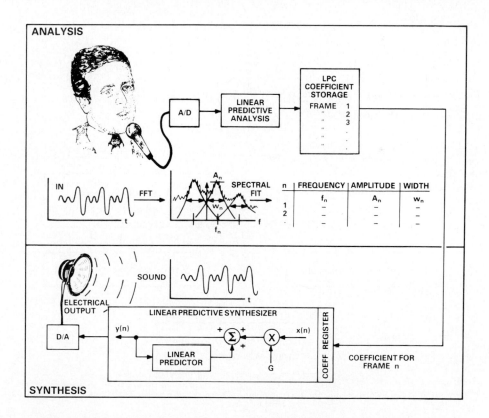

Figure 8.29. Steps in the linear predictive analysis (LPA) and linear predictive coding (LPC) of speech.

★For an excellent introduction to LPC theory, see the tutorial by Makhoul (1978).

determined, the speech will resemble the original. With LPC, the Fourier transform is not needed. Instead, the past is used to predict the future, via time series analysis, as in least-squares fits. The best-fit coefficients are put into recursive difference equations, which will be recognized as identical to those which generate IIR bandpass filters (that's why the FFT step is unnecessary). Since a typical speech utterance has only a few dominant peaks in the spectrum, relatively few parameters suffice to describe the vocal tract. That is why LPC is so compact.

The term *linear prediction* means the prediction of future results, $y(n)$, from past experience, $y(n-k)$, and present stimulus, $x(n)$, using linear difference equations:

$$y(n) = \sum_{k=1}^{p} a_k y(n-k) + G x(n) \qquad (8.35)$$

This will be recognized as the difference equation for an all-pole* filter with the transfer function:

$$H(z) = \frac{G}{1 - \sum_{k=1}^{p} a_k z^{-k}} \qquad (8.36)$$

How are the LPC coefficients, a_k, actually determined? There are two equivalent approaches:

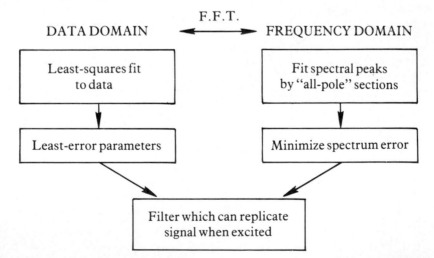

*The terminology of LPC overlaps into the terminology of spectral estimation. The following equivalents may prove useful.

Speech term	Filter	Spectral Estimation term
"all-pole"	IIR	autoregression (AR)
"all-zero"	FIR	moving average (MA)
poles & zeros	General	ARMA

The a_k are determined by minimizing the error in a fit to the previous speech interval. For example, the *autocorrelation method* follows the pathway on the left above. A least-squares fit to p past values of the signal is to be determined:

$$y_{est}(n) = \sum_{k=1}^{p} a_k\ y(n-k) \tag{8.37}$$

Minimize the total squared-error, E, of the actual $y(n)$ from this estimate over the data window of length N:

$$E = \left\{ \sum_{n-1}^{N} y(n) - \sum_{k-1}^{p} a_k\ y(n-k)^2 \right\}^2 \tag{8.38}$$

by finding the value of each coefficient, a_k, that brings to zero each derivative of error, E, with respect to a_k:

$$\frac{dE}{da_k} = 0, \qquad 0 \le k \le p \tag{8.39}$$

This gives a set of linear equations for the a_k:

$$\sum_{k=1}^{p} a_k R(i-k) = -R(i),\ \ 0 \le i \le p \tag{8.40}$$

where

$$R(k) = \sum_{n=0}^{N-k-1} y(n)\ y(n+k) \tag{8.41}$$

$R(k)$ may be recognized as the autocorrelation of the signal over the window; p, often referred to as the number of correlation lags, sets the order of the filter approximation, i.e., the number of poles in the LPC model. The autocorrelation estimate must be carefully windowed, especially for sounds in which the pitch is changing rapidly.

Solutions for the LPC coefficients from the autocorrelation estimates are found by standard linear-equation techniques, optimized for computation speed—especially if, as in the example below, the LPC model is to be used in real time for speech recognition. For instance, a method known as the Levinson-Durbin reduction (consult Makhoul for references) requires only p^2 multiplies and $2p$ storage locations—a modest requirement, since the number of poles, p, required to model speech is not large. The evaluation of p coefficients need not be the speed-limiting step, since the prior evaluation of the autocorrelation estimate takes Np multiplications for window length N, and usually $N \gg p$.

An example of LPC results calculated with high resolution appears in Figure 8.30. The reconstructed signal is shown below the natural voice input. They are not identical, since phase information is lost, but they are "similar" enough to sound alike. In practice, speech *recognition* processing can be done with a quite modest number of poles in an LPA (linear-predictive-analysis) fit, as shown in Figure 8.30b. The more-important 0-5-kHz region is fit with 14 poles, while the region above 5 kHz is approximated by only 5 poles, since the energy is low and relatively unimportant in extracting a template. The fine-structure in the actual spectrum is from the pitch-pulse excitation.

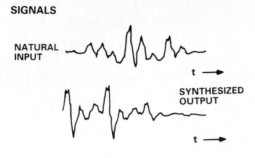

a. A speech signal and the resulting resynthesized signal.

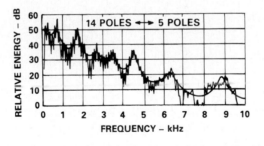

b. Spectrum and LPA fit.

Figure 8.30. Linear predictive analysis example. (Source: Makhoul, ©1975 IEEE—see Bibliography)

LPC Design Exercise

An LPC analyzer/synthesizer implemented with a TMS-32010 DSP microcomputer for a personal computer product is diagrammed in Figure 8.31. The application is a trainable voice recognizer. In the "learn" phase, the user is presented with a menu of command words to pronounce; they are recorded and LPC-analyzed. In the "command" phase, the user can speak any of the words whose template has been stored and cause the computer to respond. The voice input to be encoded is put through two analysis branches (Figure 8.31); the top branch extracts formant parameters and the bottom branch extracts pitch. Ten-vocal tract parameters are used, two for each second-order section or spectral peak. With overall gain ("energy") and pitch

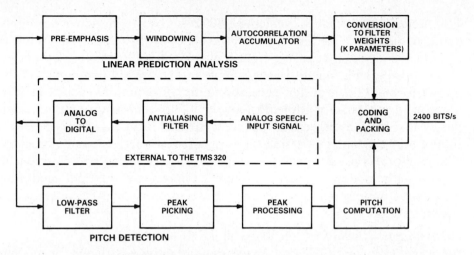

Figure 8.31. LPC analysis algorithm to extract both vocal tract resonances and pitch in the T.I. Speech-Command microcomputer peripheral. Model: 10-pole LPC, 20-ms frame period, 30-ms window length. (Source: Bucy et al, ©1984 IEEE—see Bibliography)

for the excitation, the total number of parameters is only 12. For unvoiced sounds, the noise excitation must be added but the number of spectral-peak parameters is less. The algorithm computes 10 LPC coefficients from the 11 autocorrelation coefficients ($p = 10$ in Eq. 8.34).

Pitch tracking to determine the excitation is done in the lower branch of the Figure. The signal is low-pass filtered to emphasize the pitch over the overtones, and peaks and valleys are examined to look for repetitions in time. While zero crossings are sufficient for pure tones, voice sounds require a more complex strategy. In the Gold-Rabiner (1969) pitch tracker (see Rabiner and Gold 1975 for a summary), six different pitch estimators are generated from combinations of two adjacent peak and valley heights and polled to see if a consensus exists. If no single pitch period is determined, the interval (frame) is declared unvoiced—i.e., the excitation is noise rather than a tone.

Because the peaks in the voice spectrum are so broad, recognizable—though not hi-fi—speech can be generated with coefficients of only moderate resolution. Further compression results if the same sound is repeated over more than one frame; a single bit can signal a repeat with no change in filter coefficients. This results in a modest 2,400 bits/s encoding in the TI Speech Command module:

LPC SPEECH-ENCODER PERFORMANCE

Frame length	20ms (50 frames/second)
Filter coefficients	10 at 5 bits each
Energy coefficient	1 at 4 bits
Pitch coefficient	1 at 5 bits
Total bit rate	2,950 bits/second

This modest sum is shaved further to 2,400 bits/s by a reduction of the number of bits in the higher-frequency peaks, since they are less important. While the 31 channels of pitch available from 5 bits of pitch information do not give smooth tones, and 5-bit filter coefficients can be readily distinguished from human formants, this design exercise and the computations required to carry it out were successfully implemented with a first-generation DSP processor, the TMS-32010. Synthesis of the stored sounds (the resynthesis process of Figure 8.29) is done by passing the parameters through the same processor. This design exercise accomplishes respectable voice storage, recognition, and output with modest bit-rates through compromises on pitch and filter-parameter quantization, and through the use of compact lattice-filter methods (see Bucy et al. (1984)).

Other Adaptive Predictors, Equalizers, and Echo-Cancellers

Adaptive predictors for data compression, adaptive equalizers for modems, and adaptive echo cancellers for long-distance voice communications are closely related to the least-squares extrapolation or LPC modelling of speech described above. The coefficients for an adaptive predictor in Eq. 8.37 are solved by the same method as LPC from the autocorrelation estimate of recent data. The adaptive equalizer and echo canceller are briefly described below to illustrate that LPC and predictive methods have generality far beyond speech synthesis.

A modem receiver section employs a coherent detector (as in Section 8.2) to convert the frequency-shifted or phase-shifted audio signal back to binary pulses. An *adaptive equalizer* (Figure 8.32a) is a key component of the modem receiver. Here's why: the transmission of data suffers from losses and phase shifts over the line. A pulse stream sent over a voice-grade phone line is severely distorted by the 5kHz bandwidth. Adjacent pulses at rates beyond 1,200 baud are so distorted that they can overlap one another. This effect can be reduced by convolving the received signal with a filter (equalizer) whose transfer function is the inverse of the phone line's impulse response. But the nature of the communications channel can change during a communications session; for example, it changes each time another connection is made (line loading). *Adaptive equalizers* adjust the filter transfer-function for a changing phone-line characteristic in an error-minimization loop, which restores the received pulses to optimal pulse-shape.

Other modem components may be recognized as elements in the general digital transceiver of Figure 8.1. The *Hilbert-transform* block is another name for quadrature heterodyning, to remove the negative-frequency images (Section 8.2, Figure 8.7). The *Rectangle Converter* brings the signal into clean binary form for the decision logic, which recovers the clock frequency and the (demodulated) binary data stream. The *Digital PLL* block is the same combination of digital VCO and feedback loop described above in the phase estimator portion of Section 8.3 (Figure 8.12). Beyond the principles mentioned here, further detail is avoided, since modem design is a very specialized topic; its implementation is already well-accomplished with available special-purpose DSP chips.

An echo canceller (Figure 8.32b), now implemented in large-scale ICs, performs a function which is essential over very-long-distance satellite

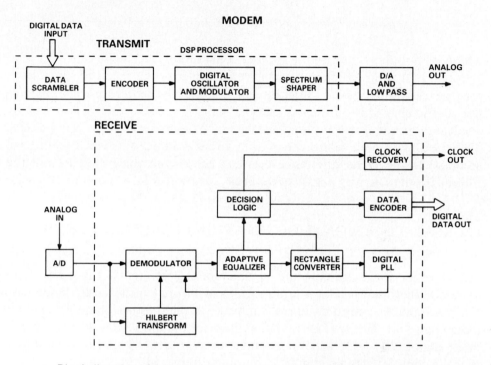

a. Block diagram of a modem, showing the use of an adaptive equalizer.

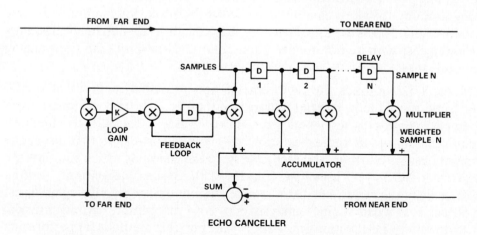

b. Adaptive echo-canceller for long-distance voice communications.

Figure 8.32. Other adaptive DSP applications.

channels and important in other long-distance telecommunications. Unintentional feedback of a receiver signal back over the transmission path leads to echoes, which become an increasingly serious impediment to conversation as the distance or echo time increases. For example, with a satellite in a stationary orbit, a 12,000-mile transmission has an echo time of

$$t(echo) = (2 \times 12,000\,\text{mi})/(180,000\,\text{mi/s}) = 0.13\,\text{s}$$

which is enough to confuse the speaker considerably. An echo canceller is an adaptive FIR filter which injects delayed samples of the received signal back into the transmission path. Since the properties of the echo path are unknown, the coefficients of the filter are automatically (i.e., adaptively) adjusted by small amounts at each sampling interval in the feedback loop of Figure 8.32b. The circuit responds adaptively within a few syllables of speech (about 1-2 seconds) to drive the difference between echo from the near end and the filtered input to zero by negative feedback, so no net echo is transmitted back.

8.5 REAL-TIME SIGNAL GENERATION AND MUSIC SYNTHESIS

8.5.1 INTRODUCTION AND SECTION OVERVIEW

Signal generators stimulate the system under test (recall Figure 8.1), or generate the tones in telecommunications or music synthesizers. Done with DSP, a signal is computed in real time and shipped to the outside world via a d/a converter. While it is easy enough to set the amplitude and frequency of an analog sine-wave generator, and sine-wave generators can be summed to make complex signals, digital techniques offer stability, accuracy, and ease of parameter variation, and they can generate complex signals with modest programming. Modern electronic music-synthesizers, for example, are now totally digital (up to the final DAC(s)); they can generate rich musical sounds at lower cost with higher quality, flexibility, and stability than their analog predecessors.

Section 8.5.2 outlines the requirements and principal methods of digital function generation: storage and readout of an arbitrary function from memory; interpolation between stored values to reduce memory requirements and increase resolution; polynomial expansion, to improve the smoothness still further, and recursive synthesis methods which simulate the system or recursively hunt for solutions of the system's equations. Section 8.5.3 covers the requirements of realistic digital electronic music synthesis; it describes what digital analysis has told us about natural musical instruments. Then Section 8.5.4 surveys the possible methods for performing the synthesis to reproduce these sounds. Section 8.5.5 discusses one successful solution, known as the parametric method, which generates harmonic-rich sounds by frequency modulation and lends itself to a real-time performing synthesizer (as opposed to an off-line composing instrument) with responsive dynamics to satisfy the musician.

8.5.2 FAST FUNCTION GENERATION

Digital Function Generation

The following section summarizes methods of digital fast function generation. It is useful to relate the digital methods to be discussed to two classic types of analog signal generators, the sine-wave oscillator and the square/triangle function generator.

(a) The traditional *sine-wave oscillator* excites a resonant circuit that has a characteristic frequency which can be changed by varying a circuit parameter; or it is tuned to feed back an output signal regeneratively at a single frequency. The digital counterparts to be discussed below perform by either *simulating* an oscillatory system or *dynamically computing* the power-series expansion of the function being generated, with enhanced frequency stability and accuracy of frequency selection while locked to a stable crystal timing system.

(b) Analog *function generators* combine integrators and switching circuits to generate square waves, ramps, and triangular waves, which may be shaped by nonlinear function-circuits into sine-wave approximations. The *digital oscillator* employs an accumulator into which a constant, the *frequency-control word* (FCW), is added each clock cycle. Since the number of counts required to accumulate to overflow (ramp time) depends on the ratio of full-scale to the value of the FCW, the frequency of oscillation depends directly on the FCW. Software switching determines the waveshape: a ramp, if overflow triggers a reset-to-zero command, or a triangle, if overflow changes the sign of the constant. A sine wave or any arbitrary waveshape may be generated by storing the waveshape in memory with addresses accessed by the accumulator.

(c) *Memory-stored waveforms* produce a point-by-point approximation which (if memory space is limited) may be too jagged or harmonic for high-resolution applications. Function-generator performance is enhanced using linear interpolation, with only a doubling of memory requirements. A small number of function values are stored in a lookup table along with the derivative values. Linear interpolation increases function and frequency resolution by several orders of magnitude at moderate computational cost.

The digital function generator in (b) has no direct analog counterpart, since waveforms that are arbitrary, dynamically modifiable, mathematically derived, and even user-defined, may be stored in memory. The general avoidance of on-line computation makes parameter variation fast and easy. The above examples are limiting cases of computation-intensive (a) and memory-intensive (b, c) solutions. When multiplication was a slow operation, (a) was avoided when possible, for the same reason that a hand-calculator takes much more time to calculate a sine function than it does to add two numbers. Avoidance of computation is no longer necessary, now that array-multiplication time is comparable to addition or memory-access time.

Other considerations dictate the choice of method: with (a), simulation may suffer from IIR roundoff errors, while power-series approximation is limited to the set of functions for which the series converges rapidly; (c) is very fast (table lookup only) but may be limited by memory size or else have high harmonic distortion, while (b) is an effective tradeoff between the computation approach of (a) and the lookup approach of (c). Finally, digital synthesis makes it easy to vary the frequency or phase of a digital oscillator dynamically, by summing with the frequency control word a small modification, whose time-dependence shifts either the phase or the frequency. This method, called *parametric synthesis*, is a computationally efficient method to generate both frequency-shift keying and phase-shift keying in the modem transmitter (Figure 8.32); and it is also the basis for the music synthesizer example below. Terminologies used for the analog instrument and its digital replacement in the following section are summarized here.

Analog Instrument	Digital Technique
Sine-wave oscillator	System simulation (biquad)
Trigonometric functions	Power-series approximation
Function generator (arbitrary)	Lookup table Newton-Raphson recursion
Frequency modulation	Parametric synthesis
Voltage-controlled oscillator	Parametric synthesis

Comparison of Function Generation Methods

Mathematical functions, such as sine, cosine, tangent, arctan, log, exponential, square, square-root, etc., are often essential to a DSP or numerical analysis computation. The problem is to compute the function "on the fly" fast enough to give a real-time output with adequate accuracy. There are three basic approaches:

Lookup table

Lookup table plus linear interpolation

Direct calculation:

(a) recursive solution

(b) polynomial expansion

The *lookup table* approach requires the most memory, while *direct calculation* requires the most computation; interpolation is intermediate. Lookup methods work for any function, even a table of numerical data, while direct calculation methods require an analytic form. While *lookup plus interpolation* requires a derivative, the function need not be analytic, since the derivative can be precalculated numerically. The balance between memory and computation is compared in Figure 8.33 for a sine wave. An automatic reduction in storage or computation results if the function has any symmetry. For the sine, only one quadrant need be stored, with the other quadrants calculated by reflection. The resolution is increased and the error cut down for a given memory size or computation time.

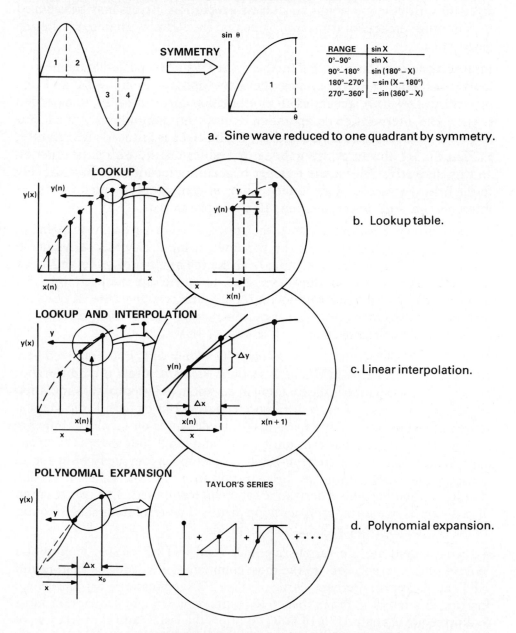

a. Sine wave reduced to one quadrant by symmetry.

b. Lookup table.

c. Linear interpolation.

d. Polynomial expansion.

Figure 8.33. Comparison of function-generation approaches.

A lookup table stores precomputed values of the function in memory. The x-input is truncated to the nearest address-value to generate the lookup address. The data at that address is the approximate value of the function. Because the same y value is output for any x-input value between adjacent addresses, direct lookup has significant error ϵ, shown to the right. To reduce the error in the lookup method, the memory size must be increased. Table lookup is therefore demanding on memory, though it is the fastest approach, since the only computation is an address-generation. Errors may be reduced by rounding x instead of truncating it, so that the address generated is the one nearest to x.

Interpolation cuts the error still further. The x-value truncated to the nearest lookup address, $x(n)$, generates the first-approximation, $y(n)$, and the increment, $(x - x(n))$, generates an interpolation correction, Δy, to be added to $y(n)$. The interpolation computation reduces throughput by about a factor of 2 compared to direct table lookup; but the error is reduced significantly, as Figure 8.33c illustrates. Even though the table must store both the function and its derivative values, the number of memory locations is considerably fewer than that needed for a table-lookup of comparable accuracy without interpolation, since fewer function values need be stored.

Polynomial expansion extrapolates the function from an origin (a point of symmetry, or the midpoint of the window), using a Taylor's series, as in Figure 8.33d. The accuracy is limited only by the number of terms and it can be quite high. Speed is slow compared to a lookup method, but fast multipliers reduce the computation burden. Stability is guaranteed, since the method is nonrecursive, but the requirements on the function or range of convergence can be restrictive.

Recursive-solution methods are of several types; they will be summarized here and discussed in detail in the next section. *Simulation* sets up a difference equation whose solution is the function in question, as in computer simulation of physical systems. Where a difference equation can be found, the method can be as fast as interpolation and makes *no* demands on memory, but care is required to avoid long-term limit cycles and amplitude errors. *Inversion* approaches, such as the Newton-Raphson method, recursively hunt for an output value which minimizes the deviation from the true value. Inversion requires a function whose derivative can be precomputed. The combination of a single Newton-Raphson iteration with a low-resolution table lookup yields very small errors with reasonable speed.

To what extent are the various methods used in DSP? Table lookup and lookup-plus-interpolation are the most common methods for oscillators and FFT applications. Perhaps 90% of all FFT applications for electronic instrumentation have fewer than 4 K points, so a straight 1-quadrant table lookup requiring only $1K \times 16$ of ROM offers the fastest and cheapest way of generating twiddle factors. For higher resolution, lookup plus interpolation, together with decimation tricks and address-generator chips, will generate functions at frequencies over the full audio bandwidth. For precision instrument standards and certain demanding scientific experiments requiring both ultra-low distortion and nearly continuous frequency selection at high speed, the recursive-simulation oscillator described below may be chosen. Many functions besides sine waves are well-suited to simulation, polynomial expansions, or recursive solutions, so these alternatives will be briefly considered.

Sine Waves Generated by Recursive Simulation

The Recursion Oscillator

The recursive method generates ultra-high-resolution sine waves by computing the projections of a rotating vector on the x- and y-axes. The advantages are:

The method is extremely fast: each value is computed with only two to four MAC steps.

Since no storage is required, recursion can rapidly generate high-resolution FFT twiddle factors, which would otherwise require unrealistic amounts of memory.

Two outputs in precise quadrature can be generated with no extra computation, to speed FFT twiddles as well as quadrature heterodyning.

The waveform has extremely low distortion, since the difference equation simulates a physical system whose solution is a perfect sine.

Recursion requires the handling of limit cycles over the long-term, but they can be made to lock on to periodic values. The two coupled 1st-order systems shown in Figure 8.34a carry out the infinitesimal rotation of a vector, $R(n)$, around a circle. The vector projections on the x and y axes generate the sine, $S(n)$, and cosine, $C(n)$. Writing the vector as a complex number,

$$R(n+1) = R(n) e^{j2\pi\alpha} \qquad (8.42)$$

with

$$R(n+1) = C(n) + jS(n)$$

$$\alpha = \frac{k}{N} \qquad \text{with } k/N \ll 1$$

The frequency is set by the phase increment, $2\pi\alpha$; for example, fixing N at 36,000 makes the phase advance in 0.01-degree increments when $k = 1$. Selecting k sets the frequency:

$$f = \frac{k}{Nt_s} \qquad (8.43)$$

QUADRATURE OSCILLATOR

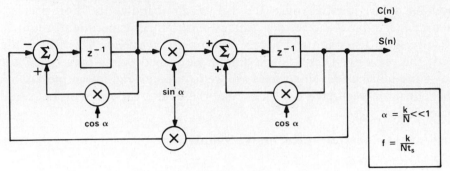

a. Two-phase quadrature oscillator.

BIQUAD OSCILLATOR

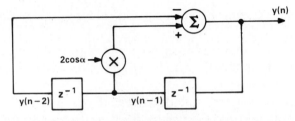

b. Single-output oscillator with one multiply per cycle.

Figure 8.34. Recursive sine-wave generator.

where t_s is the time for one iteration of the difference equations which generate the cosine, C, and sine, S, components:

$$C(n+1) = C(n)\cos\alpha - S(n)\sin\alpha$$
$$\text{(cosine of previous-angle-plus-}\alpha\text{)} \qquad (8.44)$$

$$S(n+1) = C(n)\sin\alpha + S(n)\cos\alpha$$
$$\text{(sine of previous-angle-plus-}\alpha\text{)}$$

With a given set of initial conditions, the circuit will recursively generate the sine and cosine as time evolves. The increments, $\cos\alpha$ and $\sin\alpha$, are precomputed and input *only once* as initial values, so only a pair of multiply-and-add steps are required to generate the next cosine and sine values. A program to demonstrate recursive oscillator operation is simple and convincing (see BOX).

Recursive Oscillator Performance
When the BASIC program example is run, the sine and cosine values are generated with remarkably low harmonic distortion and high frequency-

---

TWO-PHASE OSCILLATOR

The reader may explore the two-phase oscillator with the following BASIC program.

```
        REM: INITIALIZATION
    S=0
    C=1
    INPUT"number of 0.01 deg incr per step",K
    A=K/5729.577951#:REM  2πK/36000
    CA=COS(A)
    SA=SIN(A)

    START:
        TEMP=SA*C+CA*S
        C=CA*C-SA*S
        S=TEMP: PRINT S

        REM put additional printout statement here;
        REM for example, insert a conditional PRINT
        REM statement to show only when
        REM the cycle is close to a peak

    GOTO START
```

---

stability. There is some creep of amplitude values due to roundoff errors. In a test over some 20,000 iterations, the amplitude error accumulated to 0.5%, consistent with the 6-digit precision of the calculation in a particular BASIC interpreter. Roundoff in the floating-point BASIC example lets poles drift away from the unit circle in the z-plane (where they must remain for a perfectly stable oscillator). In fixed-point implementation, roundoff can be made to lock onto stable integer values, so the pole drift is eliminated and amplitudes can be as stable as the resolution of the computation. Alternatively, a simple condition test can reset one component to 1.0000 whenever the other component crosses zero (see Exercise just before Section 8.6).

Single-output sine generator with only 1 multiplication step
If the dual-output sine/cosine is not needed, a simpler (and therefore faster) sine wave can be implemented from the biquad of Chapter 4. With no zeros and the damping set to zero, the biquad transfer function becomes:

$$H(z) = \frac{1}{1 - 2\cos\alpha\, z^{-1} + z^{-2}} \qquad (8.45)$$

so a simple oscillator may be computed by solving the corresponding difference equation:

$$y(n) = 2\cos\alpha\, y(n-1) - y(n-2) \qquad (8.46)$$

This leads to the very efficient circuit shown in Figure 8.34b. There is only one parameter input, and one MAC step per output value! The angle increment and operating frequency are set as initial values.

Limitations of the Recursive Sine Generator

There are limitations to the recursive sine generator as a free-running oscillator:

How to start it? With the coupled 1st-order sine/cosine version (Eq. 8.44), the answer is simple: load the initial conditions, $C(0) =$ full-scale and $S(0) = 0$. But with the 2nd-order biquad version (eq. 8.46), the initial condition value varies with frequency if the sine amplitude is to stay constant.

The spread of available frequencies is uniform in f. This is fine for linear FFT's, but some audio applications may require a log range, spread equally over each decade of range.

Once a mistake is made, recovery is difficult, since quantization error will add up in indefinite recursion. The biquad *filter* does not suffer from this objection since it is not normally run at infinite Q.

In long-term operation, the frequency and amplitude are affected by limit cycles.

Some improvements are easy: for example, the last two objections can be eliminated by a limiter, as in analog oscillators, with an added feedback step to sense and recover from drift (see Exercises). The benefit of the recursive oscillator for generation of precise twiddle factors in an ultra-high-resolution FFT can be well worth the small extra trouble to avoid error accumulation.

Polynomial Expansions of Functions

The Taylor's Series as a DSP Method

Store a selected number of function values, $f(x_0)$, in a small lookup table, then use the Taylor's series to approximate function, $f(x)$, by its value, $f(x_0)$, at point $x = x_0$, plus corrections whose size depends upon the derivatives of $f(x)$:

$$f(x) = f(x_0) + (x - x_0)f'(x_0) + \frac{(x - x_0)^2}{2!}f''(x_0) + \ldots$$

$$= f(x_0) + \sum \frac{(x - x_0)^n}{n!}f^{(n)}(x_0) \qquad (8.47)$$

Power-series function generators work well when the number of terms in the series can be kept small, poorly for functions with rapid changes, and not at all for functions with discontinuities. Convergence is rapid as long as the derivatives are well-behaved, because $n!$ grows so fast. The interval must selected to avoid regions where any derivative grows large. Enough lookup initial values, $f(x_0)$, must be stored to speed convergence by keeping $(x - x_0) \ll 1$, so powers of $(x - x_0)$ diminish rapidly and few terms need be computed.

Example: Square-Root and Division Computations

A fast square-root function is of great utility in DSP. For example, graphics and spectral analysis both require that vector magnitudes be computed from the components. A fast divider could be as useful as a fast multiplier in, for example, vector scaling and perspective operations in graphics.

Today's high-performance square-root and division operations are performed in hardware; for example, the ADSP-3212 floating-point Multiplier/Divider performs a 32-bit single-precision divide in 300 ns and a double-precision division in 600 ns; and the AD3222 floating-point ALU performs a single-precision square-root in 1.45 μs and a double-precision square root in 2.9 μs. However, it is instructive to consider other hardware approaches for formulating approximations to square-roots and division, because the techniques are still valuable for fitting these and other functions, including arbitrary ones, at lower cost, employing simple multipliers.

The first few Taylor's series terms in the expansion of a square root function about a point, x_0, may be written:

$$y = \sqrt{x_0} \left[1 + P - \frac{P^2}{2!} \right] \qquad (8.48)$$

$$\text{where} \quad P = \frac{x - x_0}{2x_0}$$

The steps and timing in a Taylor's-series algorithm for a 16-bit computation, summarized in Figure 8.35, are:

1. Prenormalize the inputs so that x falls in the range $1/4 < x < 1$. The range near zero is avoided because the derivatives become large there. Time: up to $(8 - 2) = 6$ two-bit shifts.
2. Use the upper 4 bits of the normalized x-value as a lookup address, x_0, for precomputed values of $\sqrt{x_0}$ and $1/(2x_0)$ stored in memory.
3. Compute P (one MAC step), using $1/(2x_0)$ from the previous step and $(x - x_0)$.
4. Compute y from Eq. 8.48 above (one MAC, a subtract, and a final MPY)
5. Denormalize to undo step 1 so that y falls within $1/2 < y < 1$. Timing is the same as in normalization step.

By keeping the normalization window away from steep derivatives of the square root near $x = 0$ and by using a 4-bit first approximation to keep $(x - x_0)$ small, the rms error in this 2nd-order expansion is kept to 2 parts in 10^5 (i.e., error only in the 15th bit), at the cost of only $2 \times 12 = 24$ words

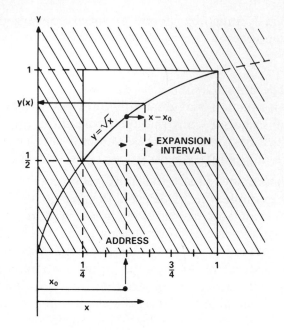

Figure 8.35. Function generation by Taylor's series, for the square-root function. Coarse values, x_0, correspond to lookup table adresses for terms in the approximation.

of lookup memory. The time is roughly:

> $6 \times 2 = 12$ shifts for norm/denorm (worst case)
> Two table lookups (address generation and data-fetch time)
> 3 MPYs or MACs
> 1 Subtraction

For computing square roots, and functions of similar shape, about 18 cycles must be allowed in the worst case, with components that allow a MPY/MAC operation in one cycle. The circuit can readily be constructed in hardware. It also provides a fast method of performing square roots using the ADSP-2100 single-chip DSP microprocessor.

Newton-Raphson Recursion
A recursive solution to the calculation of $y = 1/x$ illustrates how the Newton-Raphson method "homes in" rapidly on solutions to this and other transcendental functions. Although high-performance division and reciprocal calculations maybe performed in one step with devices such as the ADSP-3212 floating-point multiplier/divider and the ADSP-3221 floating-point ALU, the Newton-Raphson approach remains very useful. Convergence is fast: one iteration, together with a 256-entry lookup table, brings results having 14-bit accuracy. We summarize here an example of a hardware divider design (Ref: Matt Johnson, *Analog Dialogue* 18-1, 1984).

In the Newton-Raphson method, the equation to be solved is put into a form where the "correct" but unknown root is zero. The function is rearranged until it takes on the form:

$$f(y_0) = 0 = f(y(n)) + (y_0 - y(n))f'(y_0) \tag{8.49}$$

which becomes truer as $y(n)$ gets close enough to the location, y_0, where $f(y)$ crosses zero. Not knowing y_0, take a guess, $y(n)$, and compute an iteration, $y(n+1)$, from

$$0 = f(y(n)) + [y(n+) - y(n)]f'(y(n))$$

$$y(n+1) = y(n) - \frac{f(y(n))}{f'(y(n))} \qquad \text{Newton-Raphson algorithm} \tag{8.50}$$

where y_0 and the derivative of f at y_0 have been replaced by their values at the previous guess. It is clear from the geometry of Figure 8.36 that this converges rapidly to y_0 (but see Caveat $\star$).

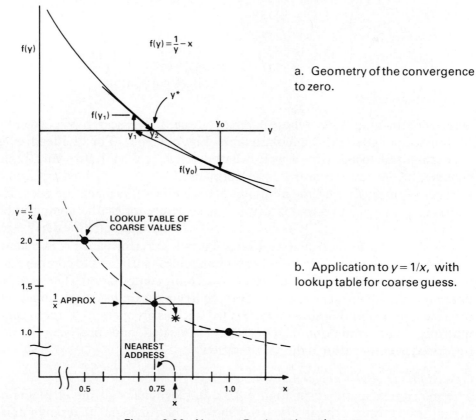

a. Geometry of the convergence to zero.

b. Application to $y = 1/x$, with lookup table for coarse guess.

Figure 8.36. Newton-Raphson iteration.

*Caveat: The function must be monotonic. If the derivative of f passes through zero within the search region, the iteration may either diverge or oscillate; this can be easily seen by making a sketch of such a function in the manner of Figure 8.36a. See the Johnson article.

Example: Fast Divider

A fast divider may be constructed from a simple multiplier, with great speed advantages over a shift-and-subtract software divider. For example, consider a quick way to compute

$$y = 1/x$$

To converge in fewer iterations, use a lookup table preloaded with coarse values of the function, so that the iteration is needed only to find the least-significant bits. This is shown for the $1/x$ function in Figure 8.36b. To apply the Newton-Raphson iteration, find the value of x such that

$$f(y) \;=\; 1/y - x \;=\; 0 \qquad\qquad (8.51)$$

Straightforward application of the Newton-Raphson algorithm gives a search strategy.

$$y(n+1) = y(n) + \frac{1/y(n) - x}{[-1/y^2(n)]}$$

$$= y(n) \; [2 - xy(n)] \qquad\qquad (8.52)$$
$$\text{feedback correction}$$

where x is the input value whose inverse is desired. The search is speeded by finding a close guess from a lookup table (Figure 8.36b). The dividend, x, is prenormalized to fall into a well-behaved range, 1/2 to 1, to avoid rapid changes in the inverse near 0; the normalized range is divided up into a number of intervals; and the midpoint of each interval is made the guess for values of x falling within that interval. The structure for the $1/x$ computation is very simple (Figure 8.37); it can be simplified further to combine the $(2 - R)$ step with the multiplication (see Johnson article). Performance is excellent: a single iteration from a 256-member lookup table is sufficient to compute the inverse to better than 14-bit accuracy in a 16-bit computation. The error is not reduced with further iterations if the recursion uses only a single-precision intermediate result, because $(x\, y(n))$ is fed back as an operand to the second multiply. The correction is effectively truncated and precision is not improved over the result of the first iteration.

Case Study: Digital Waveform Synthesizer

The instrument outlined in Figure 8.38 employs many of the elements of digital waveform synthesis discussed above. The Analogic 2020 polynomial waveform synthesizer mathematically derives its functions from expressions keyed in by the user. (For further information, see the article by Lester Brodeur in Electronic Design, May 16, 1985, pp. 155-165.) The unit combines

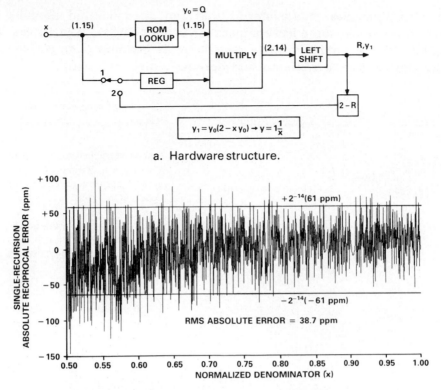

a. Hardware structure.

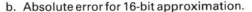

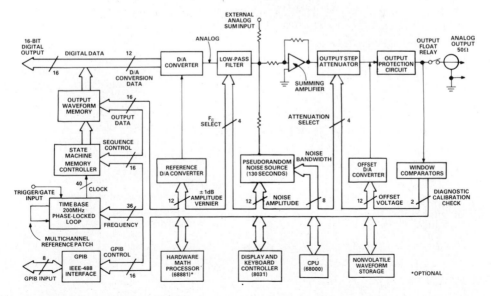

b. Absolute error for 16-bit approximation.

Figure 8.37. Fast fit to a function of the form, $y = 1/x$.

Figure 8.38. The Analogic 2020 digital polynomial waveform synthesizer derives its functions mathematically. (Copyright ©1985 Analogic Corporation and reprinted with permission)

Table 8.8. Examples of waveforms that can be generated digitally by keyboard entry and numerical computation in the Analogic 2020. Driving expressions are in similar form to equations, as the first example shows. (Copyright © 1985 Lester Brodeur and reprinted with permission)

Waveform	Typical Equation	Requirements for Conventional Generation Techniques
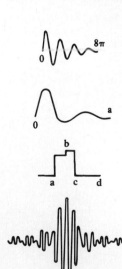	$At^2 e^{-t/b} \sin \omega t$ (2020 driving expression: $AT^2 E^{(-T/B)} \mathrm{SIN}\ WT$)	Sine-wave oscillator, phase-locked gate and carrier switch, envelope function generator (discrete circuitry), four-quadrant multiplier, and buffer.
	$0 \Big|_0^a + A \sin \omega t \Big|_0^{2\pi} + 0 \Big|_{2\pi}^b$	Sine-wave generator, amplitude limiter, digital divider, divider control circuitry, divider decoder, gate and buffer.
	$At \Big|_0^a + 0 \Big|_a^b$	Ramp generator, differentiator and clamp, divide-by-2 circuit, integrator switch, and driver.
	$A(1 - e^{-t/x}) \Big|_0^a + A e^{-t/x} \Big|_a^b$	Square-wave generator with external filter.
	$A e^{-t} \sin \omega t \Big|_0^{8\pi}$	Impulse source and tuned circuit with adjustable F_0 and damping.
	$A t e^{-Ct} \sin 2t \Big|_0^a$	Impulse source with underdamped tuned circuit and driver.
	$A \Big|_0^a + B \Big|_a^b + C \Big|_b^c + A \Big|_c^d$	Multitapped voltage divider, multipole switch, switch logic, switch driver, and buffer.
	$\dfrac{2A\, \omega_0 \omega_c}{\pi} \times \dfrac{\sin \omega_c (t - t_a)}{\omega_c (t - t_a)}$	Impulse generator, square symmetrical bandpass filter, and drives.

a math package, a PLL frequency synthesizer, real-time waveform memory, and an internal parameter file system. In addition to mathematical waveform synthesis, popular alternatives are also available: data-point entry from the front panel or over the IEEE-488 bus, linear interpolation between data points, and waveform plotting on an external oscilloscope. The mathematical approach preserves maximum vertical and horizontal resolution. A given function is typed in: to generate an exponentially damped sine wave, type $Y = A \, e^{-at} \sin(\omega t)$, etc. The notation allows concatenation of functions in adjacent time intervals. Internal computations are done with 24-bit floating-point intermediate accuracy, with waveform points stored as 16-bit values. Once a wave is precomputed, it can be strobed out in continuous, gated, or triggered modes with 10ns rise-time, 40ns settling time (to 0.2%), S/N of 72dB (12-bit output DAC), 25MHz upper frequency response, and 0.02% frequency increments down to 0.002Hz.

A wide range of complicated waveforms: modulations, tone-bursts, pulses, or quite arbitrary functions can be created, as exemplified in Table 8.8. The right-hand column shows by comparison how much hardware and circuitry would be required to generate the signal by conventional analog techniques. The relative simplicity and ready changeability during use are persuasive evidence for DSP methods in signal generation.

8.5.3 WHAT MAKES MUSIC? ANALYSIS OF MUSICAL SOUNDS

Desiderata for a Real-time Digital Music Synthesizer

While few would trade their Stradivarius violin or Steinway piano for an electronic equivalent, digital music synthesizers can provide good imitations of their sounds, and in addition: new sounds, stability, real-time performance capability, ruggedness of digital components, and multi-track recording capability—all within an inexpensive portable instrument. This has been amply demonstrated in the rock-music environment, where the wherewithal was available to pay for the development. A recent generation of superior digital instruments at moderate cost now attracts classical performers and composers as well. For example, the American Ballet Theatre has been known to use a portable synthesizer in order to avoid the uneven quality of grand pianos available on tours.

The music synthesizer as signal generator is a challenging real-time DSP application. The performer's inputs are the selected notes on a keyboard, changing instant by instant, or the modulating pressure and flowrate in a flute, etc. After substantial digital processing and intermediate RAM storage, output is generated in real time through one or more DACs. Typical digital synthesizer requirements include:

A wide choice of sounds, easily changeable *during performance* by altering parameters which determine the waveforms. Some sounds will replicate those generated by conventional instruments; others will be uniquely new; yet others will start as conventional input sounds played back in arbitrarily altered form.

Polyphonic (more than one sound simultaneously) ability for multi-note chords, often with more than one instrument sound or voice at a time.

Dynamics in the keyboard for changes in sound level or harmonic content in response to the performer's touch, emulating subtleties which give acoustic instruments much of their charm.

The human interface requires that this feeling for the nuance be transmittable from the musician to the synthesizer I/O, and be alterable by the performer (e.g., instant control of *vibrato*).

Voice memory to store good "patches" or waveform combinations.

Sequencer memory to store in RAM several pages of playing for low-distortion digital multi-track overlays of independent layers or voices.

Pitch transposition without altering the rate of playback of sequences, as well as sequence playback rate control without altering the pitch; neither feature is feasible with analog tape recorders.

Few musicians are computer-oriented, so the human interface is especially challenging. The I/O and responsiveness must resemble a natural instrument more than a computer terminal.

This section develops the DSP requirements that must be met; it examines what makes musical instruments sound as they do, surveys briefly the main methods in use, then develops in some detail the specifications and performance of one popular class of solutions, the parametric or fm synthesizer.

The Frequency Spectra of Musical Instruments

Overview of the Requirements

The basic requirements are a set of oscillators capable of being excited simultaneously at selectable pitches; their harmonic content (usually referred to as the *timbre**) and amplitude-versus-time (called the *envelope*) are both changeable to represent the different instruments. But realistic music, like realistic speech, demands a deeper understanding of the mechanisms and models of its generation in natural instruments. For example,

The piano string is not one-dimensional and not infinitely flexible; its overtones are less than integer multiples of the fundamental, particularly in the bass notes where the string is thicker. The digital synthesizer must be capable of shifts away from harmonic partials, as well as the "chorusing" of notes in unison slightly out of pitch that gives a string orchestra its richness.

The "bang" of a drum is a rapid variation of amplitude envelope. A synthesizer must be capable of amplitude changes over several-

---

*The American Standards Association defines timbre as a psychophysical quantity: if two sounds having identical pitch and sound pressure are perceived as different (e.g., flute and violin playing the same sustained note), they are said to have different *timbres*.

milisecond intervals of attack yet also have decays of several seconds for high-Q resonant instruments, a 4-order dynamic range.

The "oohwaa" of a trumpet attack is a time-variation of harmonic content during a single note. Synthesizer spectra must consequently be capable of changing over 100 ms or less.

The "bong" of a bell is a non-integer sequence of overtones. Rich synthesizers must be capable of rapid nonharmonic variation of overtone frequency and amplitude.

Comparison of Instrument Spectra

Which DSP methods work best for music synthesis? Much has been learned from analyzing the spectra of musical instruments—which are quite different from voice spectra. Consider the examples shown in Figure 8.39, taken from Moorer (1977). The amplitude envelope is characterized by distinct portions, labelled the attack, the steady state (SS), and the decay. The examples shown display a harmonic series with many terms at near-integer values.

The *trumpet* displays a pulse-like waveform much like speech, but there is little speech-like formant structure (compare Figure 8.28) in the spectrum. The attack is faster than the cello's, and the decay is slower than the clarinet's.

The *cello* displays more energy towards the fundamental frequency compared to trumpet or clarinet. There is a clear formant resonance near 1 kHz, so a cello might be modeled by LPC methods, though this would not be fruitful for the trumpet and clarinet.

The *clarinet* (and oboe) have the striking feature that at low frequencies mostly odd harmonics are present. The harmonics extend to very high order: at least 20 are significant.

Not visible on the scale of these pictures are subtleties: the noise component at the beginning of string bowing, irregularities in frequency during the trumpet attack, or modulations such as vibrato in the sustain.

Comparison with Speech Analysis and Synthesis

The most common methods of speech analysis and synthesis are not as fruitful for music synthesis. Except for the violin family, the spectra of instruments do not display as pronounced a formant structure as the voice, while the harmonic series extends far higher than the voice (compare Figure 8.39 with Figure 8.28). Thus, all-pole LPC methods are not particularly fruitful. In addition, the wider dynamic range of music (>80 dB), compared to speech, puts higher computational demands on the accuracy of the modelling—for which the recursive filters of LPC are not ideal. Waveform coding and compression are attractive, except that each note of each instrument must be pre-stored, and some means must be found to allow dynamics in performance to avoid music-box regularity. The trumpet is used below as an example to illustrate important changes in the spectrum with time—during a note and with loudness of playing.

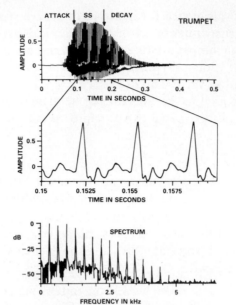

a. Trumpet.

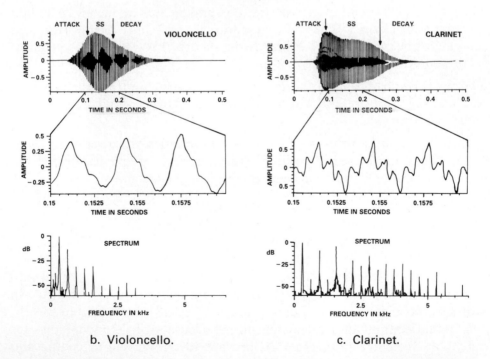

b. Violoncello.   c. Clarinet.

Figure 8.39. Waveforms and spectra of musical instruments. In all cases the top picture is the overview of a note envelope, the center picture shows the detailed waveform during the "steady-state" portion of a note, and the bottom shows the frequency spectrum at the same point. (J. A. Moorer, ©1977 IEEE—see Bibliography)

Dynamics: Envelope Function and Spectral Shifts

Change in spectrum during a note

The frequency components of the trumpet spectrum change in relative amplitude during a note, as shown in Figure 8.40. There are several characteristic features.

(1) The frequencies form a near-harmonic series. The higher harmonics die away in relative importance during the sustain and even more rapidly during the decay.

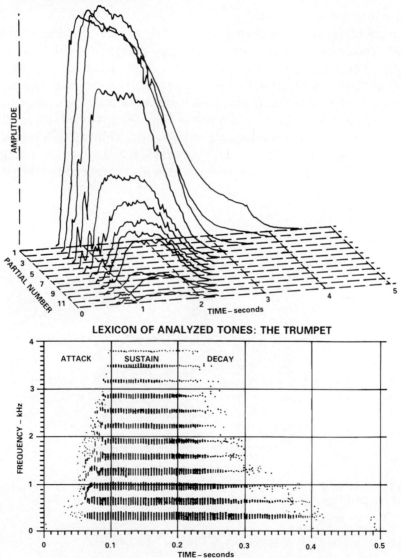

LEXICON OF ANALYZED TONES: THE TRUMPET

Figure 8.40. Amplitude function, $A_k(t)$, for the spectrum of a trumpet note derived by additive synthesis. (Reprinted from *Computer Music Journal* 2-2, James A. Moorer, John Grey, and John Strawn, "Lexicon of Analyzed Tones (Part III: The Trumpet)" by permission of the MIT Press, Cambridge, Massachusetts ©1978 the MIT Press)

(2) The attack starts out as a pure tone, with higher harmonics adding in later. The higher the frequency, the later a harmonic appears in the tone. The harmonics also exhibit a small but significant frequency modulation away from a harmonic location during the attack.

This combination gives the trumpet much of its characteristic "bite". A given envelope rises rapidly and may overshoot. The harmonic rise time is slower than the fundamental. Control of these harmonic envelopes, particularly during the attack, is crucial to a realistic sound.

Change in spectrum with loudness
The spectrum of Figure 8.40 is correct only at a particular loudness. At louder or softer sound-production intensities, the harmonic structure changes markedly. The relative importance of the higher harmonics changes with loudness as shown in Figure 8.41, because of nonlinearities in the system. For notes played softly, low-order harmonics are more important than for loud notes. At low levels (pp), the fundamental dominates, at medium levels (mp) at least 6 harmonics have comparable importance, while at high levels (fff), the tone is dominated by the higher harmonics, which gives the trumpet its "brassiness." This nuance of dynamics is essential to reproduce the trumpet satisfactorily.

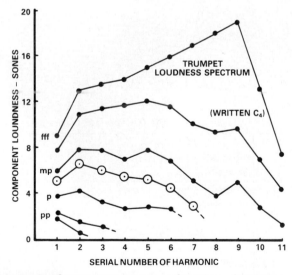

Figure 8.41. Spectrum of a trumpet tone as a function of loudness. (From *Fundamentals of Musical Acoustics,* by Arthur H. Benade. Copyright ©1976, Virginia Benade. Reprinted by permission)

8.5.4 HOW DIGITAL MUSIC SYNTHESIS IS DONE

Overview of Digital Music-Synthesis Methods
Digital electronic music synthesis illustrates the power of DSP methods in a truly real-time environment. The methods surveyed below will be found generally useful in the synthesis of other complex and rapidly changing

waveforms. Synthesis methods are introduced here grouped into four classes: additive, subtractive, compressed, and parametric.

(1) *Additive* synthesis Fourier-analyzes the frequencies and amplitudes of a recorded sample of a signal and then combines a finite sum of Fourier components to regenerate the sound.

(2) *Subtractive* synthesis begins with a harmonics-rich excitation and filters it. Human speech and analog music synthesizers use this method.

(3) *Compressed* "real" synthesis or waveform coding, as with adaptive differential PCM (Section 8.4.2), takes an accurately digitized record of a signal, compresses the data to lessen storage requirements, and then reconstructs the waveform when a note is called for in playback.

(4) *Parametric* synthesis generates a complex sound by modulation of a simple one using, for example, frequency modulation. The method is common in digital music synthesizers; Chowning (1977) showed musicians why they might want to know about Bessel functions.

Each method has its strengths and weaknesses. Method (3) is the most realistic but suffers from vocabulary (i.e., memory) limitations: each note must be prerecorded. Method (1) can make any sound, but sounds the most machine-like. Method (2) is compact in its storage demands, and has intermediate quality. Method (4) lends itself to exploration of sounds which have no analog counterpart as well as to the economical generation of realistic instrument sounds. For the natural instrument spectra of Figure 8.39, a Fourier synthesis would be recognizable, but it sounds dull and artificial.

For *real-time* generation, no single technique appears ideally suited for all instruments: Fourier synthesis requires too many components for the woodwinds, LPC would do justice only to formant-rich instruments like the cello, and filtering of a pulse train in subtractive synthesis requires too many rapid adjustments of parameters during each note. Moreover, while the spectra displayed above have overtones at near-integer frequencies, other instruments, such as bells, have overtones at much more complex intervals. Fourier or integer harmonic methods are not particularly fruitful there.

Additive, Subtractive, Compressed, and Parametric Digital Synthesis

Additive Synthesis

Additive synthesis is a Fourier-series approach. From an analysis like that shown in Figure 8.39, the amplitude-versus-time or *envelope* is extracted for each characteristic frequency. The sound is resynthesized by driving a set of oscillators as shown in Figure 8.42. While of importance in *analyzing* characteristic instrument sounds (e.g., Moorer (1977)), this method is usually limited to off-line use in *synthesis*, since a large amount of data is needed to specify a timbre adequately. Two independent envelopes are required for each

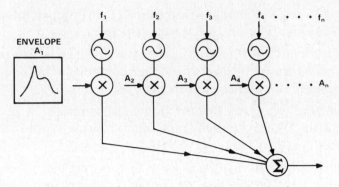

Figure 8.42. Additive digital music synthesis.

oscillator, one for amplitude, a second for frequency. The envelope may simply specify the attack rate, the sustained amplitude, and the decay rate, or a more complex form, such as that shown in Figure 8.43. When each envelope is approximated by a series of line segments, each segment requires two parameters, a rate and a final value, i.e., the rates, R, and the levels, L, at which the next R,L set is to be loaded.

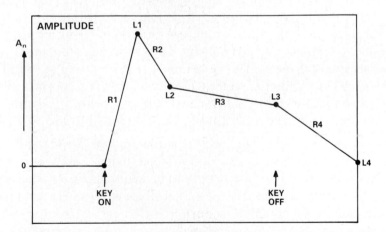

a. Amplitude envelope for a given attack, overshoot, sustain, and decay.

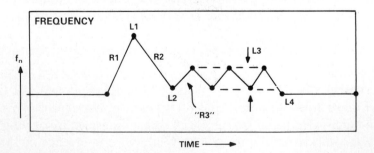

b. Frequency envelope, showing an initial overshoot of pitch, and a vibrato during the sustain.

Figure 8.43. Envelope functions for a given oscillator.

Even for such a modest envelope approximation as this, at least 8 parameters are required for a single oscillator. With a dozen oscillators required to replicate a natural spectrum, (such as in Figure 8.39), $8 \times 12 = 96$, or nearly a hundred parameters, must be supplied in sounding a single note. Polyphonic action multiplies this requirement even further. Real-time additive synthesis by sine-wave oscillators is thus very demanding of parameter storage and control. Even so, all this gives only the change with time but no timbre change with volume, as in the natural trumpet of Figure 8.41. An approximation to volume-dependence stores two sets of timbre parameters, one for "loud" and the other for "soft" playing, and interpolates between them depending on loudness called for by the key-velocity or similar input. In summary, additive synthesis using sine-wave oscillators plays somewhat the same role as fixed phoneme systems in speech synthesis: good for learning about sounds, but not satisfactory for real-time I/O.

Subtractive Synthesis

Subtractive synthesis filters a harmonic-rich waveform to leave the spectrum needed: many harmonics remaining give a harsh tone, like those of the oboe or clarinet; few harmonics give a pure tone like that of a flute. A square-wave gives the odd harmonics of reeds, while a triangular wave contains all harmonics. The subtractive method was the basis for early electronic organs and for such analog synthesizers as the original Moog. An example of the process is shown in Figure 8.44.

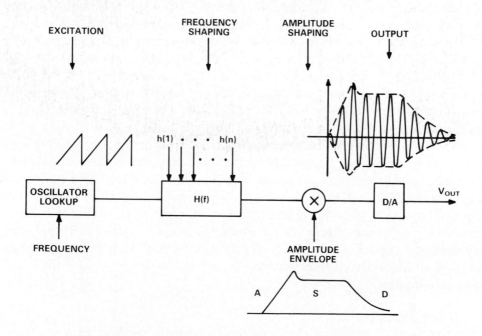

Figure 8.44. Subtractive digital music synthesis.

In spite of the success of digital adaptive filters for speech, subtractive synthesis is not very successful for digital music. FIR filters are appropriate to shape the relatively formant-less spectrum examples of Figure 8.39, but require the alteration of too many parameters in real time. IIR filters, while more compact in computation, generate enough noise due to roundoff accumulation to be readily detected as a buzz in hi-fi sound. In addition, the input signal must be harmonic-rich yet carefully band-limited in order to avoid aliasing. Digital subtractive synthesis is successful in the voice-range, e.g., for choral singing, where the noise can be band-limited out. Some very amusing effects can be created, using a symphony recording as the excitation function, but with voice formants doing the frequency shaping—a poetry reading by a talking orchestra, for example.

Few digital subtractive imitations of instruments are judged realistic. It is hard to make a subtractively synthesized violin sound like a real violin. The vibrating wooden box excited by the tuned strings driven by the hairs of the bow is too complex a system to model effectively by subtractive methods. Excitation and formant shaping seem to be less separable for instruments than for the voice; the correct excitation must be different for each instrument, unlike the voice; and our definition of "adequate" is more demanding for musical instruments than for voice communications.

Compressed "real" synthesis

This category is comparable to the compressed waveform coding (e.g., adaptive differential PCM) of speech (section 8.4), and gives comparable high quality, since a true sound is recorded and then compressed rather than being modeled. The method is only recently becoming common. For example, the Kurzweil 250 keyboard instrument sets a high standard, capturing the sound of a $30,000 Bechstein grand piano—as well as 40 other sounds. The compression techniques used stem from an earlier Kurzweil product, an optical character reader (OCR) for the blind. The two problems share a common need for data compression. The OCR must be able to "learn" to recognize a number of different typefaces. Rather than attempt to bit-map each and every image, the pattern-recognition algorithms extract and store key features that identify a particular letter or numeral. While few details have been released on the Kurzweil synthesizer's operation, the rules are stored as "sound models," which remember, for example, that a piano's high-frequency overtones decay faster than the fundamental, and that this difference increases as a function of increasing pitch. So it it not necessary to store all notes played at all loudnesses, just the rules extracted from actual digitized inputs.

Parametric Synthesis

The parametric or "fm" synthesis method modulates the frequency of a wave to generate a complicated spectrum with relatively little computation and few parameters. The method is both efficient and effective: the parameters can be dynamically altered in real-time to create quite realistic sounds. The parametric method is extensively used in commercial digital synthesizers such as the Synergy (Digital Keyboards, Inc.) or the Yamaha DX-7/DX-9. The basic building block, the digital oscillator, can be chained in parallel for additive components or in cascade for parameter modulation, or both. The success of fm synthesis lies in an insight about human perception: the evolution with time of the spectral components is of critical importance in establishing a musically satisfactory timbre. We hear this in the note of a bell or the "waah-waah" of a saxophone. Even when the time-evolution of a spectrum is too fast to follow explicitly, it is an important part of the perceived signature of a "real" (as opposed to artificial) instrument. The key to parametric synthesis is the spectrum of a frequency-modulated wave, which follows a Bessel-function series.

$$y(t) \;=\; A(t) \;\sin(\omega_c t + I(t)\sin\omega_m t) \tag{8.53}$$

$$=\; A(t) \;\sum_n \; \mathcal{J}_n(I(t))\sin(\omega_c t + n\omega_m t)$$

Here, ω_c is the carrier frequency and ω_m is the modulating frequency, $I(t)$ is the modulation amplitude or index, and the $\mathcal{J}_n$ are Bessel functions of order n. Since the envelope of a given harmonic goes as $A(t)\,\mathcal{J}_n(I(t))$, complex timbre combinations follow from setting just two envelope functions, the amplitude function, $A(t)$, and the index function, $I(t)$. If $I(t)$ is gradually increased from zero, the note begins as a pure sinewave and takes on higher harmonics as $I(t)$ increases. The higher harmonics sweep up in amplitude as I increases in size, as shown in Figure 8.45. At higher values of I, the fundamental diminishes again. For ω_c/ω_m an integer, practically any spectrum structure can be generated.

While natural instruments do not necessarily follow frequency-modulation laws, the fact that a complex timbre can be introduced with only two oscillators and a single modulation index, $I(t)$, gives fm synthesis a computational advantage. Surprisingly good sounds can be made for most orchestral instruments, including far better bells (with their non-integer overtones), than other methods. While good analytical methods for fm do not exist, hundreds of different instrument combinations have been developed empirically, using the same hardware and software structure. Pioneers include the Stanford group (see Chowning (1973), Moorer (1977)) and composers/musicians, such as W. Carlos (who earlier gained fame with "Switched on Bach" with analog synthesizers). A design study for a parametric synthesizer will be outlined below.

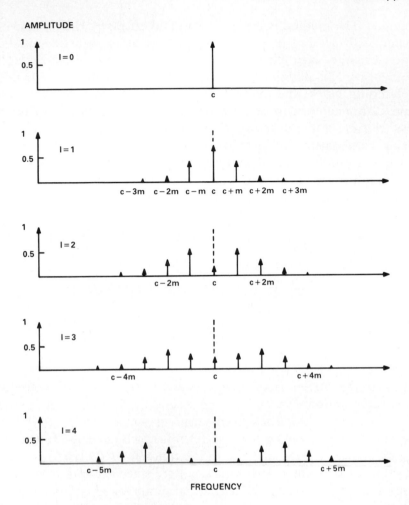

Figure 8.45. The spectrum of a sine wave that is frequency-modulated by another sine wave for various values of modulation index, *I*.

8.5.5 PARAMETRIC (FM) SYNTHESIZER DESIGN

Parametric Synthesis Requirements

The Digital Oscillator "Operator"
Figure 8.46 shows the digital-oscillator "operator" (upper figure) that carries out the frequency-modulation computation of Eq. 8.53 and is the basis for fm synthesis. The main oscillator (lower figure) includes an accumulator, or phase integrator, whose output determines the address in a waveshape lookup table, which reads out the waveform at the corresponding phase. The rate of accumulation of phase at constant clock frequency is set by the frequency-control word (FCW), which establishes the pitch. Since the waveshape table is arbitrary, the waveform is not limited to a sine-wave. The value of the FCW is the sum of independent inputs: the carrier frequency, f_c, which sets the

pitch, any pitch-shifts, Δf, and fm-timbral variations via a second *frequency modulation oscillator*, at frequency f_m, with amplitude $I(t)$. The oscillator parameters are thus set by:

Function	Label	Parameter or I/O
Pitch	f_c	carrier frequency on input port
Frequency	FCW	frequency control word
Frequency modulation	f_m	fm input port
Frequency sweep	Δf	delta frequency input port
Envelope	$A(t)$	envelope input port
	$I(t)$	modulation index

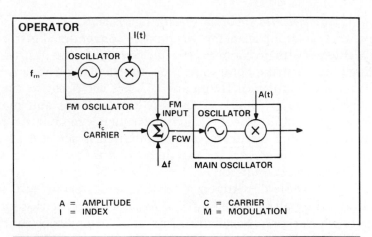

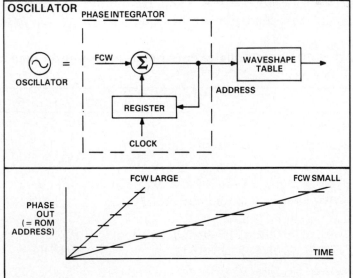

Figure 8.46. The digital fm oscillator. Upper figure: complete fm oscillator. Lower figure: digital oscillator concept.

A set of fm oscillators can be combined in various ways. Put in parallel, oscillators do additive synthesis if their frequencies are set to desired harmonics—or chorusing if the frequencies are set close to one another but randomly varying a little, as in an orchestral string section or in the multiple strings of a piano. Put in series, an oscillator can modulate the amplitude or the frequency of another oscillator, since any oscillator output can be the input to another. At the envelope input port, $A(t)$, the modultion gives *tremolo*, at an fm input port it gives *pitch bend* for a dc shift, *vibrato* if the frequency is low, or *complex fm harmonics* if the frequency is related to the carrier. The following sections detail parameter settings for specific instruments and establish throughput requirements for real-time performance.

Example: FM Parameters to Simulate Specific Instruments

A sample of fm oscillator parameters to simulate conventional instruments is shown in Table 8.9 (due to Moorer (1977)). Here, f_0 is the desired pitch of the resulting tone. To make a brass-like tone, the oscillator carrier frequency, f_c, and frequency-modulation frequency, f_m, are set equal to the desired pitch, f_0. By contrast, for the clarinet, the fm oscillator and modulation frequencies are set at multiples of f_0, and the frequency modulation produces a component at f_0. For the drum, the modulation is set at a non-integer interval, since drumhead oscillations do not form an integer series of overtones. The pitch for trumpet and clarinet can be set in the typical scale range, say 300 Hz, while the drum oscillation is set lower, say at $f_0 = 40\text{Hz}$. The amplitude, $A(n)$, and fm index, $I(n)$, envelopes are specified by ordered pairs, each of which represents one breakpoint of the envelope function (see envelope sketch in Figure 8.43). The first number specifies the time in milliseconds and the second number the value of amplitude, A, or modulation

Table 8.9 FM Digital Oscillator Parameters for Common Instruments

Interval	1	2	3	4	
Brass-like tone					
$f_c = f_m = f_0$					
$A(n)$	$(0,0)$ - -	$(50,1)$ - -	$(300,1)$ -	$(400,0)$	
$I(n)$	$(0,1)$	$(50,4.5)$		$(400,4.5)$	
Clarinet tone					
$f_c = 3f_0,$	$f_m = 2f_0$				
$A(n)$	$(0,0)$ - -	$(100,1)$ - -	$(300,1)$ -	$(400,0)$	
$I(n)$	$(0,2)$	$(100,1)$		$(400,1)$	
Drum-like tone					
$f_m = 1.414f_0,$	$f_c = f_0$				
$A(n)$	$(0,0.75)$ -	$(50,1)$ - -	$(100,0.5)$ -	$(300,0.2)$ -	$(1000,0)$
$I(n)$	$(0,1.5)$		$(100,1)$	$(300,.07)$	$(1000,0)$

index, I. For the trumpet and clarinet, the amplitude starts out at zero, builds in interval 1, is sustained at constant amplitude in interval 2, and decays back to 0 in interval 3.. The drum starts at nonzero amplitude, for the instantaneous attack of a percussive sound.

The trumpet undergoes a strong *increase* in modulation index after the attack, to give a transition to a brassy sound, while the clarinet modulation *decreases* its value at the attack, becoming more mellow in the sustain interval. The drum's modulation index decreases with each time interval, since its overtones die out more rapidly than its fundamental. These fm experiments are easily accomplished by the user by programming a DSP microprocessor to carry out the fm oscillator of Figure 8.46; they can even be tried in software on a personal computer with programs written in a compiled or assembled language (Vedanayagam and Higgins, 1981).

MIPS Requirements for a Digital Orchestra
Throughput requirements to create an all-digital "VLSI orchestra" are evaluated by taking into account (1) the number of oscillators required to synthesize a given instrument, (2) the computation time required to update a given oscillator value, and (3) the need to multiplex the above as many times as there are instruments.

(1) Oscillators per Instrument: How many fm oscillators can be simultaneously multiplexed by the same DSP processor? A set of memory locations is set aside to store current parameters for each oscillator. The DSP processor sequences through each set, calculating the updates which set frequency, amplitude, and modulation for the next time interval. The parallel outputs are digitally accumulated until all computations are done for a single output point. The number of oscillators determines the richness of the sound. It may take 3 or 4 per note, and 8 to 30 notes for a piano or orchestra chord. To prevent a previously selected oscillator from being dynamically reassigned to a new note before its sound has decayed away, cutting off a decay too early, enough independent oscillators must be available to generate the notes for several seconds of music.

Oscillators required for an easy case (flute):

$$2 \ \frac{\text{oscillators}}{\text{note}} \times 1 \ \frac{\text{note}}{\text{chord}} \times 8 \ \frac{\text{chords}}{\text{decay interval}} \ = \ 16$$

Oscillators required for moderate case (piano) :

$$4 \ \frac{\text{oscillators}}{\text{note}} \times 8 \ \frac{\text{notes}}{\text{chord}} \times 4 \ \frac{\text{chords}}{\text{decay interval}} \ = \ 128$$

An orchestra may tolerate 3 oscillators per note, since most instruments are less complex than the piano, but require four times as many instruments playing each chord, for a total of nearly 500 oscillators in parallel!

(2) *Time per Oscillator Update:* The oscillator update calculation must be complete in the sampling interval adequate for hi-fi sound, approximately $1/(30\text{kHz}) = 33\mu\text{s}$. The basic calculation for a parametric oscillator updates 8 parameter words: 3 words for phase accumulation, 3 for amplitude, one for patching oscillators to one another, one for interrupts. Each update requires at most a MAC and a memory reference. An update may take two instruction cycles—or only one if MAC and memory reference can operate in parallel.

> Steps per oscillator update:
> 8 parameter words $\times$ 2 instruction cycles each = 16 cycles
> (8 with MAC and memory reference paralleled)

Running at 125ns per instruction cycle (8 MIPS) in a DSP microprocessor such as the ADSP-2100, the update is done in 2 μs (1.0 with paralleled MAC and memory reference). A single DSP processor can therefore multiplex:

> $32\,\mu\text{s}/2\,\mu\text{s} = 16$ oscillators (32 with paralleling)

Such dynamic oscillator resource assignment can handle a monophonic instrument; however, for a complex tone such as from a piano, the designer must compromise on the number of notes playable at a time or on the decay time allowed before "stealing" a previously played oscillator. Polyphonic commercial instruments such as the Synergy (Digital Keyboards, Inc.) with 32 oscillators compromise on number of notes playable. For example, a set of voices programmed by composer W. Carlos for the Synergy has 3 oscillators per note and 10 notes playable at once when synthesizing a modest piano, but a "rich" piano requires 5 oscillators per note with only 6 notes playable before all 32 multiplexed oscillators become committed.

(3) *MIPS Requirements for the Real-time Digital Orchestra:* The throughput requirements for an all-digital real-time orchestra based on the above design criteria are very challenging:

> Speed per oscillator: 30kHz
> (33μs available between computed samples)
> Instructions per update: 8
> Instructions per second per oscillator: $30 \times 8 = 240\,\text{kHz}$
> Oscillators per instrument (average): 3
> Instructions per second per instrument: 0.72 MIPS
> Throughput, 40-instrument orchestra: $0.72\,\text{MIPS} \times 40 = 24\,\text{MIPS}$

A full orchestra is thus (at this writing) beyond the range of a single DSP microprocessor; but a quartet with voices added up digitally is within range, and parallel processing of several DSP microprocessors could give a full orchestra. This illustrates well the rapid evolution in computing performance: at the time Moorer's 1977 experiments were done, prior to DSP hardware development, a single instrument required the full resources of a PDP-10 mainframe, because the number of instructions per oscillator update was high (Von Neumann architecture, no parallelism, no pipelining) and the multiply time was much longer. By the mid-1980s, several pioneering synthesizers brought multiple-oscillator polyphonic digital synthesizers with custom DSP LSI chips in professional products. For example,

Synthesizer:	**Synergy**	**DX-7**
Manufacturer:	Digital Keyboards	Yamaha
Oscillators	32	6
Polyphonicity	2 to 16 notes	16

and a plethora of consumer synthesizers has followed rapidly.

The step to a full digital orchestra requiring close to 100 MIPS is a current DSP VLSI design challenge. Such studies have been done at IRCAM in Paris, at Stanford, and at MIT. Implementation of one particular VLSI music architecture, based on a unit called the "universal processing element," has been undertaken by Carver Mead's group at Cal Tech (Wawrzynek and Mead (1984)).

Exercise: Recursive sine-wave generator with stable amplitude.
Modify the BASIC-language sine/cosine program (Box near Figure 8.34 in Section 8.5.2) to remove long-term amplitude drift due to accumulated roundoff errors. Insert a step which searches for a zero crossing by either component; if true, it resets the other component to 1.0000. Verify by printout statements that the improvement works and that the reset glitch is invisible, since the error accumulation in one revolution is small.

8.6 IMAGE PROCESSING

8.6.1 INTRODUCTION

This section focuses on image processing as an example of two- and three-dimensional array operations with DSP components and algorithms. While the emphasis elsewhere has been on the real-time manipulation of real-world input and output *signals*, a gap would be left if no mention was made of array processing in transforming other kinds of information. Some of the content is an extension of earlier ideas; for example, image *enhancement* is digital filtering in two dimensions. Next are discussed issues in *computer vision* for the recognition of objects, as in robotic manufacturing. Finally some image *enhancement* examples are given from medical applications and image *reconstruction* as in the CAT scanner, a medically important example of image transformation from 2-D snapshots into a 3-D picture. The progression in this section is towards higher levels of abstraction and lower data rates (as discussed in the digital transceiver of Section 8.2). Image processing well illustrates both the need for fewer data points than the raw pixels, and the successively increased complexity of the programs as they progress from storage to enhancement to recognition or reconstruction.

Many DSP primitives and many DSP chips find their way into CAD workstations and specialized graphics processors for the generation and manipulation of images. The graphics generation world is divided into two areas: polygon pushers (70% of applications) and solid modelers (the rest). In the former case, images are modeled by a connected array of polygons. This "fixed" data base can give very fast display updates, since geometric transformations, clippings, shading, etc. can be handled by operations on a relatively small data base, usually within the refresh cycle of the screen. This domain is quite satisfactory for visualizing essential features such as a "walking" tour, which graphically brings the viewer through the inside of a computer-image of a DNA molecule in crystallography.

In the solid-model case, in order to achieve the very high level of resolution and realism required by its application to high-end mechanical or architectural CAD, flight simulators, and movie studios, one has to evaluate in addition complex parametric models. These include higher-order spline fits, fractals, continuous shadings and highlight reflections. Graphic image generation is at the large-system end of DSP, with a mainframe and perhaps an array processor as typical system components, since high-resolution image generation has severe computation requirements. For example, a Cray-1 supercomputer operating at 200 MFLOPS takes from 3 seconds to 1 hour of CPU time (depending on image complexity) to generate 1 second (30 frames) of high-resolution solid-model computer-graphic film images (see Demos et al. (1984)). Since little of this is real-time, we will not discuss it further below.

The challenges of image processing fall into a number of categories summarized in Table 8.10. Available space limits discussion to only a few issues; the reader is referred to the extensive literature in the Bibliography for further information.

8.6.2 MACHINE VISION: ACQUISITION, ENHANCEMENT, RECOGNITION

Parameters in Pixel Portraits

Machine vision graphically illustrates the essential need for progressing during processing towards higher levels of abstraction and concomitantly lower data rates, as the following example demonstrates. Consider an image as a "Pixel Portrait": bit-mapped representations as in a TV picture or digitized photo. The only information available is the intensity (and perhaps color) of each point in a 2-D representation; our eyes are as the eye of a camera. The image parameters are:

Resolution in number of pixels, N
(e.g., 1,024 vertical $\times$ 1,024 horizontal)

Gray or binary (black/white)?
   If gray, then how many shades (e.g., 64)
Color? If YES, then how many colors? (e.g., 16)
Repetition rate: how many frames must be processed per second
   (e.g., 30/s for TV/film)?

The product of these factors determines the number of elements to be manipulated; for complete replacement of a frame in the above example, one must compute on

$$(1{,}024 \times 1{,}024 \text{ pixels}) \times (30 \text{ frames/second}) = 31.5 \times 10^6 \text{ elements/second}$$

each of which comprises a data word of width

$$(6 \text{ bits grayscale}) + (4 \text{ bits color}) = 10 \text{ bits}$$

Image *processing* at the pixel level is thus more demanding than 1-D signals such as music or speech, since the number of 2-D elements goes as the square of the number of 1-D elements. Image processing therefore moves as soon as possible to higher-level constructs which determine key features: edges, shapes, object identification. The techniques of artificial intelligence may be called into play, since the higher-order decision strategies ("insert nut A onto bolt B") go far beyond the "modelling" of a 1-D signal source where a few poles tell all. Image enhancement, background suppression, shape extraction, and object identification reduce the number of independent pieces of information to be manipulated, as they do in human vision.

For high-resolution image *generation*, higher-level vectors or shapes are the basic construct, manipulated in this compact form until a bit-mapped image is generated at the final stage. The problem is that a large number of elements are required to create a high-resolution image: 1,000 pixels on a side give $10^3 \times 10^3 = 10^6$ data points to manipulate, for example. If the image is to be altered at TV or film rates, processing would have to proceed at 30×10^6 pixels/second.

To reduce this requirement, an image may be computed from a much smaller collection of vectors or primitive shapes, which are assembled into an object by vector operations and only converted to pixels at display time. But high-resolution manipulation of these objects calls for high-precision rotations and matrix manipulations; these generally require floating point processing. While flat rather than rounded images suffice for circuit design, CAD workstations in mechanical engineering and architecture require full 3-D operations. By contrast, *image enhancement*, discussed below, manipulates the pixels in the flat-screen image with relatively simple convolution operations, often involving only the nearest pixels to a given point; such operations can often be performed in fixed-point.

Image Operations
Consider a two-dimensional image (the extension to 3 dimensions is

straightforward, but issues such as hidden-line elimination must also be considered). Basic image operations include the following.

Image-Processing Primitives:

Rotation = Matrix × Vector
  Matrix = rotation matrix (2×2)
  Vector = (x,y) for each pixel
Translation by (X,Y) = Sum $[(x(i),y(i)] + (X,Y)$
Scale (shrink, expand) by scale factor A:
  $(x(i),y(i)) \rightarrow (Ax(i),Ay(i))$
Smoothing or other filtering (convolution)
FFT: 2-D spectral content of the image

Rotation is among the more demanding operations, because the matrix multiplication (4 multiplies) must be repeated for each pixel. On the other hand, doing a rotation of the image can speed later processing if the object has "natural" coordinates which can be aligned with x-y edge filters or oriented object-templates.

Translation is simple and fast, a repointing of the center of view. Scaling the field of view to zoom in $(A>1)$ or to reduce a bigger picture $(A<1)$ is also rapid, unless the zoom requires interpolation of points not in the original image.

Convolution is the basic image enhancement-filter operation. A 5×5 or 7×7 convolution "mask", as the filter is called, is commonly used in image enhancement algorithms, since the 25 or 49 multiplications-per-pixel filter operation must be repeated for each pixel in the image: 25 to 50 Million multiplies per frame!

FFT processing is less common than the above operations; most image processing is done in the data domain. However, an FFT can remove textural periodicities which might camouflage an object, or compress the amount of information stored. The FFT can provide insights about what information is most crucial. For example, it has been shown (Oppenheim (1981)) that phase information keeps the most important image details even if amplitude information is totally neglected.

Stages in Machine Vision
The sequence of image acquisition, enhancement, analysis, and interpretation is called *machine vision*. Feature extraction and pattern recognition progresses in stages to higher levels of abstraction with fewer pieces of data at each stage. A summary of the steps is shown in Table 8.10. The steps are symbolized by the example of frog vision in Figure 8.47; compare them with the conceptual stages in the digital receiver of Figure 8.1.

In this simple model, images in a received sequence are *enhanced* by edge-detection, followed by *simplification* of shapes to nearly black and white blobs.

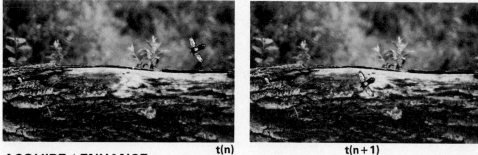

ACQUIRE / ENHANCE:

DIFFERENCE / SIMPLIFY: Image (n + 1) — Image (n)

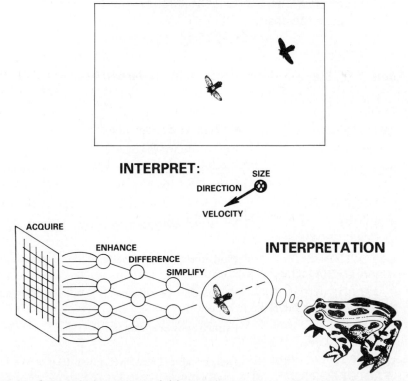

Figure 8.47. Stages in image acquisition, enhancement, analysis, and interpretation. The algorithm is a simplified model of frog vision.

Differencing adjacent time-frames to reveal only *changes* in the scene from one frame to the next reduces the amount of information enormously. *Analysis* then proceeds to model the remaining objects: size, shape, components, and to compute their velocity and trajectory, from which an extrapolation of future position can be sent—in this case to the hungry frog's tongue. Image *interpretation* may compare the object features against a template, which in this case is a dinner menu; the frog rarely decides to snap at a cow, for example. Computer vision benefits from discoveries in neurobiology. In the frog brain, it is thought that some of these steps take place in the neural links from optic nerve on the way to the brain. Distributed processing such as this permits the enormous information in a raw image to be successively reduced to transmit only the most essential features (which shadows are moving, for example) to the decision-maker, which help explain, for example, why the pea-brained frog is so good at catching flies.

The principal stages in machine vision are:

Image acquisition and display: frame grabbing and storage of image planes. Preprocessing at this stage, including summation, averaging, thresholding, and contrast enhancement, reduces the demand on later operations.

Image processing. There are two principal functions:

(1) Image enhancement: to increase signal-to-noise ratio.
(2) Image simplification: to reduce the amount of unnecessary information in the image.

Table 8.10 Stages in Image Processing (adapted from Jain (1981)).

Item	Problem	Description
1	SMOOTHING	Filter to smooth out noise
2	ENHANCEMENT	Enhance contrast, edges
3	RESTORATION	Restore degradation; deblurring, reconstruction, geometrical correction
4	COMPRESSION	Minimize bits, store/transmit at a given distortion level
5	FEATURE EXTRACTION	Extract edge/shape information
6	DETECTION AND IDENTIFICATION	Find object in scene: matched filter, pattern, segment, texture
7	INTERPOLATION & EXTRAPOLATION	Estimate image value inside or outside the sample space
8	SPECTRAL ESTIMATION	Estimate power spectrum
9	SPECTRAL FACTORIZATION	Design stable causal filter, given desired frequency response
10	SYNTHESIS	Given description or features, design system which replicates (e.g., texture synthesis)

It is convenient to divide operations at this level into two categories: neighborhood processing, and frame processing.

> *Neighborhood* operations change the state of a pixel, based on the content of a nearby region. Convolution filters and gradient filters are typical operations. The purpose is to bring out an image definition that higher-order steps later on can use optimally, such as vertical lines or contrast-enhanced edge definition.

> *Frame processing* operates on a window within the image. Frame processing operations are geometric: translation, rotation, and zoom, for example. Frame processing lets the user, perhaps with joystick control, limit the domain (hence the complexity and time) of further analysis.

Image analysis: produces descriptive information. There are two phases:

> *Segmentation*, the partitioning of the image into meaningful regions which have something in common: for example, boundaries which include a single object within them. Edge-following algorithms of considerable cleverness have been developed (see Ballard and Brown (1982)).

> *Feature extraction*—the generation of data describing the regions of interest: geometry, position, gray-scale, etc. Feature extraction includes measurements, such as area and perimeter, and projection of net intensity onto x and y axes, from which the object location can be measured.

Image interpretation uses the above data to accomplish one of several primary image functions: recognition, inspection, location, measurement, and identification.

Through these steps, the image is divided into "meaningful" regions: edges (directional and gradient), texture, color, contrast, etc. These parameters are identified through convolution operations over an entire image with 2-D convolution masks, which characterize feature parameters—for example an edge detector that pulls out lines running only in a selected direction. Mask size depends upon degree of detail in the image but is normally limited by processing speed to quite small (e.g., 3×3 to 7×7) values. Once the analysis is performed, pattern-recognition algorithms seek a match between extracted and analyzed features and a *template* in memory. The template library must contain a "vocabulary" of anticipated patterns and objects. The library is built by the analysis of simple objects by identical methods, as in the "training" of a speech-recognition processor. The goal of machine vision is to analyze TV images to make decisions. For example, a robot assembler (Figure 8.48) is told to find the nut that is not yet installed and screw it on the bolt. What is a nut? What is a bolt? Look at the template. Where are the nut and the nutless bolt in the picture? Correlate the template for each with the image and determine the *offset*, giving maximum correlation that places the template of the nut on

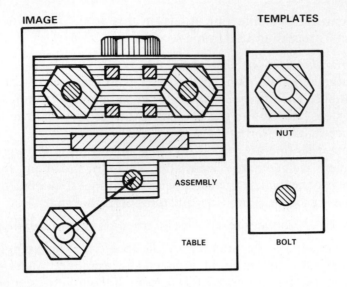

IMAGE   TEMPLATES

ASSEMBLY

NUT

TABLE   BOLT

Figure 8.48. Machine vision analyzes TV images to make decisions. In this case, find the nut remaining on the table and screw it on to the bolt which does not yet have a nut.

the table squarely over the nutless bolt in the image.

For example, consider the process of locating the position of a square box in a room. Suppose the box is located in the upper left corner of the room, at position 0,0. A box template is cross-correlated with the room image. A correlation peak should appear when the template matches up with the box's location. The correlation is the 2-D version of the product-sum convolver or correlator, with the result at a given value of template offset equal to the sum of products of image times template at that offset.

TEMPLATE FOR FINDING THE POSITION OF A BOX IN A ROOM

Template	Room Image with box (origin at top left)	Correlation	
1 1 1	1 1 0 0 0	7 4 2 x x	x = undefined
1 1 1	1 1 1 0 0	5 3 2 x x	
1 1 1	1 0 1 0 0·	2 1 5 x x	
	0 0 0 0 0	x x x x x	
	0 0 0 0 4	x x x x x	

Here, the image of the box contains noise, and there is a noise spot in the lower right corner of the room as well. The correlation in this case is carried out only over offsets that leave the template in the room, to avoid wraparound issues.

The correlation has a peak (7) at offset (0,0) as it should, but there is also a false alarm (value 5 in the center of the room) due to the noise spot.

Clearly, machine vision is a subject of deep subtlety. Only a portion of it is within the province of classic DSP; the higher levels of recognition and decision-making often involve tools of artificial-intelligence programming. We illustrate below some of the DSP operations most useful in the preprocessing of images prior to the segmentation (into shapes, etc.) and identification stages.

Image Enhancement Example: Edge Detector

Consider an edge-detection example. Each pixel's intensity is modified to a value which depends on the intensities of its neighbors. The image is convolved with a 3×3 mask. For example, the masks below

$$V = \begin{matrix} -1 & 0 & 1 \\ -2 & 0 & 2 \\ -1 & 0 & 1 \end{matrix} \qquad H = \begin{matrix} 1 & 2 & 1 \\ 0 & 0 & 0 \\ -1 & -2 & -1 \end{matrix}$$

sense vertical and horizontal edges, respectively. In the output after matrix V is "slid" across, or convolved with the image, a nonzero intensity will only be found in the neighborhood of vertical edges.

Exercise: try out the above edge detectors for an 8×8 image of the form:

IMAGE	RESULT for mask V:
0 0 0 0 0 0 0 0	0 0 0 0 0 0 0 0
0 0 0 0 0 0 0 0	0 0 0 0 0 0 0 0
0 0 0 0 0 0 0 0	0 0 0 0 0 0 0 0
0 0 0 0 0 0 0 0	−1 0 0 1 1 0 0 −1
0 0 0 0 1 1 1 1	−3 0 0 3 3 0 0 −3
0 0 0 0 1 1 1 1	−4 0 0 4 4 0 0 −4
0 0 0 0 1 1 1 1	−4 0 0 4 4 0 0 −4
0 0 0 0 1 1 1 1	−3 0 0 3 3 0 0 −3

To carry out the operation, one may choose to assume that the image is periodically repeated (edge artifacts are present but clearly recognizable as such in more-typical images, which are generally larger than the above example.) The vertical edge is clearly brought out by the vertical edge filter V in this example. Filter output values are highest where the gradient is steep; if the image had largely the same intensity, whether dark or light, the gradient mask would bring out an image with an overall low intensity. Edge-detection algorithms of much higher sophistication than the above have been developed, since edge-detection is a key step towards image simplification in machine vision (see Baxes (1984)) or Ballard and Brown (1982)).

Example of Enhancement Alternatives

A well-defined set of standard masks has evolved in image-enhancement literature (see Baxes (1984), for example). The choice of enhancement strategy depends on the objectives.

Low-pass filtering combines information from adjacent pixels to give a smoother version of a noisy image, at the expense of some blurring.

High-pass filtering is like a derivative, sensing changes only, and is one way to bring out edges.

High-emphasis filtering weights the high-frequencies preferentially, but leaves some low-frequency information.

The *spatial-frequency* components of most interest can be enhanced by a recursive filter or a shaped FIR bandpass.

Histogram equalization transforms the gray scale in each region to cover the full dynamic range available. This enhances contrast of an image which shows too little variation of grayscale.

Background subtraction removes uniform gray levels to enhance smaller variations. Besides removing the obviously constant or linear background, more-sophisticated approaches, such as a spline filter, using piecewise-linear polynomials, can identify and remove background levels that vary arbitrarily (but slowly) over the image.

Many of these techniques have been applied in the medical-imaging example of Figure 8.49, to be discussed below.

a. Computed tomography; " + " marks the tumor. The complete presentation includes pictures at different angles to target the tumor. Courtesy of Sloan-Kettering Cancer Center, New York.

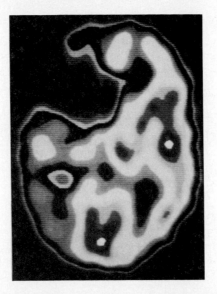

b. Radioisotope imaging; positron-emission tomography (PET) scan. Darkened area indicates damage from a stroke. Courtesy of Division of Radiation Sciences, Mallinckrodt Institute of Radiology, Washington University School of Medicine, St. Louis. See also Figure 8.54.

Figure 8.49. Examples of advanced imaging techniques in medicine. Source: *The National Geographic* 171-1, January, 1987.

8.6.3 MEDICAL IMAGE-PROCESSING

Comparison of Medical-Imaging Examples (Figure 8.49)

Ultrasonic imaging, digital radiography, and various forms of tomography—CAT, PET, MRI*—illustrate DSP benefits in medical instrumentation. The outline of the three digital processes is shown in Figure 8.50.

Ultrasonic imaging sends non-damaging ultrasonic waves through the body via an array of transducers. The ultrasound wavelength is small enough to resolve millimeter-scale features. The technique is particularly valuable in prenatal examination of a fetus. DSP algorithms are employed in various ways:

Digitize the reflected waves from a "phased array" of transducers.

Combine the images from different transducers, with proper corrections for delays and attenuation.

Time-average to enhance sensitivity.

Scan convert, to transform the detector images, expressed in polar coordinates of angle and distance relative to the detector, to *x-y* Cartesian coordinates for combining and viewing.

*CAT: computer-aided tomography; PET: positron-emission tomography; MRI: Magnetic resonance imaging (formerly NMR (nuclear magnetic resonance) imaging).

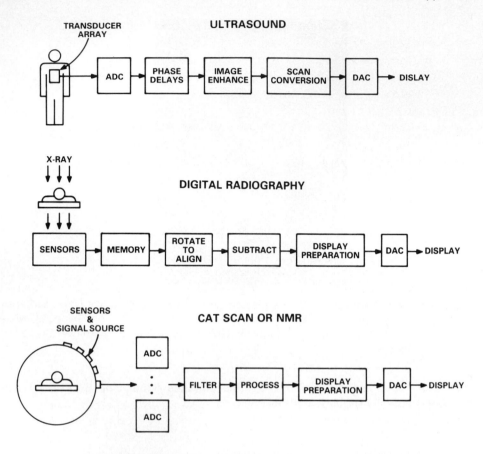

Figure 8.50. Comparison of three medical imaging techniques.

Spectral analysis (FFT) to get blood-flow velocity from the Doppler shift of the returned acoustic wave; arterial integrity and clot buildup can be viewed.

Freeze heartbeats and freeze-frame scan the heartbeat cycle.

Digital radiography is a form of x-ray imaging that lessens the exposure to x-rays required to acquire an adequate image, and enhances the diagnostic repertoire by using differential techniques to bring out anomalies. The transmitted x-ray signal is digitized by an *x-y* sensor array and put into memory. DSP algorithms are employed to:

Enhance the image, or bring out outlines, or suppress regions of less interest.

Display the difference between two pictures, the second made after an absorber is injected to bring out contrast.

Time-average to lessen the dose.

Reconstructive tomography differs from the above two techniques in the same sense that a hologram differs from a conventional image. An array of sensors, arranged in a circle around the body, picks up transmitted x-rays (CAT) or emitted positrons (PET) or spin-echo signals (NMR). Each transducer sees an average in time and space from a macroscopic region of the body. These records are combined to deconvolute the information into a 3-D data base viewed in 2-D slices. DSP techniques perform:

Multiple-detector digitizing and storage.

Filtering of each sensor, or tuning for particular wavelengths or frequencies of information.

Deconvolution or inversion algorithms to reconstruct the intensity at single points in the body.

Display preparation, combining point information into 2-D slice images.

Differencing of two scans (before-and-after contrast-enhancing dyes in CAT, under different psychophysical stimulation in PET, during a metabolic process in MRI, etc.)

The key DSP operations in ultrasonic imaging and digital radiography are similar to those already discussed in *machine vision* above. A quite new and widely useful DSP technique, image *reconstruction*, emerges in CAT scanner instrumentation; it is discussed below.

CAT, PET, and MRI Image Reconstruction

Overview

Image reconstruction, best known from CAT scanners in medicine, has a long history which, like many DSP algorithms, includes independent discoveries in several fields. Reconstruction was used as early as the 1950's in radio astronomy; to build a picture of the shape of an interstellar radio source. Then medicine adopted it, driven by the necessity of early cancer diagnosis. Reconstruction has recently been extended to map the earth's core from correlated seismic observations on the earth's surface. For a review of the subject, with emphasis on x-rays, emission, and ultrasound sources, see Kak (1979). Some highlights of reconstruction algorithms are traced below and related to DSP software and hardware systems; they include a few recent applications, ranging from diagnosis of Alzheimer's disease to learning what parts of the brain are most active when listening to music.

Comparison of X-ray, MRI, and PET Imaging

Many alternative sources (x-rays, gamma rays, electrons, positrons, ultrasound, electric currents, magnetic fields) are potential reconstruction sources. Three are compared here:

X-ray CAT (computer-assisted tomography) scanning uses an x-ray tube in conventional transmission, but the signal is recorded electronically. The output is a measure of point density, which may be enhanced by differential use of high-absorption materials ingested or injected into regions under diagnosis.

MRI tomography is an emission method, sensing the spins on protons (within the body), which have been excited by a pulse of RF energy. The method can either record proton density (and hence permit organ mapping), or, by looking for chemical shifts in the magnetic resonance lines, sense metabolic activity.

PET scanning is also an emission method. A mildly radioactive glucose containing a positron-emitting source is ingested and, while the glucose is being metabolized, the most metabolically active regions are sensed by the array of positron detectors surrounding the body.

In *x-ray CAT scanning*, a transmission is recorded as the projection of the body's x-ray absorption onto the plane of observation. If enough different angles of projection are accumulated, the reconstruction algorithm below is the recipe to transform 2-D projections into a 3-D point map of interior density.

In *MRI tomography*, a clever trick is used to single out one slice of the body at a time, recording the signal produced only by that slice and reconstructing the point image by a sequence of slices at different angles. Magnetic-Resonance Imaging requires a very uniform magnetic field, since the spin resonance linewidth is quite narrow and will easily broaden below detectability. This is used to advantage: a gradient coil on top of the steady dc field is swept to create in the body a local region of high field-homogeneity which contributes the dominant signal. The MRI image emphasizes water-containing living tissue, rather than the highest-density material—such as bones in x-rays. Since the excitation is a weak rf field, there is little if any ionizing radiation risk. Changes due to changes in chemical uptake can also be monitored by measuring the shift of the magnetic resonance line of protons chemically bound to various organic molecules.

PET scanning has, in addition to its sensitivity to metabolic activity, two key features that aid data-gathering:

Electronic collimation: coincidence detection locates the emitting source within the body on the line between two detectors.

Absorption losses between emission and detection are easily corrected for.

Electronic collimation in PET scans follows from the physics of the emission. A positron is unstable, combining with an electron within a few microseconds. Since there are many electrons in matter, the positron will recombine within a few millimeters of its origin; this distance sets the spatial-resolution limit. *Two* photons are emitted in the recombination, traveling with equal speeds in opposite directions (momentum conservation). If two detectors are set up in coincidence, counting only when a photon arrives at both detectors, then the only coincidences will come from sources along a line between the two detectors (Figure 8.51). Source A emits two photons. If one is received by detector D1, the other will necessarily miss detector D2. Only sources such as B, along the line between two detectors, will be recorded. A typical detector array, shown in Figure 8.51b, sets up a number of coincidence lines for each detector pair; detector outputs are fed to fast pulse-detecting digital electronics and thence to digital-reconstruction processing.

Correcting for absorption of photons in body tissue is also easier for PET than for single-photon processes, such as x-ray CAT. Single-photon methods have to contend with an absorption correction from each interior point, *L*, and

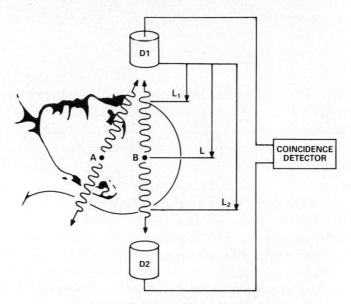

a. The principle of electronic collimation.

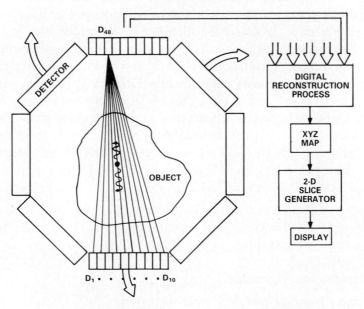

b. Detector array in the positron camera.

Figure 8.51. PET-scanner detection geometry.

ambiguities can lead to ghost and artifact images. The coincidence probability in PET, however, is independent of the point of origin, L, of the photon pair:

$$P(\text{coincidence}) = P(L1,t)\,P(L2,t) \tag{8.58}$$

$$= \left\{ \exp\left[-\int_{L}^{L_1} \mu(x)dx \right] \right\} \left\{ \exp\left[-\int_{L_2}^{L} \mu(x)dx \right] \right\}$$

$$= \exp\left[\int_{L_2}^{L_1} \mu(x)dx \right]$$

$P(L1)$ and $P(L2)$ are signals attenuated by passing from the point of photon origin, B, in Figure 8.51a to the edge of the body, $L1$ or $L2$. While $P(L1)$ and $P(L2)$ both contain (unknown) interior position L, through the distance to $L1$ or $L2$, L cancels in the joint probability because the exponents add in the product of the probabilities.

Where's the DSP?

Tomography exercises practically every key DSP algorithm and introduces a new one, the reconstruction algorithm. Tomography data analysis demands filtering, Fourier transforms, and function generation. The sampled nature of the data (limited by finite time, money, or dose tolerance) demands careful anti-alias treatment and the use of circular convolutions. The time and frequency variables are replaced by their spatial equivalents: the data is a function of position and the transforms specify the spatial frequency. Tomography *data gathering* is similar to that in image enhancement, collecting intensities on pixel planes. Because data from many different angles must be collected, each with pixel resolution adequate for the millimeter scale needed for medical diagnosis, the data-collection step alone imposes an enormous computational burden, which could be improved by clever use of parallelism. Tomography *analysis* benefits from the new DSP architectures, since many multiplications are involved in each step. While a typical installation of the mid-1980's depends upon a mainframe computer and array processor, optimized hardware—using the techniques of earlier chapters and high-speed VLSI components for DSP—is enhancing tomography and reducing its currently enormous cost (a Megabuck—$\$10^6$—per installation). More patient-care facilities can now afford the benefits that tomography provides.

The following examples analyze 2-D objects by 1-D projections, for simpler understanding (and drawing). The extension to 3-D, though harder to visualize, is straightforward.

The Reconstruction Algorithm

Reconstruction Algorithm from the Fourier Transform

This is the key concept in image reconstruction: the data consists of a sampled set of transmissions through (or emission from) an object over a set of angles, θ, or "points of view" of radiation source and detector array. From the data, one can reconstruct an image of the interior with points of view impossible to achieve directly, such as slices through the human brain. The reconstruction algorithm does not violate information theory, just recombines it.

The idea is shown by a two-dimensional example in Figure 8.52. The three objects are identical circles as seen from the outside. Only the detected intensity projections are shown; the beam passes through the object at right angles to the plane of projection. In an x-ray projection, $p(x(\theta)$ at a given angle, holes on the inside show up as regions of lower intensity. By rotating the plane of the projection, a data base builds up from which a model of the interior could be constructed.

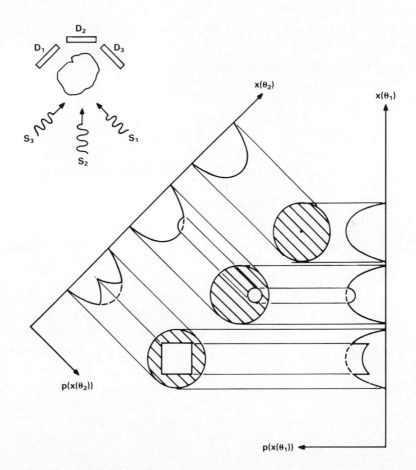

Figure 8.52. 2-D example of projection image-collection. The transmitted intensity, $p(x(\theta))$, is collected on the plane of projection, x, as the x-ray source-and-detector array is rotated at angular increments around the object. INSET: The source and detector are rotated stepwise as shown to collect the data array. Only the projected intensity at detector array, D1, is shown in the main figure.

The projection theorem shows why the reconstruction can be exact: *The FT of a projection of an object is equal to the central section of the FT of the object.* The process is pictured in Figure 8.53. Suppose the density of an object, $f(x,y)$, is projected as $p_y(x)$ on the x-axis. The Fourier transform of $p_y(x)$ is the section of the object's FT, $F(u,v)$, which passes through the frequency origin, and is labeled the *central section*, $P(u) = F(u,0)$. Rotating the plane of the projection, as in Figure 8.52, rotates the plane of the FT section. With enough fractional rotations to build up an adequate collection of central sections, the internal density of the object itself can be obtained from an inverse FT, as shown below.

The math corresponding to the above statements starts with writing the 2-D FT of the object's density, $f(x,y)$, and then examining the projection.

$$F(u,v) = \iint f(x,y)\, e^{-j2\pi(ux + vy)}\, dx\, dy \qquad (8.59)$$

Project the density of the object onto the x-axis.

$$p_y(x) = \int f(x,y)\, dy \qquad (8.60)$$

The FT of the projection is:

$$p_y(u) = \iint f(x,y)\, e^{-j2\pi ux} dx\, dy \qquad (8.61)$$

$$= F(u,0)$$

which, by inspection, is just the section of the FT that passes through the origin and is parallel to the plane of the projection—the so-called *central section*. Each projection builds up a section in the spectrum, a cut through the object's density spectrum, $F(u,v)$, at a single angle. Since all sections pass through the origin, a complete picture of the spectrum is not built up until enough angles of projection have been taken to sweep out frequency space with no big gaps, as is evident from Figure 8.53b. It should be clear from this picture that:

> The number of angles of data collection sets the high-frequency limit on knowledge of the object, since the spectrum is filled in densely towards $f = 0$ (dc limit)—but gaps open up radially at any angle.

> No partial image follows from a partial collection of sections; the data is a collection of 1-D F.T.'s, but a 2-D inverse FT is required to re-constitute the image.

> For a given number of sections, each projection must be filtered to avoid the aliasing of high-frequency artifacts back into the central region of the spectrum.

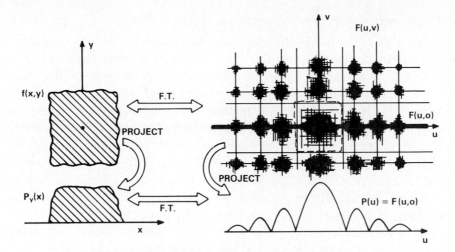

a. An object, its spectrum, and their projections.

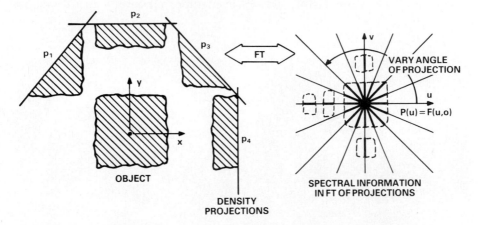

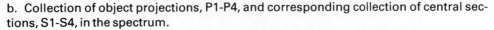

b. Collection of object projections, P1-P4, and corresponding collection of central sections, S1-S4, in the spectrum.

Figure 8.53. Fourier-transform theory of reconstruction via the central-section theorem.

This Fourier-transform method of image reconstruction is carried out by various DSP coordinate transformations and interpolations. The 2-D inverse FT for each slice imposes large computation demands, of order $(N \log_2 N)^2$, with an FFT, or 4×10^{-6} multiplications—even for a 256×256 moderate-resolution image. A computationally more-efficient alternative (but equivalent) approach, called *direct reconstruction via backprojection*, is used in practice (see Kak (1979) or Hall (1979) for details). Backprojection has an advantage over the central section method; all its F.T.'s are one-dimensional, so the computations go as $N \log_2 N$ rather than as the square.

A PET-Scan Example: Brain-Function Mapping

The PET scan is not used solely to detect brain anomalies. Physiologists can profit from its ability to determine non-invasively what regions of the brain are active in everyday activities of a normal living subject. A Brookhaven PET-scan experiment involving the perception of music is an elegant example.* Figure 8.54 shows a comparison of two slices of the same region of the brain taken under different conditions. The subject, a pianist, was reading the score for a Beethoven piano sonata during the half-hour of uptake of radioactive tracer-tagged glucose. A PET scan (left) was performed while the ears were covered to block any aural activity; it shows which brain regions were most metabolically active while reading music. The right-hand slice is a repetition of the first, except that during the tracer uptake the subject was *listening* to an excellent recording of the same sonata. There is a dramatic global increase in metabolic rate, as marked by the increase in areas marked A and B in the image, and areas dealing with the auditory processing of the musical input are quite visibly active (even more strikingly in the original color image).

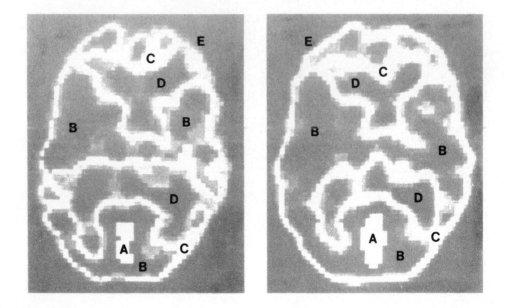

Figure 8.54. PET scan reveals areas of the brain active in processing aural input of music. (Left) Reading a musical score in silence with ears covered. (Right) Rereading the same score while listening to an excellent recording of the same piece. In this black-and-white rendering, areas A-E map the original colors indicating increasing glucose metabolic rate: $E = 0$, $A = max$ (60μmoles/min per 100 cm^3 of tissue). (Courtesy of Jerome Russell, Brookhaven National Laboratory)

*The author is indebted to Dr. Jerome Russell and the Cyclotron PET Group, Brookhaven National Laboratory, for this illuminating example.

PET-scan processing is computationally demanding; it benefits greatly from the enhanced speed of modern DSP components. The DSP process steps are:

Acquire PET Count by pixels

Process: reconstruct slices

Analysis: ratioing, kinetic modelling, etc.

In this experiment, the computations were done with a dedicated VAX. There was no array processor. With 14 slices and a resolution of 256×256 pixels per slice, the requisite time for data acquisition was of the order of one-half hour, a speed fortuitously matched to the peak in metabolic uptake; but it would have been impossible to follow faster changes. Reconstruction was not done in real-time. Since the computations above are so regular, PET image reconstruction could benefit greatly from special purpose architectures or from algorithms matched to the superior architecture of DSP chips.

As a further step, physiologists can translate this mapping of radioactive counts into a picture of brain function, in the kinetics and biological modelling of glucose uptake. The computational requirements are formidable. Each image pixel is a tiny experimental cell for chemistry. Each cell in the kinetic model communicates with each other cell. The processing of the kinetic modelling in the Brookhaven experiment for one slice alone takes 13 hours of VAX time! The regularity of the computational structure could benefit from parallel processing, or—at the very least—digital signal-processing in VLSI.

Bibliography

Alles, Harold G., "Music Synthesis Using Real-Time Digital Techniques," *Proc. IEEE*, 68, No. 4, 1978 (436-449). Overview of digital music methods, hardware, and software.

Alles, H. G., "An Inexpensive Digital Sound Synthesizer," *Computer Music J.*, 3, No. 3, 1978 (28-). The digital fm oscillator: 32 simultaneous waves generated from a lookup table, with modulation gives complex fm timbres.

————, and R. J. Higgins, "Software-Emulated Multichannel Lock-in Detector," *Rev. Sci. Instrum.*, 44, 1973 (1646-1650). Multi-channel lock-in in software.

————, and R. J. Higgins, "de Haas-van Alphen Measurements of the Exchange Energy and Spin-Scattering Anisotropy in Cu-Fe," *Phys. Rev.*, B9, 1974 (158-173). Spectral estimation by iterative peak fitting and subtraction.

Analog Devices, Inc., *ADSP-2100 Applications Handbook,* Volume 1 (1987) and Volume 2 (1988). Volume 1's topics include arithmetic, filters, FFTs, LPC, modem algorithms; Volume 2 covers graphics, multirate filters, PCM & ADPCM, and DTMF. These applications handbooks and the following publications from a manufacturer of DSP and data-acquisition products exemplify the depth of help usually available to the user in the form of manufacturers' literature (from this company and others).

————, *ADSP-2100 Cross-Software Manual—Programming Reference*, 1988. Complete programmer's reference, including optional C compiler, for a DSP microprocessor.

————, *ADSP-2100 Emulator Manual*, 1988. User's Manual for an in-circuit emulator.

————, *ADSP-2100 Evaluation Board Manual*, 1988. A guide to the evaluation board including schematics for prototyping.

————, *ADSP-2100 User's Manual—Architecture*, 1988. Complete description of architecture and system interface for a DSP microprocessor.

————, *DSP Products Databook*, 1989. Data sheets, selection guides, application notes, and a wealth of background information on components for number crunching and digital signal-processing.

Andrews, H. C. and B. R. Hunt, *Digital Image Restoration*. Englewood Cliffs, NJ: Prentice Hall, 1977. Accessible introduction, illustrated with many reconstructed-image examples.

Baggeroer, A. B., "Sonar Signal Processing," in: *Applications of Digital Signal Processing*, A. V. Oppenheim, Ed. Englewood Cliffs, NJ: Prentice Hall, 1978. The correlator receiver, for S/N enhancement (active sonar). Clear explanations of spectral estimation (passive sonar).

Ballard, Dana H., and Christopher M. Brown, *Computer Vision*. Englewood Cliffs, NJ: Prentice Hall, 1982. A superb introduction; accessible yet not lightweight.

Bates, R. H. T., Kathryn L. Garden, and Terence M. Peters, "Overview of Computerized Tomography with Emphasis on Future Developments," *Proc. IEEE*, 71, No. 3, 1983 (356-372). Nonmathematical survey of reconstruction of medical images by many techniques.

Baxes, Gregory A., *Digital Image Processing: a Practical Primer*. Englewood Cliffs, NJ: Prentice Hall, 1984. An approachable introduction with nice examples.

Benade, A. H., *Fundamentals of Musical Acoustics*. New York & London: Oxford University Press, 1976. Non-mathematical but perceptive survey.

Bendat, Julius, and Allan Piersol, *Engineering Applications of Correlation and Spectral Analysis*. New York: John Wiley & Sons, 1980.

Berglund, Glenn D., "A Radix-Eight Fast Fourier Transform Subroutine for Real-Valued Series," *IEEE Trans. Audio and Electroacoustics*, AU-17, No. 2, 1969 (138-144).

Blesser, B., and J. M. Kates, "Digital Processing of Audio Signals" in: *Applications of Digital Signal Processing*, A. V. Oppenheim, ed. Englewood Cliffs, NJ: Prentice Hall, 1978. One of the best overview surveys of audio DSP.

Boddie, J. R., et al., "Adaptive Differential Pulse-Code Modulation Coding," *Bell System Technical Journal*, 60, 1981 (1547-1561). ADPCM using a Bell Labs programmable DSP microprocessor.

Bogert, B. P., M. J. R. Healey, and J. W. Tukey, "The Quefrency Alanysis of Time Series for Echoes: Cepstrum, Pseudo-Autocovariance, Cross-Cepstrum, and Saphe Cracking," *Proc. Symp. Time Series Analysis*, M. Rosenblatt, ed. New York: John Wiley & Sons, Inc., 1963 (209-243).

Bracewell, R. N., *The Fourier Transform and its Applications*. New York: McGraw-Hill, 1965. A classic, distinguished for its drawings of signals, spectra, and convolutions.

Brigham, E. O., *The Fast Fourier Transform and Its Application*. Englewood Cliffs, NJ: Prentice Hall, 1988. One of the best on FFT and DFT.

Browne, James C., "Parallel Architecture for Computer Systems," *Physics Today*, May, 1984 (28-35). Introductory survey for non-architects.

Brukert, Edward, Martin Minow, and Walter Tetschner, "Three-Tiered Software and VLSI Aid Developmental System to Read Text Aloud," *Electronics*, April 21, 1983 (133-138). Phonemic dictionary and translation rules plus transition smoothing give excellent text-to-speech output from formant synthesis.

Bucy, J. Fred, et al., "Ease-of-Use Features in the TI Professional Computer," *Proc. IEEE*, 72, No. 3, 1984 (269-282). Speech analysis, synthesis, compression. Summary of issues in feature extraction, similarity measure, dynamic programming, decision strategy.

Buric, M. R., K. Kohut, and J. P. Olive, "Speech Synthesis," *Bell System Technical Journal*, 60, No. 7, 1981(1621-31). Brief survey, with an application of a Bell Labs DSP programmable microprocessor.

Capello, Peter R., et al., eds., *VLSI Signal Processing*. New York: IEEE Press, 1985. Papers from a 1984 workshop: Image Analysis, Algorithmic Processors, VLSI Architectures, Telecommunications, Advances in Algorithms, CAD, Languages and Software, Speech, Systolic Architectures.

Carayannis, George, et al., "Fast Recursive Algorithms for a Class of Linear Equations," *IEEE Trans. Acoustics, Speech, and Signal Processing*, ASSP-30, No. 2, 1982 (227-239). Wide class of linear equations solvable with reduced computation. Includes FORTRAN codes.

Chamberlin, Hal, *Musical Applications of Microprocessors*. Rochelle Park, NJ: Hayden Book Co., 1980. Survey at introductory level includes dynamic range issues; sine-wave lookup; reverb, polynomial interpolation; spectrographs, linear prediction, homomorphic (cepstrum), pitch detectors, time-domain methods.

Chowning, J. M., "The synthesis of complex audio spectra by means of frequency modulation," *J. Audio Eng. Soc.*, 21, 7, 1973 (526-534). Key article on fm synthesis of complex sounds. Also reprinted in *Computer Music J.*, 1, 1977 (46-54).

Cooley, J. W., P. A. Lewis and P. D. Welch, "The FFT Algorithm: Programming Considerations in the Calculation of Sine, Cosine, and Laplace Transforms, *J. Sound and Vibration*, 12, 1970 (315-337). FFT program enhancements for special classes of data.

Cooley, James W., and John W. Tukey, "An Algorithm for the Machine Calculation of Complex Fourier Series," *Math. of Computation*, 19, 1965 (297-301). The article that got the FFT started.

Crochiere, R. E., and J. L. Flanagan, "Current Perspectives in Digital

Speech," *IEEE Communications*, Jan., 1983 (32-40). Tutorial overview of key ideas (modeling, compression) and applications.

Crochiere, Ronald E., and Lawrence R. Rabiner, "Interpolation and Decimation of Digital Signals—a Tutorial Review," *Proc. IEEE*, 69, No. 3, 1981 (300-331).

Crowell, C. R., and S. Alipanhi, "Transient Distortion and nth Order Filtering in Deep Level Transient Spectroscopy," *Solid State Electronics*, 24, 1981 (25-36). Optimum correlator waveforms for exponential decays.

Delves, L. M., and A. Samba, "Deconvolution of Seismic Data," *Computer Physics Communications*, 26, 1982 (473-6). A new optimum algorithm for separating seismic echoes from noise, implemented on an array processor system.

Demos, Gary, Maxine D. Brown, and Richard A. Weinberg, "Digital Scene Simulation: The Synergy of Computer Technology and Human Creativity," *Proc. IEEE*, 72, No. 1, 1984(22-31). Throughput issue starkly defined: it takes a dedicated Cray1 operating at 200 MFLOPS from 3 s to 10 hours to generate 1 s of high-resolution computer graphics!

Dintersmith, Ted, and Paul Toldalagi, "Apply Modern Control Theory to Optimize Digital Systems," *EDN*, April 28, 1983 (165-179).

Doi, Toshi, "Error Correction in the Compact Disk System," *Proc. 73rd Convention Audio Engineering Society*, Repr. 1991 (B4), 1983. See *Audio*, April, 1984, (24), for popular version. Error measurement, error detection. Principles of error correction: Supermarket sales ticket model. Error concealment and correction methods.

Drake, Frank, John H. Wolfe, and Charles L. Seeger, *SETI Science Working Group Report*, NASA Technical Paper 2224, 1983. See for SETI background; also, Ch. 5 on the NASA multiband filter and Appendix C on signal/noise issues.

DSP Committee of IEEE Acoustics, Speech, and Signal Processing Society, *Programs for Digital Signal Processing*. New York: IEEE Press, 1983. Listings of FORTRAN programs for DFT, power spectrum, correlation, convolution, linear prediction, FIR and IIR filters, cepstrum, interpolation, decimation.

Duff, I. S., and J. K. Reid, "Experience of Sparse Matrix Codes on the Cray1," *Computer Physics Communications*, 26, 1982(293-302). How to avoid defeating a vector architecture when computationally efficient sparse matrices are used.

Dusek, Lee, Thomas B. Shalk, and Michael McMahan, "Voice Recognition Joins Speech on Programmable Board," *Electronics*, April 21, 1983 (128-132). Hardware and algorithm design for an inexpensive trainable voice recognizer for PCs.

Eldon, John, "Digital signal processing hits stride with 64-bit correlator IC," *Electronics*, July 14, 1981 (118-123). 15-MHz correlator chip.

————, "Digital Correlator Defends Signal Integrity with Multibit Precision," *Electronic Design*, May 17, 1984.

Elliott, Douglas F., and K. Ramamohan Rao, *Fast Transforms*. New York: Academic Press, 1982. Combines DFT/FFT (including filter applications) with binary Walsh-Hadamard transforms.

Fichtner, Wolfgang, et al, "The Impact of Supercomputers on IC Technology Development and Design," *Proc. IEEE*, 72, No. 1, 1984 (96-112). Modeling of IC process and device performance. See re-*sparse matrix* definition and key references.

Franz, Gene A and Richard H. Wiggins, "Design Case History: Speak and Spell Learns to Talk," *IEEE Spectrum*, Feb., 1982 (45-49).

Friedlander, Benjamin, "Lattice Filters for Adaptive Processing," *Proc. IEEE*, 79, No. 8, 1982 (829-865). Tutorial review, relates to adaptive prediction.

Gonzales, Raphael C., and Paul Wintz, *Digital Image Processing*. Reading, MA: Addison-Wesley, 1977. Fundamentals, transforms, enhancement, restoration, encoding, segmentation, description.

Griffeths, Peter R., "Fourier Transform Infrared Spectrometry," *Science*, 222, Oct. 21, 1983 (297-302). Introductory survey. F.T. method lets signals get encoded 1,000 times faster than monochromator method.

Hall, Ernest L., *Computer Image Processing and Recognition*. New York: Academic Press, 1979. Accessible introductory survey by a pioneer in digital image enhancement in radiography.

Hamming, R. W., *Digital Filters*. Second Edition. Englewood Cliffs, NJ: Prentice Hall, 1983. Slim and elegant introduction (still only 257 pages in the second edition), from a numerical-analysis guru.

Harris, Fredric J., "On the Use of Windows for Harmonic Analysis with the Discrete Fourier Transform," *Proc. IEEE*, 66, No. 1, 1978 (51-83). Exhaustive compilation with critical analysis of spectral separation criteria and computed examples.

Helms, Howard D., and Lawrence R. Rabiner, *Literature in Digital Signal Processing*. New York: IEEE Press, 1973. Reprints the Rabiner (1972) Terminology article plus extensive bibliography of cornerstone works.

Higgins, Richard J., *Electronics with Digital and Analog Integrated Circuits*. Englewood Cliffs, NJ: Prentice Hall, 1983. User's guide to IC applications in instrumentation and signal-processing.

Higgins, R. J., "Fast Fourier transform: An introduction with some minicomputer experiments," *Amer. J. Physics*, 44, No. 8, 1976 (766-773). Tutorial introduction.

————, and D. H. Lowndes, "Waveshape Analysis in the dHvA Effect," in: *Electrons on the Fermi Surface*, M. Springford, ed. Cambridge and New York: Cambridge University Press, 1980 (394-478). Harmonic content analysis in a solid-state physics application.

Hockney, R. W., and C. R. Jesshope, *Parallel Computers.* Bristol, England: Hilger Publ., 1981.

Horowitz, Paul, et al., "Ultranarrowband Searches for Extraterrestrial Intelligence with Dedicated Signal Processing Hardware," *Icarus*, 67, 1986 (525-539). FFT spectrum analyzer with more than 100,000 channels and 1-in-10^{12} resolution.

—————, John Forster, and Ivan Linscott, "The 8-Million Channel Narrowband Analyzer," in *The Search for Extraterrestrial Life, proc. of IAU Symposium*, #112, Michael D. Papagiannis, ed., Dordrecht, Netherlands: D. Reidel, Publishers, 1985 (361-371). Signal processing in the search for extraterrestrial intelligence.

Hoste, Daniel, and M. Grinde, "High Performance Spectrum Analyzer Using Am29500 FFT Address Generator." Advanced Micro Devices, Application Note 0579A, January, 1985.

Hunt, B. R. ,"Digital Image Processing" in: *Applications of Digital Signal Processing*, A. V. Oppenheim, ed. Englewood Cliffs, NJ: Prentice Hall, 1978. Accessible intro, though somewhat eclipsed by recent progress.

IEEE Acoustics, Speech, and Signal Processing Society: Digital Signal Processing Committee, eds., *Programs for Digital Signal Processing*. New York: IEEE Press, 1979. FORTRAN listings (magnetic tape available) for FFT, power spectrum, linear prediction, filters.

IEEE Task P754 Group, "A Proposed Standard for Binary Floating Point Arithmetic," *IEEE Computer*, 14, No. 3, 1981 (51-62). Draft 8.0 of the famous IEEE-754 standard.

IEEE Transactions on Audio/Electroacoustics. (1969) and (1967). Key collection of papers from two seminal workshops when the FFT was young and new versions flowed freely. Many excellent papers not otherwise cited explicitly are to be found here.

Isaksson, Anders, Arne Wennberg, and Lars H. Zetterberg, "Computer Analysis of EEG Signals with Parametric Models," *Proc. IEEE*, 69, No. 4, 1981 (451-461). Spectral estimation clearly surveyed in the context of brain waves.

Jack, Mervyn A., Peter M. Grant, and Jeffrey H. Collins, "Theory, Design, and Applic.of SAW Fourier Transform Processors," *Proc. IEEE*, 68, No. 4, 1980 (450-). Analog F.T. processing at 100-MHz speeds, using chirp filter; application to radar and sonar.

Jain, Anil K., "Advances in Mathematical Models for Image Processing," *Proc. IEEE*, 69, No. 5, 1981 (502-528). Tutorial review of key image issues and models.

Jain, Anil K., "Image Compression: A Review," *Proc. IEEE*, 69, No. 3, 1981 (349-389). PCM, DPCM, LPC, etc., with particular application to image processing.

Jayant, N. S., and Peter Noll, *Digital Coding of Waveforms*. Englewood Cliffs, NJ: Prentice Hall, 1984. Compression survey, with applications to speech and video. The derivations are relegated to appendices to help the flow. Includes computer experiments.

Jesshope, C. R., "Programming with a High Degree of Parallelism in FORTRAN," *Computer Physics Communications*, 26, 1982 (237-246). Parallel extensions to FORTRAN generate fast compact code which lets vector architecture operate; application examples in matrix multiplication and FFT.

Johnson, Matt, "Approximating Functions with Digital Signal Processing ICs," *Analog Dialogue*, 18, No. 1, 1984 (14-17). See for Newton-Raphson iteration and digital divider.

———, "Implement Stable IIR Filters Using Minimal Hardware," *EDN*, April 14, 1983 (153-165). Accessible introduction.

Jordan, Harry F., "Experience with Pipelined Multiple Instruction Streams," *Proc. IEEE*, 72, No. 1, 1984 (113-123). SISD, SIMD, MIMD machines, where S=single, M=multiple, I=instruction, D=data.

Juang, Biing-Hwang, Lawrence R. Rabiner, and Jay G. Wilpon, "On the Use of Bandpass Liftering in Speech Recognition," *IEEE Trans. Acoustics, Speech, and Signal Processing*, ASSP-35, No. 7, 1987 (947-54)

Kaiser, J. F., and W. A. Reed, "Data smoothing using low-pass digital filters," *Rev. Sci. Instr.*, 48, No. 11, 1977 (1447-1457). Beginner's guide to the Kaiser method of filter design.

Kaiser, James F., and Ronald W. Schafer, "On the Use of the I_0 Sinh Window for Spectrum Analysis," *IEEE Trans. Acoustics, Speech, and Signal Processing*, ASSP-28, No. 1, 1980 (105-107). How to optimize the Kaiser window parameters for spectral analysis rather than for digital filtering.

Kaiser, L. R., and W. A. Reed, "Bandpass (bandstop) Digital Filter Design Routine," *Rev. Sci. Instrum.*, 49, No. 8, 1978 (1103-1106). Complementary to Kaiser's low-pass article; another beginner's guide.

Kak, Avinash C., "Computerized Tomography with X-ray, Emission, and Ultrasound Sources," *Proc. IEEE*, 67, No. 9, 1979 (1245-1272). Image reconstruction: major advances reviewed.

Kay, Steven M., and Stanley Lawrence Marple, Jr., "Spectrum Analysis—a Modern Perspective," *Proc. IEEE*, 69, No. 11, 1981 (1380-1419). How do modern spectral estimation methods (12 in all) actually compare for real data examples?

Langston, J. Leland, "μC Chip Implements High-speed Modems Digitally," *Electronic Design*, June 24, 1982 (109-116). How key modem algorithms could be implemented on TMS320—conceptually.

Lin, Zse-Cherng, and Glen Wade, "Signal Processing and Image Reconstruction in Microscopic Acoustical Holography." *IEEE Trans. Acoustics, Speech, and Signal Processing*, ASSP-35, No. 7, July, 1987 (1037-45).

Lindsey, William C., and Chak Ming Chie, "A Survey of Digital Phase-Locked Loops," *Proc. IEEE*, 69, No. 4, 1981 (410-431). An analog signal processing component with performance improved by going digital.

Makhoul, J., "Linear Prediction: a Tutorial Review," *Proc. IEEE*, 63, 1975 (561-580). Elegant and lucid LPC intro; equivalence of time-prediction and spectral fitting.

Mansfield, P., and P. G. Morris, "NMR Imaging in Biomedicine," in: *Advances in Magnetic Resonance*, John S. Waugh, ed., New York: Academic Press, 1982. See Ch. 3, Basic Imaging Principles, and Ch. 4, NMR Imaging Methods.

Markel, J. D., and A. H. Gray, *Linear Prediction of Speech*. New York: Springer Verlag, 1976. The bible of linear prediction.

Matheson, T. G., and R. J. Higgins, "Microprocessor-Assisted Real-Time Harmonic Analysis by Minicomputer," *Rev. Sci. Instrum.*, 49, 1978 (1694-1697). Multiple-channel lock-in with true sine-wave mixer.

McClellan, J. H., and R. J. Purdy, "Applications of DSP to Radar," in: *Applications of Digital Signal Processing*. A. V. Oppenheim, ed., Englewood Cliffs, NJ: 1978, Prentice Hall. The radar environment demands optimum filters for noise improvement, and uses FFTs for high-speed filtering.

Mead, Carver, and Lynn Conway, *Introduction to VLSI Systems*. Reading, MA: Addison-Wesley, 1980. The bible for VLSI design beginners.

Melen, Roger D., Albert Macovski, and James D. Meindl, "Applications of Integrated Electronics to Ultrasonic Medical Instruments," *Proc. IEEE*, 67, No. 9, 1979 (1274-1285). See re-multiple-transducer array recombination.

Mick, John, and Jim Brick, *Bit-Slice Microprocessor Design*. New York: McGraw-Hill, 1980. A bible for microprogramming fundamentals and specific Bit-Slice chip utilization.

Moorer, James A., "Signal Processing Aspects of Computer Music: A Survey," *Proc. IEEE*, 65, No. 8, 1977 (1108-1137). Waveforms and spectra of various instruments. Additive synthesis (Fourier analysis), subtractive synthesis (time-varying filtering), fm and related parametric methods.

——, and John Grey, "Lexicon of Analyzed Tones," *Computer Music J.*, 1, 1, 1977 (39-45). Part I Violin tone is here; see also 1, 12-29, June 1977 (Clarinet and Oboe); and 2, 23-31, 1978 (Trumpet).

——, John Grey, and John Strawn, *Computer Music Journal* 2, 2, 1978 (12).

Nagle, H. T., Jr., and V. P. Nelson, "Digital Filter Implementation on 16-bit Microcomputers," *IEEE Micro*, Feb., 1981 (23-41). Survey, strong on IIR, with specific application programs and benchmarks on 8086 microprocessor.

Neves, Kenneth W., "Mathematical Software Libraries for Vector Computers,"*Computer Physics Communications*, 26, 1982 (303-310). Optimization of performance for vector (Cray1) and array processor (FPS AP120B).

Oppenheim, A. V., "Digital Processing of Speech," in: *Applications of Digital Signal Processing*, A. V. Oppenheim, ed. Englewood Cliffs, NJ: Prentice Hall, 1978. Excellent introduction to speech analysis and synthesis.

————, and Jae S. Lim, "The Importance of Phase in Signals," *Proc. IEEE*, 69, No. 5, 1981 (529-541). Phase alone (magnitude set to one) can reconstruct essential features of an image.

————, and R. W. Schafer, *Digital Signal Processing*. Englewood Cliffs, NJ: Prentice Hall, 1975. The bible; authoritative but not easy reading.

————, ed, *Applications of Digital Signal Processing*. Englewood Cliffs, NJ: Prentice Hall, 1978. Excellent collection; specific items listed separately in this bibliography.

Oppenheim, Alan V. and Clifford J. Weinstein, "Effects of Finite Register Length in Digital Filtering and the Fast Fourier Transform," *Proc. IEEE*, 60, No. 8, 1972 (957-976). Very comprehensible survey.

————., A. S. Willsky, and I. T. Young, *Signals and Systems*. Englewood Cliffs, NJ: Prentice Hall, 1983. Background on signals, both analog and digital, at an undergraduate level, from a discrete perspective.

Oxaal, John, "Temporal averaging techniques reduce image noise," *EDN*, March 17, 1983 (211-215). The signal averager, with DSP enhancements such as exponential smoothing.

————, "DSP hardware improves multiband filters," *EDN*, March 31, 1983 (193-197). Heterodyning and decimation filtering.

Parikh, Dakshesh, Susan Charbonneau, and Susan Barber, "Single-chip speech synthesizers speak well of their algorithms,"*Electronics*, July 28, 1983 (135-139). Parametric coding: linear prediction vs. formant coding.

Peled, Abraham, and Bede Liu, *Digital Signal Processing*. New York: Wiley, 1976. A practical introduction (filters, FFT) with a collection of computer programs and good treatment of throughput and accuracy.

Rabiner, L. R., and B. Gold, *Theory and Application of Digital Signal Processing*. Englewood Cliffs, NJ: Prentice Hall, 1975.

————, and C. M. Rader, eds., *Digital Signal Processing*. New York: IEEE Press, 1972. A reprint volume with several cornerstones: Steiglitz on Equivalence of Analog and Digital Signal Proc. and Rader & Gold: Digital Filter Design in the Frequency Domain.

————, and R. W. Schafer, *Digital Processing of Speech Signals*. Englewood Cliffs, NJ: Prentice Hall, 1978.

————, et al., "Terminology in Digital Signal Processing," *IEEE Trans. Audio and Electroacoustics*, AU-20, No. 5, 1972 (322-337). A basic review; more than just terminology.

Robinson, Enders A., "A Historical Perspective of Spectrum Estimation," *Proc. IEEE*, 70, No. 9, 1982 (885-907). What did Pythagoras, Fourier, Schrödinger, Von Neumann, Wiener, Tukey, et al. have in common? The search for periodicities.

Rosenfeld, Azriel, and A. C. Kak, *Digital Picture Processing* (2nd Ed.). Academic Press, 1982. Key image-processing reference.

Schwartz, Mischa, and Leonard Shaw, *Signal Processing: Discrete Spectral Analysis, Detection, and Estimation*. New York: McGraw-Hill, 1975. Intelligible introduction, with real-world examples such as lab measurements and traffic flow.

SETI Science Working Group Report (see: Drake et al, eds., 1983.)

Sheahen, Thomas P., "Use of Chirping to Compensate for Nonlinearities in the Fourier Spectroscopy," *J. Opt. Soc. Amer.*, 64, No. 4, 1974 (485-494). Good non-specialized chirping background here.

Sheingold, Daniel H., ed., *Analog-Digital Conversion Handbook*. Englewood Cliffs, NJ: Prentice Hall, 1986. More than 700 pages of basic information about data converters and systems that use them—for engineers and scientists on both sides of the interface.

————, ed., *Nonlinear Circuits Handbook—designing with analog function modules and ICs*. Norwood, MA: Analog Devices, 1974 & 1976. A guide to analog signal processing using analog multipliers, logarithmic circuits, and other elements based on semiconductor junction characteristics.

————, ed., *Transducer Interfacing Handbook—a guide to analog signal conditioning*. Norwood, MA: Analog Devices, 1980. A practical designers' guide to how analog signals from transducers are processed before entering the digital realm.

Skordalakis, E., "Meta-assemblers," *IEEE Micro*, April, 1983 (6-16). Code-generation for microcoded systems. See table of commercial products and vendors.

Steiglitz, Kenneth, "Computer-Aided Design of Recursive Digital Filters," *IEEE Trans. Audio and Electroacoustics*, AU-18, No. 2, 1970 (123-129). Pivotal article on Fletcher-Powell optimization.

Stockham, Thomas G., Thomas M. Cannon, and Robert B. Ingebretsen, "Blind Deconvolution through Digital Signal Processing," *Proc. IEEE* 63, 1975 (678-692).

Swingler, D. N., "A Differential Technique for the Fourier Transform Processing of Multicomponent Exponentials," *IEEE Trans. Biomed. Eng.*, BME-24, No. 4, 1977 (408-410). Brings out rate constants from multi-rate processes via a transform method.

Templeton, Clive, "Fast Methods on Parallel and Vector Machines," *Computer Physics Communications*, 26, 1982 (331-334). FFT benchmark runs a Cray1 at 4 to 87 MFLOPS, depending on how the code is vectorized. A Cyber operating in parallel fashion competes at 50 MFLOPS.

Templeton, I. M. , "Real-time harmonic analysis by minicomputer," *Rev. Sci. Instrum.*, 46, No. 4, 1975 (301-305). True sine-wave lock-in, multichannel.

Trussel, H. Joel, "Processing of X-Ray Images," *Proc. IEEE*, 69, No. 5, 1981 (615-627). Short review of tomography (reconstruction), coded-aperture imaging, and radiographic image enhancement.

Tyson, J. Anthony and R. P. Giffard, "Gravitational Wave Astronomy," *Ann. Rev. Astronomy & Astrophysics*, 16, 1978 (521-554). Optimum digital filtering application to an astrophysics problem.

Wawrzynek, John, and Carver Mead, "A VLSI Architecture for Sound Synthesis," in: *VLSI Signal Processing: A Bit-Serial Approach*, Peter Denyer and David Renshaw, eds., Reading, MA: Prentice Hall Publishers, 1986. Novel Bit-Slice universal processing element optimized for music difference equations.

Whalen, Anthony D., *Detection of Signals in Noise*. New York: Academic Press, 1971. Matched and optimum filters and receivers.

Yianilos, Peter N., "Dedicated Comparator Matches Symbol Strings Fast and Intelligently," *Electronics*, Dec. 1, 1983 (113-117). Chip and algorithm design to look for similarities between strings.

Index